Molecular Symmetry and Spectroscopy

Molecular Symmetry and Spectroscopy

Philip R. Bunker

Herzberg Institute of Astrophysics
National Research Council
Ottawa, Canada

ACADEMIC PRESS   New York   San Francisco   London   1979
A Subsidiary of Harcourt Brace Jovanovich, Publishers

6189-7243

PHYSICS

ACADEMIC PRESS, INC.
111 Fifth Avenue, New York, New York 10003

*212-
800-346-8648
Harcourt Brace*

United Kingdom Edition published by
ACADEMIC PRESS, INC. (LONDON) LTD.
24/28 Oval Road, London NW1   7DX

Library of Congress Cataloging in Publication Data

Bunker, Philip R
    Molecular symmetry and spectroscopy.

    Bibliography: p.
    Includes index.
    1.   Molecular structure.   2.   Molecular spectra.
I.   Title.
QD461.B958   541'.28   78−51240
ISBN 0−12−141350−0

PRINTED IN THE UNITED STATES OF AMERICA

79 80 81 82   9 8 7 6 5 4 3 2 1

To Éva and Alex

Contents

vii

3. ROTATION GROUPS AND POINT GROUPS

4. REPRESENTATIONS OF GROUPS

5. THE USE OF REPRESENTATIONS FOR LABELING MOLECULAR ENERGY LEVELS

6. THE MOLECULAR HAMILTONIAN AND ITS TRUE SYMMETRY

7. THE COORDINATES IN THE ROVIBRONIC SCHRÖDINGER EQUATION

Contents

8. THE ROVIBRONIC WAVEFUNCTIONS

9. THE DEFINITION OF THE MOLECULAR SYMMETRY GROUP

10. THE CLASSIFICATION OF MOLECULAR WAVEFUNCTIONS IN THE MOLECULAR SYMMETRY GROUP

11. NEAR SYMMETRY, PERTURBATIONS, AND OPTICAL SELECTION RULES

12. LINEAR MOLECULES AND NONRIGID MOLECULES

Appendix A.   THE CHARACTER TABLES

Appendix B.   THE CORRELATION TABLES

REFERENCES

Preface

This is a book about the use of group theory in quantum mechanics with particular reference to problems in molecular spectroscopy. There are so many books, so many good books, on this subject that it is hard to believe that there can be any justification for writing another. The justification lies in the fact that whereas the existing literature is concerned with the use of the molecular point group, whose elements consist of rotations and reflections of vibronic variables, the present volume discusses the use of the molecular symmetry group, whose elements consist of permutations of identical nuclei with or without the inversion. The molecular symmetry group is of more general use than the molecular point group since the effect of molecular rotation and the effect of tunneling due to nonrigidity (such as inversion tunneling in the ammonia molecule) are allowed for. Also, because of the fundamental nature of its elements, the molecular symmetry group provides a very good pedagogical vehicle for teaching group theory and the application of group theory to problems in molecular spectroscopy.

The book is aimed at the serious student of molecular spectroscopy, and although knowledge of the postulatory basis of quantum mechanics is assumed, group theory is developed from first principles. The idea of the molecular symmetry group is introduced early in the book (in Chapter 2) after the definition of the concept of a group by using permutations. This is followed by a discussion of point groups and rotation groups. The definition of the representations of a group and a general account of the use of representations in classifying (or labeling) molecular states are given in Chapters 4 and 5. In Chapter 6 the symmetry of the exact molecular Hamiltonian is considered and the role of identical particle permutation and overall rotation is stressed. In order to classify molecular states it is necessary to obtain appropriate approximate molecular wavefunctions and to understand how the wavefunctions transform under the effects of symmetry operations; these topics are discussed in Chapters 7, 8, and 10, and the coordinates occurring in the approximate molecular wavefunctions (in particular the Euler angles and vibrational normal coordinates) are discussed in detail. Chapter 9 is devoted to a detailed discussion of the definition of the molecular symmetry group and to the application of the definition in a variety of circumstances. Near symmetry is defined in Chapter 11, and the use of the energy level classification scheme obtained by using near symmetry groups (such as the molecular point group), as well as true symmetry groups (such as the molecular symmetry group), in understanding perturbations, optical selection rules, forbidden transitions, the Stark effect, and the Zeeman effect is described. The special problems present in the application of these symmetry ideas to linear molecules, to nonrigid molecules, and to molecules in which spin–orbit coupling is strong are also discussed. The electron spin double groups of molecular symmetry groups are derived for the first time (in Chapter 10 and in Problem 12-1) and used for the classification by symmetry of molecular wavefunctions for half-integral angular momentum states. In the Appendixes the character tables and correlation tables of some molecular symmetry groups are given.

A reader already conversant with abstract group theory, the use of point groups, and the detailed form of molecular wavefunctions can jump straight from the end of Chapter 2 to Chapter 9 (in which the molecular symmetry group is defined in detail) and then to Chapters 10–12 in which the applications of the molecular symmetry group are given. Chapter 11 is the central chapter of the book and the one in which the relationship of the molecular symmetry group to the molecular point group is discussed in detail (see, in particular, Figs. 11-3–11-5). In this chapter the usefulness of the molecular symmetry group for labeling the states of rigid molecules (i.e. molecules that do not tunnel between different conformations) is stressed.

I hope that the reader will gain from this book a good understanding of the place of the molecular symmetry group, in relation to the molecular point group and the molecular rotation group, in the application of group theory to problems in molecular spectroscopy. To aid in this understanding I have included many applications of the ideas as they arise and many figures to show the effect of symmetry operations. I have also included problems, followed by worked solutions, in the body of the text; as a result of this the reader can "self pace" his reading by either (a) omitting to read the problems and solutions, (b) solving the problems as they arise (and then reading the solution in the text as a check), or (c) reading the problems and solutions and treating them as a continuing part of the text.

Acknowledgments

I am very grateful to D. M. Bishop, J. M. Brown, R. E. Moss, T. Nakagawa, and M. Vernon who each read the entire book in manuscript form and gave much helpful advice in order to improve it. I am also grateful to H. H. Günthard, J. Hardwick, G. Hills, J. T. Hougen, V. Laurie, I. M. Mills, A. R. W. McKellar, M. T. Riggin, and S. Novick for their advice on particular points. M. Herman, J. K. G. Watson, and B. P. Winnewisser each read and corrected the proofs of the book, and my wife Éva, together with M. B. Wadsworth and A. M. Lyyra, helped me to compare the proofs with the typewritten manuscript; I am very grateful for their help.

I would especially like to thank Helen Letaif who typed the entire final draft of the manuscript and made a special effort to ensure a well produced and consistent product. Assisting in the typing of the first draft and in typing the modifications that I made to the second draft were Denise Charette, Gloria Dumoulin, Marilyn Nadon, and Lorette Ernst; I very much appreciate all their help.

Introduction

A molecule possesses structural symmetry that is of the same type as that of a macroscopic body, and it can be described in terms of rotation axes and reflection planes. For example, a methane molecule and a macroscopic tetrahedron both have the same structural symmetry. This symmetry is specified by saying that the molecule belongs to a certain *point group*, and the group consists of a definite set of rotation and reflection operations (or elements); for the methane molecule the point group is called T_d. In molecular physics we make great use of symmetry in order to label (or classify) molecular energy levels. Point group symmetry is not the only symmetry present in a molecule, and in this volume the various types of symmetry are discussed at length. The use of these symmetries in labeling molecular states and the application of these labels in understanding molecular processes are also described.

The treatment given in this volume concerning the symmetry classification of the states of polyatomic molecules can be understood by reference to the symmetry classification scheme used for the states of diatomic molecules. The reader unfamiliar with this scheme will have to take the next two sentences on trust, but the scheme will be explained fully in Chapters 7 and 12. For a homonuclear diatomic molecule such as H_2 the vibronic (vibration–electronic) states are classified or labeled using the particular molecular point group $D_{\infty h}$. On the other hand the rovibronic (rotation–vibration–electronic) energy levels (sometimes called rotational levels) are labeled simply as $+s$, $+a$, $-s$, or $-a$ depending on whether the rovibronic wavefunction is invariant $(+)$ or changed in sign $(-)$ by the inversion of the molecular coordinates in the center of mass, and on whether it is invariant (s) or changed in sign (a) by the interchange or permutation of the two identical nuclei. Two such classification schemes are also possible for the states of polyatomic molecules, but although the use of the molecular point group to classify vibronic states is the subject of a vast textbook literature, the use of inversion and nuclear permutation symmetry to classify rovibronic states is but little described. Both classification schemes are useful in understanding and categorizing molecular interactions, and in this book, for a change, emphasis is given to the use of inversion and nuclear permutation symmetry; these symmetry elements constitute the molecular symmetry group.

The molecular point group, whose elements are rotations and reflections of vibronic variables, is used when studying the vibronic levels of a molecule in a given electronic state that has a unique equilibrium configuration with no observable tunneling between configurations (i.e., a rigid molecule). This group is useful in understanding, for example, the infrared and Raman activity of vibrational fundamentals, the terms that can occur in the molecular potential function, and the atomic orbital functions that can be included in a particular molecular orbital. Although the molecular point group is introduced and defined in Chapter 3 here, and its use discussed in Chapter 11, the reader is referred to the texts listed at the end of the chapter for a more exhaustive discussion.

The molecular symmetry group, whose elements consist of the permutations of identical nuclei with or without inversion, is used when studying rovibronic levels of a molecule and the presence of a unique equilibrium configuration is immaterial. The molecular symmetry group has its origins in the work of Christopher Longuet-Higgins and Jon Hougen. The importance of the molecular symmetry group lies not only in its use for studying nonrigid molecules such as ammonia that have large amplitude vibrations, or for studying electronic transitions in which there are changes in the nuclear geometry, but also in its application to rovibronic as well as vibronic states.

Apart from the limitations on using the molecular point group imposed by the need for there to be a unique equilibrium configuration, it is very important to realize that the molecular point group cannot be used if we wish to allow for molecular rotation. The operations of the molecular point group are rotations of the vibronic variables about molecule fixed axes and reflections of the vibronic variables in planes defined by the molecule fixed axes. For a rotating molecule the molecule fixed axes are not inertial axes, and in this molecule fixed frame there will be centrifugal and Coriolis forces. As a result the operations of the molecular point group are only true symmetry operations (in the sense of not affecting the molecular energy) for the nonrotating molecule. If we wish to allow for the effects of molecular rotation (i.e., for the effects of centrifugal distortions and Coriolis forces) we cannot use the molecular point group but must use the molecular symmetry group instead.

A simple example of the effect of molecular rotation is provided by the methane molecule. The methane molecule has a tetrahedral equilibrium geometry in its electronic ground state and we use the T_d point group to classify the vibrational states. On the basis of point group symmetry considerations, we would say that this molecule has no electric dipole moment and no allowed electric dipole rotational spectrum. However centrifugal distortion in the rotating molecule can give rise to a nonvanishing electric dipole moment so that the molecule does have a rotational spectrum. The molecular symmetry group of the methane molecule helps us to understand which rovibration states can interact as a result of centrifugal distortion and to determine which rotational transitions can occur in the spectrum.

The most important general idea that this book is designed to convey is that in molecular physics we use two types of symmetry: true symmetry and near symmetry. The molecular symmetry group is a group of true symmetry operations for an isolated molecule, whereas the molecular point group is a group of near symmetry operations. True symmetry is symmetry that remains when every detail in the molecule is considered, and near symmetry is the symmetry when certain details are neglected; for the molecular point group the small details that are neglected are the effects of molecular rotation. A true symmetry group is not "better" than a near symmetry group in molecular applications, and it should be emphasized that the two types of groups complement each other. However, the molecular symmetry group provides a more fundamental and simpler vehicle for teaching group theory and its use in molecular spectroscopy than does the molecular point group.

BIBLIOGRAPHICAL NOTES

Diatomic Molecules

Herzberg (1950a).[1]   The symmetry of a diatomic molecule is discussed on pages 128–131 and 212–218.

---

[1] References are given in full at the end of the book.

Point Groups

Herzberg (1945).   Pages 82–131.
Herzberg (1966).   Pages 563–579.
Tinkham (1964).   Chapter 4.
Wilson, Decius, and Cross (1955).   See particularly Section 5-5 in which the effect of a point group operation on a molecule is described.

The Molecular Symmetry Group

Longuet-Higgins (1963).   The birth of the molecular symmetry group; its general definition is given.

Hougen (1962b).   The labor pains before the birth. In this paper the "full molecular point group" for a symmetric top molecule is defined by combining molecular point group operations and rotations. It is shown that the elements of the group are permutations of identical nuclei in the molecule with or without inversion. This group is, in fact, the molecular symmetry group of a symmetric top molecule.

Hougen (1963).   This extends Hougen (1962b) to all nonlinear rigid molecules.

Bunker and Papousek (1969).   This extends the definition of the molecular symmetry group to linear molecules and introduces the *extended* molecular symmetry group.

Bunker (1975).   This review paper is devoted to the molecular symmetry group and many nonrigid molecules are discussed.

Hougen (1976).   This review paper discusses the coordinates and the symmetry of the methane molecule.

1

Permutations
and Permutation Groups

In this chapter permutation operations are introduced and the result of multiplying (or combining) permutations together is defined. We then define a group and in particular we define the complete nuclear permutation group of a molecule. The methyl fluoride and ethylene molecules are used as examples. The effect of nuclear permutations on nuclear coordinates, and on functions of the coordinates, is explained.

PERMUTATIONS

Permutations change the order of an ordered set of numbers. For example, the numbers 1, 2, and 3 can be ordered as

$$\underline{123}, \underline{132}, \underline{213}, \underline{231}, \underline{312}, \text{ or } \underline{321}, \tag{1-1}$$

and permutations take us from one ordering to another. The permutation of the numbers 1 and 3 changes the order from $\underline{213}$ to $\underline{231}$, for example, and introducing the notation (13) for the operation of permuting 1 and 3 we write

$$(13)\underline{213} = \underline{231}. \tag{1-2}$$

6

To change the number order from $\underline{321}$ to $\underline{132}$, the number 1 is replaced by 2, 2 is not replaced by 1 but by 3, and 3 is replaced by 1; we define this as the permutation (123) so that

$$(123)\underline{321} = \underline{132}. \tag{1-3}$$

In the notation used throughout this book $(abcd \cdots yz)$ is a permutation that replaces a by b, b by c, c by d, ..., y by z, and z by a. Other possible permutations of the numbers 1, 2, and 3 are (12), (23), and (132). The permutations (12), (23), and (13) are called *transpositions* or *interchanges*, and (123) and (132) are called *cycles*. If we permute more than three numbers we can have longer cycles. The following identities hold

$$(12) \equiv (21), \qquad (23) \equiv (32), \qquad (13) \equiv (31),$$
$$(123) \equiv (231) \equiv (312), \qquad (132) \equiv (321) \equiv (213). \tag{1-4}$$

An alternative notation for permutations can be used in which a permutation that replaces 1 by α_1, 2 by α_2, ..., n by α_n is denoted

$$\begin{pmatrix} 1 & 2 & \cdots & n \\ \alpha_1 & \alpha_2 & \cdots & \alpha_n \end{pmatrix}. \tag{1-5}$$

A permutation that we would write as $(abcd \cdots yz)$ would be written in the notation of Eq. (1-5) as

$$\begin{pmatrix} a & b & c & \cdots & y & z \\ b & c & d & \cdots & z & a \end{pmatrix}. \tag{1-6}$$

In this notation the order of the columns is immaterial. We can write the five permutations of the numbers 1, 2, and 3 as

$$(12) \equiv \begin{pmatrix} 123 \\ 213 \end{pmatrix}, \qquad (23) \equiv \begin{pmatrix} 123 \\ 132 \end{pmatrix}, \qquad (13) \equiv \begin{pmatrix} 123 \\ 321 \end{pmatrix},$$
$$(123) \equiv \begin{pmatrix} 123 \\ 231 \end{pmatrix}, \qquad \text{and} \qquad (132) \equiv \begin{pmatrix} 123 \\ 312 \end{pmatrix}. \tag{1-7}$$

Problem 1-1.   Which of the five distinct permutations in Eq. (1-4) changes the number order (a) from $\underline{123}$ to $\underline{132}$, (b) from $\underline{132}$ to $\underline{321}$, and (c) from $\underline{123}$ to $\underline{321}$?

Answer.   (a) (23), (b) (132), and (c) (13).

Problem 1-2.   If we label the protons in a CH_3F molecule with the numbers 1, 2, and 3 (see Fig. 1-1a, for example) we find that there are two distinct labeled forms: one with an anticlockwise labeling arrangement (looking down the C → F direction) and one, as in Fig. 1-1, with a clockwise labeling arrangement. On applying the five distinct permutations of Eq. (1-4) to the labels, what do you notice about the effect of (12), (23), or (13) when compared to the effect of (123) or (132)?

Answer. The permutations (12), (23), and (13) all interconvert the clockwise and anticlockwise labeled forms whereas (123) and (132) preserve the form.

Problem 1-2 is the first example we have met of permuting nuclei in a molecule and it requires further discussion. The permutation (12) is an operation that interchanges the nuclei labeled 1 and 2 and as a result, if the coordinates of nuclei 1 and 2 in a space fixed (X, Y, Z) axis system are initially (X_1, Y_1, Z_1) and (X_2, Y_2, Z_2), respectively, then after applying (12) they become (X_2, Y_2, Z_2) and (X_1, Y_1, Z_1), respectively; the nuclei are interchanged and thereby gain each other's coordinates. We can write

$$(12)[\underbrace{X_1, Y_1, Z_1}_{①}, \underbrace{X_2, Y_2, Z_2}_{②}, \ldots] = [\underbrace{X_1, Y_1, Z_1}_{②}, \underbrace{X_2, Y_2, Z_2}_{①}, \ldots] \quad (1\text{-}8)$$

$$\equiv [\underbrace{X_2, Y_2, Z_2}_{①}, \underbrace{X_1, Y_1, Z_1}_{②}, \ldots], \quad (1\text{-}9)$$

where the circled numbers represent the nuclei and the braces indicate their coordinates. Equation (1-8) shows the exchange of the nuclei 1 and 2, and Eq. (1-9) is just a reordering so that the coordinates of nucleus 1 are given first. An abbreviated notation in which the nuclear labels (i.e., the numbers in circles) are omitted but in which the coordinates of nucleus 1 are always given first, those of 2 second, etc., can be used so that we have

$$(12)[X_1, Y_1, Z_1, X_2, Y_2, Z_2, W] = [X_1', Y_1', Z_1', X_2', Y_2', Z_2', W']$$
$$\equiv [X_2, Y_2, Z_2, X_1, Y_1, Z_1, W]. \quad (1\text{-}10)$$

In Eq. (1-10) (X_i, Y_i, Z_i) are the initial coordinates of nucleus i, W represents the other coordinates in the molecule, and (X_i', Y_i', Z_i') are the coordinates of nucleus i after the permutation has been performed; clearly $W' \equiv W$.

Problem 1-3. Write down equations similar to Eqs. (1-8)–(1-10) for the effect of (123) on the spatial coordinates of the labeled protons in a CH_3F molecule.

Answer. Let us initially give protons 1, 2, and 3 the coordinates (X_1, Y_1, Z_1), (X_2, Y_2, Z_2), and (X_3, Y_3, Z_3), respectively, in the space fixed (X, Y, Z) system. The permutation (123) replaces proton 1 by proton 2 so that proton 2 then has as new coordinates the coordinates that proton 1 had initially, i.e., (X_1, Y_1, Z_1). Similarly (123) replaces 2 by 3 so that 3 then has the coordinates that 2 had, i.e., (X_2, Y_2, Z_2), and finally 1 ends up with the coordinates

(X_3, Y_3, Z_3) that 3 had. We can write

$$(123)[\underbrace{X_1, Y_1, Z_1}_{①}, \underbrace{X_2, Y_2, Z_2}_{②}, \underbrace{X_3, Y_3, Z_3}_{③}, \ldots]$$

$$= [\underbrace{X_1, Y_1, Z_1}_{②}, \underbrace{X_2, Y_2, Z_2}_{③}, \underbrace{X_3, Y_3, Z_3}_{①}, \ldots] \qquad (1\text{-}11)$$

$$\equiv [\underbrace{X_3, Y_3, Z_3}_{①}, \underbrace{X_1, Y_1, Z_1}_{②}, \underbrace{X_2, Y_2, Z_2}_{③}, \ldots], \qquad (1\text{-}12)$$

or omitting the circled numbers,

$$(123)[X_1, Y_1, Z_1, X_2, Y_2, Z_2, X_3, Y_3, Z_3, W]$$
$$= [X_1', Y_1', Z_1', X_2', Y_2', Z_2', X_3', Y_3', Z_3', W']$$
$$\equiv [X_3, Y_3, Z_3, X_1, Y_1, Z_1, X_2, Y_2, Z_2, W], \qquad (1\text{-}13)$$

where, here, (X_i', Y_i', Z_i') are the coordinates of proton i after having made the permutation (123). Figure 1-1 shows the effect of the permutation (123).

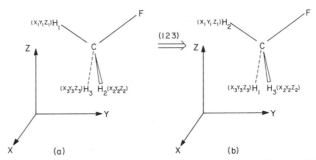

Fig. 1-1. (a) A proton labeled CH_3F molecule with (X, Y, Z) coordinates of H_1, H_2, and H_3 in space given by (X_1, Y_1, Z_1), (X_2, Y_2, Z_2), and (X_3, Y_3, Z_3), respectively. The nuclei H_1, C, and F are in the plane of the page, H_3 is below it and H_2 above it. (b) The same after having performed the permutation (123), i.e., H_1 replaced by H_2, H_2 by H_3, and H_3 by H_1. Now H_2, C, and F are in the plane of the page with H_1 below it and H_3 above it. The coordinates of H_1, H_2, and H_3 are now (X_3, Y_3, Z_3), (X_1, Y_1, Z_1), and (X_2, Y_2, Z_2), respectively.

It will be useful to define the effect of a nuclear permutation operation on a function of the nuclear coordinates (i.e., on a molecular wavefunction). To introduce the definition that we will adopt for this effect let us consider the following function of the X coordinates of the three protons in a CH_3F molecule:

$$f(X_1, X_2, X_3) = X_1 + 2X_2 + 3X_3. \qquad (1\text{-}14)$$

Using Eq. (1-13) we see that if initially the (X_1, X_2, X_3) nuclear coordinates are (a, b, c) then after applying the permutation (123) they are (c, a, b). The function defined in Eq. (1-14) has the value $a + 2b + 3c$ at the point (a, b, c) in (X_1, X_2, X_3) "space" (called *configuration space*) and the value $c + 2a + 3b$ at the point (c, a, b). Therefore, the effect of (123) is to change the value of the function at each point (X_1, X_2, X_3) into its value at the point $(X_1', X_2', X_3') = (X_3, X_1, X_2)$. We now define the effect of (123) on the function by writing

$$(123)f(X_1, X_2, X_3) = (123)[X_1 + 2X_2 + 3X_3]$$
$$= X_1' + 2X_2' + 3X_3' = X_3 + 2X_1 + 3X_2, \quad (1\text{-}15)$$

where we have introduced a new function of the nuclear coordinates, i.e., $X_3 + 2X_1 + 3X_2$. This new function has been defined so that its value at the point (X_1, X_2, X_3) is the same as the value of the original function, $X_1 + 2X_2 + 3X_3$, at the point $(X_1', X_2', X_3') = (X_3, X_1, X_2)$. We denote this new function by

$$f^{(123)}(X_1, X_2, X_3) = X_3 + 2X_1 + 3X_2, \quad (1\text{-}16)$$

and write

$$(123)f(X_1, X_2, X_3) = f^{(123)}(X_1, X_2, X_3). \quad (1\text{-}17)$$

With this convention the general equation defining the effect of a permutation P on a function is

$$Pf(X_1, Y_1, Z_1, \ldots, X_l, Y_l, Z_l) = f(X_1', Y_1', Z_1', \ldots, X_l', Y_l', Z_l')$$
$$= f^P(X_1, Y_1, Z_1, \ldots, X_l, Y_l, Z_l), \quad (1\text{-}18)$$

where (X_i', Y_i', Z_i') are the coordinates of nucleus i after applying the permutation P; $f^P(\ )$ is a new function generated from $f(\ )$ by applying P, and $f^P(\ )$ is such that its value at the point $(X_1, Y_1, Z_1, \ldots, X_l, Y_l, Z_l)$ is the same as the value of $f(\ )$ at the point $(X_1', Y_1', Z_1', \ldots, X_l', Y_l', Z_l')$. Equation (1-18) taken with the equation $P(X_i, Y_i, Z_i) = (X_i', Y_i', Z_i')$, for which examples are given in Eqs. (1-10) and (1-13), defines the effect of the nuclear permutation operator P as it is used in this book.

THE SUCCESSIVE APPLICATION OF PERMUTATIONS

The solution to Problem 1-1 can be represented pictorially as

$$\underline{123} \xrightarrow{(23)} \underline{132} \xrightarrow{(132)} \underline{321}.$$
$$\underline{\qquad\qquad\qquad} \;(13)\; \uparrow$$

This shows how the successive application of first (23) and then (132) to the number order $\underline{123}$ is equivalent to performing the single operation (13); we write

$$(132)(23)\underline{123} = (13)\underline{123}. \tag{1-19}$$

The left hand side of this equation can be expanded as

$$(132)[(23)\underline{123}] = (132)\underline{132} = \underline{321} \tag{1-20}$$

in order to make it clear why we chose to write the operations in the order $(132)(23)$, and not $(23)(132)$, when (23) is applied first. The successive application of first (23) and then (132) to any of the number arrangements in Eq. (1-1) is always equivalent to the effect of (13), as the reader can verify, and as a result we can write the *operator equation*

$$(132)(23) = (13). \tag{1-21}$$

We say that the *product* of (132) and (23) (or the result of *multiplying* them together) in this order is *equal* (or *equivalent*) to (13); they have the same effect on a number order.

In general the product of two permutations is independent of the number order chosen for the determination of the product [in Eq. (1-19) the number order $\underline{123}$ is used to determine $(132)(23)$], and with some practice it is possible to determine the product of two permutations without applying them to a number order. As an example let us determine the product $(23)(132)$; this is the same as the product in Eq. (1-21) but taken the other way around. The effect of (132) is to replace 1 by 3, 3 by 2, and 2 by 1; the effect of (23) is then to replace 3 by 2, 2 by 3, and 1 by 1. Overall doing first (132) and then (23) we have the sequences of replacements 1 by 3 by 2, 3 by 2 by 3, and 2 by 1 by 1, i.e., the overall replacements are 1 by 2, 3 by 3, and 2 by 1; this is the effect of (12). Hence

$$(23)(132) = (12). \tag{1-22}$$

Comparing Eqs. (1-21) and (1-22) we see that permutation multiplication is not necessarily commutative, i.e., if A and B are two permutations it is not necessarily true that $AB = BA$.

The result of multiplying permutations together using the notation of Eq. (1-5) can be understood by extending that notation a little. We can write Eq. (1-22) as

$$(23)(132) = \begin{pmatrix} 123 \\ 132 \end{pmatrix} \begin{pmatrix} 123 \\ 312 \end{pmatrix} = \begin{pmatrix} 312 \\ 213 \end{pmatrix} \begin{pmatrix} 123 \\ 312 \end{pmatrix} = \begin{pmatrix} 123 \\ 312 \\ 213 \end{pmatrix} = \begin{pmatrix} 123 \\ 213 \end{pmatrix} = (12), \tag{1-23}$$

where the "three-level" parenthesis indicates the successive replacements 1 by 3 by 2, 2 by 1 by 1, and 3 by 2 by 3 caused by the successive application of the two permutations.

The reader should now determine the result of multiplying all pairs of distinct permutations from Eq. (1-4) together. Three of the permutation products obtained are

$$(123)(123) = (132), \qquad (1\text{-}24)$$

$$(12)(23) = (123), \qquad (1\text{-}25)$$

and

$$(13)(23) = (132). \qquad (1\text{-}26)$$

From Eqs. (1-25) and (1-26) we see that the cycles can be written as products of transpositions, and it turns out that any permutation can be written as the product of a sequence of transpositions. For example, if we are interested in permuting the first seven integers we can have

$$(15432)(67) = (15)(54)(43)(32)(67). \qquad (1\text{-}27)$$

In general the breaking down of a permutation into a product of a series of transpositions is not unique, but what is unique is whether there is an odd or even number of transpositions in the product. A permutation is called *even* or *odd* depending on whether there is an even or odd number of transpositions in its transposition product. From Eqs. (1-25) and (1-26) we see that the permutations (123) and (132) are even, and from Eq. (1-27) we see that the permutation (15432)(67) is odd. The importance of determining whether a permutation is even or odd will become apparent when we consider Bose–Einstein and Fermi–Dirac statistical formulas in Chapter 6.

Problem 1-4. Determine whether the following permutations are even or odd: (146)(2357), (17)(23456), (1462357), and (14)(27)(36).

Answer. We can write these permutations in many ways as the product of transpositions but one possible way for each is now given:

$$\begin{aligned}
(146)(2357) &= (14)(46)(23)(35)(57), \\
(17)(23456) &= (17)(23)(34)(45)(56), \\
(1462357) &= (14)(46)(62)(23)(35)(57), \\
(14)(27)(36) &= (14)(27)(36).
\end{aligned} \qquad (1\text{-}28)$$

The third one is even and the rest are odd.

It is presumed that on multiplying all the permutations of Eq. (1-4) together in pairs the reader will have attempted

$$(123)(132) = ? \qquad (1\text{-}29)$$

and have wondered about what to give as the answer. Obviously doing first (132) and then (123) has the overall effect of leaving the number order unchanged, and we simply have to invent a symbol to mean "doing nothing" (the equivalent of 1 in algebraic multiplication). The symbol we use is E and we write

$$(123)(132) = E. \tag{1-30}$$

Also we can write equations such as

$$(12)E = (12). \tag{1-31}$$

E is called the *identity operation.*

Having defined the identity operation we can now define the *reciprocal,* or *inverse,* of an operation. We define the reciprocal of an operation A, say, as that operation which when multiplied on the right of A gives the identity. For example (132) is the reciprocal of (123) from the result in Eq. (1-30); i.e.,

$$(123)(123)^{-1} = E \tag{1-32}$$

defines $(123)^{-1}$, and from Eq. (1-30) we have

$$(123)^{-1} = (132). \tag{1-33}$$

Similarly

$$(132)^{-1} = (123), \tag{1-34}$$

and all the transpositions are self-reciprocal, i.e.,

$$(12)^{-1} = (12), \tag{1-35}$$

$$(23)^{-1} = (23), \tag{1-36}$$

and

$$(13)^{-1} = (13). \tag{1-37}$$

Using the notation of Eq. (1-5) the inverse of a permutation is obtained by simply turning the permutation "upside down," that is,

$$(abcd \cdots xyz)^{-1} = \begin{pmatrix} abc \cdots xyz \\ bcd \cdots yza \end{pmatrix}^{-1} = \begin{pmatrix} bcd \cdots yza \\ abc \cdots xyz \end{pmatrix}$$

$$\equiv \begin{pmatrix} azy \cdots dcb \\ zyx \cdots cba \end{pmatrix}$$

$$= (azyx \cdots dcb). \tag{1-38}$$

Hence, for example, we have,

$$(1423756)^{-1} = (1657324). \tag{1-39}$$

Problem 1-5. Determine the effect of the successive applications of first (132) and then (23) on the coordinates of the three protons in CH_3F and on the function $[X_1 + 2X_2 + 3X_3]$.

Answer. The effect of the successive application of these permutations on the proton coordinates is best understood by using the notation of Eqs. (1-8) and (1-11). In this case we write

$$(23)(132)[\underbrace{X_1,Y_1,Z_1}_{①}, \underbrace{X_2,Y_2,Z_2}_{②}, \underbrace{X_3,Y_3,Z_3}_{③}]$$

$$= (23)[\underbrace{X_1,Y_1,Z_1}_{③}, \underbrace{X_2,Y_2,Z_2}_{①}, \underbrace{X_3,Y_3,Z_3}_{②}]$$

$$= [\underbrace{X_1,Y_1,Z_1}_{②}, \underbrace{X_2,Y_2,Z_2}_{①}, \underbrace{X_3,Y_3,Z_3}_{③}]. \qquad (1\text{-}40)$$

Thus proton 3 ends up in the same place in space as it started and protons 1 and 2 are interchanged; this is in accordance with Eq. (1-22). A function such as $[X_1 + 2X_2 + 3X_3]$, which is to be read as "the X coordinate of proton 1 plus twice the X coordinate of proton 2 plus three times the X coordinate of proton 3," transforms under (23)(132) as

$$(23)(132)[X_1 + 2X_2 + 3X_3] = (23)[X_2 + 2X_3 + 3X_1]$$
$$= [X_2 + 2X_1 + 3X_3]. \qquad (1\text{-}41)$$

After applying (132) the X coordinates of protons 1, 2, and 3 are X_2, X_3, and X_1, respectively so that (23) permutes X_3 and X_1 in the function (the coordinates of 2 and 3). This is perhaps better understood if we introduce the notation that X_i' is the X coordinate of proton i after having applied (132) so that

$$(23)(132)[X_1 + 2X_2 + 3X_3] = (23)[X_1' + 2X_2' + 3X_3']$$
$$= [X_1' + 2X_3' + 3X_2']$$
$$= [X_2 + 2X_1 + 3X_3]. \qquad (1\text{-}42)$$

We see that if

$$f(X_1,X_2,X_3) = X_1 + 2X_2 + 3X_3,$$

then

$$f^{(23)(132)}(X_1,X_2,X_3) = X_2 + 2X_1 + 3X_3 \qquad (1\text{-}43)$$

and

$$f^{(23)(132)}(X_1,X_2,X_3) = f^{(12)}(X_1,X_2,X_3). \qquad (1\text{-}44)$$

PERMUTATION GROUPS

By considering all permutations of the integers 1, 2, and 3, and the results of all possible products of them, we have constructed the following set of distinguishable permutation operations:

$$\{E,(12),(23),(13),(123),(132)\}. \tag{1-45}$$

Having defined how to multiply these operations together in pairs we can construct the *multiplication table* of this set of elements and this is given in Table 1-1. In Table 1-1 each entry C represents the result of first applying the operation B at the head of its column and then applying the operation A at the beginning of its row, i.e., $C = AB$. The set of operations in Eq. (1-45) is, in fact, a *group* because it satisfies the following *group axioms*:

1. We can multiply (i.e., successively apply) the operations (or elements) together in pairs and the result is a member of the group.
2. One of the operations in the group is the identity operation E.
3. The reciprocal of each operation is a member of the group.
4. Multiplication of the operations is associative; that is, in a multiple product the answer is independent of how the operations are associated in pairs, e.g.,

$$(12)(123)(23) = (12)\underbrace{[(123)(23)]}_{(12)} = \underbrace{[(12)(123)]}_{(23)}(23) = E. \tag{1-46}$$

The fact that the group axioms 1, 2, 3, and 4 are satisfied by the set in Eq. (1-45) can be verified by inspecting the multiplication table (Table 1-1), and thus the set is a group. We will generally enclose the members of a group in braces { }.

Table 1-1

The multiplication table of the S_3 group[a]

	E	(12)	(23)	(13)	(123)	(132)
E:	E	(12)	(23)	(13)	(123)	(132)
(12):	(12)	E	(123)	(132)	(23)	(13)
(23):	(23)	(132)	E	(123)	(13)	(12)
(13):	(13)	(123)	(132)	E	(12)	(23)
(123):	(123)	(13)	(12)	(23)	(132)	E
(132):	(132)	(23)	(13)	(12)	E	(123)

[a] Each entry is the product of first applying the permutation at the top of the column and then applying the permutation at the left end of the row.

This particular group is called S_3, the permutation group (or symmetric group) of degree 3, and it consists of all permutations of three objects. There are six elements in S_3 and the group is said to have *order* six. In general, the permutation group S_n (all permutations of n objects) has order $n!$.

Problem 1-6. Prove that if $AB = E$, i.e., $A^{-1} = B$ from Eq. (1-32), then $BA = E$, i.e., $B^{-1} = A$ also, if A and B are in a group. (Hint: Evaluate BAB two different ways using axiom 4.)

Answer. The triple product BAB can be evaluated as $B[AB]$ or $[BA]B$. The first way of associating the elements gives the answer $BE = B$, since $AB = E$. The second way of associating the elements must give the same answer (by group axiom 4), i.e., $[BA]B = B$. This can only be true if $BA = E$ which was what we set out to prove. In S_3 we have, for example,

$$(123)(132) = (132)(123) = E. \qquad (1\text{-}47)$$

Axiom 4 can be used to prove that

$$(AB)^{-1} = B^{-1}A^{-1}. \qquad (1\text{-}48)$$

We can write

$$(AB)(B^{-1}A^{-1}) = A(BB^{-1})A^{-1} = AEA^{-1} = AA^{-1} = E, \qquad (1\text{-}49)$$

but

$$(AB)(AB)^{-1} = E, \qquad (1\text{-}50)$$

thus

$$(AB)^{-1} = B^{-1}A^{-1}.$$

In general the reciprocal of a product is the product of the reciprocals taken in the reverse order, e.g.,

$$(ABCD)^{-1} = D^{-1}C^{-1}B^{-1}A^{-1}. \qquad (1\text{-}51)$$

Problem 1-7. Three of the elements of S_3 taken by themselves satisfy the group axioms and hence form a group. Which three elements are they? Such a group is called a *subgroup* of the S_3 group.

Answer. It is clear that E has to be in the subgroup in order for it to form a group (by axiom 2). Perhaps the set $\{E,(12),(23)\}$ is a subgroup? This set satisfies axiom 3 by virtue of Eqs. (1-35) and (1-36). However, $(12)(23) = (123)$, and (123) is not in the set; thus by axiom 1 this set is not a group. By studying Table 1-1 the subgroup is found to be $\{E,(123),(132)\}$.

The only condition for a subset of the elements of a group to form a subgroup is that the subset contain all products of its members. The subgroup discussed in Problem 1-7 has order 3 and it is called an *Abelian group* since multiplication within it is commutative [see the first equality in Eq. (1-47)]. Other subgroups of S_3 are $\{E,(12)\}$, $\{E,(23)\}$, and $\{E,(13)\}$, each having order 2, and $\{E\}$ of order 1.

THE COMPLETE NUCLEAR PERMUTATION GROUP OF A MOLECULE

Labeling the protons in the CH_3F molecule as 1, 2, and 3, the group S_3 of Eq. (1-45) contains all possible permutations of identical nuclei in the molecule. We call this group the *complete nuclear permutation* (CNP) group of the CH_3F molecule. The CNP group of a molecule having n identical nuclei of one type, and no sets of other identical nuclei, will be the group S_n of permutations of these nuclei.

If a molecule has more than one set of identical nuclei the definition of the CNP group is more complicated and we will illustrate this by considering the ethylene molecule C_2H_4. Labeling the protons in the molecule 1 to 4, the group of all proton permutations is the group S_4. Numbering the carbon nuclei 5 and 6 the group of all carbon nuclei permutations is the group $S_2 = \{E,(56)\}$. We will denote these two nuclear permutation groups $S_4^{(H)}$ and $S_2^{(C)}$. The group of all possible permutations of identical nuclei in the molecule (the CNP group) will therefore consist of all 4! elements of the group $S_4^{(H)}$ *and* of all of these elements taken in combination with (56); $2 \times 4!$ elements in all. The element (56) will commute with all the elements of the $S_4^{(H)}$ group, since these two groups involve permutations of different types of nuclei. This CNP group is called the *direct product* of the groups $S_4^{(H)}$ and $S_2^{(C)}$ and is written

$$G^{CNP} = S_4^{(H)} \otimes S_2^{(C)}. \tag{1-52}$$

In general the direct product of a group $A = \{A_1 \equiv E, A_2, \ldots, A_n\}$ and a group $B = \{B_1 \equiv E, B_2, \ldots, B_m\}$, where these are different types of groups so that all A_i commute with all B_j, is the set of $n \times m$ elements $A_iB_j(= B_jA_i)$ where $i = 1$ to n and $j = 1$ to m. One element in this set, the element A_1B_1, will be the identity, and the product of any two elements in the set will give another element of the set. This latter result follows by considering

$$(A_iB_j)(A_kB_l) = A_iA_kB_jB_l = A_pB_q, \tag{1-53}$$

where

$$A_p = A_iA_k \quad \text{in the group } A,$$
$$B_q = B_jB_l \quad \text{in the group } B,$$

and A_pB_q must be in the set of direct product elements since it is defined to contain all products. The inverse of each element is present since

$$(A_i B_j)^{-1} = B_j^{-1} A_i^{-1} \tag{1-54}$$

and B_j^{-1} is an element in B and A_i^{-1} in A, since A and B are groups. The associative law of multiplication must hold for the direct product elements since it holds within the groups A and B. Thus the set of $n \times m$ elements $A_i B_j$ forms a group and it is called the direct product group. In this definition it is crucial that each element in A commute with each element in B.

The CNP group of a molecule containing l identical nuclei of one set, m of another, n of another, and so on, is the direct product group

$$G^{\text{CNP}} = S_l \otimes S_m \otimes S_n \otimes \cdots, \tag{1-55}$$

and the order of the group is $l! \times m! \times n! \times \cdots$.

Problem 1-8. Write down the elements of the CNP group of the molecule CHOCOOH.

Answer. The CNP group is the direct product group

$$
\begin{aligned}
G^{\text{CNP}} &= S_3^{(O)} \otimes S_2^{(H)} \otimes S_2^{(C)} \\
&= \{E, (12), (23), (13), (123), (132)\} \otimes \{E, (45)\} \otimes \{E, (67)\}, \tag{1-56}
\end{aligned}
$$

where we label the oxygen nuclei 1, 2, and 3; the protons 4 and 5; the carbon nuclei 6 and 7. The complete list of $(3! \times 2! \times 2! = 24)$ elements is

$$
\begin{aligned}
\{E, &(12), (23), (13), (123), (132), \\
&(45), (12)(45), (23)(45), (13)(45), (123)(45), (132)(45), \\
&(67), (12)(67), (23)(67), (13)(67), (123)(67), (132)(67), (45)(67), \\
&(12)(45)(67), (23)(45)(67), (13)(45)(67), (123)(45)(67), (132)(45)(67)\}. \tag{1-57}
\end{aligned}
$$

Problem 1-9. We label the protons of an ethylene molecule 1 to 4, and the carbon nuclei 5 and 6 as shown in Fig. 1-2. There are 11 other distinct

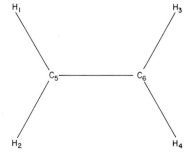

Fig. 1-2. An ethylene molecule with the nuclei labeled 1 to 6.

ways of putting the labels 1 to 6 on the nuclei so that the protons are labeled 1 to 4 and the carbon nuclei 5 and 6; write down these other distinct ways (distinct means those not simply obtained by rigidly rotating the molecule in space). If we only include permutations from the CNP group of ethylene that do not convert our chosen labeled form into any of the other 11 forms what set of permutations is obtained? Prove that this set of permutations forms a group.

Answer.   The twelve distinct nuclear labeled forms of ethylene are shown in Fig. 1-3. The set of permutations that does not convert the form in Fig. 1-3a into any other form is

$$\{E, (12)(34), (13)(24)(56), (14)(23)(56)\},\qquad\qquad (1\text{-}58)$$

and this set is a group, a subgroup of the CNP group, since any product of the elements of the set gives an element that is in the set.

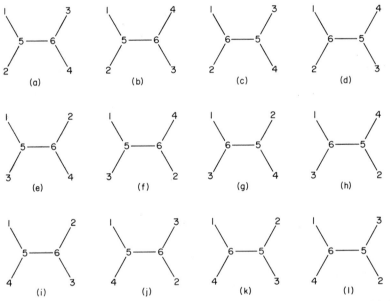

Fig. 1-3.   The 12 distinct forms of an ethylene molecule in which the protons are labeled 1 to 4 and the carbon nuclei 5 and 6. Distinct forms cannot be interconverted by simply rigidly rotating them in space.

In Problem 1-9 we obtained a subgroup of the CNP group of the ethylene molecule by considering only elements that do not interconvert some distinct nuclear labeled forms of the molecule. Notice that the subgroup of S_3, of order 3, obtained as answer to Problem 1-7, contains all elements of the CNP group of CH_3F that do not interconvert the two distinct (clockwise and anticlockwise) labeled forms of the molecule. The idea of distinct nuclear labeled forms of a molecule and of a subgroup of elements that does not interconvert them will prove to be very important when we come to define the molecular symmetry group.

BIBLIOGRAPHICAL NOTES

Hamermesh (1962).   Chapter 1.

Tinkham (1964).   Chapter 2.

Wigner (1959).   Permutations are discussed on pages 64–65 and on pages 124–126. A group is defined on pages 58–59. The effect of an operation on a function is discussed on pages 104–106 and a different convention is used by Wigner from that adopted here; this is discussed further in the Bibliographical Notes to Chapter 5.

2

The Inversion Operation and Permutation Inversion Groups

In this chapter the inversion operation E^* is defined and its effect on a molecule is discussed; its use allows us to determine the parity of a molecular wavefunction. The effect of the successive application of E^* and of a nuclear permutation operation is also considered, and as a result we are able to define the complete nuclear permutation inversion group of a molecule. The idea (but not a rigorous definition) of the molecular symmetry group is introduced, and the detailed effect of an element of the group on the coordinates of the nuclei and electrons in a molecule is presented.

THE INVERSION OPERATION AND PARITY

As well as permuting the coordinates of identical nuclei it will be necessary for us to consider the effect of inverting the spatial coordinates of all particles (nuclei and electrons) in a molecule through the origin of the space fixed axis system. This operation involves changing the sign of the Cartesian coordinates in space of all the particles in the molecule. Since the translational motion of a molecule will not be of any concern to us, it will prove convenient to refer the coordinates of the electrons and nuclei to the molecular center of mass; this involves *the separation of the translational motion*, which will be discussed more fully in Chapter 6. From now on when we discuss the spatial coordinates of the nuclei and electrons in a molecule we will usually use an (X, Y, Z) axis system parallel to the space fixed (X, Y, Z) axis system but with origin at the molecular center of mass. The inversion operation E^*, when applied to a molecule, is defined as the operation of inverting the spatial coordinates of all the nuclei and electrons through the molecular center of mass. Using the (X, Y, Z) coordinates of a nucleus or electron we can write

$$E^*[X_i, Y_i, Z_i] = [X_i', Y_i', Z_i'] = [-X_i, -Y_i, -Z_i]. \qquad (2\text{-}1)$$

Problem 2-1. Write down the effect of E^* on the spatial coordinates of the five nuclei in a CH_3F molecule. Draw a picture of a CH_3F molecule and draw the effect of E^*.

Answer. By definition E^* simply changes the sign of the (X, Y, Z) coordinates of all the nuclei, i.e.,

$$E^*[X_1, Y_1, Z_1, \ldots, X_F, Y_F, Z_F] = [-X_1, -Y_1, -Z_1, \ldots, -X_F, -Y_F, -Z_F]. \qquad (2\text{-}2)$$

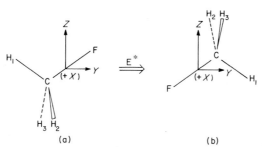

Fig. 2-1. The effect of E^* on a CH_3F molecule. The $(+X)$ indicates that the X axis is pointing up out of the plane of the page.

In Fig. 2-1a a methyl fluoride molecule is arbitrarily oriented with its center of mass at the origin of the (X, Y, Z) axis system. The operation E^* inverts the coordinates of all particles in the origin of the (X, Y, Z) system, and the result is shown in Fig. 2-1b; the (X, Y, Z) axis system is unaffected by E^* by definition. In Fig. 2-2 the effect of E^* is given without showing the (X, Y, Z) axes and in future figures the axes will usually be omitted. Figures 2-1 and 2-2 are simply pictorial representations of Eq. (2-2). Although not mentioned in Eq. (2-2) and not explicitly shown in Figs. 2-1 and 2-2, the coordinates of the electrons are also inverted by E^*.

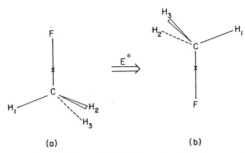

(a)   (b)

Fig. 2-2.   The effect of E^* on CH_3F without drawing the space fixed axes.

We define the effect of E^* on a function of the nuclear and electronic co-ordinates in the same way as we did for a nuclear permutation in Eq. (1-18), i.e.,

$$E^*f(X_1, Y_1, Z_1, \ldots, X_n, Y_n, Z_n) = f(-X_1, -Y_1, -Z_1, \ldots, -X_n, -Y_n, -Z_n)$$
$$= f^{E^*}(X_1, Y_1, Z_1, \ldots, X_n, Y_n, Z_n). \qquad (2\text{-}3)$$

The value of the new function $f^{E^*}()$ at each point $(X_1, Y_1, Z_1, \ldots)$ is the same as the value of the original function $f()$ at each point $(-X_1, -Y_1, -Z_1, \ldots)$. If a function $f()$ is such that its values at the points $(X_1, Y_1, Z_1, \ldots)$ and $(-X_1, -Y_1, -Z_1, \ldots)$ are the same then

$$f^{E^*}() = f() \qquad (2\text{-}4)$$

and the function is said to have *positive parity*. If on the other hand the values of the function at the points $(X_1, Y_1, Z_1, \ldots)$ and $(-X_1, -Y_1, -Z_1, \ldots)$ are the negative of each other then

$$f^{E^*}() = -f() \qquad (2\text{-}5)$$

and the function has *negative parity*. For a function of positive parity we can write

$$E^*f() = f(), \qquad (2\text{-}6)$$

and for a function of negative parity we can write

$$E^*f() = -f(). \qquad (2\text{-}7)$$

COMBINING PERMUTATIONS WITH THE INVERSION

Figure 2-2 shows the effect of the inversion operation E^* on a methyl fluoride molecule. We could easily apply a permutation operation, (12) say, after doing E^* (i.e., to Fig. 2-2b), and the resulting combined operation would be written $(12)E^*$. The effect of $(12)E^*$ on the coordinates of the nuclei is given by

$$(12)E^*[X_1, Y_1, Z_1, X_2, Y_2, Z_2, X_3, Y_3, Z_3, X_C, Y_C, Z_C, X_F, Y_F, Z_F]$$
$$= [-X_2, -Y_2, -Z_2, -X_1, -Y_1, -Z_1, -X_3, -Y_3, -Z_3,$$
$$-X_C, -Y_C, -Z_C, -X_F, -Y_F, -Z_F]. \tag{2-8}$$

The effect of a permutation operation P is to interchange the nuclear coordinates, and the inversion changes the sign of all coordinates. It is clearly immaterial whether we change the sign of the coordinates with E^* before or after interchanging them with P, and thus E^* and P will commute. We can write

$$(12)E^* = E^*(12) = (12)^*, \tag{2-9}$$

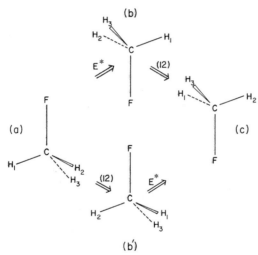

Fig. 2-3. Drawings to show the equivalence of $E^*(12)$ $[a \rightarrow b' \rightarrow c]$ and $(12)E^*$ $[a \rightarrow b \rightarrow c]$ for a CH_3F molecule.

where we introduce the notation of P^* for $PE^* = E^*P$. In Fig. 2-3 the results of $(12)E^*$ and $E^*(12)$ are drawn to show their equivalence. The transformation from Fig. 2-3a to Fig. 2-3c is a pictorial representation of Eq. (2-8), and it should be remembered that the electron coordinates are inverted by any permutation inversion operation P^* such as this. We will be concerned with the effects of both P and P^* operations on nuclear and electron coordinates and on functions of the coordinates.

Problem 2-2.   By drawing pictures such as Fig. 2-3 verify that E^* commutes with all the elements of the complete nuclear permutation group of CH_3F, where this group is

$$S_3^{(H)} = \{E,(12),(23),(13),(123),(132)\}. \tag{2-10}$$

Which of the operations in the direct product group $S_3^{(H)} \otimes \{E, E^*\}$, i.e., the group

$$\{E,(12),(23),(13),(123),(132), E^*,(12)^*,(23)^*,(13)^*,(123)^*,(132)^*\}, \tag{2-11}$$

does not interconvert clockwise and anticlockwise labeled CH_3F?

Answer.   The set of operations that does not interconvert the clockwise and anticlockwise labeled forms is the group

$$\{E,(123),(132),(12)^*,(23)^*,(13)^*\}. \tag{2-12}$$

For reasons that will emerge we call this group $C_{3v}(M)$.

The group in Eq. (2-11) is the *complete nuclear permutation inversion* (CNPI) group of CH_3F; the CNPI group of a given molecule contains all possible permutations of identical nuclei in the molecule with and without inversion. Therefore, the CNPI group of a molecule is the direct product of the complete nuclear permutation group [as introduced in Eq. (1-55)] and the *inversion group* $\mathscr{E} = \{E, E^*\}$; the CNPI group contains twice as many elements as the complete nuclear permutation group.

The group in Eq. (2-12) is the molecular symmetry group of the CH_3F molecule in its ground electronic state. The definition of the molecular symmetry group will be discussed in detail in Chapter 9 but it is appropriate to give an outline of the definition here. The molecular symmetry group is a subgroup of the CNPI group and it results from the fact that for many molecules it is possible in practice to define *distinctly different nuclear labeled forms*. These forms are such that they do not interconvert (or "tunnel") on a time scale short enough to be experimentally detected (i.e., no energy level splittings or shifts resulting from the tunneling are observed). For example, in CH_3F the two distinctly different nuclear labeled forms are the clockwise and anticlockwise labeled forms. The molecular symmetry group is that subgroup of the CNPI group consisting of all elements that do not interconvert distinctly different nuclear labeled forms.

Problem 2-3.   What is the complete nuclear permutation inversion group of the ethylene molecule using the nuclear labeling convention of Fig. 1-2? Which of the elements do not convert the form in Fig. 1-3a into any of the other 11 distinct nuclear labeled forms of Fig. 1-3?

Answer. The complete nuclear permutation inversion group is the direct product

$$S_4^{(H)} \otimes S_2^{(C)} \otimes \mathscr{E}, \qquad (2\text{-}13)$$

and it has $4! \times 2! \times 2 = 96$ elements. Typical elements are (132), $(13)(56)$, $(1234)(56)$, $(12)(34)^*$, $(124)(56)^*$, etc. The subgroup that does not convert the form in Fig. 1-3a into any of the other distinct nuclear labeled forms is

$$\{E, (12)(34), (13)(24)(56), (14)(23)(56), E^*, (12)(34)^*,$$
$$(13)(24)(56)^*, (14)(23)(56)^*\}, \qquad (2\text{-}14)$$

and this is the molecular symmetry group of the rigidly planar ethylene molecule; it is called $D_{2h}(M)$. Because E^* commutes with permutations we can write, for example,

$$(12)(34)^* = (12)(34)E^* = (12)E^*(34) = E^*(12)(34) = (12)^*(34). \quad (2\text{-}15)$$

Problem 2-4. Draw the effect of separately applying $(12)(34)^*$ and $(13)(24)(56)$ to the ethylene molecule in Fig. 1-2. Is there any difference between the effects of these operations?

Answer. Yes. If the nuclei of the ethylene molecule are in the equilibrium configuration (as they are in Fig. 1-2) then $(12)(34)^*$ and $(13)(24)(56)$ have the same effect on the nuclear coordinates, but $(12)(34)^*$ inverts the electron coordinates whereas $(13)(24)(56)$ does not. If the nuclei are not in the equilibrium configuration then the effects of these operations on the nuclear coordinates will usually be different, and this is discussed in more detail in the next section.

THE DETAILED EFFECTS OF P AND P^* OPERATIONS

In the preceding section we have discussed the effect of a nuclear permutation operation P, and the effect of a nuclear permutation operation accompanied by an inversion, P^*, several times. However, in the figures we have always considered the molecule to be in its equilibrium configuration. This was not necessary but it made it easier to draw the figures. Also we have given figures *and* equations, and it is a good idea to use both in order to have a full understanding of and ability to use the molecular symmetry group. In this section we are going to look at the effect of each of the two operations used in Problem 2-4 in more detail, both with figures and equations, and we are going to apply them to an ethylene molecule that is not in its equilibrium configuration.

In Fig. 2-4 the shaded circles show the instantaneous positions of the four hydrogen nuclei and two carbon nuclei in an ethylene molecule that is

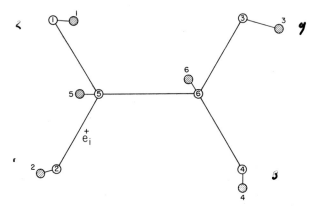

Fig. 2-4. An ethylene molecule. The shaded circles represent an instantaneous nuclear arrangement (i.e., a "snapshot" of the rotating–vibrating molecule), and the open circles the appropriate equilibrium nuclear arrangement. $\overset{+}{e}_i$ represents an electron i above the plane of the page.

rotating and vibrating in space; the open circles show the appropriate equilibrium nuclear positions, and $\overset{+}{e}_i$ represents electron i above the plane of the page ($\overset{-}{e}_i$ would represent electron i below the plane of the page). For simplicity we have chosen to keep all the nuclei in the plane of the page. We can write the instantaneous coordinates of the nuclei and electrons in the space fixed (X, Y, Z) axis system (the origin is at the common center of mass of the instantaneous and equilibrium configuration molecules) as $[X_1, Y_1, Z_1, \ldots, X_6, Y_6, Z_6; X_i, Y_i, Z_i]$, where the electrons are labeled i.

We wish to consider the effects of the two operations $(12)(34)^*$ and $(13)(24)(56)$ of the molecular symmetry group of ethylene. Equation (2-16) and Fig. 2-5 show the effect of $(12)(34)^*$, and Eq. (2-17) and Fig. 2-6 show the effect of $(13)(24)(56)$, on the nuclear and electronic coordinates.

$$(12)(34)^*[X_1, Y_1, Z_1, X_2, Y_2, Z_2, X_3, Y_3, Z_3, X_4, Y_4, Z_4, X_5, Y_5, Z_5,$$
$$X_6, Y_6, Z_6; X_i, Y_i, Z_i] = [-X_2, -Y_2, -Z_2, -X_1, -Y_1, -Z_1,$$
$$-X_4, -Y_4, -Z_4, -X_3, -Y_3, -Z_3,$$
$$-X_5, -Y_5, -Z_5, -X_6, -Y_6, -Z_6;$$
$$-X_i, -Y_i, -Z_i], \qquad (2\text{-}16)$$

and

$$(13)(24)(56)[X_1, Y_1, Z_1, X_2, Y_2, Z_2, X_3, Y_3, Z_3, X_4, Y_4, Z_4, X_5, Y_5, Z_5,$$
$$X_6, Y_6, Z_6; X_i, Y_i, Z_i] = [X_3, Y_3, Z_3, X_4, Y_4, Z_4, X_1, Y_1, Z_1,$$
$$X_2, Y_2, Z_2, X_6, Y_6, Z_6, X_5, Y_5, Z_5;$$
$$X_i, Y_i, Z_i]. \qquad (2\text{-}17)$$

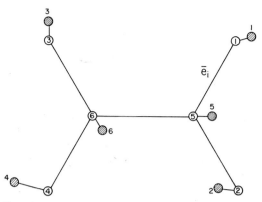

Fig. 2-5. The effect of applying the operation (12)(34)* to the ethylene molecule of Fig. 2-4. $\bar{e}_i$ represents electron i below the plane of the page.

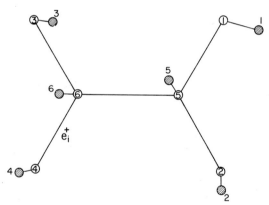

Fig. 2-6. The effect of applying the operation (13)(24)(56) to the ethylene molecule of Fig. 2-4.

If the molecule were in its equilibrium configuration then because of the centrosymmetric nature of the molecule we would have

$$X_1 = -X_4, \quad Y_1 = -Y_4, \quad Z_1 = -Z_4, \quad X_2 = -X_3, \quad Y_2 = -Y_3,$$
$$Z_2 = -Z_3, \quad X_5 = -X_6, \quad Y_5 = -Y_6, \quad \text{and} \quad Z_5 = -Z_6, \tag{2-18}$$

so that the effects of (12)(34)* and (13)(24)(56) on the nuclear coordinates would be the same. This can be seen by inserting Eq. (2-18) into the right hand side of Eq. (2-16) and comparing with Eq. (2-17), or by looking at the positions of the open circles in Figs. 2-5 and 2-6. The electronic coordinates are inverted by (12)(34)* but not by (13)(24)(56).

SUMMARY

In this chapter and the previous one the complete nuclear permutation (CNP) group and the complete nuclear permutation inversion (CNPI) group of a molecule have been defined. The idea of the molecular symmetry (MS) group has also been introduced. The effects of the elements of these groups, and of products of these elements, on nuclear and electronic coordinates in space (and on functions of these coordinates) has been explained.

The reader who has some idea of how symmetry groups are used, and who wants to understand the detailed definition of the MS group without further ado, could now read Chapter 9 before reading Chapter 3.

3

Rotation Groups
and Point Groups

In this chapter the geometrical symmetries of some three-dimensional objects are considered in order to define rotation groups and point groups. The application of these groups to molecules is discussed in an introductory fashion only.

ROTATIONAL SYMMETRY AND THE ROTATION GROUP

In Fig. 3-1a we show an equilateral triangular prism, and in Fig. 3-1b a box, the "prism box," into which the prism exactly fits when the lid is closed. There are six ways in which the prism can be placed to fit into the prism box depending on which of the two triangular faces is down and on which of the three triangle vertices is in a given corner of the box; there are thus six equivalent spatial orientations of an equilateral triangular prism. Similarly a solid cube can be placed inside a "cube box" that exactly fits it in any one of 24 different ways. The number of different ways that a solid body can be placed into a box that exactly fits it is a measure of the rotational symmetry of the object, and a cube has more rotational symmetry than an equilateral triangular prism.

To be more specific about the rotational symmetry of an object we introduce *rotational symmetry axes* and *rotational symmetry operations*. As

29

(a)   (b)

Fig. 3-1. (a) An equilateral triangular prism and (b) the "prism box" into which it exactly fits when the lid is closed.

drawn in Fig. 3-1 the equilateral prism can be dropped straight into the prism box. If the prism is rotated by $2\pi/3$ or $4\pi/3$ radians about an axis passing vertically through the center of the triangular faces it can again be dropped straight into the prism box; such an axis is termed a rotational symmetry axis, and the rotations through $2\pi/3$ or $4\pi/3$ radians about the axis are rotational symmetry operations. A rotational symmetry operation of a solid object is a rotation of the object, about an axis passing through its center of mass, that leaves the object in an equivalent spatial orientation; the axis about which the rotation takes place is called a rotational symmetry axis. To specify the rotational symmetry axes of the prism we label the vertices and introduce a, b, c, and d axes as shown in Fig. 3-2. Each of the a, b, c, and d axes is a rotational symmetry axis and the rotational symmetry operations are

C_{2a} a rotation through π radians about the a axis,
C_{2b} a rotation through π radians about the b axis,
C_{2c} a rotation through π radians about the c axis,
C_{3d} a rotation through $2\pi/3$ radians in a right handed sense about the d axis (this takes vertex 1 to where vertex 3 was), and
C_{3d}^2 a rotation through $4\pi/3$ radians in a right handed sense about the d axis.

Including the identity operation E (that of doing no rotation) with these five operations, and defining multiplication of these operations to be their successive application, we obtain the rotational symmetry group $\boldsymbol{D}_3$:

$$\boldsymbol{D}_3 = \{E, C_{2a}, C_{2b}, C_{2c}, C_{3d}, C_{3d}^2\}. \tag{3-1}$$

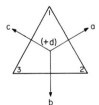

Fig. 3-2. The axis and vertex labeling convention used for an equilateral triangular prism. The $(+d)$ indicates that the d axis is pointing up out of the page.

D_3 is the rotational symmetry group of an equilateral triangular prism. The multiplication table for the elements of D_3 is given in Table 3-1. The reader is urged to check that this table is correct, and to use it to prove that D_3 is a group by testing that the four group axioms are obeyed. As an example of an entry in the table we show, in Fig. 3-3, that (we define the axes to move with the prism when it is rotated)

$$C_{3d}^2 C_{2a} = C_{2b}. \qquad (3-2)$$

The application of each of the six operations of the D_3 group to the equilateral triangular prism produces the six different ways that the prism can be placed into the prism box. The operations C_{2a}, C_{2b}, and C_{2c} are *twofold rotation operations*, since if done twice they are equivalent to E, and the operation C_{3d} is a threefold rotation operation. The a, b, and c axes are called twofold rotation axes and the d axis is a threefold rotation axis; rotation axes must pass through the center of mass of the object. An equilateral triangular prism is said to *have* D_3 rotational symmetry or to *belong* to the D_3 rotational symmetry group.

The rotational symmetry of an object is obtained by determining the number and type of rotational symmetry axes that are present. An object

Table 3-1
The multiplication table of the rotation group D_3[a]

	E	C_{2a}	C_{2b}	C_{2c}	C_{3d}	C_{3d}^2
E:	E	C_{2a}	C_{2b}	C_{2c}	C_{3d}	C_{3d}^2
C_{2a}:	C_{2a}	E	C_{3d}	C_{3d}^2	C_{2b}	C_{2c}
C_{2b}:	C_{2b}	C_{3d}^2	E	C_{3d}	C_{2c}	C_{2a}
C_{2c}:	C_{2c}	C_{3d}	C_{3d}^2	E	C_{2a}	C_{2b}
C_{3d}:	C_{3d}	C_{2c}	C_{2a}	C_{2b}	C_{3d}^2	E
C_{3d}^2:	C_{3d}^2	C_{2b}	C_{2c}	C_{2a}	E	C_{3d}

[a] Each entry represents the result of first applying the operation at the top of the column and then applying the operation at the left end of the row.

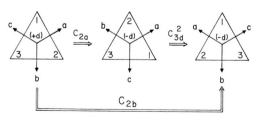

Fig. 3-3. Diagram to show that $C_{3d}^2 C_{2a} = C_{2b}$ in the D_3 rotation group of Eq. (3-1). The $(-d)$ indicates that the d axis is pointing down into the page.

having one n-fold rotational symmetry axis, and no other rotational symmetry axes, has C_n rotational symmetry. For example, a square based pyramid has C_4 rotational symmetry and the C_4 rotation group has elements $\{E, C_4, C_4{}^2, C_4{}^3\}$, where the rotations are about the vertical axis of the pyramid. An object having one n-fold rotation axis, and n twofold rotation axes perpendicular to it, has D_n rotational symmetry. The rotation group of a regular tetrahedron has order 12 and is called T, and that of a cube has order 24 and is called O.

The rotation group of a sphere consists of all rotations about any axis that passes through the center of mass of the sphere. We can specify a given rotation operation by specifying the direction of the axis of rotation relative to the space fixed axis system, for which we need two angles (α and β, say), and by specifying the amount of the rotation about this axis, for which we need another angle (γ, say). To each rotation there corresponds a set of values of these parameters, where we have the restrictions

$$0 \le \alpha \le 2\pi, \qquad 0 \le \beta \le \pi, \qquad 0 \le \gamma \le \pi. \qquad (3\text{-}3)$$

The restrictions follow since a rotation through $\varepsilon + 2\pi$ radians about an axis is identical to a rotation through ε about the axis. This group is called the *three-dimensional pure rotation group* K, and it has an infinite number of elements, i.e., it is an *infinite group*. An infinite group whose elements are specified by continuously variable parameters, as is the case here (the parameters being α, β, and γ) is said to be a *continuous group*. Two other continuous rotation groups are the C_∞ and D_∞ groups. The C_∞ group is the rotation group of a cone and the D_∞ group is the rotation group of a cylinder.

REFLECTION SYMMETRY AND THE POINT GROUP

In Fig. 3-4 a pyramid with an equilateral triangular base is drawn; this three-dimensional figure will be used to introduce the concept of a point group. The pyramid has threefold rotational symmetry about the d axis, and

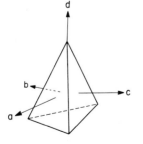

Fig. 3-4.   An equilateral triangular based pyramid with the axes used in describing its symmetry.

it also has *reflection symmetry* in the *ad*, *bd*, and *cd* planes. A reflection symmetry operation for a three-dimensional object is a reflection of the object through a plane (a reflection symmetry plane) that leaves the object in an equivalent spatial orientation; the plane will pass through the center of mass of the object and this center point will be common to all rotational symmetry axes and reflection symmetry planes (hence the name *point* group). The point group of a three-dimensional object consists of all rotation symmetry operations, all reflection symmetry operations, and of all possible products of such operations (although the individual rotation and reflection operations that make up a rotation–reflection product symmetry operation do not necessarily have to be symmetry operations). The point group of the pyramid in Fig. 3-4 is called C_{3v}, and it consists of the rotation operations E, C_{3d}, and C_{3d}^2, plus the operations of reflection in the planes *ad*, *bd*, and *cd*; we will call these reflection symmetry operations σ_{ad}, σ_{bd}, and σ_{cd}. The operations E, C_{3d}, and C_{3d}^2 form the rotational subgroup of C_{3v}; the rotational subgroup of the point group of an object consists of E and of all the rotational symmetry operations in the group (it is the rotational symmetry group of the object).

In naming the point groups we use the *Schoenflies notation*, and the possible names in this notation are:

$$C_s(= C_{1v} = C_{1h} = S_1), C_i(= S_2), C_n, D_n, C_{nv}, C_{nh}(= S_n \text{ with } n \text{ odd}),$$
$$D_{nd}, D_{nh}, S_n(n \text{ even}), T, T_d, O, O_h, I, I_h, \text{ and } K_h,$$

where n is an integer (for continuous point groups it is infinity). Each of these point groups consists of a certain definite set of symmetry operations (rotations, reflections, and rotation–reflections), and any three-dimensional object can be classified as *belonging* to one of these point groups. The point group symmetry of an object is determined from the following:

C_n one *n*-fold rotation axis,

C_{nv} one *n*-fold rotation axis and *n* reflection planes containing this axis,

C_{nh} one *n*-fold rotation axis and one reflection plane perpendicular to this axis,

D_n one *n*-fold rotation axis and *n* twofold rotation axes perpendicular to it,

D_{nd} those of D_n plus *n* reflection planes containing the *n*-fold rotation axis and bisecting the angles between the *n* twofold rotation axes,

D_{nh} those of D_n plus a reflection plane perpendicular to the *n*-fold rotation axis,

S_n one alternating axis of symmetry (about which rotation by $2\pi/n$ radians followed by reflection in a plane perpendicular to the axis is a symmetry operation).

The point groups T_d, O_h, and I_h consist of all rotation, reflection, and rotation–reflection symmetry operations of a regular tetrahedron, cube, and icosahedron, respectively (the rotational subgroups of these point groups are called T, O, and I, respectively). The point group K_h consists of all the rotation, reflection, and rotation–reflection symmetry operations of a sphere.

THE POINT GROUP SYMMETRY OF MOLECULES

If a molecule has a unique equilibrium nuclear configuration then we can determine the point group symmetry of the equilibrium molecular structure. For example, the nuclei of a CH_3F molecule in the equilibrium configuration of its electronic ground state form a structure with C_{3v} point group symmetry; the C–F bond being the threefold rotation axis and the H–C–F planes the reflection planes. The CH_3F molecule is said to have C_{3v} point group symmetry in its electronic ground state.

Problem 3-1.   What is the point group symmetry of the ethylene molecule in its electronic ground state (see Fig. 1-2)?

Answer.   The equilibrium molecular structure of the ethylene molecule has three twofold rotation axes: one (the z axis) along the C–C bond, one (the y axis) perpendicular to the z axis and in the molecular plane, and one (the x axis) that is perpendicular to both the z and y axes. The xy, yz, and xz planes are reflection symmetry planes so that the point group symmetry is D_{2h}. The elements in the D_{2h} point group of ethylene are (where i is the product $C_{2z}\sigma_{xy}$)

$$\{E, C_{2x}, C_{2y}, C_{2z}, i, \sigma_{xy}, \sigma_{yz}, \sigma_{xz}\}. \tag{3-4}$$

Examples of some common molecular point group symmetries are:

C_s the HDO molecule (a monodeuterated water molecule),
C_{2v} the H_2O molecule,
C_{2h} the trans $C_2H_2F_2$ molecule (*trans*-1,2-difluoroethylene),
C_{3v} the CH_3F molecule,
$C_{\infty v}$ the HCN molecule,
D_{2d} the CH_2CCH_2 molecule (allene),
D_{2h} the CH_2CH_2 molecule (ethylene),
D_{3d} the CH_3CH_3 molecule in its "staggered" conformation (ethane),
D_{3h} the BF_3 molecule,
D_{6h} the C_6H_6 molecule (benzene),
$D_{\infty h}$ the CO_2 molecule,
T_d the CH_4 molecule (methane), and
O_h the SF_6 molecule.

Although it is easy to appreciate the effect of point group operations on a molecule in its equilibrium configuration, the application of the point group to molecular wavefunctions is different [see Section 5-5 in Wilson, Decius, and Cross (1955)]. The elements of the point group of a molecule are interpreted as rotating and/or reflecting the vibronic coordinates (the electronic coordinates, and the vibrational displacement coordinates away from equilibrium) with respect to the molecule fixed axes; these axes remaining fixed. With this interpretation of the elements we will call the group *the molecular point group* in order to distinguish it from the simple point group of a three-dimensional object in which the operations are the bodily rotation or reflection of the object. The molecular point group is used to classify the vibronic states of a molecule, and the effects of the elements of the group are discussed in detail in Chapter 11.

One point worthy of special discussion concerns the point group operation called i. This operation is the product of a C_2 rotation and a reflection in a plane perpendicular to the C_2 axis; its effect is to "invert" the object at its center. The effect of this operation in the molecular point group is to invert the vibronic coordinates in the origin of the molecule fixed axis system. This operation is not the same as the (spatial) inversion operation E^*, and it is important to realize that it is E^* and not i that determines the parity of a state; this is more fully discussed in Chapter 11 [see Eqs. (11-12)–(11-16)]. The operation i determines the g or u label of the state.

THE ROTATION GROUP SYMMETRY OF MOLECULES

In the application of rotation groups (as distinct from point groups) to molecules only three rotation groups are used: D_2, D_∞, and K. The group K is used in two ways: either as *the molecular three-dimensional pure rotation group*, which we denote K(mol), or as *the spatial three-dimensional pure rotation group* K(spatial). The group K(mol) consists of all rotations of a molecule about all axes passing through the molecular center of mass and fixed to the molecule, and the group K(spatial) consists of all rotations of a molecule about all axes passing through the molecular center of mass and fixed in space. These two groups are different and are used in different ways for labeling molecular states.

The detailed effect of rotational symmetry operations on molecular coordinates will be discussed in Chapter 11, after we have introduced the definition of molecule fixed axes about which the rotations take place, but there is one exception to this, however, and that involves the use of the group K(spatial). The energy of any isolated molecule in free space (where there are no external electric or magnetic fields) is not altered by rotating all the electrons and nuclei in the molecule about an axis fixed in space that passes through the center of mass of the molecule; the energy of an isolated molecule is therefore invariant to all the operations of the group K(spatial). With an extended definition of the use of the phrase "symmetry group" to mean a group of elements that do not change the energy of a molecule, we say that the group K(spatial) is a rotational symmetry group for any isolated molecule in field free space. The use of this group is the basis for our labeling of molecular energy levels with the total angular momentum quantum number J or F, and this is looked into further in Chapter 6. As far as rotation groups consisting of rotation operations about molecule fixed axes (which we call *molecular rotation groups*) are concerned we will need to use the group K(mol) for classifying the rotational states of spherical top molecules (such as methane), the group D_∞ for classifying the rotational states of symmetric top molecules (such as methyl fluoride), and the group D_2 (sometimes called V) for classifying the rotational states of all other molecules (asymmetric top molecules such as water). When applied to a molecule each of the operations of the appropriate molecular rotation group is a bodily rotation of all the nuclei and electrons about a molecule fixed axis.

DISCUSSION

The three types of groups that we have now discussed (the molecular symmetry group, the molecular point group, and the molecular rotation group) are all very important in helping us to understand molecules and their dynamical behavior. In discussing point groups, rotation groups, permutation groups, and inversion (E^*) symmetry we should appreciate that different types of "symmetry" are involved. Point groups and rotation groups are the symmetry groups of macroscopic three-dimensional bodies, where these bodies have a certain geometrical (or structural) symmetry represented by the rotation axes and reflection planes that are present. The application of these two groups to molecules hinges on a very important fact about molecules, namely that in a molecule the nuclei generally form a rigid framework which can be thought of as a classical *structure*. We can talk about the equilibrium nuclear structure of CH_3F as being pyramidal, and we say that it has C_{3v} point group symmetry. The molecular point group can be used, as we shall see, in studying the nuclear vibrational states and electronic states in a molecule, whereas the molecular rotation group is used in studying the rotational states of a molecule. The effects of molecular rotation and reflection symmetry operations on molecular coordinates will be discussed later, since to understand their effect in detail we must define a "molecule fixed" axis system, and this is done in Chapter 7.

The electron arrangement in a molecule (or atom) does not exhibit any classical structure or structural symmetry, and we do not use point groups based on the structural symmetry of some electron arrangement. Instead we use electron permutation symmetry and electron permutation groups. This symmetry results from the identity of the electrons rather than from any structural symmetry of the spatial arrangement of the electrons. In a similar way there is nuclear permutation symmetry in a molecule when identical nuclei are present. Using this type of symmetry we can develop nuclear permutation groups.

The third type of symmetry, inversion symmetry (this is the symmetry under the effect of the E^* operation discussed in Chapter 2, and not the point group operation i which only occurs for centrosymmetric molecules), does not depend on the presence of any structural symmetry in the molecule or on the presence of the identical particles; its use results from the fact that the energy of a system of nuclei and electrons in field-free space is invariant to the effect of E^* just as it is to the operations of the spatial three-dimensional pure rotation group K(spatial). In the end it turns out that in all our uses of symmetry groups for molecules it is the invariance of the energy of a molecule that is more important than any structural symmetry it may have.

BIBLIOGRAPHICAL NOTES

Rotation Groups

Hamermesh (1962).   Chapter 9.
Tinkham (1964).   Chapter 5.
Wigner (1959).   Chapter 14.

Point Groups and Their Application in Molecular Spectroscopy

Herzberg (1945).   Pages 82–131.
Wilson, Decius, and Cross (1955).   Chapter 5.

Electron Permutation Groups

Matsen (1964).
Ruedenberg and Poshusta (1972).
Walker and Musher (1974).

4

Representations
of Groups

In this chapter matrices and matrix groups are introduced, and then the concepts of isomorphism and homomorphism are defined. Examples of isomorphic and homomorphic groups are discussed by using the groups $D_3, S_3,$ S_2 and various matrix groups. The matrix groups involved form what are called representations of the D_3 and S_3 groups, and we focus particular attention on the definition of irreducible and inequivalent representations. We also discuss the class structure and character table of a group.

MATRICES AND MATRIX GROUPS

It is necessary to understand what matrices are and to be able to multiply them together in order to be able to work with matrix groups, and we summarize some important definitions here. A matrix is an array of numbers (called elements) arranged in rows and columns; for example

$$D = \begin{bmatrix} 2 & 3 \\ 4 & 5 \end{bmatrix} \tag{4-1}$$

is a matrix. The matrix in Eq. (4-1) has two rows and two columns, and since it has the same number of rows as columns it is said to be a *square matrix*;

matrices are not necessarily square. An $n \times n$ square matrix (having n rows and n columns) is said to be n-dimensional. In a general matrix, A say, the element occurring at the intersection of the ith row and jth column is called A_{ij}. In the matrix in Eq. (4-1) we have

$$D_{11} = 2, \quad D_{12} = 3,$$
$$D_{21} = 4, \quad D_{22} = 5. \tag{4-2}$$

In applying group theory to molecules it will often be necessary to multiply matrices together, and we must explain how the product of two matrices is defined. The product of an $n \times m$ matrix A (having n rows and m columns) and an $m \times q$ matrix B in the order AB is an $n \times q$ matrix C where the ijth element of C is given by

$$C_{ij} = \sum_{k=1}^{m} A_{ik}B_{kj}. \tag{4-3}$$

Notice that for the multiplication AB to be defined the number of columns in A must be equal to the number of rows in B. The product of an $n \times n$ square matrix A with a column matrix (to the right of A) of n elements is another column matrix with n elements. The product of an $n \times n$ square matrix A with a row matrix (to the left of A) of n elements is another row matrix with n elements. Three examples follow to help in the understanding of this:

A square matrix times a square matrix,

$$\begin{bmatrix} 1 & 2 \\ 3 & 4 \end{bmatrix}\begin{bmatrix} 5 & 6 \\ 7 & 8 \end{bmatrix} = \begin{bmatrix} 1 \times 5 + 2 \times 7 & 1 \times 6 + 2 \times 8 \\ 3 \times 5 + 4 \times 7 & 3 \times 6 + 4 \times 8 \end{bmatrix} = \begin{bmatrix} 19 & 22 \\ 43 & 50 \end{bmatrix}. \tag{4-4}$$

A square matrix times a column matrix,

$$\begin{bmatrix} 1 & 2 \\ 3 & 4 \end{bmatrix}\begin{bmatrix} 5 \\ 7 \end{bmatrix} = \begin{bmatrix} 1 \times 5 + 2 \times 7 \\ 3 \times 5 + 4 \times 7 \end{bmatrix} = \begin{bmatrix} 19 \\ 43 \end{bmatrix}. \tag{4-5}$$

A row matrix times a square matrix,

$$\begin{bmatrix} 1 & 2 \end{bmatrix}\begin{bmatrix} 5 & 6 \\ 7 & 8 \end{bmatrix} = \begin{bmatrix} 1 \times 5 + 2 \times 7 & 1 \times 6 + 2 \times 8 \end{bmatrix} = \begin{bmatrix} 19 & 22 \end{bmatrix}. \tag{4-6}$$

We can also take the product in Eq. (4-4) the other way round to get:

$$\begin{bmatrix} 5 & 6 \\ 7 & 8 \end{bmatrix}\begin{bmatrix} 1 & 2 \\ 3 & 4 \end{bmatrix} = \begin{bmatrix} 5 \times 1 + 6 \times 3 & 5 \times 2 + 6 \times 4 \\ 7 \times 1 + 8 \times 3 & 7 \times 2 + 8 \times 4 \end{bmatrix} = \begin{bmatrix} 23 & 34 \\ 31 & 46 \end{bmatrix}. \tag{4-7}$$

We see that matrix multiplication, like permutation multiplication or the multiplication of the elements of rotation groups or point groups, is not necessarily commutative. However, matrix multiplication is associative.

If we multiply an n-dimensional square matrix A by an n-dimensional square matrix E having 1 in all diagonal positions ($E_{ii} = 1$), and zero in all off-diagonal positions ($E_{ij} = 0$ for $i \neq j$), the result is still A. That is the matrix E plays the role in matrix multiplication that unity plays in the ordinary algebraic multiplication of numbers, and such a matrix is called an n-dimensional *unit matrix*. Borrowing the language of ordinary algebraic multiplication we say that if the product of two $n \times n$ matrices is the n-dimensional unit matrix then one matrix is the inverse of the other. For example,

$$
\begin{bmatrix} -\dfrac{1}{2} & \dfrac{\sqrt{3}}{2} \\ -\dfrac{\sqrt{3}}{2} & -\dfrac{1}{2} \end{bmatrix}
\begin{bmatrix} -\dfrac{1}{2} & -\dfrac{\sqrt{3}}{2} \\ \dfrac{\sqrt{3}}{2} & -\dfrac{1}{2} \end{bmatrix}
= \begin{bmatrix} 1 & 0 \\ 0 & 1 \end{bmatrix},
\tag{4-8}
$$

and the matrices on the left hand side in this equation are, therefore, the inverse of each other. These two matrices are also equal to the *transpose* of each other, where the transpose of a matrix A, say, is obtained by interchanging each element A_{ij} with the element A_{ji}, and the matrix is written $\tilde{A}$. Thus from Eq. (4-1) we have

$$
\tilde{D} = \begin{bmatrix} 2 & 4 \\ 3 & 5 \end{bmatrix}.
\tag{4-9}
$$

If a matrix is equal to its transpose then the matrix is *symmetric*, and if a matrix is equal to the inverse of its transpose then the matrix is *orthogonal*. Each of the two matrices on the left hand side in Eq. (4-8) is orthogonal.

The *hermitian conjugate* (or *conjugate transpose*) $A^\dagger$ of a matrix A is obtained by taking the complex conjugate of the transpose of the matrix. Thus

$$
A^\dagger = (\tilde{A})^*
\tag{4-10}
$$

and

$$
(A^\dagger)_{ij} = A_{ji}^*.
\tag{4-11}
$$

A matrix that is equal to its hermitian conjugate is *hermitian* and a matrix that is equal to the inverse of its hermitian conjugate is *unitary*.

The sum of the diagonal elements of a square matrix is the *trace* of the matrix, and the Greek letter chi (χ) is used for it; for example, from Eqs. (4-1) and (4-9) we have

$$
\chi(D) = \chi(\tilde{D}) = 7.
\tag{4-12}
$$

Table 4-1

Summary of some important definitions in
matrix algebra

Notation	Meaning
Transpose $\tilde{A}$	$(\tilde{A})_{ij} = A_{ji}$
Hermitian conjugate $A^\dagger$	$(A^\dagger)_{ij} = A^*_{ji} = (\tilde{A})^*_{ij}$
Unit matrix E	$E_{ij} = 0$ if $i \neq j$
(E is square)	$E_{ij} = 1$ if $i = j$
Inverse A^{-1}	$AA^{-1} = A^{-1}A = E$
Symmetric matrix	$A = \tilde{A}$
Orthogonal matrix	$A^{-1} = \tilde{A}$
Hermitian matrix	$A = A^\dagger$
Unitary matrix	$A^{-1} = A^\dagger$
Trace χ of matrix A	$\chi = \sum_i A_{ii}$
Kronecker delta δ	$\delta_{ij} = 0$ if $i \neq j$
	$\delta_{ij} = 1$ if $i = j$
Diagonal matrix	$A_{ij} = 0$ if $i \neq j$

In Table 4-1 we summarize these important definitions.

Problem 4-1.   Consider the following two-dimensional square matrix,

$$M = \begin{bmatrix} \dfrac{1}{\sqrt{2}} & \dfrac{i}{\sqrt{2}} \\ -\dfrac{i}{\sqrt{2}} & \dfrac{1}{\sqrt{2}} \end{bmatrix}, \tag{4-13}$$

where $i^2 = -1$. Is this matrix unitary? What is the trace of this matrix?

Answer.   To test if M is unitary we must see if

$$MM^\dagger = E, \tag{4-14}$$

where E is the 2×2 unit matrix. To form $M^\dagger$ we must form the transpose of M and then take the complex conjugate.

$$\tilde{M} = \begin{bmatrix} \dfrac{1}{\sqrt{2}} & -\dfrac{i}{\sqrt{2}} \\ \dfrac{i}{\sqrt{2}} & \dfrac{1}{\sqrt{2}} \end{bmatrix} \tag{4-15}$$

and

$$M^\dagger = (\tilde{M})^* = \begin{bmatrix} \dfrac{1}{\sqrt{2}} & \dfrac{i}{\sqrt{2}} \\ -\dfrac{i}{\sqrt{2}} & \dfrac{1}{\sqrt{2}} \end{bmatrix}. \tag{4-16}$$

We see that

$$M^\dagger = M \tag{4-17}$$

and therefore M is hermitian. Testing Eq. (4-14) we have

$$MM^\dagger = \begin{bmatrix} \dfrac{1}{\sqrt{2}} & \dfrac{i}{\sqrt{2}} \\ -\dfrac{i}{\sqrt{2}} & \dfrac{1}{\sqrt{2}} \end{bmatrix}\begin{bmatrix} \dfrac{1}{\sqrt{2}} & \dfrac{i}{\sqrt{2}} \\ -\dfrac{i}{\sqrt{2}} & \dfrac{1}{\sqrt{2}} \end{bmatrix} = \begin{bmatrix} 1 & i \\ -i & 1 \end{bmatrix}. \tag{4-18}$$

Thus

$$MM^\dagger \neq \begin{bmatrix} 1 & 0 \\ 0 & 1 \end{bmatrix}, \tag{4-19}$$

and M is not unitary. The trace of M is given by taking the sum of the diagonal elements, i.e.,

$$\chi(M) = \sum_j M_{jj} = 1/\sqrt{2} + 1/\sqrt{2} = \sqrt{2}. \tag{4-20}$$

We can now define a *matrix group*. A matrix group is a set of square matrices (all with the same dimension) that obey the group axioms; in a matrix group the operation of multiplication is matrix multiplication and the identity is the unit matrix. The following set of six matrices is an example of a two-dimensional matrix group, and we call the group Γ_3 (the subscript 3 has no special significance but it is a useful notation in this chapter):

$$\left\{ \begin{bmatrix} 1 & 0 \\ 0 & 1 \end{bmatrix}, \begin{bmatrix} 1 & 0 \\ 0 & -1 \end{bmatrix}, \begin{bmatrix} -\dfrac{1}{2} & \dfrac{\sqrt{3}}{2} \\ \dfrac{\sqrt{3}}{2} & \dfrac{1}{2} \end{bmatrix}, \begin{bmatrix} -\dfrac{1}{2} & -\dfrac{\sqrt{3}}{2} \\ -\dfrac{\sqrt{3}}{2} & \dfrac{1}{2} \end{bmatrix}, \begin{bmatrix} -\dfrac{1}{2} & \dfrac{\sqrt{3}}{2} \\ -\dfrac{\sqrt{3}}{2} & -\dfrac{1}{2} \end{bmatrix}, \begin{bmatrix} -\dfrac{1}{2} & -\dfrac{\sqrt{3}}{2} \\ \dfrac{\sqrt{3}}{2} & -\dfrac{1}{2} \end{bmatrix} \right\}. \tag{4-21}$$

The reader can form the multiplication table of this group and confirm that the four group axioms are satisfied. From Eq. (4-8) we see that the last two matrices in this group are the inverse (or reciprocal) of each other. The first four matrices in the group are symmetric, and since they are also self-reciprocal they are orthogonal. In this matrix group, therefore, all the matrices are orthogonal.

ISOMORPHISM AND FAITHFUL REPRESENTATIONS

In Chapters 1 and 3 we have introduced the two groups S_3 (a permutation group) and D_3 (a rotation group and a point group). These groups, together with the matrix group Γ_3 of Eq. (4-21), will be used to explain the concept of isomorphism. We can make the following one-to-one correspondences between the elements of the S_3 and D_3 groups

$$E\text{--}E,$$

$$(12)\text{--}C_{2a}, \quad (23)\text{--}C_{2b}, \quad (13)\text{--}C_{2c}, \qquad (4\text{-}22)$$
$$(123)\text{--}C_{3d}, \quad (132)\text{--}C_{3d}^2.$$

Each element in S_3 has a partner in the D_3 group according to Eq. (4-22). We have chosen the partners in Eq. (4-22) in a special way so that if we replace each element of S_3 in the multiplication table for S_3 (Table 1-1) by its partner in the D_3 group [according to Eq. (4-22)] then we obtain the multiplication table of the D_3 group (see Table 3-1). These two multiplication tables have the same structure, and the S_3 and D_3 groups are therefore said to be isomorphic. For example, in the S_3 group we have the product [see Eq. (1-21)]

$$(132)(23) = (13), \qquad (4\text{-}23)$$

from which, by replacing each element by its partner according to Eq. (4-22), we obtain

$$C_{3d}^2 C_{2b} = C_{2c}, \qquad (4\text{-}24)$$

and this is true (see Table 3-1). Formally two groups are isomorphic if the elements of one $(A, B, C, \ldots)$ correspond to (or can be *mapped* onto) the elements of the other $(\bar{A}, \bar{B}, \bar{C}, \ldots)$ in a unique one-to-one fashion $(A\text{--}\bar{A}, B\text{--}\bar{B}, C\text{--}\bar{C}, \ldots)$ so that from $AB = C$ it can be inferred that $\bar{A}\bar{B} = \bar{C}$, etc. Clearly isomorphic groups must have the same order, but groups of the same order are not necessarily isomorphic. The best way to test if two groups are isomorphic is to look at their multiplication tables and to check that everywhere A occurs in the one, $\bar{A}$ occurs in the other, similarly for B and $\bar{B}$, etc.

The matrix group Γ_3 introduced in Eq. (4-21) is isomorphic to both the S_3 and D_3 groups, and the mapping is as follows:

$$S_3: \quad E \qquad (12) \qquad (23) \qquad (13) \qquad (123) \qquad (132)$$
$$D_3: \quad E \qquad C_{2a} \qquad C_{2b} \qquad C_{2c} \qquad C_{3d} \qquad C_{3d}^2$$

$$\Gamma_3: \begin{bmatrix} 1 & 0 \\ 0 & 1 \end{bmatrix} \begin{bmatrix} 1 & 0 \\ 0 & -1 \end{bmatrix} \begin{bmatrix} -\dfrac{1}{2} & \dfrac{\sqrt{3}}{2} \\ \dfrac{\sqrt{3}}{2} & \dfrac{1}{2} \end{bmatrix} \begin{bmatrix} -\dfrac{1}{2} & -\dfrac{\sqrt{3}}{2} \\ -\dfrac{\sqrt{3}}{2} & \dfrac{1}{2} \end{bmatrix} \begin{bmatrix} -\dfrac{1}{2} & \dfrac{\sqrt{3}}{2} \\ -\dfrac{\sqrt{3}}{2} & -\dfrac{1}{2} \end{bmatrix} \begin{bmatrix} -\dfrac{1}{2} & -\dfrac{\sqrt{3}}{2} \\ \dfrac{\sqrt{3}}{2} & -\dfrac{1}{2} \end{bmatrix}.$$

$$(4\text{-}25)$$

A matrix group that is isomorphic to another group is said to form a faithful representation of that group. Thus the matrix group Γ_3 forms a faithful representation of the S_3 and D_3 groups. Using the mapping of Eq. (4-25), we see that Eqs. (4-23) and (4-24) map onto the product

$$\begin{bmatrix} -\dfrac{1}{2} & -\dfrac{\sqrt{3}}{2} \\ \dfrac{\sqrt{3}}{2} & -\dfrac{1}{2} \end{bmatrix} \begin{bmatrix} -\dfrac{1}{2} & \dfrac{\sqrt{3}}{2} \\ \dfrac{\sqrt{3}}{2} & \dfrac{1}{2} \end{bmatrix} = \begin{bmatrix} -\dfrac{1}{2} & -\dfrac{\sqrt{3}}{2} \\ -\dfrac{\sqrt{3}}{2} & \dfrac{1}{2} \end{bmatrix}, \qquad (4\text{-}26)$$

and Eq. (1-22) maps onto the product

$$\begin{bmatrix} -\dfrac{1}{2} & \dfrac{\sqrt{3}}{2} \\ \dfrac{\sqrt{3}}{2} & \dfrac{1}{2} \end{bmatrix} \begin{bmatrix} -\dfrac{1}{2} & -\dfrac{\sqrt{3}}{2} \\ \dfrac{\sqrt{3}}{2} & -\dfrac{1}{2} \end{bmatrix} = \begin{bmatrix} 1 & 0 \\ 0 & -1 \end{bmatrix}, \qquad (4\text{-}27)$$

in the matrix group. The multiplication table of this matrix group can be obtained by replacing each element in the multiplication table of the S_3 group in Table 1-1 (or of the D_3 group in Table 3-1) by its partner in the matrix group using the mapping of Eq. (4-25).

HOMOMORPHISM AND UNFAITHFUL REPRESENTATIONS

Homomorphism is similar to isomorphism except that instead of the correspondence being a one-to-one correspondence between elements of groups having the same order, it is a many-to-one correspondence between elements of groups having a different order. The larger group is said to be homomorphic onto the smaller group. For example, the group D_3 is homomorphic onto the S_2 group $\{E,(12)\}$ with the following correspondences:

$$D_3: \quad \underbrace{E \quad C_{3d} \quad C_{3d}^2} \quad \underbrace{C_{2a} \quad C_{2b} \quad C_{2c}}$$
$$S_2: \quad\quad E \quad\quad\quad\quad (12) \tag{4-28}$$

If, in the multiplication table of D_3 (Table 3-1), the elements E, C_{3d}, and C_{3d}^2 are replaced by E (of S_2), and C_{2a}, C_{2b}, and C_{2c} by (12), we obtain the multiplication table of S_2 (nine times over).

We can also make many-to-one mappings of S_3 and D_3 onto two one-dimensional matrix groups. These two groups are

$$\Gamma_1: \quad \{1\} \tag{4-29}$$

and

$$\Gamma_2: \quad \{1, -1\}. \tag{4-30}$$

These are one-dimensional matrix groups, and matrix multiplication is simply ordinary algebraic multiplication in this case. The homomorphic mapping of S_3 and D_3 onto these matrix groups is shown below.

$$\begin{array}{lcccccc} D_3: & E & C_{3d} & C_{3d}^2 & C_{2a} & C_{2b} & C_{2c} \\ S_3: & E & (123) & (132) & (12) & (23) & (13) \\ \\ \Gamma_2: & 1 & 1 & 1 & -1 & -1 & -1 \\ \Gamma_1: & 1 & 1 & 1 & 1 & 1 & 1 \end{array} \tag{4-31}$$

If a group is homomorphic onto a matrix group then the matrix group is said to form an unfaithful representation of the group; thus Γ_1 and Γ_2 each form unfaithful representations of S_3 and D_3.

EQUIVALENT AND IRREDUCIBLE REPRESENTATIONS

Now we have introduced the three representations Γ_1, Γ_2, and Γ_3 of the groups S_3 and D_3. We can use these representations in two different ways to construct others, but these three are special in that they are *irreducible* and also all other irreducible representations of S_3 and D_3 are *equivalent* to these. We must now define the terms irreducible and equivalent, and show how other representations can be formed from these three.

Equivalent Representations

One way of forming a new representation from the Γ_3 representation involves forming an equivalent representation. We could subject each of the matrices of the representation Γ_3 to the same *similarity transformation,* i.e., post multiply each matrix of Γ_3 by a 2×2 matrix, A say, and then premultiply each result by the inverse of A; thus a matrix M_r in Γ_3 would be changed to $A^{-1}M_rA$ and, since matrix multiplication is associative,

$$(A^{-1}M_rA)(A^{-1}M_sA) = A^{-1}M_r(AA^{-1})M_sA = A^{-1}M_rM_sA. \quad (4\text{-}32)$$

The new matrix group will be isomorphic to Γ_3 with the mapping of M_r onto $A^{-1}M_rA$, etc. The new set of matrices will, therefore, also be isomorphic to S_3 and D_3 and form a faithful representation of them. Two representations related by a similarity transformation are said to be equivalent, and for our applications in molecular physics can usually be considered to be the same representation (as we will see).

Problem 4-2.   Post multiply each matrix of Γ_3 by the matrix

$$A = \begin{bmatrix} 1 & 1 \\ 3 & 4 \end{bmatrix}$$

and premultiply by

$$A^{-1} = \begin{bmatrix} 4 & -1 \\ -3 & 1 \end{bmatrix}.$$

Test that the new representation is isomorphic to Γ_3. What do you notice about the traces of each of the matrices in the new representation compared to those of the matrices in the group Γ_3?

Answer.   The new matrix group obtained from Γ_3 in the order that the matrices are given in Eq. (4-25) is

$$\left\{ \begin{bmatrix} 1 & 0 \\ 0 & 1 \end{bmatrix}, \begin{bmatrix} 7 & 8 \\ -6 & -7 \end{bmatrix}, \begin{bmatrix} -\dfrac{7}{2}+\dfrac{11\sqrt{3}}{2} & -4+\dfrac{15\sqrt{3}}{2} \\ 3-4\sqrt{3} & \dfrac{7}{2}-\dfrac{11\sqrt{3}}{2} \end{bmatrix}, \right.$$

$$\begin{bmatrix} -\dfrac{7}{2}-\dfrac{11\sqrt{3}}{2} & -4-\dfrac{15\sqrt{3}}{2} \\ 3+4\sqrt{3} & \dfrac{7}{2}+\dfrac{11\sqrt{3}}{2} \end{bmatrix}, \begin{bmatrix} -\dfrac{1}{2}+\dfrac{13\sqrt{3}}{2} & \dfrac{17\sqrt{3}}{2} \\ -5\sqrt{3} & -\dfrac{1}{2}-\dfrac{13\sqrt{3}}{2} \end{bmatrix},$$

$$\left. \begin{bmatrix} -\dfrac{1}{2}-\dfrac{13\sqrt{3}}{2} & -\dfrac{17\sqrt{3}}{2} \\ 5\sqrt{3} & -\dfrac{1}{2}+\dfrac{13\sqrt{3}}{2} \end{bmatrix} \right\}. \quad (4\text{-}33)$$

These matrices look complicated compared to those that make up the representation Γ_3, but it is straightforward (although tedious) to show by working out the multiplication table, that this matrix group is isomorphic to the group Γ_3. This matrix group, therefore, forms a representation of the S_3 and D_3 groups. Equation (1-22), for example, maps onto

$$\begin{bmatrix} -\dfrac{7}{2} + \dfrac{11\sqrt{3}}{2} & -4 + \dfrac{15\sqrt{3}}{2} \\[3mm] 3 - 4\sqrt{3} & \dfrac{7}{2} - \dfrac{11\sqrt{3}}{2} \end{bmatrix} \begin{bmatrix} -\dfrac{1}{2} - \dfrac{13\sqrt{3}}{2} & -\dfrac{17\sqrt{3}}{2} \\[3mm] 5\sqrt{3} & -\dfrac{1}{2} + \dfrac{13\sqrt{3}}{2} \end{bmatrix} = \begin{bmatrix} 7 & 8 \\[3mm] -6 & -7 \end{bmatrix}.$$

$$(4\text{-}34)$$

Part of the reason why the matrices in Γ_3 are more simple looking than those in the new representation of Eq. (4-33) is that each of the matrices in Γ_3 is orthogonal. In general it is possible to choose a similarity transformation so that all the matrices in a representation are unitary and such a representation is said to be a *unitary representation*. We will always use unitary representations in our applications of representation theory to molecules.

The trace of a matrix in a representation is called the *character* of the matrix. As to the characters of the matrices in this new representation, we should notice that each matrix has the same character as its partner in the group Γ_3. It is left as an exercise for the reader to show that a similarity transformation does not change the character of a matrix, i.e., that the characters of M_r and $A^{-1}M_rA$ are identical. As a result we see that equivalent representations have the same characters.

Reducible Representations

Another way of forming a new representation of S_3 or D_3 from the representations Γ_1, Γ_2, and Γ_3 is by combining them to form what is called a *reducible* representation. For example, the representations Γ_2 and Γ_3 can be combined to form the following reducible three-dimensional representation [where the matrices are in the same order as in Eq. (4-25)]:

$$\begin{bmatrix} 1 & 0 & 0 \\ 0 & 1 & 0 \\ 0 & 0 & 1 \end{bmatrix}, \begin{bmatrix} 1 & 0 & 0 \\ 0 & -1 & 0 \\ 0 & 0 & -1 \end{bmatrix}, \begin{bmatrix} -\dfrac{1}{2} & \dfrac{\sqrt{3}}{2} & 0 \\[2mm] \dfrac{\sqrt{3}}{2} & \dfrac{1}{2} & 0 \\[2mm] 0 & 0 & -1 \end{bmatrix}, \begin{bmatrix} -\dfrac{1}{2} & -\dfrac{\sqrt{3}}{2} & 0 \\[2mm] -\dfrac{\sqrt{3}}{2} & \dfrac{1}{2} & 0 \\[2mm] 0 & 0 & -1 \end{bmatrix},$$

$$\begin{bmatrix} -\dfrac{1}{2} & \dfrac{\sqrt{3}}{2} & 0 \\[2mm] -\dfrac{\sqrt{3}}{2} & -\dfrac{1}{2} & 0 \\[2mm] 0 & 0 & 1 \end{bmatrix}, \begin{bmatrix} -\dfrac{1}{2} & -\dfrac{\sqrt{3}}{2} & 0 \\[2mm] \dfrac{\sqrt{3}}{2} & -\dfrac{1}{2} & 0 \\[2mm] 0 & 0 & 1 \end{bmatrix}.$$

$$(4\text{-}35)$$

Each of the matrices in the representation given in Eq. (4-35) is said to be *block diagonal*. A block diagonal matrix is such that the nonzero elements occur in blocks along the diagonal of the matrix, and in Eq. (4-35) each matrix consists of an upper left 2×2 block and a lower right 1×1 block, with zeros in the off diagonal positions that connect the elements in one block with those of the other. Block diagonal matrices can consist of more than two blocks along the diagonal, for example the matrix

$$\begin{bmatrix} 1 & 2 & 3 & 0 & 0 & 0 & 0 \\ 0 & 5 & 6 & 0 & 0 & 0 & 0 \\ 7 & 4 & 8 & 0 & 0 & 0 & 0 \\ 0 & 0 & 0 & 6 & 5 & 0 & 0 \\ 0 & 0 & 0 & 4 & 3 & 0 & 0 \\ 0 & 0 & 0 & 0 & 0 & 8 & 9 \\ 0 & 0 & 0 & 0 & 0 & 1 & 2 \end{bmatrix}$$

is also block diagonal and consists of a 3×3 block and two 2×2 blocks. The reader can use Eq. (4-3) to prove that the product of two block diagonal matrices having the same block structure produces a third block diagonal matrix that has the same block structure. The representation in Eq. (4-35) is said to be in *block diagonal form*, and in such a representation all the matrices must have the same block structure as each other.

The representation of S_3 or D_3 in Eq. (4-35) has characters that are the sum of the characters of the Γ_2 and Γ_3 representations. If we were to subject the representation to a similarity transformation, to form an equivalent representation, the characters would be the same. If we know the characters of the representations Γ_2 and Γ_3, then by inspecting the characters of this reducible representation (or one that is equivalent to it) we are able to determine how it has been built up.

The fact that the representation in Eq. (4-35) (which we will call Γ_s) is made up by combining the Γ_2 and Γ_3 representations is symbolically expressed by writing

$$\Gamma_s = \Gamma_2 \oplus \Gamma_3. \tag{4-36}$$

For an element R, say, of the D_3 or S_3 group the character in Γ_s is given by the sum of the characters of the Γ_2 and Γ_3 representations, i.e.,

$$\chi^{\Gamma_s}[R] = \chi^{\Gamma_2}[R] + \chi^{\Gamma_3}[R], \tag{4-37}$$

where $\chi^{\Gamma_i}[R]$ is the character in the representation Γ_i of the matrix $D^{\Gamma_i}[R]$ that maps onto the element R of the symmetry group.

A representation that cannot be brought into block diagonal form by any similarity transformation applied simultaneously to all the matrices of the representation is said to be irreducible, and for a finite group there are a finite number of inequivalent irreducible representations. It can be shown that the matrix elements $D^{\Gamma_i}[R]_{mn}$ and $D^{\Gamma_j}[R]_{m'n'}$ of the matrices in the irreducible representations Γ_i and Γ_j of any group satisfy the following *orthogonality relation*:

$$\sum_R D^{\Gamma_i}[R]_{mn}^* \sqrt{\frac{l_i}{h}} D^{\Gamma_j}[R]_{m'n'} \sqrt{\frac{l_j}{h}} = \delta_{ij} \delta_{mm'} \delta_{nn'}, \tag{4-38}$$

where h is the order of the group, l_i and l_j are the dimensions of the irreducible representations Γ_i and Γ_j, respectively, and the sum is over all elements R in the group. From Eq. (4-38) it follows that the characters of irreducible representations are orthogonal, i.e.,

$$\sum_R \chi^{\Gamma_i}[R]^* \chi^{\Gamma_j}[R] = h \delta_{ij} \tag{4-39}$$

for two irreducible representations Γ_i and Γ_j.

It turns out that the representations Γ_1, Γ_2, and Γ_3, in Eqs. (4-31) and (4-25), are the only inequivalent irreducible representations of the groups D_3 and S_3 [see the discussion following Eq. (4-48)]. Therefore, any representation Γ of D_3 or S_3 can be written

$$\Gamma = a_1 \Gamma_1 \oplus a_2 \Gamma_2 \oplus a_3 \Gamma_3, \tag{4-40}$$

and the characters satisfy

$$\chi^{\Gamma}[R] = a_1 \chi^{\Gamma_1}[R] + a_2 \chi^{\Gamma_2}[R] + a_3 \chi^{\Gamma_3}[R], \tag{4-41}$$

where the a_i are integers.

REDUCTION OF A REPRESENTATION

Irreducible (and inequivalent) matrix representations play a special role in molecular physics since we are able to use them to label molecular states. This provides a very useful characterization of the state, and in applications we frequently obtain a reducible representation which we need to reduce to its irreducible components. The way that a given representation of a group is reduced to its irreducible components depends only on the characters of the matrices in the representation and on the characters of the matrices in the irreducible representations of the group. For most of the groups that we are interested in the characters of the irreducible representations have been tabulated; such a table is called *the character table* of the group.

To reduce a representation to its irreducible components we can sometimes guess the answer by using Eq. (4-41), but in general we proceed as

follows. Suppose that the reducible representation is Γ and that the irreducible representations of the group involved are Γ_1, Γ_2, Γ_3, ... We
desire to find the integral coefficients a_i, where

$$\Gamma = a_1\Gamma_1 \oplus a_2\Gamma_2 \oplus a_3\Gamma_3 \oplus \cdots \qquad (4\text{-}42a)$$

and

$$\chi^\Gamma[R] = \sum_j a_j \chi^{\Gamma_j}[R], \qquad (4\text{-}42b)$$

with the sum running over all the irreducible representations of the group.
Multiplying Eq. (4-42b) on the right by $\chi^{\Gamma_i}[R]^*$ and summing over R it
follows from the character orthogonality relation [Eq. (4-39)] that the a_i
are given by

$$a_i = \frac{1}{h} \sum_R \chi^\Gamma[R] \chi^{\Gamma_i}[R]^*, \qquad (4\text{-}43)$$

where h is the order of the group and R runs over all the elements of the
group. Since the reduction of a representation depends only on the characters
this means that representations having the same characters must be equivalent and that the characters serve to distinguish inequivalent representations.
    If we put $i = j$ in Eq. (4-39) we see that for an irreducible representation

$$\sum_R |\chi^{\Gamma_i}[R]|^2 = h, \qquad (4\text{-}44)$$

which can be used as a test for the irreducibility of a representation. Another
result of importance is that the sum of the squares of the dimensions of the
irreducible representations of a group is equal to the order of the group, i.e.,

$$\sum_i l_i^2 = h, \qquad (4\text{-}45)$$

where the sum is over all irreducible representations of the group; this
equation can be used to test if we have all the irreducible representations of a
group.

CONJUGATE ELEMENTS AND CLASSES

    The elements of a group can be divided up into a number of classes, and this
important idea will now be defined using a matrix group. In a matrix group we
can take one matrix, M_r say, and form the product

$$M_s^{-1} M_r M_s = M_p \qquad (4\text{-}46)$$

with each of the other matrices M_s in the group. A few moments thought
shows that each of the matrices M_p obtained will also be in the group;
from the group axioms both $M_r M_s$ and M_s^{-1} are in the group and hence

$M_s^{-1}(M_r M_s)$ must also be in the group. Further, since a similarity transformation, such as is performed on M_r in Eq. (4-46), does not change the character of a matrix, then M_r and M_p will have the same character. Two matrices in a matrix group that are related, as are M_r and M_p in Eq. (4-46), by a similarity transformation involving another matrix (M_s here) of the group are said to be *conjugate* and are said to be in the same *class*.

In any group, whether a matrix group or not, we can divide the elements up into classes. Two elements A and B in a group are said to be in the same class if there exists an element, C say, in the group such that

$$C^{-1}AC = B. \qquad (4\text{-}47)$$

In the S_3 group there are three classes,

$$E, \quad [(12), (23), (13)], \quad \text{and} \quad [(123), \quad (132)]. \qquad (4\text{-}48)$$

Because of the isomorphism [resulting in "parallel" versions of Eq. (4-47)], the class structures of D_3 and Γ_3 are the same as that of S_3. It can be shown that there are the same number of irreducible (and inequivalent) representations of a group as there are classes in the group. There are, therefore, only three irreducible representations for the groups D_3 or S_3, and these are Γ_1, Γ_2, and Γ_3 given in Eqs. (4-25) and (4-31).

Problem 4-3. The internal displacement coordinates (bond stretching and angle bending coordinates) of the BF_3 molecule can be shown to generate (in a way to be explained in the next chapter) a representation Γ of the S_3 group (this group contains all fluorine nuclei permutations and we label the fluorine nuclei 1, 2, and 3). The characters of this representation are as follows

$$
\begin{array}{lcccccc}
R: & E & (12) & (23) & (13) & (123) & (132) \\
\chi^\Gamma[R]: & 7 & 1 & 1 & 1 & 1 & 1
\end{array}
\qquad (4\text{-}49)
$$

Use Eq. (4-43) to reduce this representation into the irreducible representations of the S_3 group. You can check that the answer is correct by testing to see if Eq. (4-42b) is satisfied.

Answer. Using Eq. (4-43) with

$$\Gamma = a_1 \Gamma_1 \oplus a_2 \Gamma_2 \oplus a_3 \Gamma_3$$

we have

$$
\begin{aligned}
a_1 &= \tfrac{1}{6}\{\chi^\Gamma[E]\chi^{\Gamma_1}[E]^* + \chi^\Gamma[(12)]\chi^{\Gamma_1}[(12)]^* + \chi^\Gamma[(23)]\chi^{\Gamma_1}[(23)]^* + \cdots\} \\
&= \tfrac{1}{6}(7 \times 1 + 1 \times 1 + 1 \times 1 + 1 \times 1 + 1 \times 1 + 1 \times 1) \\
&= 2, \qquad (4\text{-}50) \\
a_2 &= \tfrac{1}{6}[7 \times 1 + 1 \times (-1) + 1 \times (-1) + 1 \times (-1) + 1 \times 1 + 1 \times 1] \\
&= 1, \qquad (4\text{-}51)
\end{aligned}
$$

and

$$a_3 = \tfrac{1}{6}[7 \times 2 + 1 \times 0 + 1 \times 0 + 1 \times 0 + 1 \times (-1) + 1 \times (-1)]$$
$$= 2. \tag{4-52}$$

Thus

$$\Gamma = 2\Gamma_1 \oplus \Gamma_2 \oplus 2\Gamma_3. \tag{4-53}$$

We now test to see if this is correct by using Eq. (4-42b). For the operation E Eq. (4-42b) gives

$$\chi^\Gamma = 7 = 2 \times 1 + 1 \times 1 + 2 \times 2 \tag{4-54}$$

which is correct. For the operations (12), (23), or (13) Eq. (4-42b) gives

$$\chi^\Gamma = 1 = 2 \times 1 + 1 \times (-1) + 2 \times 0 \tag{4-55}$$

which is correct, and for the operations (123) and (132) Eq. (4-42b) gives

$$\chi^\Gamma = 1 = 2 \times 1 + 1 \times 1 + 2 \times (-1) \tag{4-56}$$

which is also correct.

The class structure of a permutation group or of a CNPI group can be determined by inspection. All permutations or permutation inversions of the same shape (i.e., consisting of the same number of independent transpositions, independent cycles of three, independent cycles of four, etc.) are in the same class; the proof of this is left as an exercise for the reader. For example the 120 elements of the permutation group S_5 divide up into seven classes, and the elements in these classes have the following shapes:

$$E, (xx), (xxx), (xxxx), (xxxxx), (xx)(xx), \text{ and } (xx)(xxx). \tag{4-57}$$

The number of classes in the permutation group S_n is given by the *partition number* of n, i.e., the number of ways of writing n as the sum of integers. For example, for $n = 5$ we can write

$$5 = 1 + 1 + 1 + 1 + 1, \quad 2 + 1 + 1 + 1, \quad 3 + 1 + 1, \quad 4 + 1, \quad 5,$$
$$2 + 2 + 1, \quad \text{or} \quad 2 + 3, \tag{4-58}$$

so that the partition number of 5 is 7. The parallel between this partitioning of the number 5 and the shapes of the permutations in each class in Eq. (4-57) is obvious. Since the number of irreducible representations in a group is equal to the number of classes in the group, the group S_5 has seven irreducible representations. The class structure of a CNPI group follows in a similar fashion, with E^* in a class on its own. For example, the class structure of the

CNPI group of CH_3F [see Eq. (2-11)] is

$$
\begin{array}{llllll}
E & (12) & (123) & E^* & (12)^* & (123)^* \\
  & (23) & (132) &   & (23)^* & (132)^* \\
  & (13) &   &   & (13)^*
\end{array}
\qquad (4\text{-}59)
$$

and this group, therefore, has six irreducible representations.

The class structure of a subgroup of a permutation group, or a subgroup of a CNPI group (such as an MS group), is not so easily determined and Eq. (4-47) must be used. However, in an Abelian group multiplication is commutative (as discussed after Problem 1-7) so that each element has to be in a class on its own. When multiplication within a group is commutative the right hand side of Eq. (4-47) can never be anything else except A, i.e., if A and C commute

$$
C^{-1}AC = C^{-1}CA = A. \qquad (4\text{-}60)
$$

In Problem 1-7 the subgroup $\{E, (123), (132)\}$ of the group S_3 was obtained and this is an Abelian group; hence each of the three elements is in a class of its own. Similarly the MS group of ethylene $D_{2h}(M)$, in Eq. (2-14), is Abelian so that this group has eight classes and eight irreducible representations.

For a given group the characters of the irreducible representations are tabulated in a character table. In a character table the elements in the same class are grouped together, since they all have the same character in each irreducible representation, and only one element in each class is given but the number of elements in each class is indicated. The character table of the S_3 group is given in this way in Table 4-2 and the irreducible representations are labeled Γ_1, Γ_2, and Γ_3 following our earlier usage. Groups that are isomorphic to each other have the same character table; the groups S_3, D_3, C_{3v}, and $C_{3v}(M)$ are isomorphic to each other and, therefore, they all have the same character table. The character tables of many MS groups are

Table 4-2
The character table of the S_3 group[a]

S_3:	E	(12)	(123)
	1	3	2
Γ_1:	1	1	1
Γ_2:	1	-1	1
Γ_3:	2	0	-1

[a] One representative element in each class is given, and the number written below each element is the number of elements in the class.

given in Appendix A. For all groups there is one irreducible representation, such as Γ_1 in S_3, that has character $+1$ for all operations of the group. This representation is called the *totally symmetric representation* and is denoted $\Gamma^{(s)}$ in the text; it is variously called A, A_g, A', A_1 etc. in the character tables in Appendix A.

The character tables of some groups are determined in Chapters 10 and 12 using the following general procedure. Knowing the multiplication table and the class structure of a group it is possible to multiply all the elements in one class C_i by (on the right) all the elements in another class C_j, and it is always so that the resultant elements can be arranged into classes. Let us suppose that the set of elements in the class C_k occurs c_{ijk} times in the product so that we can write

$$C_i C_j = \sum_k c_{ijk} C_k, \tag{4-61}$$

where the sum is over all classes in the group and each C_l means all the elements in that class. Having determined the coefficients c_{ijk}, the characters can be found from the relation

$$r_i r_j \chi_i \chi_j = \chi[E] \sum_k c_{ijk} r_k \chi_k \tag{4-62}$$

which holds for every representation of the group where r_i is the number of elements in class C_i. The examples given in Chapters 10 and 12 should make the application of this technique clear [see, in particular, Eqs. (10-74)–(10-81)].

BIBLIOGRAPHICAL NOTES

Matrices

Margenau and Murphy (1956).   In Chapter 10 matrices and matrix algebra are discussed.

Representation Theory

Hamermesh (1962).   Chapter 7 discusses permutation groups and their representations.
Murnaghan (1938).   Equations (4-61) and (4-62) in this chapter are proved on page 83.
Tinkham (1964).   Representations are discussed in Chapter 3.
Wigner (1959).   The orthogonality relation [Eq. (4-38) in this chapter] is proved and discussed in Chapter 9 [see Eq. (9-32)], and the proof that any matrix representation can be transformed into a unitary matrix representation is given in Eqs. (9-4)–(9-7). Equation (4-45) of this chapter is proved on page 115 of Wigner's book, and the equality of the number of classes in a group to the number of irreducible representations of the group is discussed on page 84.

5

The Use of Representations for Labeling Molecular Energy Levels

In this chapter an approximate time independent Schrödinger equation for a molecule is introduced, and the effect of a nuclear permutation or the inversion E^* on this equation is discussed. Symmetry operations and symmetry groups are defined with respect to the molecular Hamiltonian, and it is shown how molecular wavefunctions, and products of them, generate representations of a symmetry group of the Hamiltonian. As a result we can label energy levels with representation labels and use these labels to distinguish energy levels. The general use of these labels in determining which pairs of levels can interact with each other in certain circumstances is discussed, and the vanishing integral rule is derived.

A MOLECULAR SCHRÖDINGER EQUATION
IN (X, Y, Z) COORDINATES

With some simplifying approximations we can write the classical non-relativistic energy of a molecule in free space as the sum of the kinetic energies of the nuclei and electrons and of the electrostatic interactions between them. Neglecting the translational motion of the molecule and expressing the energy in terms of the coordinates and momenta (rather than in terms of the coordinates and velocities) we obtain this approximate classical energy in *Hamiltonian form* H^0; using the (X, Y, Z) coordinate system (with origin at the molecular center of mass) we have:

$$H^0 = \frac{1}{2} \sum_r \frac{P_r^2}{m_r} + \sum_{r<s} \frac{C_r C_s e^2}{R_{rs}}, \tag{5-1}$$

where r, s run over all the nuclei and electrons in the molecule, particle r has mass m_r and charge $C_r e$ (the mass of an electron is m_e and its charge is $-e$), and the momentum of particle r is P_r; we can write

$$P_r^2 = m_r^2(\dot{X}_r^2 + \dot{Y}_r^2 + \dot{Z}_r^2) = P_{X_r}^2 + P_{Y_r}^2 + P_{Z_r}^2, \tag{5-2}$$

where $\dot{X}_r = dX_r/dt$ and t is time. In Eq. (5-1) the distance between the particles r and s is written R_{rs} and

$$R_{rs} = + [(X_r - X_s)^2 + (Y_r - Y_s)^2 + (Z_r - Z_s)^2]^{1/2}. \tag{5-3}$$

In Eq. (5-1) we use $C_r e$ for the charge rather than the more customary $Z_r e$ to avoid confusion with the coordinate Z_r.

From the basic postulates of quantum mechanics the allowed stationary state energies of a molecule with classical energy given by Eq. (5-1) are the *eigenvalues* E_n in the time independent Schrödinger equation

$$\hat{H}^0 \Psi_n = E_n \Psi_n, \tag{5-4}$$

where the *eigenfunctions* Ψ_n are single valued functions of the coordinates of the nuclei and electrons in the molecule, and the quantum mechanical *Hamiltonian operator* $\hat{H}^0$ is obtained from H^0 in Eq. (5-1) by replacing the Cartesian momenta P_{X_r}, etc., by the operators $\hat{P}_{X_r}$, etc., according to the prescription

$$\hat{P}_{X_r} = -i\hbar \, \partial/\partial X_r, \qquad \text{etc.,} \tag{5-5}$$

where $\hbar = h/2\pi$ and h is Planck's constant. We obtain

$$\hat{H}^0 = -\frac{\hbar^2}{2} \sum_r \frac{\nabla_r^2}{m_r} + \sum_{r<s} \frac{C_r C_s e^2}{R_{rs}}, \tag{5-6}$$

where

$$\nabla_r^2 = (\partial^2/\partial X_r^2 + \partial^2/\partial Y_r^2 + \partial^2/\partial Z_r^2). \tag{5-7}$$

The effect of the electron spin magnetic moment, and of the possible magnetic and electric moments of each nucleus, will be to add further electric and magnetic interaction terms to the energy in Eq. (5-1) and hence to the Hamiltonian operator in Eq. (5-6). Even without worrying about these terms the use of an axis system with origin at the molecular center of mass leads to a slightly more complicated Hamiltonian than $\hat{H}^0$ after we discard the translational energy, and this is discussed in Chapter 6. For the general arguments we wish to make in this chapter we neglect all these extra terms. Equation (5-6) is an approximate molecular Hamiltonian operator, and in this chapter we use representation theory to label the energy levels (eigenstates) of this particular approximate Hamiltonian in order to show how the process of labeling energy levels works.

THE EFFECTS OF NUCLEAR PERMUTATIONS AND THE INVERSION ON THE SCHRÖDINGER EQUATION

We will consider the Schrödinger equation, Eq. (5-4), for a molecule in which nuclei 1 and 2 are identical, having mass m and charge Ce. In these circumstances we can write the Hamiltonian operator of Eq. (5-6) as

$$\hat{H}^0 = \hat{H}^0(1,2) + \hat{H}^0(\text{rest}),\qquad(5\text{-}8)$$

where

$$\hat{H}^0(1,2) = -\frac{\hbar^2}{2m}(\nabla_1^2 + \nabla_2^2) + \frac{C^2 e^2}{R_{12}} + \sum_{r \neq 1,2} CC_r e^2 \left(\frac{1}{R_{1r}} + \frac{1}{R_{2r}}\right). \quad(5\text{-}9)$$

$\hat{H}^0(\text{rest})$, therefore, does not involve the coordinates or momenta of nuclei 1 and 2. The Schrödinger equation, Eq. (5-4), for the molecule is

$$[\hat{H}^0(1,2) + \hat{H}^0(\text{rest})]\Psi_n(X_1, Y_1, Z_1, X_2, Y_2, Z_2, W)$$
$$= E_n \Psi_n(X_1, Y_1, Z_1, X_2, Y_2, Z_2, W),\qquad(5\text{-}10)$$

where W stands for the coordinates of all the other nuclei and of all the electrons in the molecule.

We wish to examine the effect of permuting nuclei 1 and 2, i.e., the effect of the operation (12), on the Schrödinger equation given in Eq. (5-10). The effect of (12) on the right hand side of the Schrödinger equation can be written

$$(12)E_n \Psi_n(X_1, Y_1, Z_1, X_2, Y_2, Z_2, W)$$
$$= E_n(12)\Psi_n(X_1, Y_1, Z_1, X_2, Y_2, Z_2, W)\qquad(5\text{-}11)$$
$$= E_n \Psi_n(X_2, Y_2, Z_2, X_1, Y_1, Z_1, W)\qquad(5\text{-}12)$$
$$= E_n \Psi_n^{(12)}(X_1, Y_1, Z_1, X_2, Y_2, Z_2, W).\qquad(5\text{-}13)$$

Eq. (5-11) simply results from the fact that E_n is a constant, independent of the coordinates of nuclei 1 and 2, and hence unaffected by (12). Equation (5-12) shows the effect of (12) on Ψ_n, and the value of the function at the point $(X_1, Y_1, Z_1, X_2, Y_2, Z_2, W)$ in configuration space is changed to its value at the point $(X_2, Y_2, Z_2, X_1, Y_1, Z_1, W)$. Equation (5-13) introduces

$\Psi_n^{(12)}$ as a new function whose value at the point $(X_1, Y_1, Z_1, X_2, Y_2, Z_2, W)$ is the same as the value of Ψ_n at the point $(X_2, Y_2, Z_2, X_1, Y_1, Z_1, W)$ [see Eqs. (1-14)–(1-18)].

The effect of (12) on the left hand side of Eq. (5-10) can be written

$$(12)\left\{-\frac{\hbar^2}{2m}(\nabla_1{}^2 + \nabla_2{}^2) + \frac{C^2e^2}{R_{12}} + \sum_{r\neq 1,2} CC_r e^2\left(\frac{1}{R_{1r}} + \frac{1}{R_{2r}}\right) + \hat{H}^0(\text{rest})\right\}$$

$$\times \Psi_n(X_1, Y_1, Z_1, X_2, Y_2, Z_2, W) = \left\{-\frac{\hbar^2}{2m}(\nabla_2{}^2 + \nabla_1{}^2) + \frac{C^2e^2}{R_{21}}\right.$$

$$\left. + \sum_{r\neq 1,2} CC_r e^2\left(\frac{1}{R_{2r}} + \frac{1}{R_{1r}}\right) + \hat{H}^0(\text{rest})\right\}$$

$$\times (12)\Psi_n(X_1, Y_1, Z_1, X_2, Y_2, Z_2, W), \tag{5-14}$$

permutation of identical nuclei affects Ψ not Ĥ

that is,

$$(12)[\hat{H}^0(1,2) + \hat{H}^0(\text{rest})]\Psi_n(X_1, Y_1, Z_1, X_2, Y_2, Z_2, W)$$
$$= [\hat{H}^0(2,1) + \hat{H}^0(\text{rest})](12)\Psi_n(X_1, Y_1, Z_1, X_2, Y_2, Z_2, W). \tag{5-15}$$

Since nuclei 1 and 2 have the same mass and charge $\hat{H}^0(2,1)$ is identical to $\hat{H}^0(1,2)$, and we can write

$$(12)[\hat{H}^0(1,2) + \hat{H}^0(\text{rest})]\Psi_n(X_1, Y_1, Z_1, X_2, Y_2, Z_2, W)$$
$$= [\hat{H}^0(1,2) + \hat{H}^0(\text{rest})](12)\Psi_n(X_1, Y_1, Z_1, X_2, Y_2, Z_2, W). \tag{5-16}$$

The right hand side of Eq. (5-16) becomes

$$[\hat{H}^0(1,2) + \hat{H}^0(\text{rest})]\Psi_n(X_2, Y_2, Z_2, X_1, Y_1, Z_1, W)$$
$$= [\hat{H}^0(1,2) + \hat{H}^0(\text{rest})]\Psi_n^{(12)}(X_1, Y_1, Z_1, X_2, Y_2, Z_2, W). \tag{5-17}$$

Equation (5-16) shows that the effect of $(12)\hat{H}^0$ on any function Ψ_n is the same as the effect of $\hat{H}^0(12)$ when 1 and 2 are identical. Thus the operation (12) commutes with this particular Hamiltonian operator $\hat{H}^0$ and we can write the operator equation

$$(12)\hat{H}^0 = \hat{H}^0(12). \tag{5-18}$$

In general the molecular Hamiltonian operator given in Eq. (5-6) will commute with any operation that is a permutation of identical nuclei (and similarly with any permutation of the electrons). As a direct result of Eq. (5-18) we have [by equating the right hand side of Eq. (5-17) with Eq. (5-13)]

$$\hat{H}^0\Psi_n^{(12)}(X_1, Y_1, Z_1, X_2, Y_2, Z_2, W) = E_n\Psi_n^{(12)}(X_1, Y_1, Z_1, X_2, Y_2, Z_2, W), \tag{5-19}$$

and the operation (12) has converted Ψ_n to a new function $\Psi_n^{(12)}$ which is also an eigenfunction of $\hat{H}^0$ with eigenvalue E_n. In general we can say that any operation R that commutes with the molecular Hamiltonian will convert an eigenfunction of the Hamiltonian to a new function having the same eigenvalue and such an operation is called a *symmetry operation* of the Hamiltonian. Symmetry operations are restricted to be unitary or anti-unitary operators [see p. 107 of Wigner (1959)]. A *symmetry group* of a Hamiltonian is a group of symmetry operations of the Hamiltonian. We sometimes say that the Hamiltonian is *invariant* to a symmetry operation in the sense that if R is a symmetry operation the effect of R on $\hat{H}$ (not considering any function that $\hat{H}$ acts on) is to leave $\hat{H}$ unchanged.

We can easily see that the inversion E^* will also commute with the Hamiltonian operator given in Eq. (5-6). This depends on results such as those now given in Eqs. (5-20) and (5-23).

$$E^* \frac{\partial^2}{\partial X_1^{\,2}} \Psi_n(X_1, Y_1, Z_1, \ldots) = \frac{\partial^2}{\partial X_1^{\,2}} E^* \Psi_n(X_1, Y_1, Z_1, \ldots) \tag{5-20}$$

$$= \frac{\partial^2}{\partial X_1^{\,2}} \Psi_n(-X_1, -Y_1, -Z_1, \ldots) \tag{5-21}$$

$$= \frac{\partial^2}{\partial X_1^{\,2}} \Psi_n^{E^*}(X_1, Y_1, Z_1, \ldots) \tag{5-22}$$

and

$$E^* R_{rs}^{-1} \Psi_n(X_1, Y_1, Z_1, \ldots) = R_{rs}^{-1} E^* \Psi_n(X_1, Y_1, Z_1, \ldots) \tag{5-23}$$

$$= R_{rs}^{-1} \Psi_n(-X_1, -Y_1, -Z_1, \ldots) \tag{5-24}$$

$$= R_{rs}^{-1} \Psi_n^{E^*}(X_1, Y_1, Z_1, \ldots). \tag{5-25}$$

Equation (5-20) follows from the fact that $\partial^2/\partial(-X_i)^2 = \partial^2/\partial X_i^{\,2}$, and Eq. (5-23) follows from the fact that R_{rs} [see Eq. (5-3)] depends only on the squares of the differences of the Cartesian coordinates and does not change if the signs of all the coordinates change (obviously the distances between the particles are unaffected by inverting the molecule).

Since any permutation of identical nuclei P commutes with $\hat{H}^0$, and E^* commutes with $\hat{H}^0$, it must be that $PE^* = E^*P = P^*$ commutes with H, i.e.,

$$P^*\hat{H}^0 = PE^*\hat{H}^0 = P\hat{H}^0E^* = \hat{H}^0PE^* = \hat{H}^0P^*. \tag{5-26}$$

Thus all the elements of the CNPI group of a molecule will commute with the molecular Hamiltonian given in Eq. (5-6) and the CNPI group is therefore a symmetry group of that Hamiltonian. Any element R of the CNPI group of a molecule will convert an eigenfunction Ψ_n of $\hat{H}^0$ into a new eigenfunction

Ψ_n^R of $\hat{H}^0$ having the same eigenvalue as Ψ_n. To proceed further we discuss the cases of nondegenerate and degenerate energy levels separately. A molecular state with energy E is said to be degenerate if more than one linearly independent eigenfunction of the molecular Hamiltonian has eigenvalue E.

Nondegenerate Levels

If E_n is a nondegenerate eigenvalue of $\hat{H}^0$, belonging to eigenfunction Ψ_n, then we see from Eq. (5-19) that $\Psi_n^{(12)} = (12)\Psi_n$ is also an eigenfunction of $\hat{H}^0$ with eigenvalue E_n. For this to be true we must have

$$\Psi_n^{(12)} = c\Psi_n, \tag{5-27}$$

where c is a constant, since Ψ_n is the *only* eigenfunction of $\hat{H}^0$ with eigenvalue E_n (it is nondegenerate by definition). We can evaluate c as follows: Writing

$$(12)\Psi_n = c\Psi_n, \tag{5-28}$$

and applying (12) to both sides, we obtain

$$(12)^2\Psi_n = (12)c\Psi_n. \tag{5-29}$$

Since $(12)^2$ is the identity and (12) commutes with c (a constant) this equation can be rewritten as

$$\Psi_n = c(12)\Psi_n \tag{5-30}$$

$$= c^2\Psi_n, \tag{5-31}$$

i.e., $c^2 = 1$, and therefore

$$c = \pm 1. \tag{5-32}$$

A nondegenerate eigenfunction of $\hat{H}^0$ for a molecule in which 1 and 2 are identical nuclei is thus either invariant ($c = +1$) to (12) or is changed in sign ($c = -1$) by (12). Alternatively we can say that for a molecule in which 1 and 2 are identical nuclei the value of a nondegenerate eigenfunction Ψ_n at each point $(X_1, Y_1, Z_1, X_2, Y_2, Z_2, W)$ in configuration space is either equal to, or equal to the negative of, its value at each point $(X_2, Y_2, Z_2, X_1, Y_1, Z_1, W)$. Some of the nondegenerate eigenfunctions will be unaffected by (12), i.e., symmetric under (12), and all the others will be changed in sign by (12), i.e., antisymmetric under (12). We cannot determine which will be symmetric and which antisymmetric except by looking at the functions in detail.

A simple example of this is afforded by the function $\sin(X_1 - X_2)$ which is antisymmetric under (12):

$$(12)\sin(X_1 - X_2) = \sin(X_2 - X_1) = -\sin(X_1 - X_2). \tag{5-33}$$

Alternatively $\cos(X_1 - X_2)$ is symmetric under (12). The functions $\sin(X_1 - X_2)$ and $\cos(X_1 - X_2)$, and the functions used in Problems 5-1, 5-2, and 5-3 that follow, are not eigenfunctions of a molecular Hamiltonian; these functions are introduced simply as convenient functions for studying transformation properties.

To introduce some more of the definitions that are used we can take (12) together with the identity E to make the S_2 group. The S_2 group has two irreducible representations, which we call Γ_1 and Γ_2, and the character table of S_2 is given in Table 5-1. A function that is symmetric under (12) is said to *generate* (or *form a basis for*) the representation Γ_1 of the S_2 group, since application of either E or (12) to a symmetric function produces $(+1)$ times the function, and the representation Γ_1 has $(+1)$ as its elements both under E and (12). A function that is antisymmetric under (12) generates the numbers $(+1)$ and (-1) under the operations E and (12), respectively, and thus generates the representation Γ_2. Symmetric functions would then be said to be of Γ_1 *symmetry* in the S_2 group, while antisymmetric functions are of Γ_2 symmetry.

Table 5-1
The character table
of the S_2 group

S_2:	E	(12)
Γ_1:	1	1
Γ_2:	1	-1

In deriving Eq. (5-32) from Eq. (5-28) for the effect of (12) on Ψ_n it was important that $(12)^2 = E$. An arbitrary operation R that commutes with the Hamiltonian need not be self-reciprocal like this and in general we have $R^m = E$ where m need not be 2. In this case we determine that for a non-degenerate eigenfunction

$$R^m \Psi_n = c^m \Psi_n = \Psi_n, \qquad (5\text{-}34)$$

i.e.,

$$c = \sqrt[m]{1}. \qquad (5\text{-}35)$$

For example, the operation (123) is such that $(123)^3 = E$, and nondegenerate eigenfunctions of a Hamiltonian that commutes with (123) will be multiplied by 1, $\omega = \exp(2\pi i/3)$, or ω^2 by the effect of (123).

In general any nondegenerate eigenfunction of the Hamiltonian will generate a one-dimensional representation of a symmetry group of the Hamiltonian (as is proved in Appendix 5-1 at the end of this chapter) and we say that we can *classify* the nondegenerate eigenfunction according to the one-dimensional representations of the symmetry group. We pay special attention to the effect of E^* by saying that an eigenfunction that is symmetric under it is of positive parity whereas one that is antisymmetric under it is of negative parity.

Problem 5-1. Consider the following functions for the water molecule, where 1 and 2 label the protons,

$$\Psi_1 = \sin(X_1 - X_2 + Y_1 - Y_2 + Z_1 - Z_2),$$
$$\Psi_2 = \cos(X_1 - X_2 + Y_1 - Y_2 + Z_1 - Z_2),$$

and

$$\Psi_3 = \sin(X_1 + X_2 + Y_1 + Y_2 + Z_1 + Z_2).$$

Each of these functions generates a one-dimensional representation of the CNPI group of the water molecule; determine which representations are generated. In Table 5-2 the character table of the CNPI group of the water molecule is given. We call this group $C_{2v}(M)$ and label the irreducible representations A_1, A_2, B_1, and B_2 (the reason for this notation and the similar notation used in Problem 5-2 will become apparent later).

Table 5-2

The character table of the CNPI group
$C_{2v}(M)$ of the water molecule[a]

$C_{2v}(M)$:	E	(12)	E^*	$(12)^*$
A_1:	1	1	1	1
A_2:	1	1	-1	-1
B_1:	1	-1	-1	1
B_2:	1	-1	1	-1

[a] The protons are labeled 1 and 2.

Answer. To determine the representation for which each Ψ_i forms a basis we must determine the coefficients c_i^R, where

$$E\Psi_i = c_i^E \Psi_i, \tag{5-36}$$

$$(12)\Psi_i = c_i^{(12)} \Psi_i, \tag{5-37}$$

$$E^*\Psi_i = c_i^{E^*} \Psi_i, \tag{5-38}$$

and

$$(12)^*\Psi_i = c_i^{(12)^*} \Psi_i. \tag{5-39}$$

The representation generated is then $[c_i^E, c_i^{(12)}, c_i^{E*}, c_i^{(12)*}]$. Clearly

$$c_i^E = 1, \tag{5-40}$$

since E does nothing to the functions. Also we have

$$c_i^{(12)*} = c_i^{(12)} c_i^{E*}, \tag{5-41}$$

since the c_i^R form a one-dimensional representation. As a result of Eqs. (5-40) and (5-41) we only have to determine $c_i^{(12)}$ and c_i^{E*} for each Ψ_i in order to determine the representation that the function generates.

We look first at Ψ_1.

$$\begin{aligned}(12)\Psi_1 &= (12)\sin(X_1 - X_2 + Y_1 - Y_2 + Z_1 - Z_2) \\ &= \sin(X_2 - X_1 + Y_2 - Y_1 + Z_2 - Z_1) \\ &= (-1)\Psi_1 \end{aligned} \tag{5-42}$$

and

$$\begin{aligned}E^*\Psi_1 &= E^* \sin(X_1 - X_2 + Y_1 - Y_2 + Z_1 - Z_2) \\ &= \sin(-X_1 + X_2 - Y_1 + Y_2 - Z_1 + Z_2) \\ &= (-1)\Psi_1. \end{aligned} \tag{5-43}$$

Collecting the results for Ψ_1 we have

$$c_1^E = +1, \tag{5-44a}$$

$$c_1^{(12)} = -1, \tag{5-44b}$$

$$c_1^{E*} = -1, \tag{5-44c}$$

and

$$c_1^{(12)*} = c_1^{(12)} c_1^{E*} = +1, \tag{5-44d}$$

so that the representation generated by Ψ_1 is $[+1, -1, -1, +1]$, i.e., B_1, in the CNPI group of the water molecule.

We next look at the function Ψ_2:

$$(12)\Psi_2 = \cos(X_2 - X_1 + Y_2 - Y_1 + Z_2 - Z_1) = (+1)\Psi_2 \tag{5-45}$$

and

$$E^*\Psi_2 = \cos(-X_1 + X_2 - Y_1 + Y_2 - Z_1 + Z_2) = (+1)\Psi_2, \tag{5-46}$$

and the function Ψ_2 generates the A_1 representation of the CNPI group.

Finally we look at Ψ_3.

$$(12)\Psi_3 = \sin(X_2 + X_1 + Y_2 + Y_1 + Z_2 + Z_1) = (+1)\Psi_3 \tag{5-47}$$

and

$$E^*\Psi_3 = \sin(-X_1 - X_2 - Y_1 - Y_2 - Z_1 - Z_2) = (-1)\Psi_3, \tag{5-48}$$

and the function Ψ_3 generates the representation A_2. The functions Ψ_1, Ψ_2, and Ψ_3 are thus of symmetry B_1, A_1, and A_2, respectively; Ψ_1 and Ψ_3 are of negative parity whereas Ψ_2 is of positive parity.

Degenerate Levels

If E_n is an l-fold degenerate eigenvalue of the molecular Hamiltonian with eigenfunctions Ψ_{n1}, Ψ_{n2}, ..., Ψ_{nl}, then the effect of a symmetry operation R on one of these functions will be to convert it to a linear combination of these l functions. This must be true since the function that results by applying R to any one of these functions still has eigenvalue E_n [see Eq. (5-19) and the discussion after it], and the most general function of this type is a linear combination of these functions.

From the above remarks we see that the effect of R can be written, using matrix notation, as

$$R\Psi_{ni} = \sum_{j=1}^{l} D[R]_{ij}\Psi_{nj}, \qquad (5\text{-}49)$$

where $i = 1, 2, \ldots, l$. For example, choosing $i = 1$, we have the effect of R on Ψ_{n1} as:

$$R\Psi_{n1} = D[R]_{11}\Psi_{n1} + D[R]_{12}\Psi_{n2} + \cdots + D[R]_{1l}\Psi_{nl}. \qquad (5\text{-}50)$$

The $D[R]_{ij}$ are numbers and $D[R]$ is a matrix of these numbers; the matrix $D[R]$ is *generated* by the effect of R on the l functions Ψ_{ni}. We can visualize Eq. (5-49) as a column matrix $R\Psi_n$ being equal to the product of a square matrix $D[R]$ and a column matrix Ψ_n, i.e.,

$$R\begin{bmatrix} \Psi_n \end{bmatrix} = \begin{bmatrix} D[R] \end{bmatrix}\begin{bmatrix} \Psi_n \end{bmatrix}. \qquad (5\text{-}51)$$

Each operation in a symmetry group of the Hamiltonian will generate such an $l \times l$ matrix, and it is shown in Appendix 5-1 that if three operations of the group P_1, P_2, and P_{12} are related by

$$P_1 P_2 = P_{12} \qquad (5\text{-}52)$$

then the matrices generated by application of them to the Ψ_{ni} [as described by Eq. (5-49)] will satisfy

$$D[P_1]D[P_2] = D[P_{12}]. \qquad (5\text{-}53)$$

Thus the matrices will have a multiplication table with the same structure as the multiplication table of the symmetry group and hence will form an l-dimensional representation of the group.

A given l-fold degenerate state may generate a reducible or an irreducible l-dimensional representation of the symmetry group considered. If the representation is irreducible then the degeneracy is said to be *necessary*, i.e., imposed by the symmetry of the Hamiltonian. However, if the representation is reducible then the degeneracy between different irreducible representations is said to be accidental, and it is not imposed by the symmetry of the Hamiltonian.

Suppose that, for the given l-fold degenerate energy level E_n, we choose a different set of l linearly independent functions, i.e., use Φ_{nk} instead of Ψ_{ni}, where

$$\Phi_{nk} = \sum_{i=1}^{l} A_{ki}\Psi_{ni} \qquad (5\text{-}54)$$

and A is an orthogonal matrix. Such a transformation is called an orthogonal transformation, and if the Ψ_{ni} are eigenfunctions of the Hamiltonian with eigenvalue E_n then the Φ_{nk} will also be eigenfunctions having the same eigenvalue. The eigenfunctions for a given degenerate energy level are therefore arbitrary to the extent of an orthogonal transformation. The question arises: What representation of the symmetry group of the Hamiltonian will the new functions Φ_{nk} generate? The matrix representing the operation R that is generated by the new functions Φ_{nk} will be given by

$$R\Phi_{nk} = \sum_{r=1}^{l} \bar{D}[R]_{kr}\Phi_{nr}. \qquad (5\text{-}55)$$

Applying R to both sides of Eq. (5-54) we obtain

$$R\Phi_{nk} = \sum_{i=1}^{l} A_{ki}R\Psi_{ni}. \qquad (5\text{-}56)$$

Using Eq. (5-49) we can write Eq. (5-56) as

$$R\Phi_{nk} = \sum_{i=1}^{l} A_{ki} \sum_{j=1}^{l} D[R]_{ij}\Psi_{nj}. \qquad (5\text{-}57)$$

The inverse of Eq. (5-54) can be written as

$$\Psi_{nj} = \sum_{r=1}^{l} (A^{-1})_{jr}\Phi_{nr} \qquad (5\text{-}58)$$

and substituting this equation into Eq. (5-57) we obtain

$$R\Phi_{nk} = \sum_{i=1}^{l} A_{ki} \sum_{j=1}^{l} D[R]_{ij} \sum_{r=1}^{l} (A^{-1})_{jr}\Phi_{nr} \qquad (5\text{-}59)$$

$$= \sum_{i,j,r} A_{ki}D[R]_{ij}(A^{-1})_{jr}\Phi_{nr}. \qquad (5\text{-}60)$$

From Eqs. (5-60) and (5-55) we obtain the matrix equation

$$\bar{D}[R] = AD[R]A^{-1}. \tag{5-61}$$

Thus the matrix representation generated by the Φ_n, containing $\bar{D}$, is obtained from that generated by the Ψ_n by performing the similarity transformation of Eq. (5-61) with the matrix A, and these representations are, therefore, equivalent. This means that the representation generated by the eigenfunctions of a particular degenerate energy level is unique (apart from a similarity transformation) and can be unambiguously reduced to its irreducible components. Therefore, each energy level can be labeled according to the irreducible representations of the symmetry group, and this is an important characteristic of use in distinguishing the energy levels.

Using l mutually orthogonal functions[1] to describe an l-fold degenerate level it can be shown that the matrix representation obtained will be unitary, i.e.,

$$D[R^{-1}]_{ij} = D[R]^*_{ji}. \tag{5-62}$$

We always use orthonormal wavefunctions and hence always obtain unitary representations of the symmetry group.

We frequently consider a set of functions, $\Phi_1, \ldots, \Phi_r$, say, which generates a reducible representation Γ of a symmetry group, i.e., for a particular operation R of the group we have

$$R\Phi_n = \sum_{m=1}^{r} D^\Gamma[R]_{nm}\Phi_m, \tag{5-63}$$

where the $D^\Gamma[R]$ are the matrices of the reducible representation Γ. We wish to determine which linear combinations of the functions Φ_n transform according to the irreducible representations Γ_i that make up Γ. To do this we use a *projection operator*, and this is a particular combination of the operations in the group defined by the equation

$$P^{\Gamma_i} = \sum_R \chi^{\Gamma_i}[R]^*R, \tag{5-64}$$

where the sum is over all operations in the group. There is a different projection operator for each irreducible representation of the group, and we need to know the characters of the representations in order to be able to construct the projection operators. If we apply the projection operator P^{Γ_i} to Φ_n, which

---

[1] Two functions Ψ_{ni} and Ψ_{nj} are orthogonal if the product $\Psi_{ni}^*\Psi_{nj}$, integrated over all configuration space, vanishes. A function Ψ is normalized if the product $\Psi^*\Psi$ integrated over all configuration space is unity. An orthonormal set contains functions that are normalized and orthogonal to each other.

is one of the functions $\Phi_1, \ldots, \Phi_r$, the result is

$$P^{\Gamma_i}\Phi_n = \sum_R \chi^{\Gamma_i}[R]^* R\Phi_n$$

$$= \sum_{R,k,m} D^{\Gamma_i}[R]^*_{kk} D^{\Gamma}[R]_{nm}\Phi_m. \tag{5-65}$$

We thus obtain a linear combination of the r functions Φ_1 to Φ_r, and by use of the orthogonality relation [Eq. (4-38)] it can be shown (see Appendix 5-2) that the linear combination obtained transforms according to the irreducible representation Γ_i. If Γ_i has dimension l_i then it will be necessary to apply P^{Γ_i} to l_i of the functions Φ_n in order to obtain l_i functions that generate Γ_i; it will usually be necessary to form linear combinations of these l_i functions in a special way, using *Schmidt orthogonalization* [see Eq. (5-83)], in order to obtain l_i orthogonal functions of species Γ_i. Some examples of the application of projection operators are given in Problems 5-2 and 5-3. If in the reduction of the representation Γ generated by the functions Φ_n the irreducible representation Γ_i does not occur then the effect of P^{Γ_i} on Φ_n will be to give zero; i.e., the projection operator *annihilates* Φ_n. The reader can test this by applying the operator P^{A_1} to any of the three functions Ψ_a, Ψ_b, or Ψ_c in Problem 5-2 which follows.

Problem 5-2. The character table of the MS group $C_{3v}(M)$ of the CH_3F molecule is given in Table 5-3. We label the irreducible representations A_1, A_2, and E. Suppose that Ψ_a, Ψ_b, and Ψ_c are threefold degenerate orthonormal eigenfunctions for CH_3F where

$$\Psi_a = X_1, \qquad \Psi_b = X_2, \qquad \text{and} \qquad \Psi_c = X_3,$$

where the protons are labeled 1, 2, and 3. What representation of the $C_{3v}(M)$ group do these three functions generate? Reduce the representation obtained to its irreducible components, and find the combinations of the functions

Table 5-3
The character table of the MS group
$C_{3v}(M)$ of the methyl fluoride
molecule[a]

$C_{3v}(M)$:	E	(123) (132)	(12)* (23)* (13)*
A_1:	1	1	1
A_2:	1	1	-1
E:	2	-1	0

[a] The protons are labeled 1, 2, and 3.

that generate these irreducible representations by using the projection operator technique. Equations (1-12) and (2-8) will be of use in solving this problem. In this artificial example we are, for the sake of clarity, taking the functions X_i (the nuclear coordinates) as being normalized; it would be more correct to take three identical (normalized) *functions* of X_1, X_2, and X_3 instead of X_1, X_2, and X_3, respectively.

Answer. We must determine the 3×3 matrices $D[R]$ generated by the effect of the operations R of the $C_{3v}(M)$ group on the three functions (Ψ_a, Ψ_b, Ψ_c). We consider (123) and (12)* in detail with the help of Eqs. (1-12) and (2-8).

$$(123)\Psi_a = (123)X_1 = X_3 = \Psi_c, \tag{5-66}$$

$$(123)\Psi_b = (123)X_2 = X_1 = \Psi_a, \tag{5-67}$$

and

$$(123)\Psi_c = \Psi_b, \tag{5-68}$$

so that

$$(123)\begin{bmatrix} \Psi_a \\ \Psi_b \\ \Psi_c \end{bmatrix} = \begin{bmatrix} \Psi_c \\ \Psi_a \\ \Psi_b \end{bmatrix} = \underbrace{\begin{bmatrix} 0 & 0 & 1 \\ 1 & 0 & 0 \\ 0 & 1 & 0 \end{bmatrix}}_{D[(123)]}\begin{bmatrix} \Psi_a \\ \Psi_b \\ \Psi_c \end{bmatrix}. \tag{5-69}$$

Also

$$(12)^*\Psi_a = (12)^*X_1 = -X_2 = -\Psi_b, \tag{5-70}$$

$$(12)^*\Psi_b = -\Psi_a, \tag{5-71}$$

and

$$(12)^*\Psi_c = -\Psi_c, \tag{5-72}$$

so that

$$(12)^*\begin{bmatrix} \Psi_a \\ \Psi_b \\ \Psi_c \end{bmatrix} = \begin{bmatrix} -\Psi_b \\ -\Psi_a \\ -\Psi_c \end{bmatrix} = \underbrace{\begin{bmatrix} 0 & -1 & 0 \\ -1 & 0 & 0 \\ 0 & 0 & -1 \end{bmatrix}}_{D[(12)^*]}\begin{bmatrix} \Psi_a \\ \Psi_b \\ \Psi_c \end{bmatrix}. \tag{5-73}$$

Similarly the operations (132), (23)*, and (13)* generate the matrices

$$\begin{bmatrix} 0 & 1 & 0 \\ 0 & 0 & 1 \\ 1 & 0 & 0 \end{bmatrix}, \quad \begin{bmatrix} -1 & 0 & 0 \\ 0 & 0 & -1 \\ 0 & -1 & 0 \end{bmatrix}, \quad \text{and} \quad \begin{bmatrix} 0 & 0 & -1 \\ 0 & -1 & 0 \\ -1 & 0 & 0 \end{bmatrix}, \tag{5-74}$$

respectively, and E generates the 3×3 unit matrix. These matrices form a representation of the $C_{3v}(M)$ group as the reader can test by forming their multiplication table. The characters of this representation are

$$\chi[E] = 3, \quad \chi[(123)] = \chi[(132)] = 0,$$
$$\chi[(12)^*] = \chi[(23)^*] = \chi[(13)^*] = -1. \tag{5-75}$$

By inspection of Table 5-3, or by using Eq. (4-43), we see that this representation reduces to $A_2 \oplus E$. Since we are only interested in the characters of the representation generated by the set of functions we really need only have considered one operation from each class in the group.

The combinations of the functions Ψ_a, Ψ_b, and Ψ_c that are of species A_2 and E are found by using projection operators. An unnormalized combination of species A_2 is given by

$$\Phi'(A_2) = \left\{ \sum_R \chi^{A_2}[R]^* R \right\} \Psi_a, \tag{5-76}$$

where the operator in the braces is the A_2 projection operator P^{A_2}. Operating with P^{A_2} on Ψ_a we have

$$\begin{aligned}
\Phi'(A_2) &= (+1) \times E\Psi_a + (+1) \times (123)\Psi_a + (+1) \times (132)\Psi_a \\
&\quad + (-1) \times (12)^*\Psi_a + (-1) \times (23)^*\Psi_a + (-1) \times (13)^*\Psi_a \\
&= (+1)\Psi_a + (+1)\Psi_c + (+1)\Psi_b + (-1)(-\Psi_b) \\
&\quad + (-1)(-\Psi_a) + (-1)(-\Psi_c) \\
&= 2(\Psi_a + \Psi_b + \Psi_c).
\end{aligned} \tag{5-77}$$

Normalizing we obtain

$$\Phi(A_2) = (\Psi_a + \Psi_b + \Psi_c)/\sqrt{3} = (X_1 + X_2 + X_3)/\sqrt{3}. \tag{5-78}$$

The same function would be obtained if we applied P^{A_2} to Ψ_b or Ψ_c. The reader should notice that P^{A_1} annihilates Ψ_a, Ψ_b, or Ψ_c since the representation generated by these functions does not contain A_1.

Starting from the function Ψ_a and using the projection operator P^E, where

$$P^E = \sum_R \chi^E[R]^* R, \tag{5-79}$$

we obtain the function

$$\Phi_a'(E) = (+2)\Psi_a + (-1)\Psi_c + (-1)\Psi_b, \tag{5-80}$$

which on normalizing gives

$$\Phi_a(E) = (2X_1 - X_2 - X_3)/\sqrt{6}. \tag{5-81}$$

Starting from Ψ_b and applying P^E we obtain the normalized function

$$\Phi_b''(E) = (2X_2 - X_3 - X_1)/\sqrt{6}, \tag{5-82}$$

which is not orthogonal to $\Phi_a(E)$. We can use a procedure called *Schmidt orthogonalization* to obtain from Φ_b'' the function Φ_b that is orthogonal to Φ_a; this procedure gives

$$\Phi_b' = \Phi_b''(E) - \left[\int \Phi_a(E)^* \Phi_b''(E) \, dX_1 \, dX_2 \, dX_3 \right] \Phi_a(E). \qquad (5\text{-}83)$$

With $\int X_i X_j \, dX_1 \, dX_2 \, dX_3 = \delta_{ij}$ we obtain

$$\Phi_b' = \Phi_b''(E) - (-\tfrac{1}{2})\Phi_a(E) = \sqrt{3}(X_2 - X_3)/(2\sqrt{2}), \qquad (5\text{-}84)$$

which on normalization gives

$$\Phi_b(E) = (X_2 - X_3)/\sqrt{2}. \qquad (5\text{-}85)$$

Thus this set of functions would be considered as an accidentally threefold degenerate set of states $\Phi(A_2)$ and $[\Phi_a(E), \Phi_b(E)]$, where $\Phi_a(E)$ and $\Phi_b(E)$ are necessarily twofold degenerate from symmetry. The reader can test that these three functions transform separately as A_2 and E by applying the operations of the $C_{3v}(M)$ group to them (see Appendix 5-3).

THE SYMMETRY OF A PRODUCT

Given an s-fold degenerate state of energy E_n and symmetry Γ_n, with eigenfunctions $\Phi_{n1}, \Phi_{n2}, \ldots, \Phi_{ns}$, and an r-fold degenerate state of energy E_m and symmetry Γ_m, with eigenfunctions $\Phi_{m1}, \Phi_{m2}, \ldots, \Phi_{mr}$, we wish to determine the symmetry Γ_{mn} of the set of functions $\Psi_{ij} = \Phi_{ni}\Phi_{mj}$, where $i = 1, 2, \ldots, s$ and $j = 1, 2, \ldots, r$. There will be $s \times r$ functions of the type Ψ_{ij}. The matrices $D^{(n)}$ and $D^{(m)}$ in the representations Γ_n and Γ_m, respectively, are obtained from

$$R\Phi_{ni} = \sum_{k=1}^{s} D^{(n)}[R]_{ik}\Phi_{nk} \qquad (5\text{-}86)$$

and

$$R\Phi_{mj} = \sum_{l=1}^{r} D^{(m)}[R]_{jl}\Phi_{ml}, \qquad (5\text{-}87)$$

where R is an operation of the symmetry group. To obtain the matrices in the representation Γ_{nm} we write

$$R[\Phi_{ni}\Phi_{mj}] = \sum_{k=1}^{s} \sum_{l=1}^{r} D^{(n)}[R]_{ik}D^{(m)}[R]_{jl}\Phi_{nk}\Phi_{ml} \qquad (5\text{-}88)$$

and we can write this as

$$R\Psi_{ij} = \sum_{k,l=1,1}^{s,r} D^{(nm)}[R]_{ij,kl}\Psi_{kl}. \qquad (5\text{-}89)$$

From this we see that the $s \times r$ dimensional representation Γ_{nm} generated by the $s \times r$ functions Ψ_{ij} has matrices with elements given by

$$D^{(nm)}[R]_{ij,kl} = D^{(n)}[R]_{ik}D^{(m)}[R]_{jl}, \tag{5-90}$$

where each element of $D^{(nm)}$ is indexed by a row label ij and a column label kl, each of which runs over $s \times r$ values. The ij,ij diagonal element is given by

$$D^{(nm)}[R]_{ij,ij} = D^{(n)}[R]_{ii}D^{(m)}[R]_{jj}, \tag{5-91}$$

and the character of the matrix is given by

$$\chi^{\Gamma_{nm}}[R] = \sum_{i,j=1,1}^{s,r} D^{(nm)}[R]_{ij,ij} = \sum_{i,j=1,1}^{s,r} D^{(n)}[R]_{ii}D^{(m)}[R]_{jj}$$

$$= \chi^{\Gamma_n}[R]\chi^{\Gamma_m}[R]. \tag{5-92}$$

We can therefore calculate the character, under a symmetry operation R, in the representation generated by the product of two sets of functions, by multiplying together the characters under R in the representations generated by each of the sets of functions. We write Γ_{nm} symbolically as

$$\Gamma_{nm} = \Gamma_n \otimes \Gamma_m, \tag{5-93}$$

where the characters satisfy Eq. (5-92) in which usual algebraic multiplication is used. Knowing the character in Γ_{nm} from Eq. (5-92) we can then reduce the representation to its irreducible components using Eq. (4-43). Suppose Γ_{nm} can be reduced to irreducible representations Γ_1, Γ_2, and Γ_3 according to

$$\Gamma_n \otimes \Gamma_m = 3\Gamma_1 \oplus \Gamma_2 \oplus 2\Gamma_3. \tag{5-94}$$

In this circumstance we say that Γ_{nm} *contains* Γ_1, Γ_2, and Γ_3; since $\Gamma_n \otimes \Gamma_m$ contains Γ_1, for example, we write

$$\Gamma_n \otimes \Gamma_m \supset \Gamma_1. \tag{5-95}$$

Problem 5-3.   Determine the representation generated by the product $\Phi(A_2) \times [\Phi_a(E), \Phi_b(E)]$ where these functions are given in Eqs. (5-78), (5-81), and (5-85). If we introduce

$$\theta_a(E) = (2Y_1 - Y_2 - Y_3)/\sqrt{6} \quad \text{and} \quad \theta_b(E) = (Y_2 - Y_3)/\sqrt{2}$$

what is the symmetry of the product

$$[\Phi_a(E), \Phi_b(E)] \times [\theta_a(E), \theta_b(E)]?$$

Also determine the symmetry of the product

$$[\Phi_a(E), \Phi_b(E)] \times [\Phi_a(E), \Phi_b(E)].$$

Answer.   The two functions formed by multiplying $\Phi(A_2)$ and $[\Phi_a(E), \Phi_b(E)]$ are $\Phi(A_2)\Phi_a(E)$ and $\Phi(A_2)\Phi_b(E)$. These will transform according to a

representation $\Gamma^{(1)}$, say, where by using Eq. (5-92) we deduce that the characters in $\Gamma^{(1)}$ are as follows:

$$\chi^{\Gamma^{(1)}}[E] = \chi^{A_2}[E] \times \chi^{E}[E] = 1 \times 2 = 2, \qquad (5\text{-}96)$$

$$\chi^{\Gamma^{(1)}}[(123)] = \chi^{A_2}[(123)] \times \chi^{E}[(123)] = 1 \times (-1) = -1, \qquad (5\text{-}97)$$

and

$$\chi^{\Gamma^{(1)}}[(12)^*] = \chi^{A_2}[(12)^*] \times \chi^{E}[(12)^*] = (-1) \times 0 = 0. \qquad (5\text{-}98)$$

We have chosen one element from each class in the $C_{3v}(M)$ group and we see
that the representation generated is the irreducible representation E. We
write

$$A_2 \otimes E = E. \qquad (5\text{-}99)$$

We deduce that $[\Phi(A_2)\Phi_a(E), \Phi(A_2)\Phi_b(E)]$ transforms according to the E
representation of the $C_{3v}(M)$ group.

When we multiply $[\Phi_a(E), \Phi_b(E)]$ and $[\theta_a(E), \theta_b(E)]$ together we obtain
the four functions $[\Phi_a\theta_a, \Phi_a\theta_b, \Phi_b\theta_a, \Phi_b\theta_b]$ and these functions will form
the basis for the representation $\Gamma^{(2)}$, say, where [using Eq. (5-92)] the characters in this representation are given by

$$\chi^{\Gamma^{(2)}}[E] = \chi^{E}[E] \times \chi^{E}[E] = 4, \qquad (5\text{-}100)$$

$$\chi^{\Gamma^{(2)}}[(123)] = \chi^{E}[(123)] \times \chi^{E}[(123)] = 1, \qquad (5\text{-}101)$$

and

$$\chi^{\Gamma^{(2)}}[(12)^*] = \chi^{E}[(12)^*] \times \chi^{E}[(12)^*] = 0. \qquad (5\text{-}102)$$

Thus $\Gamma^{(2)}$ has characters $[4, 1, 0]$. This reduces to $A_1 \oplus A_2 \oplus E$ and we write

$$E \otimes E = A_1 \oplus A_2 \oplus E. \qquad (5\text{-}103)$$

It is left as an exercise for the reader to show, by using projection operators,
that the combinations of the product functions that transform irreducibly are

$$(\Phi_a\theta_a + \Phi_b\theta_b): \quad A_1, \qquad (5\text{-}104)$$

$$(\Phi_a\theta_b - \Phi_b\theta_a): \quad A_2, \qquad (5\text{-}105)$$

and

$$[(\Phi_a\theta_a - \Phi_b\theta_b), (\Phi_a\theta_b + \Phi_b\theta_a)]: \quad E. \qquad (5\text{-}106)$$

The three *symmetric product functions* $\Phi_a\theta_a$, $\Phi_b\theta_b$, and $(\Phi_a\theta_b + \Phi_b\theta_a)$
generate the representation $A_1 \oplus E$, and the *antisymmetric product function*,
defined as $(\Phi_a\theta_b - \Phi_b\theta_a)$, generates the representation A_2.

We now consider the product $[\Phi_a(E), \Phi_b(E)] \times [\Phi_a(E), \Phi_b(E)]$. This product gives us the functions $[\Phi_a\Phi_a, \Phi_a\Phi_b, \Phi_b\Phi_a, \Phi_b\Phi_b]$ and we might expect,

from Eq. (5-103), that these four functions would transform as $A_1 \oplus A_2 \oplus E$. However, $\Phi_a\Phi_b = \Phi_b\Phi_a$, and there are only three independent functions in the product. From Eq. (5-105) we see that the antisymmetric product function vanishes, and the three independent functions in the product (of necessity symmetric product functions) transform as $A_1 \oplus E$.

We can extend the product by introducing the functions

$$\Psi_a = (2Z_1 - Z_2 - Z_3)/\sqrt{6} \qquad (5\text{-}107)$$

and

$$\Psi_b = (Z_2 - Z_3)/\sqrt{2}, \qquad (5\text{-}108)$$

and we can form the product $[\Phi_a, \Phi_b] \times [\theta_a, \theta_b] \times [\Psi_a, \Psi_b]$. These eight functions generate the representation

$$E \otimes E \otimes E = E \otimes (A_1 \oplus A_2 \oplus E) = A_1 \oplus A_2 \oplus 3E. \qquad (5\text{-}109)$$

On the other hand if we form the product $[\Phi_a, \Phi_b] \times [\Phi_a, \Phi_b] \times [\Phi_a, \Phi_b]$ we obtain the four independent functions $\Phi_a\Phi_a\Phi_a$, $\Phi_a\Phi_a\Phi_b$, $\Phi_a\Phi_b\Phi_b$, and $\Phi_b\Phi_b\Phi_b$; these four functions transform as $A_1 \oplus A_2 \oplus E$.

The solution to Problem 5-3 provides examples of the *symmetric product representation* and the *antisymmetric product representation* of a representation with itself. The symmetric product of a doubly degenerate representation, E, say, of any group is the representation generated by the functions $\Phi_a\theta_a$, $\Phi_b\theta_b$ and $(\Phi_a\theta_b + \Phi_b\theta_a)$, where (Φ_a, Φ_b) and (θ_a, θ_b) are each of species E; the antisymmetric product is generated by $(\Phi_a\theta_b - \Phi_b\theta_a)$. In the group $C_{3v}(M)$ the product representation $E \otimes E$ is reducible into the sum of $A_1 \oplus E$ and A_2, where the former is the symmetric product and the latter is the antisymmetric product. We write the symmetric product as

$$[E]^2 = [E \otimes E] = A_1 \oplus E \qquad (5\text{-}110)$$

and the antisymmetric product as

$$\{E\}^2 = \{E \otimes E\} = A_2. \qquad (5\text{-}111)$$

In general the product of any doubly degenerate representation E with itself is reducible to the sum of the symmetric product representation $[E \otimes E]$ and the antisymmetric product representation $\{E \otimes E\}$ where the characters in the symmetric product are given by

$$\chi^{[E \otimes E]}[R] = \tfrac{1}{2}((\chi^E[R])^2 + \chi^E[R^2]), \qquad (5\text{-}112)$$

and the characters in the antisymmetric product are given by

$$\chi^{\{E \otimes E\}}[R] = \tfrac{1}{2}((\chi^E[R])^2 - \chi^E[R^2]). \qquad (5\text{-}113)$$

The characters in the symmetric nth power of E (i.e., the symmetry of the set of $n + 1$ independent functions obtained by taking the nth power of a pair of E functions) can be obtained from the characters in the symmetric $(n - 1)$th power of E by using (where E is doubly degenerate)

$$\chi^{[E]^n}[R] = \tfrac{1}{2}(\chi^E[R]\chi^{[E]^{n-1}}[R] + \chi^E[R^n]).$$ (5-114)

An example with $n = 3$ is given at the end of the solution to Problem 5-3. In Eqs. (5-112)–(5-114), $\chi^E[R^n]$ is the number obtained by determining the character in E under the operation $P = R^n$. The symmetric nth power of a representation is used in determining the species of vibrational wavefunctions in Chapter 10 [see Eqs. (10-33) and (10-35)].

In considering the symmetry of a product there is one especially important result, namely that the product representation Γ of two irreducible representations[2] $\Gamma_n{}^*$ and Γ_m contains the totally symmetric representation $\Gamma^{(s)}$ once if $\Gamma_n = \Gamma_m$, otherwise the product does not contain $\Gamma^{(s)}$. This can be proved as follows. The character under the operation R in the representation Γ is given by

$$\chi^\Gamma[R] = \chi^{\Gamma_n}[R]^*\chi^{\Gamma_m}[R]$$ (5-115)

from Eq. (5-92). The number of times the totally symmetric representation occurs is given by

$$a^{\Gamma^{(s)}} = \frac{1}{h}\sum_R \chi^\Gamma[R]$$ (5-116)

from Eq. (4-43) since $\chi^{\Gamma^{(s)}}[R] = 1$ for all R. Substituting Eq. (5-115) into Eq. (5-116) we obtain

$$a^{\Gamma^{(s)}} = \frac{1}{h}\sum_R \chi^{\Gamma_n}[R]^*\chi^{\Gamma_m}[R].$$ (5-117)

Using Eq. (4-39) we derive

$$a^{\Gamma^{(s)}} = \delta_{nm},$$ (5-118)

i.e.,

$$a^{\Gamma^{(s)}} = 1 \quad \text{if} \quad \Gamma_n = \Gamma_m$$

or

$$a^{\Gamma^{(s)}} = 0 \quad \text{if} \quad \Gamma_n \neq \Gamma_m.$$

[2] $\Gamma_n{}^*$ is the irreducible representation whose matrices $D^n[R]^*$ are the complex conjugates of the matrices of Γ_n. If Γ_n is an irreducible representation then so is $\Gamma_n{}^*$, although if the $D^n[R]$ are real then $\Gamma_n \equiv \Gamma_n{}^*$.

As an example with imaginary characters, the character table of the C_3 group is given in Table 5-4 and we see that

$$E_a^* \otimes E_a = A_1 \tag{5-119}$$

but

$$E_a \otimes E_a = E_b, \tag{5-120}$$

since $\omega^* = \omega^2$ and $(\omega^2)^* = \omega$. Two representations (such as E_a and E_b in C_3) which have imaginary characters, and whose characters are the complex conjugates of each other, are said to be *separably degenerate* for reasons that we will appreciate in Chapter 6 when we discuss the operation of time reversal.

Table 5-4
The character table of the
C_3 group^a

C_3:	E	C_3	C_3^2
A_1:	1	1	1
E_a:	1	ω	ω^2
E_b:	1	ω^2	ω

^{a} $\omega = \exp(2\pi i/3)$ and $\omega^3 = 1$.

THE USE OF SYMMETRY LABELS AND THE VANISHING INTEGRAL RULE

As a result of the discussion given after Eq. (5-61) we see that we can label molecular energy levels according to the irreducible representations of a symmetry group of the molecular Hamiltonian. The use of the symmetry labels that we put on energy levels is that they enable us to tell which of the levels can *interact* with each other as a result of adding some previously unconsidered term $\hat{H}'$ to the molecular Hamiltonian $\hat{H}^0$. The term $\hat{H}'$ may be part of the exact Hamiltonian that was ignored initially for convenience, or it may be the result of applying an external perturbation such as an electric or magnetic field or electromagnetic radiation.

Let us suppose that the Hamiltonian $\hat{H}^0$ ($\hat{H}'$ having been neglected) has normalized eigenfunctions Ψ_m^0 and Ψ_n^0, with eigenvalues E_m^0 and E_n^0, respectively, and that $\hat{H}^0$ commutes with the group of symmetry operations $G = \{R_1, R_2, \ldots, R_h\}$. $\hat{H}^0$ will transform as the totally symmetric representation $\Gamma^{(s)}$ of G, and we let Ψ_m^0, Ψ_n^0, and $\hat{H}'$ generate the representations Γ_m, Γ_n, and Γ' of G, respectively. The complete set of eigenfunctions of $\hat{H}^0$ form a *basis set* for determining the eigenfunctions and eigenvalues of the

Hamiltonian $\hat{H} = (\hat{H}^0 + \hat{H}')$, and we can define the *Hamiltonian matrix* **H** in this basis set to be a matrix with *matrix elements* H_{mn} given by the integrals

$$H_{mn} = \int \Psi_m^{0*}(\hat{H}^0 + \hat{H}')\Psi_n^{\,0}\, d\tau_S = \delta_{mn}E_n^{\,0} + H'_{mn}, \qquad (5\text{-}121)$$

where

$$H'_{mn} = \int \Psi_m^{0*}\hat{H}'\Psi_n^{\,0}\, d\tau_S \qquad (5\text{-}122)$$

and $d\tau_S$ is the volume element in the space of the coordinates $X_1, Y_1, Z_1, \ldots$ of the particles in the molecule (i.e., $d\tau_S = dX_1\, dY_1\, dZ_1 \cdots$). The eigenvalues E of $\hat{H}$ can be determined (as we will see in the next section) from the Hamiltonian matrix by solving the *secular equation*

$$|H_{mn} - \delta_{mn}E| = 0. \qquad (5\text{-}123)$$

In solving the secular equation (this is called *diagonalizing* the Hamiltonian matrix) it is important to know which of the off-diagonal matrix elements H'_{mn} vanish since this will enable us to simplify the equation.

We can use the symmetry labels Γ_m and Γ_n on the levels $E_m^{\,0}$ and $E_n^{\,0}$, together with the symmetry Γ' of $\hat{H}'$, to determine which H'_{mn} elements must vanish. We write the integrand of H'_{mn} as

$$f(S) = \Psi_m^{0*}\hat{H}'\Psi_n^{\,0}, \qquad (5\text{-}124)$$

where S is a general point with coordinates $(X_1, Y_1, Z_1, \ldots)$. Suppose that the operation R_q of **G** moves S to a point S' with coordinates $(X_1', Y_1', Z_1', \ldots)$; the volume element at S', $d\tau_{S'} = dX_1'\, dY_1'\, dZ_1' \cdots$, must be equal to the volume element $d\tau_S$ at S for such a coordinate transformation. From Eq. (1-18) we have

$$R_q f(S) = f(S') = f^{R_q}(S), \qquad (5\text{-}125)$$

and since integration is carried out over all points S or S' (and since $d\tau_{S'} = d\tau_S$) we have

$$\int f(S)\, d\tau_S = \int f(S')\, d\tau_{S'} = \int f^{R_q}(S)\, d\tau_S = \int [R_q f(S)]\, d\tau_S. \qquad (5\text{-}126)$$

Thus

$$h \int f(S)\, d\tau_S = \int \left[\sum_q R_q f(S) \right] d\tau_S, \qquad (5\text{-}127)$$

where there are h operations R_q in the group **G**. We can therefore write

$$H'_{mn} = \int \Psi_m^{0*}\hat{H}'\Psi_n^{\,0}\, d\tau_S = h^{-1} \int \left[\sum_q R_q(\Psi_m^{0*}\hat{H}'\Psi_n^{\,0}) \right] d\tau_S. \qquad (5\text{-}128)$$

We will use this equation to determine if H'_{mn} must vanish. The function $\Psi_m^{0*}\hat{H}'\Psi_n^0$ generates the product representation $\Gamma_m^* \otimes \Gamma' \otimes \Gamma_n = \Gamma'_{mn}$ (Ψ_m^* has symmetry Γ_m^*), and we consider the situation in which this does not contain the totally symmetric representation $\Gamma^{(s)}$, i.e.,

$$\Gamma_m^* \otimes \Gamma' \otimes \Gamma_n \not\supset \Gamma^{(s)}. \qquad (5\text{-}129)$$

The $\Gamma^{(s)}$ projection operator in the group G is [see Eq. (5-64)]

$$P^{\Gamma^{(s)}} = \sum_q R_q, \qquad (5\text{-}130)$$

and if Eq. (5-129) is satisfied then $\Gamma^{(s)}$ will not occur in $\Gamma_m^* \otimes \Gamma' \otimes \Gamma_n$ so that

$$P^{\Gamma^{(s)}}(\Psi_m^{0*}\hat{H}'\Psi_n^0) = 0, \qquad (5\text{-}131)$$

i.e.,

$$\sum_q R_q(\Psi_m^{0*}\hat{H}'\Psi_n^0) = 0. \qquad (5\text{-}132)$$

Combining Eqs. (5-132) and (5-128) we see that the integral H'_{mn} will vanish if Eq. (5-129) is satisfied, i.e., if Γ'_{mn} does not contain $\Gamma^{(s)}$. Thus

$$\int \Psi_m^{0*}\hat{H}'\Psi_n^0 \, d\tau = 0$$

if

$$\Gamma_m^* \otimes \Gamma' \otimes \Gamma_n \not\supset \Gamma^{(s)}. \qquad (5\text{-}133)$$

This is *the vanishing integral rule*. If $\hat{H}'$ is totally symmetric in G then H'_{mn} will vanish if

$$\Gamma_m^* \otimes \Gamma_n \not\supset \Gamma^{(s)},$$

i.e. [from Eq. (5-118)], if

$$\Gamma_m \neq \Gamma_n. \qquad (5\text{-}134)$$

It can happen that H'_{mn} vanishes although $\Gamma'_{mn} \supset \Gamma^{(s)}$, and this is said to be an accident.

DIAGONALIZING THE HAMILTONIAN MATRIX

We will now consider the secular equation [Eq. (5-123)] in more detail and demonstrate the importance of the vanishing integral rule. The eigenfunctions Ψ_n^0 and eigenvalues E_n^0 of the Hamiltonian operator $\hat{H}^0$ in Eq. (5-121) are such that

$$\hat{H}^0\Psi_n^0 = E_n^0\Psi_n^0, \qquad (5\text{-}135)$$

and eigenfunctions belonging to different eigenvalues must be orthogonal to each other. We choose the degenerate eigenfunctions so that they are mutually orthogonal and we assume that all functions are normalized. In these circumstances we can set up the Hamiltonian matrix H^0 using the eigenfunctions Ψ^0, and the matrix elements will be given by (where we write the volume element simply as $d\tau$ here)

$$H_{mn}^0 = \int \Psi_m^{0*} \hat{H}^0 \Psi_n^0 \, d\tau = \delta_{mn} E_n^0. \tag{5-136}$$

The H^0 matrix is diagonal in the eigenfunctions Ψ^0 of $\hat{H}^0$, and the diagonal elements are the eigenvalues of $\hat{H}^0$.

Let us suppose that the Ψ^0 functions introduced above are not eigenfunctions of the Hamiltonian operator $\hat{H} = \hat{H}^0 + \hat{H}'$. This means that

$$\hat{H}\Psi_n^0 \neq \varepsilon \Psi_n^0, \tag{5-137}$$

where ε is a constant, but rather that

$$\hat{H}\left(\sum_n C_{jn}\Psi_n^0\right) = E_j\left(\sum_n C_{jn}\Psi_n^0\right), \tag{5-138}$$

where the Ψ^0 functions form a complete set of basis functions. We wish to solve the set of simultaneous differential equations, Eq. (5-138), involving all E_j to obtain the eigenvalues E_j and eigenfunctions

$$\Psi_j = \sum_n C_{jn}\Psi_n^0 \tag{5-139}$$

of the Hamiltonian $\hat{H}$. Multiplying Eq. (5-138) on the left by Ψ_m^{0*} and integrating over configuration space we obtain

$$\sum_n C_{jn}H_{mn} = E_j \sum_n C_{jn} \int \Psi_m^{0*}\Psi_n^0 \, d\tau = E_j \sum_n C_{jn}\delta_{mn}, \tag{5-140}$$

where

$$H_{mn} = \int \Psi_m^{0*}\hat{H}\Psi_n^0 \, d\tau. \tag{5-141}$$

We can rewrite Eq. (5-140) as

$$\sum_n (H_{mn} - \delta_{mn}E_j)C_{jn} = 0. \tag{5-142a}$$

The matrix of elements C_{jn} is orthogonal (since the eigenfunctions Ψ_j are also required to be mutually orthogonal) so that $C_{jn} = (C^{-1})_{nj}$, and we can write Eq. (5-142a) as the matrix product

$$\sum_n (H_{mn} - \delta_{mn}E_j)(C^{-1})_{nj} = 0. \tag{5-142b}$$

Since the determinants of matrices multiply together in the same way as the matrices themselves we see that, apart from the useless solution of all $(C^{-1})_{nj}$ being zero, we obtain the solution of Eqs. (5-142) as the secular equation

$$|H_{mn} - \delta_{mn}E_j| = 0. \tag{5-143}$$

In general an l-dimensional Hamiltonian matrix leads to a secular equation with l eigenvalues. Substituting the eigenvalues E_j one at a time into Eq. (5-142b) gives l simultaneous equations for the $(C^{-1})_{nj}$ ($m = 1$ to l), and we obtain the elements in the jth column of the matrix C^{-1}. Since $(C^{-1})_{nj} = C_{jn}$ these coefficients form the jth row of the C matrix and are the coefficients of the basis functions $\Psi_n{}^0$ in the eigenfunction Ψ_j.

Let us introduce the diagonal matrix Λ where $\Lambda_{jj} = E_j$ and $\Lambda_{ij} = 0$ if $i \neq j$. Substituting this into Eq. (5-142b) we obtain the equation

$$\sum_n H_{mn}(C^{-1})_{nj} = (C^{-1})_{mj}\Lambda_{jj} \tag{5-144}$$

which in matrix notation is

$$HC^{-1} = C^{-1}\Lambda,$$

i.e.,

$$CHC^{-1} = \Lambda. \tag{5-145}$$

Thus the similarity transformation of the Hamiltonian matrix H using the matrix of eigenfunction coefficients C produces the diagonal matrix Λ of eigenvalues of H.

The value of the vanishing integral theorem is that it allows the matrix H to be block diagonalized. This occurs if we order the eigenfunctions $\Psi_n{}^0$ according to their symmetry when we set up H. Let us initially consider the case when $\Gamma' = \Gamma^{(s)}$. In this case all off-diagonal matrix elements between $\Psi_n{}^0$ basis functions of different symmetry will vanish, and the Hamiltonian matrix will block diagonalize with there being one block for each symmetry type of $\Psi_n{}^0$ function. To diagonalize a block diagonal matrix according to Eq. (5-145) it follows from the definition of matrix multiplication that the C and C^{-1} matrices must be block diagonal in the same fashion. From the fact that the C matrix is block diagonal like the H matrix we see that each eigenfunction of $\hat{H}$ will only be a linear combination of $\Psi_n{}^0$ functions having the same symmetry in G (G being the symmetry group of $\hat{H}^0$). Thus the symmetry of each eigenfunction Ψ_j of $\hat{H}$ in the group G will be the same as the symmetry of the $\Psi_n{}^0$ basis functions that make it up (G is a symmetry group of $\hat{H}$ when $\Gamma' = \Gamma^{(s)}$), and each block of a block diagonal matrix can be diagonalized separately, which is a great simplification. The symmetry of the Ψ_j functions can be obtained from the symmetry of the $\Psi_n{}^0$ functions without worrying about the details of $\hat{H}'$ and this is frequently very useful. When $\Gamma' \neq \Gamma^{(s)}$ all off-diagonal matrix elements between Ψ^0 functions of symmetry Γ_m and Γ_n will vanish if Eq. (5-129) is satisfied, and there will also be a block diagonalization of H (it may be necessary to rearrange the rows and columns of H, i.e., to rearrange the order of the Ψ_n^0 functions, to obtain H in block diagonal form). However, now nonvanishing matrix elements occur in H that connect Ψ_n^0 functions of different symmetry in G, and as a result the eigenfunctions of $\hat{H}$ may not contain only functions of one symmetry type of G; when $\Gamma' \neq \Gamma^{(s)}$ the group G is not a symmetry group of $\hat{H}$ and its eigenfunctions Ψ_j cannot be classified in G. However, the classification of the basis functions Ψ_n^0 in G will still allow a simplification of the Hamiltonian matrix.

APPENDIX 5-1: PROOF THAT THE MATRICES $D[R]$ GENERATED IN EQ. (5-49) ARE REPRESENTATIONS

Consider two symmetry operations P_1 and P_2, and an l-fold degenerate energy level E_n with eigenfunctions $\Psi_{n1}, \Psi_{n2}, \ldots, \Psi_{nl}$. We can write [see Eq. (1-18)]

$$P_2\Psi_{nk}(X_1, X_2, \ldots) = \Psi_{nk}(X_1', X_2', \ldots) = \Psi_{nk}^{P_2}(X_1, X_2, \ldots), \quad (5\text{-}146)$$

where

$$P_2 X_i = X_i'. \quad (5\text{-}147)$$

However $\Psi_{nk}^{P_2}(X_1, X_2, \ldots)$ is also an eigenfunction of $\hat{H}$ with eigenvalue E_n so we can write it as a linear combination of the initial l functions, i.e.,

$$\Psi_{nk}(X_1', X_2', \ldots) = \Psi_{nk}^{P_2}(X_1, X_2, \ldots) = \sum_{j=1}^{l} D[P_2]_{kj}\Psi_{nj}(X_1, X_2, \ldots). \quad (5\text{-}148)$$

Using these results we can write [see Eq. (1-42), for example]

$$P_1 P_2 \Psi_{ni}(X_1, X_2, \ldots) = P_1 \Psi_{ni}(X_1', X_2', \ldots) \quad (5\text{-}149)$$

$$= \sum_{k=1}^{l} D[P_1]_{ik}\Psi_{nk}(X_1', X_2', \ldots) \quad (5\text{-}150)$$

$$= \sum_{k=1}^{l} D[P_1]_{ik} \sum_{j=1}^{l} D[P_2]_{kj}\Psi_{nj}(X_1, X_2, \ldots) \quad (5\text{-}151)$$

$$= \sum_{j=1}^{l} \sum_{k=1}^{l} D[P_1]_{ik}D[P_2]_{kj}\Psi_{nj}(X_1, X_2, \ldots). \quad (5\text{-}152)$$

The operation P_{12} is also a symmetry operation, where

$$P_{12} = P_1 P_2, \qquad (5\text{-}153)$$

so that

$$P_1 P_2 \Psi_{ni}(X_1, X_2, \ldots) = P_{12} \Psi_{ni}(X_1, X_2, \ldots) = \sum_{j=1}^{l} D[P_{12}]_{ij} \Psi_{nj}(X_1, X_2, \ldots).$$

$$(5\text{-}154)$$

Equating the right hand sides of Eqs. (5-152) and (5-154) we have

$$D[P_{12}]_{ij} = \sum_{k=1}^{l} D[P_1]_{ik} D[P_2]_{kj}. \qquad (5\text{-}155)$$

This means that the matrices D obtained in this way form a representation of the group. (If they are one-dimensional matrices this still holds, and a one-dimensional representation is obtained.) This is discussed further in the Bibliographical Notes at the end of this chapter.

APPENDIX 5-2: PROJECTION OPERATORS

We consider the r functions $\Phi_1, \ldots, \Phi_r$ that transform according to the reducible representation Γ which can be reduced to the sum of irreducible representations $\Gamma_1, \Gamma_2, \Gamma_3$, etc. Since Γ is reducible we can block diagonalize all its matrices $D^{\Gamma}[R]$ by the same similarity transformation, and we let the orthogonal $r \times r$ matrix required be A such that

$$AD^{\Gamma}[R]A^{-1} = D^{\Gamma_1}[R] \oplus D^{\Gamma_2}[R] \oplus D^{\Gamma_3}[R] \oplus \cdots, \qquad (5\text{-}156)$$

where the right hand side of this equation is shorthand for an $r \times r$ matrix block diagonalized into irreducible blocks $D^{\Gamma_i}[R]$. Using Eqs. (5-54)–(5-61) we deduce that we can write each function Φ_n as a linear combination of functions Ψ_l^i that transform irreducibly (the superscript denotes the irreducible representation Γ_i that a particular Ψ_l generates) by

$$\Phi_n = \sum_l A_{nl} \Psi_l, \qquad (5\text{-}157)$$

and the inverse is

$$\Psi_l^i = \sum_n [A^{-1}]_{ln} \Phi_n. \qquad (5\text{-}158)$$

The problem we face is: Given the r functions Φ_n transforming as the reducible representation Γ, how do we determine the linear combinations Ψ_l^i that transform irreducibly? In the remarks after Eq. (5-65) we stated that the projection operator P^{Γ_i} forms such a linear combination (or *projects out* from Φ_n the $\Psi_l{}^i$) and we now show this. We write

$$P^{\Gamma_i}\Phi_n = \sum_R \chi^{\Gamma_i}[R]^* R\Phi_n \qquad (5\text{-}159)$$

$$= \sum_R \chi^{\Gamma_i}[R]^* R \sum_l A_{nl}\Psi_l{}^j \qquad (5\text{-}160)$$

$$= \sum_{R,k} D^{\Gamma_i}[R]_{kk}^* \sum_l A_{nl} \sum_m D^{\Gamma_j}[R]_{lm}\Psi_m{}^j \qquad (5\text{-}161)$$

$$= \sum_{k,l,m} A_{nl}\Psi_m{}^j \sum_R D^{\Gamma_i}[R]_{kk}^* D^{\Gamma_j}[R]_{lm}, \qquad (5\text{-}162)$$

where Eq. (5-160) follows from Eq. (5-157) and $\Psi_l{}^j$ transforms irreducibly according to Γ_j. Substituting the orthogonality relation Eq. (4-38) into Eq. (5-162) we obtain

$$P^{\Gamma_i}\Phi_n = \frac{h}{l_i} \sum_k A_{nk}\Psi_k{}^i, \qquad (5\text{-}163)$$

where l_i is the dimension of Γ_i, and we have produced a function that is a linear combination only of functions of symmetry Γ_i, i.e., we have produced a function of symmetry Γ_i. Note that if there are no functions of symmetry Γ_i in the set of Ψ_l functions to which Φ_n reduces then

$$P^{\Gamma_i}\Phi_n = 0. \qquad (5\text{-}164)$$

Even when Γ contains Γ_i Eq. (5-164) can be true for some of the r functions Φ_n but if Γ does not contain Γ_i then Eq. (5-164) is true for *all* of the r functions Φ_n.

APPENDIX 5-3: ADDENDUM TO PROBLEM 5-2

To determine the representation of $C_{3v}(M)$ generated by the functions $\Phi_a(E)$ and $\Phi_b(E)$ in Problem 5-2 we proceed as follows:

$$(123)\Phi_a(E) = (2X_3 - X_1 - X_2)/\sqrt{6}, \qquad (5\text{-}165)$$

and we write

$$(2X_3 - X_1 - X_2)/\sqrt{6} = c_{aa}\Phi_a(E) + c_{ab}\Phi_b(E), \qquad (5\text{-}166)$$

which can be solved to give

$$c_{aa} = -\tfrac{1}{2} \quad \text{and} \quad c_{ab} = -\sqrt{3}/2 \qquad (5\text{-}167)$$

Similarly

$$(123)\Phi_b(E) = c_{ba}\Phi_a(E) + c_{bb}\Phi_b(E) \qquad (5\text{-}168)$$

which gives

$$c_{ba} = \sqrt{3}/2 \quad \text{and} \quad c_{bb} = -\tfrac{1}{2}. \qquad (5\text{-}169)$$

The character of the matrix of the c_{ij} generated by (123) is thus given by

$$\chi[(123)] = c_{aa} + c_{bb} = -1. \qquad (5\text{-}170)$$

We similarly determine that for $[\Phi_a(E), \Phi_b(E)]$

$$\chi[(12)^*] = 0, \qquad (5\text{-}171)$$

so that this pair of functions is indeed of E symmetry.

BIBLIOGRAPHICAL NOTES

Margenau and Murphy (1956). The postulates of quantum mechanics are discussed on pages 337–344, and these were used (see particularly page 338) in obtaining Eqs. (5-4)–(5-7) in this chapter. The effect of a coordinate transformation on a volume element [see Eq. (5-126) here in which $d\tau_S = d\tau_{S'}$] is described on pages 172–173. The technique adopted above [see Eq. (5-49) and Appendix 5-1] for obtaining a matrix representation follows the procedure used on page 563 [see Eqs. (15-24)–(15-28)]. Determinants are defined on page 302.

Hamermesh (1962). The symmetric and antisymmetric product of a representation are discussed on page 132, and projection operators are discussed on pages 111–113.

Wigner (1959). Projection operators are discussed on pages 112–118, and the proof that mutually orthogonal wavefunctions generate a unitary representation is given on page 111. The technique used in Wigner's book (and in many other texts) to generate matrix representations is different from that used in the present book. Having defined the effect of the operation R on coordinates by $RX_i = X_i'$ in Eq. (11-18a) Wigner defines another operation P_R in Eq. (11-19) by $P_R f(X_i') = f(X_i)$. Comparing with the definition used in the present book [see Eq. (1-18)] for the effect of R on a function we see that $P_R = R^{-1}$. Wigner generates representation matrices $D^W[R]$ (W for Wigner) by using [see Eq. (11-23)]

$$P_R \Psi_{ni} = \sum_j D^W[R]_{ji} \Psi_{nj}$$

(note the order of the subscripts) where $D^W[R]$ is the matrix representing the operation R in the group of R operations. We note that, as far as the effect on a function is concerned, $P_R^{-1} = R$ so that the matrices $D[R]$, obtained in the present book [see Eq. (5-49)], are related to the $D^W[R]$ by

$$D^W[R^{-1}]_{ji} = D[R]_{ij}.$$

Since the representations are unitary this means that

$$D^W[R]_{ij}^* = D[R]_{ij}$$

and the representation matrices obtained here are the complex conjugates of the $D^W[R]$. This change is immaterial as long as one is consistent in only using one of these two possible ways of generating representation matrices, and the labeling of pairs of separably degenerate states would be reversed by these two techniques. This points out the arbitrariness of the absolute symmetry labels on a pair of separably degenerate states.

Wigner introduces the group of operations P_R, which transform functions, as being distinct from the group of operations R which change the values of nuclear coordinates and the values of functions of them. The groups $\{P_R\}$ and $\{R\}$ are isomorphic. We choose to define the nuclear permutation operations as permuting nuclei *and* transforming functions according to Eq. (1-18).

6

The Molecular
Hamiltonian
and Its True Symmetry

In this chapter the expression for the complete molecular Hamiltonian is discussed, and several groups of operations that commute with the Hamiltonian are examined. The energy levels of the Hamiltonian can be symmetry labeled by using the irreducible representations of these groups. The symmetry labels are called *true symmetry labels* since they involve the use of the true symmetry groups of the Hamiltonian. It is shown how the symmetry labels on basis functions (*basis symmetry labels*) can be used to determine the true symmetry labels for the complete wavefunction and to determine nuclear spin statistical weights and electron spin multiplicity restrictions.

THE MOLECULAR HAMILTONIAN

We consider a molecule to be a collection of nuclei and electrons held together by certain forces and obeying the laws of quantum mechanics. A classical expression for the energy of the molecule can be derived, and the postulates of quantum mechanics used to obtain the appropriate quantum mechanical Hamiltonian and Schrödinger equation. In general terms the molecular Hamiltonian operator $\hat{H}$ involves the properties of each of the nuclei and electrons in the molecule; the properties occurring are:

m_r the mass of each particle ($m_r = m_e$ for electrons),

$C_r e$ the charge of each particle ($C_r = -1$ for electrons),

g and s_i the g-factor and spin of each electron i (giving the electron spin magnetic moment),

g_α and I_α the g-factor and spin of each nucleus α (giving the nuclear spin magnetic moment),

$Q_{ab}^{(\alpha)}$ the components of the electric quadrupole moment of each nucleus α,

and the higher electric and magnetic moments of the nuclei (when present). To this list of particle properties we could add others such as the polarizability of each nucleus. The time-independent Schrödinger equation $\hat{H}\Phi = E\Phi$ provides a prescription for getting the molecular energies E and wavefunctions Φ from the properties of the nuclei and electrons in the molecule, as given above. As well as the particle properties the Hamiltonian involves the coordinates R_r and momentum operators $\hat{P}_r$ for all the particles in the molecule, and we write the Hamiltonian as

$$\hat{H} = (\underbrace{\hat{T}_{CM} + \hat{T}^0 + \hat{T}'}_{\hat{T}}) + V + \hat{H}_{es} + \hat{H}_{hfs}. \tag{6-1}$$

We will now discuss the meaning of each of these terms.

We first look at the nonrelativistic kinetic energy operator $\hat{T}$ in the Hamiltonian, and this operator is given by

$$\hat{T} = -\frac{\hbar^2}{2} \sum_{r=1}^{l} \frac{(\partial^2/\partial X_r^2) + (\partial^2/\partial Y_r^2) + (\partial^2/\partial Z_r^2)}{m_r}, \tag{6-2}$$

where r runs over all the nuclei and electrons in the molecule (the total number of particles is l), and the coordinates are in an arbitrary space fixed (X, Y, Z) axis system. We introduce the (X, Y, Z) axis system with origin at the molecular center of mass, and parallel to the (X, Y, Z) axis system, so that

$$X_r = X_r + X_0, \tag{6-3}$$

$$Y_r = Y_r + Y_0, \tag{6-4}$$

and

$$Z_r = Z_r + Z_0, \tag{6-5}$$

where (X_0, Y_0, Z_0) are the coordinates of the molecular center of mass in the (X, Y, Z) axis system. To separate the translational kinetic energy in Eq. (6-2) we write $\hat{T}$ in terms of the $3l$ coordinates

$$X_0, Y_0, Z_0, X_2, Y_2, Z_2, \ldots, X_l, Y_l, Z_l, \tag{6-6}$$

where we have eliminated X_1, Y_1, and Z_1 since

$$X_1 = -\frac{1}{m_1} \sum_{r=2}^{l} m_r X_r, \tag{6-7}$$

with similar equations for Y_1 and Z_1. To make the coordinate change we use the *chain rule*. To explain the chain rule we consider for convenience the coordinate change from the coordinates $(Q_1, Q_2, \ldots, Q_n)$ to the coordinates $(q_1, q_2, \ldots, q_n)$, where $q_i = f_i(Q_1, Q_2, \ldots, Q_n)$. For this coordinate change the chain rule gives

$$\frac{\partial}{\partial Q_j} = \sum_i \frac{\partial q_i}{\partial Q_j} \frac{\partial}{\partial q_i}, \tag{6-8}$$

and by further differentiation we deduce that

$$\frac{\partial^2}{\partial Q_j^2} = \sum_{k,i} \left(\frac{\partial q_k}{\partial Q_j}\right)\left(\frac{\partial q_i}{\partial Q_j}\right)\frac{\partial^2}{\partial q_k \partial q_i} + \sum_i \left(\frac{\partial^2 q_i}{\partial Q_j^2}\right)\frac{\partial}{\partial q_i}. \tag{6-9}$$

We use the chain rule in order to obtain $\hat{T}$ in the coordinates of Eq. (6-6) from the expression for $\hat{T}$ given in Eq. (6-2). We obtain the results that

$$\frac{\partial^2}{\partial X_1^2} = \left(\frac{m_1}{M}\right)^2\left(\frac{\partial^2}{\partial X_0^2} - 2\sum_{s=2}^{l}\frac{\partial^2}{\partial X_0 \partial X_s} + \sum_{s,t=2}^{l}\frac{\partial^2}{\partial X_s \partial X_t}\right), \tag{6-10}$$

and for $r \neq 1$

$$\frac{\partial^2}{\partial X_r^2} = \left(\frac{m_r}{M}\right)^2\left(\frac{\partial^2}{\partial X_0^2} - 2\sum_{s=2}^{l}\frac{\partial^2}{\partial X_0 \partial X_s} + \sum_{s,t=2}^{l}\frac{\partial^2}{\partial X_s \partial X_t}\right)$$
$$+ 2\left(\frac{m_r}{M}\right)\left(\frac{\partial^2}{\partial X_0 \partial X_r} - \sum_{s=2}^{l}\frac{\partial^2}{\partial X_r \partial X_s}\right) + \frac{\partial^2}{\partial X_r^2}, \tag{6-11}$$

where M is the total mass of all the electrons and nuclei in the molecule. Substituting Eqs. (6-10) and (6-11) and the similar equations for the Y and Z derivatives into Eq. (6-2) we obtain

$$\hat{T} = \hat{T}_{\text{CM}} + \hat{T}^0 + \hat{T}', \tag{6-12}$$

where

$$\hat{T}_{\text{CM}} = -(\hbar^2/2M)\mathbf{V}_{\text{CM}}^2, \tag{6-13}$$

$$\hat{T}^0 = -(\hbar^2/2)\sum_{r=2}^{l}(\mathbf{V}_r^2/m_r), \tag{6-14}$$

$$\hat{T}' = (\hbar^2/2M)\sum_{r,s=2}^{l}\mathbf{V}_r \cdot \mathbf{V}_s, \tag{6-15}$$

$$\mathbf{V}_{\text{CM}}^2 = \partial^2/\partial X_0^2 + \partial^2/\partial Y_0^2 + \partial^2/\partial Z_0^2, \tag{6-16}$$

$$\mathbf{V}_r^2 = \partial^2/\partial X_r^2 + \partial^2/\partial Y_r^2 + \partial^2/\partial Z_r^2, \tag{6-17}$$

and

$$\mathbf{V}_r \cdot \mathbf{V}_s = \partial^2/\partial X_r \partial X_s + \partial^2/\partial Y_r \partial Y_s + \partial^2/\partial Z_r \partial Z_s. \quad (6\text{-}18)$$

The next term in the molecular Hamiltonian, Eq. (6-1), after the kinetic energy operator, is the electrostatic potential energy term V, and this is given by

$$V = \sum_{r<s=1}^{l} \frac{C_r C_s e^2}{R_{rs}}. \quad (6\text{-}19)$$

This term involves the interparticle distances R_{rs} [see Eq. (5-3)] and the electrostatic charges of the particles.

The term $\hat{H}_{es}$ in the molecular Hamiltonian arises from the interaction of each of the electron spin magnetic moments with

(a) the magnetic moments generated by the orbital motions of the electrons (interaction with its own orbital motion is the most important and is called *the* electron spin–orbit interaction),

(b) the magnetic moments generated by the orbital motions of the nuclei, and

(c) the spin magnetic moments of the other electrons (the electron spin–spin interaction). The third term in Eq. (6-22) involves the Dirac delta function and represents a Fermi-contact-type interaction.

We write

$$\hat{H}_{es} = \frac{g\mu_B}{m_e c} \sum_{j>i} (-e)R_{ij}^{-3}\left[(\mathbf{R}_i - \mathbf{R}_j) \times \left(\frac{\hat{\mathbf{P}}_i}{2} - \hat{\mathbf{P}}_j\right)\right] \cdot \mathbf{s}_i \quad (6\text{-}20)$$

$$+ \frac{g\mu_B}{c} \sum_{\alpha,i} (C_\alpha e)R_{i\alpha}^{-3}\left[(\mathbf{R}_i - \mathbf{R}_\alpha) \times \left(\frac{\hat{\mathbf{P}}_i}{2m_e} - \frac{\hat{\mathbf{P}}_\alpha}{m_\alpha}\right)\right] \cdot \mathbf{s}_i \quad (6\text{-}21)$$

$$+ g^2\mu_B^2 \sum_{j>i}\left\{(\mathbf{s}_i \cdot \mathbf{s}_j)R_{ij}^{-3} - 3[\mathbf{s}_i \cdot (\mathbf{R}_i - \mathbf{R}_j)][\mathbf{s}_j \cdot (\mathbf{R}_i - \mathbf{R}_j)]R_{ij}^{-5}\right.$$

$$\left. - \frac{8\pi}{3}\delta(\mathbf{R}_i - \mathbf{R}_j)(\mathbf{s}_i \cdot \mathbf{s}_j)\right\}, \quad (6\text{-}22)$$

where α labels the nuclei, and i and j label the electrons; μ_B is the Bohr magneton ($\mu_B = he/2m_e c$), and c is the velocity of light.

$\hat{H}_{hfs}$ is the last term that we consider in the molecular Hamiltonian and it results from the interactions of the magnetic and electric moments of the nuclei with the other electric and magnetic moments in the molecule. We call it the nuclear hyperfine structure term. Nuclei with spin $\frac{1}{2}$ or greater have a nonvanishing magnetic dipole moment. The term $\hat{H}_{ns}$, which is part of $\hat{H}_{hfs}$ and which arises from the interaction of this nuclear spin magnetic moment with the other magnetic moments in the molecule, can be obtained from the expression for $\hat{H}_{es}$ given above[1] by replacing the electron labels i and j by the nuclear labels α and β, replacing α by i, interchanging $-e$ and $C_\alpha e$, introducing I_α and I_β instead of s_i and s_j, introducing the nuclear g-factor g_α instead of g, and, finally, introducing $m_p \mu_N / m_\alpha$ instead of μ_B where μ_N is the nuclear magneton ($\mu_N = \hbar e / 2 m_p c$ where m_p is the proton mass); both g_α and $1/m_\alpha$ must then be put inside the summation over α. Also included in $\hat{H}_{ns}$ is the nuclear spin–electron spin coupling term, and this is analogous to Eq. (6-22). Nuclei with spin 1 or greater also have a nonvanishing electric quadrupole moment and nuclei with spin $\frac{3}{2}$ or greater have a magnetic octupole moment as well; nuclei with higher spins can have even higher electric and magnetic moments. The electric moments arise from the non-spherical charge distribution over the finite volume of each nucleus, and the magnetic moments from the motions of the charges within each nucleus. All of these higher electric and magnetic moments interact with the charge distribution and charge motion within the molecule to give rise to further terms in $\hat{H}_{hfs}$. Neglecting all but the quadrupole moment term $\hat{H}_{quad}$ of these higher terms we can write the nuclear hyperfine term as

$$\hat{H}_{hfs} = \hat{H}_{ns} + \hat{H}_{quad}, \tag{6-23}$$

and

$$\hat{H}_{quad} = -\frac{1}{6} \sum_{\alpha,a,b} Q_{ab}^{(\alpha)} V_{ab}^{(\alpha)}, \tag{6-24}$$

where the sum over α is over all nuclei and the sum over a and b is over the three molecule fixed coordinate directions; $Q_{ab}^{(\alpha)}$ is a component of the electric quadrupole moment of nucleus α, and $V_{ab}^{(\alpha)}$ is a component of the gradient of the electric field at nucleus α from all the other charges in the molecule. Nuclei with spin zero do not contribute to $\hat{H}_{hfs}$ since such nuclei have no electric or magnetic moments.

From Eq. (6-1) we write the molecular Hamiltonian as

$$\hat{H} = \hat{T} + V + \hat{H}_{es} + \hat{H}_{hfs} \tag{6-25}$$

$$= \hat{T}_{CM} + \hat{H}^0 + \hat{T}' + \hat{H}_{es} + \hat{H}_{hfs}, \tag{6-26}$$

where $\hat{H}^0$ is given in Eq. (5-6) [but the coordinates X_1, Y_1, and Z_1 are redundant because of Eq. (6-7)]. The terms in $\hat{H}$ are summarized in Table 6-1. For applications to the study of the energy level spacings for singlet electronic ground states of molecules with unresolved nuclear hyperfine structure we can satisfactorily approximate $\hat{H}$ to $\hat{H}^0$. For the general case we cannot make such a sweeping approximation.

---

[1] See Eq. (1) in Gunther-Mohr, Townes, and Van Vleck (1954).

Table 6-1

The terms in the molecular Hamiltonian $\hat{H}$ given in Eq. (6-25)

$\hat{T}$	Kinetic energy	Eq. (6-2)
$\hat{T}_{CM}$	Kinetic energy of the center of mass	Eq. (6-13)
$\hat{T}^0$	Intramolecular kinetic energy	Eq. (6-14)
$\hat{T}'$	Cross terms in intramolecular kinetic energy	Eq. (6-15)
V	Electrostatic potential energy	Eq. (6-19)
$\hat{H}_{es}$	Interaction energy of the electron spin magnetic moments	
$\hat{H}_{so}$	Electron spin–electron orbit interaction	Eq. (6-20)
$\hat{H}_{sr}$	Electron spin–nuclear motion interaction	Eq. (6-21)
$\hat{H}_{ss}$	Electron spin–electron spin interaction	Eq. (6-22)
$\hat{H}_{hfs}$	Interaction energy of the nuclear magnetic and electric moments	Eq. (6-23)
$\hat{H}_{ns}$	Interaction energy of the nuclear spin magnetic moments	
$\hat{H}_{nso}$	Nuclear spin–electron orbit interaction	
$\hat{H}_{nsr}$	Nuclear spin–nuclear motion interaction	
$\hat{H}_{nss}$	Nuclear spin–nuclear spin interaction	
$\hat{H}_{nses}$	Nuclear spin–electron spin interaction	
$\hat{H}_{quad}$	Interaction energy of the nuclear electric quadrupole moments with the electric field gradients	Eq. (6-24)

THE FULL SYMMETRY GROUP OF THE MOLECULAR HAMILTONIAN

The complete molecular Hamiltonian $\hat{H}$ of Eq. (6-1) will commute with (or be invariant to) all of the following operations:

(a) any translation of the molecule along a space fixed direction,

(b) any rotation of the molecule about a space fixed axis passing through the center of mass of the molecule,

(c) any permutation of the space and spin coordinates of the electrons,

(d) any permutation of the space and spin coordinates of identical nuclei,

(e) the inversion of the coordinates of all the particles (nuclei and electrons) in the center of mass of the molecule, and

(f) time reversal.

The invariance of $\hat{H}$ to these operations is not axiomatic but follows in a general way from the form of $\hat{H}$, and we will now indicate how this is so. It must be realized however that we are dealing with the molecule in the idealized circumstances of there being no intermolecular interactions and of there being no external fields, i.e., a completely isolated molecule. Note that in the above list of symmetry operations the operations of the molecular point group (i.e., rotations and reflections of the vibronic variables in the molecule fixed axis system) do not appear. The reason for this is that these operations do not commute with the molecular Hamiltonian; the molecular point group is not a symmetry group of the molecular Hamiltonian. This point will be discussed further in Chapter 11 when we discuss near symmetry.

From the fact that the free space in which the isolated molecule moves is everywhere the same (uniform) we infer that the Hamiltonian is invariant to translation. The invariance of $\hat{H}$ to a translation of the whole molecule through A, say (giving the direction and distance of the shift), results from the fact that $\hat{H}$ is unchanged if we add A to the position vectors of all particles in the molecule. The terms $\hat{T}_{CM}$, $\hat{T}^0$, $\hat{T}'$, V, $\hat{H}_{es}$, and $\hat{H}_{hfs}$ are unchanged by the addition of A to the R_i and R_α, and the molecular Hamiltonian is therefore invariant to the translation through A. Since A can be arbitrary the Hamiltonian is invariant to any translation and hence to all the elements of the translational group G_T. This is an infinite group consisting of all translations of the whole molecule by any amount along any direction in space.

The invariance of the Hamiltonian to rotation follows from the fact that space is isotropic. The Hamiltonian is unchanged if we rotate the particles about any space fixed axis passing through the center of mass of the molecule and the generality of this argument is explained in the next paragraph. Such an operation does not alter interparticle distances. As a result the molecular Hamiltonian is invariant to all the elements of the spatial three-dimensional pure rotation group K(spatial) introduced in Chapter 3.

In the above discussion of the invariance of the Hamiltonian to translation and rotation operations we have used the so-called *active* picture in which the operations are interpreted as involving a translation or rotation of the whole molecule within the space fixed reference frame. The effect of these operations on the Hamiltonian is exactly duplicated if we keep the molecule fixed and translate or rotate the space fixed axes in the opposite sense; this latter way of interpreting the effect of the operations is called the *passive* picture, and it leads to a clearer understanding of the invariance of the Hamiltonian. For example, instead of moving all the particles so that their spatial positions are changed from R_r to $R_r + A$ we could move the axes through $-A$ when the same coordinate changes within the Hamiltonian will occur. It is immediately clear that a molecule is unaffected if we walk away from it, taking the space fixed axes with us, or if we walk around the molecule, twisting the space fixed axes as we go. The molecular Hamiltonian must be invariant to all translations or rotations of the space fixed axis system. Hence the Hamiltonian is invariant to the translation or rotation of the molecule within the space fixed axis system.

The molecular Hamiltonian is invariant to any permutation of the electrons. This operation merely involves interchanging the electron subscripts $i, j, k, \ldots$ on the vectors $\boldsymbol{R}$, $\hat{\boldsymbol{P}}$, and s in $\hat{H}$ where these vectors have the same coefficients (these coefficients involve the mass, charge, and g-factor of the electron). The invariance of $\hat{H}$ to a permutation of the electrons follows immediately from the fundamental fact that all the electrons in a molecule are identical and indistinguishable. The molecular Hamiltonian is therefore invariant to all the operations of the complete electron permutation group which we call $S_n^{(e)}$ (supposing there to be n electrons in the molecule).

In the same manner as the Hamiltonian is invariant to any permutation of electrons it is invariant to any permutation of identical nuclei. This just involves interchanging the nuclear labels on the quantities $\boldsymbol{R}$, $\hat{\boldsymbol{P}}$, $\boldsymbol{I}$, Q_{ab}, and V_{ab}. The invariance of $\hat{H}^0$ to any operation of permuting identical nuclei was proved in Chapter 5, and the invariance of the rest of the terms in $\hat{H}$ is easy to appreciate. This follows from the indistinguishability of identical nuclei. The Hamiltonian will therefore be invariant to the elements of the complete nuclear permutation group G^{CNP} [see Eq. (1-55)].

Now we must consider the inversion operation E^*. The effect of E^* is to change all position vectors $\boldsymbol{R}$ to $-\boldsymbol{R}$ and momentum vectors $\hat{\boldsymbol{P}}$ to $-\hat{\boldsymbol{P}}$ (these are polar vectors), within the (X, Y, Z) system, but not to change the spin vectors $\boldsymbol{I}$, and s; since they are axial vectors. An axial vector transforms under E^* like the angular momentum vector $\boldsymbol{R} \times \hat{\boldsymbol{P}}$; since $\boldsymbol{R}$ and $\hat{\boldsymbol{P}}$ are reversed by E^* the vector product is invariant. The reader can see that, for example, $\hat{H}_{es}$ and $\hat{H}_{ns}$ are not altered if $\boldsymbol{R} \to -\boldsymbol{R}$, $\hat{\boldsymbol{P}} \to -\hat{\boldsymbol{P}}$, $\boldsymbol{I} \to \boldsymbol{I}$ and $s \to s$. The invariance of $\hat{H}^0$ to E^* was discussed in Chapter 5. Examination of the electric quadrupole term in the Hamiltonian shows it to be invariant under E^* as are all terms that arise from intramolecular electromagnetic interactions. The above molecular Hamiltonian is invariant to the operation E^* and hence to the group $\mathscr{E} = \{E, E^*\}$. One should not be dogmatic about the invariance of the molecular Hamiltonian to E^*. It may be that the weak interaction force contributes to the interaction forces between electrons and nuclei in a molecule (or atom) and if this is the case then the term in the Hamiltonian that that effect contributes will not be invariant to E^*. This effect will be very small indeed, and the presence of this term has yet to be experimentally demonstrated. In all normal spectroscopic work the effect of this possible term can certainly be neglected and the Hamiltonian taken as being invariant to E^*.

The time reversal operation θ reverses all momenta ($\hat{\boldsymbol{P}}$) including spin angular momenta (s and $\boldsymbol{I}$), but not the coordinates ($\boldsymbol{R}$). It would be less enigmatic sounding (but longer) to call it the "reversal of momenta and spins" operation. This operation leaves the Hamiltonian invariant (for

example, $\hat{H}_{es}$ and $\hat{H}_{ns}$ are invariant if $R \to R$, $\hat{P} \to -\hat{P}$, $I \to -I$, and $s \to -s$).
It turns out that including θ in any symmetry group of the Hamiltonian
does not lead to any new labels on the energy levels beyond that provided
by the initial symmetry group. For this reason we will generally ignore this
symmetry operation. The operation θ can, however, be responsible for
extra degeneracies. If in the original symmetry group there is a pair of
irreducible representations, Γ and Γ^*, say, that are the complex conjugates
of each other then, as a result of time reversal symmetry, an energy level of
symmetry Γ will always coincide with an energy level of symmetry Γ^*. For
this reason such a pair of irreducible representations of a symmetry group
can be considered to be degenerate and is called separably degenerate. The
representations E_a and E_b of the C_3 group (see Table 5-4) are separably
degenerate. The character table of such a group can be condensed by adding
the characters of each pair of separably degenerate irreducible representations, and such a condensed character table for the C_3 group is given in
Table 6-2. The operations C_3 and $C_3{}^2$ are not really in the same class but
in the condensed character table that is how they appear. The fact that E is
the sum of separably degenerate irreducible representations is indicated by
writing sep in the character table. Several examples of separably degenerate
representations occur in the character tables given in Appendix A.

Table 6-2
The condensed character table of
the C_3 group[a]

C_3:	E	$C_3, C_3{}^2$	
A:	1	1	
E:	2	-1	sep

[a] The full character table is given in
Table 5-4, and the separably degenerate irreducible representations E_a and E_b of that
group are added to give E.

As a result of the above discussion we see that the molecular Hamiltonian
is invariant to the elements of the five groups $G_{\dot{T}}$, K(spatial), $S_n^{(e)}$, G^{CNP}, and $\mathscr{E}$.
Each of these groups is a true symmetry group of the molecular Hamiltonian.
The full Hamiltonian group G_{full} will therefore consist of the elements of
each of these groups and of all possible products of the elements. Thus we
can write G_{full} as the direct product of these groups:

$$G_{full} = G_T \otimes K(\text{spatial}) \otimes S_n^{(e)} \otimes G^{CNP} \otimes \mathscr{E} \qquad (6\text{-}27)$$

$$= G_T \otimes K(\text{spatial}) \otimes S_n^{(e)} \otimes G^{CNPI}, \qquad (6\text{-}28)$$

where G^{CNPI} is the complete nuclear permutation inversion group introduced in Chapter 2. Symmetry labels (i.e., irreducible representation labels) obtained by using G_{full} (or a subgroup of it) will be called true symmetry labels since they are obtained by using a group of elements that all commute with the exact molecular Hamiltonian. In particular the symmetry labels obtained by using the group G^{CNPI} or the molecular symmetry group (a subgroup of G^{CNPI}) are true symmetry labels. In applications we generally do not use the complete group G_{full} as it stands but rather we use the various subgroups of it separately and we will now discuss this.

The Translational Group

The molecular Hamiltonian $\hat{H}$ [see Eq. (6-26)] can be written as

$$\hat{H} = \hat{T}_{CM} + \hat{H}_{int}, \tag{6-29}$$

where the internal Hamiltonian $\hat{H}_{int}$ does not involve the coordinates (or momenta) of the molecular center of mass. Because of this separation of coordinates in the Hamiltonian we can write the eigenfunctions as

$$\Phi = \Phi_{CM}(X_0, Y_0, Z_0)\Phi_{int}(X_2, Y_2, Z_2, \ldots, X_l, Y_l, Z_l), \tag{6-30}$$

where

$$\hat{T}_{CM}\Phi_{CM} = E_{CM}\Phi_{CM}, \tag{6-31}$$

$$\hat{H}_{int}\Phi_{int} = E_{int}\Phi_{int}, \tag{6-32}$$

and the total energy is given by

$$E = E_{CM} + E_{int}. \tag{6-33}$$

Thus we can completely separate the translational motion from the internal motion. The translational wavefunction Φ_{CM} is obtained from Eq. (6-31) as

$$\Phi_{CM} = e^{ik \cdot R_0}, \tag{6-34}$$

where R_0 is the position vector of the center of mass [with components $X_0, Y_0,$ and Z_0 in the (X, Y, Z) axis system], $k\hbar = P_{CM}$ the classical translational momentum, and

$$k^2 = 2ME_{CM}/\hbar^2. \tag{6-35}$$

A translational operation R_T which changes R_0 to $R_0 + A$, say, will not affect Φ_{int} and we can write

$$R_T\Phi = R_T\Phi_{CM}\Phi_{int} = \Phi_{int}R_T\Phi_{CM}. \tag{6-36}$$

Thus to determine the effect of R_T on Φ we need only look at the effect of R_T on Φ_{CM}. From Eq. (6-34) we deduce that

$$R_T\Phi_{CM} = e^{ik\cdot(R_0 + A)} \tag{6-37}$$

$$= e^{ik\cdot A}\Phi_{CM}, \tag{6-38}$$

so that the effect of a translational operation is determined solely by the k vector. Labeling the states according to the effect of the operations of G_T is equivalent to labeling the states by the k vector (or translational momentum). We generally do not worry about the translational momentum of molecular states but when necessary translational states are discussed by using the k vector and by using the law of conservation of momentum rather than by making explicit use of G_T. From now on we will neglect the molecular translational motion and concentrate on $\hat{H}_{int}$ and Φ_{int}.

The Spatial Three-Dimensional Pure Rotation Group

The group K(spatial), introduced in Chapter 3, is an infinite (continuous) group with an infinite number of classes and therefore an infinite number of irreducible representations. This group is discussed in many books (see the Bibliographical Notes at the end of this chapter), and it is only necessary to give a brief summary here.

Each operation of the group K(spatial) in the passive picture can be viewed as rotating the axes from (X, Y, Z) to a new orientation (X', Y', Z') and can be specified by the values of the three angles α, β, and γ in Fig. 6-1 that relate the orientation of (X', Y', Z') to that of (X, Y, Z). These angles are called Euler angles and they are restricted according to

$$0 \le \alpha \le 2\pi, \qquad 0 \le \beta \le \pi, \qquad 0 \le \gamma \le 2\pi. \tag{6-39}$$

The angles α and γ are measured in the positive (right handed) sense about the axes Z and Z', respectively. The angle γ is measured from the positive half of the node line ON which marks the intersection of the X, Y and X', Y' planes. The positive sense of ON is defined so that on rotating a right handed

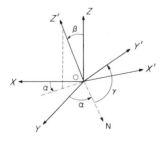

Fig. 6-1. The definition of the Euler angles (α, β, γ) that relate the orientation of the (X', Y', Z') axis system to that of the (X, Y, Z) axis system.

screw from Z to Z' through the angle β the screw travels in the positive sense along ON. Any operation of the group K(spatial) can be written as $[\alpha, \beta, \gamma]$ and this notation specifies completely the axis rotation that results from application of the operation. One must be careful to specify whether one is using the active or passive picture of the rotation operation in explicitly writing down the effect of the operation.

The irreducible representations of the group K(spatial) are written $D^{(0)}$ (the totally symmetric representation), $D^{(1)}$, $D^{(2)}$, etc., and in general we write $D^{(j)}$. The matrix under the operation $[\alpha, \beta, \gamma]$ in the representation $D^{(j)}$ is written $D^{(j)}([\alpha, \beta, \gamma])$ and the dimension of each matrix in the representation $D^{(j)}$ is $(2j + 1)$. We label the rows and columns of each matrix with $m_j = -j$, $-j + 1, \ldots, +j$. The product of two representations of K(spatial) is given by the following rule:

$$D^{(j_1)} \otimes D^{(j_2)} = D^{(j_1+j_2)} \oplus D^{(j_1+j_2-1)} \oplus \cdots \oplus D^{(|j_1-j_2|)}. \quad (6\text{-}40)$$

Extending this we can write the product of three representations as

$$D^{(j_1)} \otimes D^{(j_2)} \otimes D^{(j_3)}$$
$$= D^{(j_1+j_2+j_3)} \oplus D^{(j_1+j_2+j_3-1)} \oplus \cdots \oplus D^{(|(j_1+j_2)-j_3|)}$$
$$\oplus D^{(j_1+j_2-1+j_3)} \oplus D^{(j_1+j_2-1+j_3-1)} \oplus \cdots \oplus D^{(|(j_1+j_2-1)-j_3|)}$$
$$\oplus \cdots$$
$$\oplus D^{(|j_1-j_2|+j_3)} \oplus D^{(|j_1-j_2|+j_3-1)} \oplus \cdots \oplus D^{(||j_1-j_2|-j_3|)}. \quad (6\text{-}41)$$

For example

$$D^{(1)} \otimes D^{(1)} \otimes D^{(2)} = (D^{(2)} \oplus D^{(1)} \oplus D^{(0)}) \otimes D^{(2)}$$
$$= D^{(4)} \oplus D^{(3)} \oplus D^{(2)} \oplus D^{(1)} \oplus D^{(0)}$$
$$\oplus D^{(3)} \oplus D^{(2)} \oplus D^{(1)}$$
$$\oplus D^{(2)}. \quad (6\text{-}42)$$

In using the group K(spatial) with the vanishing integral theorem, Eq. (5-133), it is important to know that

$$D^{(j'')} \otimes D^{(1)} \otimes D^{(j')} \supset D^{(0)} \quad (6\text{-}43)$$

is only true if $j'' = j'$ (except $j' = j'' = 0$) or $j'' = j' \pm 1$.

We are interested in the effect of an operation $[\alpha, \beta, \gamma]$ of K(spatial) on the molecular wavefunction Φ_{int}. The molecular Hamiltonian $\hat{H}_{\text{int}}$ can be shown to commute with the operations $\hat{F}^2$ (the square of the total angular momentum, including the nuclear and electron spin angular momenta) and $\hat{F}_Z$ (the Z component of the total angular momentum); see, for example, Chapters IV and VIII in Landau and Lifshitz (1958). The Hamiltonian commutes with the operations $[\alpha, \beta, \gamma]$ and these operations commute with $\hat{F}^2$ and $\hat{F}_Z$. We can write

$$\hat{F}^2\Phi_{int} = F(F + 1)\hbar^2\Phi_{int} \tag{6-44}$$

and

$$\hat{F}_Z\Phi_{int} = m_F\hbar\Phi_{int} \tag{6-45}$$

and hence label the Φ_{int} with the quantum numbers F and m_F, i.e., $\Phi_{int}(F, m_F)$. It can be shown that the molecular wavefunction $\Phi_{int}(F, m_F)$ transforms in the group K(spatial) according to the row labeled m_F in the representation $D^{(F)}$. Hence, the use of K(spatial) to symmetry label the molecular wavefunctions is equivalent to labeling the states with the quantum numbers F and m_F.

Fermi–Dirac and Bose–Einstein Statistical Formulas

Before considering the electron permutation group, or the complete nuclear permutation group, we must consider a law of nature concerning the interchange of identical particles in a molecule. For this purpose it is convenient to divide particles (i.e., nuclei and electrons) into two types: fermions, having half integer spin, and bosons, having integer spin. Thus electrons, protons, and ^{13}C, ^{11}B, and ^{17}O nuclei are fermions (having spins of $\frac{1}{2}, \frac{1}{2}, \frac{1}{2}$, $\frac{3}{2}$, and $\frac{5}{2}$, respectively), and deuterons and ^{12}C, ^{14}N, and ^{16}O nuclei are bosons (having spins of 1, 0, 1, and 0, respectively). The law of nature of importance to us here is that a molecular wavefunction is unaffected if we exchange a pair of identical bosons in the molecule but it is changed in sign if we exchange a pair of identical fermions in the molecule. This law is usually taken to be axiomatic (i.e., empirical) although in the paper by Pauli cited at the end of the chapter a proof using relativistic arguments is advanced. The fact that the wavefunction is changed in sign by the permutation of a pair of identical fermions is called the *Pauli exclusion principle*. Fermions are said to obey Fermi–Dirac statistical formulas and bosons to obey Bose–Einstein statistical formulas.

The importance of these statistical formulas is that the molecular wavefunction will be invariant to any permutation of identical bosons in the molecule and to any even permutation of identical fermions, but will be changed in sign by an odd permutation of identical fermions. An even (odd) permutation can be expressed as an even (odd) number of successively applied pair transpositions (see Problem 1-4).

The Electron Permutation Group

In the group $S_n^{(e)}$ of order $n!$ there are m irreducible representations where m is the partition number of n [see Eq. (4-57) and remarks after it]. One of the irreducible representations in $S_n^{(e)}$ is called the *antisymmetric* representation, $\Gamma^{(e)}(A)$, say, and it has character $(+1)$ under all even permutations and (-1)

under all odd permutations. Since electrons are fermions they obey Fermi–Dirac statistical formulas (i.e., the Pauli exclusion principle) and the molecular wavefunction Φ_{int} must be changed in sign by an odd permutation of the electrons. Thus Φ_{int} can only transform according to the representation $\Gamma^{(e)}(A)$ of $S_n^{(e)}$. As a result of the Pauli exclusion principle all energy levels gain the same label $\Gamma^{(e)}(A)$ from the use of the group $S_n^{(e)}$, and it would appear that this group is of no use in helping us either to distinguish between energy levels or to determine which energy levels can interact with one another. However, in the section on basis function symmetry that follows we will appreciate the special use to which this group is put.

The Complete Nuclear Permutation Group

In a molecule some of the nuclei may have integer spin and obey Bose–Einstein statistical formulas and some of the nuclei may have half integer spin and obey Fermi–Dirac statistical formulas. For the group G^{CNP} of a particular molecule there will be one irreducible representation, which we call $\Gamma^{CNP}(A)$, that has character $(+1)$ for all nuclear permutations except those that involve an odd permutation of fermion nuclei for which the character is (-1). From Bose–Einstein and Fermi–Dirac statistical formulas we deduce that the wavefunction Φ_{int} can only transform according to the representation $\Gamma^{CNP}(A)$ of the group G^{CNP}. This group, like $S_n^{(e)}$, appears to be of no help for labeling energy levels but when we come to consider basis function symmetry we will appreciate the use to which this group is put.

The Inversion Group

The group $\mathscr{E}$ has two representations which we call $+$ or $-$ depending on whether the character under E^* is $+1$ or -1. The wavefunction Φ_{int} can be $+$ or $-$ depending on the effect of E^* and we label the states with this parity label.

From the above considerations we see that we can label the translational states Φ_{CM} according to their momentum using G_T, and usefully label the internal states Φ_{int} with the true symmetry labels $(F, m_F, \pm)$ from the groups K(spatial) and $\mathscr{E}$. The true symmetry labels obtained by using the permutation groups are determined completely by the spin statistical formulas and all states Φ_{int} for a particular molecule gain the same labels $(\Gamma^{(e)}(A), \Gamma^{CNP}(A))$. The useful true symmetry labels from G_{full} on a molecular energy level, E_{int} (as opposed to the label on the wavefunction Φ_{int}), are $(F, \pm)$, and each level is $2F + 1$ fold degenerate as $m_F = -F, -F + 1, \ldots, +F$. The true symmetry groups of the Hamiltonian of an isolated molecule are summarized in Table 6-3.

Table 6-3

The true symmetry groups of the Hamiltonian of an isolated molecule[a]

Fundamental concept	Symmetry group	Symmetry operation Active picture	Passive picture	Symmetry label[b]
Uniform space	Translation group G_T	Translation of the molecule along a space fixed direction	Translation of the space fixed axes	Linear momentum vector k
Isotropic space	Spatial rotation group K(spatial)	Rotation of the molecule about a space fixed axis	Rotation of the space fixed axes	Angular momentum quantum numbers F, m_F
Indistinguishability of electrons	Electron permutation group $S_n^{(e)}$	Permutation of the electrons	Permutation of the numbering of electrons	The antisymmetric representation $\Gamma^{(e)}(A)$
Indistinguishability of identical nuclei	Complete nuclear permutation group G^{CNP}	Permutation of the identical nuclei	Permutation of the numbering of identical nuclei	The antisymmetric representation $\Gamma^{\mathrm{CNP}}(A)$
Conservation of parity[c]	Inversion group $\mathscr{E}$	Inversion of all particles through the molecular center of mass	Inversion of space fixed axis system through the molecular center of mass	Parity $\pm$

[a] Time reversal symmetry is neglected.

[b] The symmetry of the complete wavefunctions in $S_n^{(e)}$ and G^{CNP} is determined by Fermi–Dirac and Bose–Einstein statistical formulas to be $\Gamma^{(e)}(A)$ and $\Gamma^{\mathrm{CNP}}(A)$, respectively.

[c] This is a property of the strong and electromagnetic interaction forces and it is not a property of the weak interaction force; it is not a property of space.

BASIS FUNCTIONS AND BASIS FUNCTION SYMMETRY

We now look at the problem of determining the true symmetry labels on the eigenstates of $\hat{H}_\mathrm{int}$, where from Eqs. (6-26) and (6-29) $\hat{H}_\mathrm{int}$ is given by

$$\hat{H}_\mathrm{int} = \hat{H}^0 + \hat{T}' + \hat{H}_\mathrm{es} + \hat{H}_\mathrm{hfs}. \qquad (6\text{-}46)$$

By judicious choice of the $(3l - 3)$ coordinates in $\hat{H}^0$ we can express the major part of $\hat{H}_\mathrm{int}$ in a form $(\hat{H}_\mathrm{int}^0)$ that is separable into the sum of five terms each of which involves different coordinates. The eigenfunctions Φ_int^0 of the separable (and approximate) Hamiltonian $\hat{H}_{int}^0$ can be determined and these form basis functions for the diagonalization of $\hat{H}_{int}$. These basis functions are each the product of five separate functions that can be classified in the true symmetry groups of the molecular Hamiltonian to give *basis function symmetry labels*. The symmetry species of the exact states can be determined from the basis function symmetry species by using the vanishing integral rule. However, the basis function symmetry labels often turn out to be very useful energy level labels in their own right, and they also enable us to determine electron spin multiplicity restrictions and nuclear spin statistical weights.

In an appropriate set of $(3l - 3)$ coordinates we can write

$$\hat{H}_{int} = \hat{H}_{int}^0 + \hat{H}', \tag{6-47}$$

where

$$\hat{H}_{int}^0 = \hat{H}_{rot} + \hat{H}_{vib} + \hat{H}_{elec} + \hat{H}_{ss} + \hat{H}_{nss}. \tag{6-48}$$

The terms in the zero order Hamiltonian $\hat{H}_{int}^0$ are, respectively, the rotational, vibrational, electron orbital, electron spin–spin, and nuclear spin–spin Hamiltonian operators. The rest of the Hamiltonian, i.e., $\hat{H}'$, contains all the terms that spoil the separation of coordinates. As a result of the separation of coordinates in $\hat{H}_{int}^0$ we can write

$$\hat{H}_{int}^0 \Phi_{int}^0 = E_{int}^0 \Phi_{int}^0, \tag{6-49}$$

where

$$\Phi_{int}^0 = \Phi_{rot} \Phi_{vib} \Phi_{elec} \Phi_{espin} \Phi_{nspin}, \tag{6-50}$$

and

$$E_{int}^0 = E_{rot} + E_{vib} + E_{elec} + E_{espin} + E_{nspin} \tag{6-51}$$

with

$$\hat{H}_{rot} \Phi_{rot} = E_{rot} \Phi_{rot}, \tag{6-52}$$

$$\hat{H}_{vib} \Phi_{vib} = E_{vib} \Phi_{vib}, \tag{6-53}$$

$$\hat{H}_{elec} \Phi_{elec} = E_{elec} \Phi_{elec}, \tag{6-54}$$

$$\hat{H}_{ss} \Phi_{espin} = E_{espin} \Phi_{espin}, \tag{6-55}$$

and

$$\hat{H}_{nss} \Phi_{nspin} = E_{nspin} \Phi_{nspin}. \tag{6-56}$$

In obtaining separable electron spin and nuclear spin parts in $\hat{H}_{int}^0$ we must neglect all the coupling terms $\hat{H}_{so}$, $\hat{H}_{sr}$, $\hat{H}_{nso}$, $\hat{H}_{nsr}$, $\hat{H}_{nses}$, and $\hat{H}_{quad}$ (see Table 6-1) in which the electron spin coordinates and the nuclear spin coordinates each occur in combination with other molecular coordinates. All these terms are contained in $\hat{H}'$. The terms $\hat{H}_{\text{rot}}$, $\hat{H}_{\text{vib}}$, and $\hat{H}_{\text{elec}}$ are obtained from $\hat{H}^0$ in Eq. (6-46) by making an appropriate coordinate choice and some approximations. Among the approximations the term $\hat{T}'$ is neglected and is therefore also contained in $\hat{H}'$.

The rotation, vibration, and electronic wave equations of Eqs. (6-52)–(6-54) will be discussed in Chapters 7 and 8, and we will take the eigenfunctions Φ_{rot}, Φ_{vib}, and Φ_{elec} as known in the rest of this chapter. As we shall see each of these functions is a function of the nuclear coordinates but only Φ_{elec} is a function of the electronic coordinates. Thus Φ_{rot} and Φ_{vib} are unaffected by any permutation of the electrons and must transform as the totally symmetric irreducible representation of the group $S_n^{(e)}$.

The eigenfunctions of $\hat{H}_{\text{int}}$ can be determined by diagonalizing the matrix of $\hat{H}_{\text{int}}$ in the basis set of eigenfunctions Φ_{int}^0 of $\hat{H}_{\text{int}}^0$. Thus each eigenfunction of $\hat{H}_{\text{int}}$ can be written as a linear combination of the complete set of functions Φ_{int}^0 [see Eq. (5-139)]. The term $\hat{H}'$, being part of $\hat{H}_{\text{int}}$, must be totally symmetric in the group G_{full} (or in any subgroup of G_{full}) and hence, from the vanishing integral rule, can only have nonvanishing matrix elements between Φ_{int}^0 functions of the same symmetry in G_{full} (or in any subgroup of G_{full}). The symmetry labels Γ_{int} on the eigenstates Φ_{int} of $\hat{H}_{\text{int}}$ will, therefore, be the same as those on the eigenstates Φ_{int}^0 of $\hat{H}_{\text{int}}^0$, and we need only symmetry classify the basis functions Φ_{int}^0 in order to determine the Γ_{int}. In practice this is only satisfactory if $\hat{H}_{\text{int}}^0$ is a reasonable approximation to $\hat{H}_{\text{int}}$. In other words the energy level pattern of the eigenstates of $\hat{H}_{\text{int}}^0$ and the experimental energy levels (i.e., eigenstates of $\hat{H}_{\text{int}}$) must be similar so that a one-to-one identification of the levels can be made. For example, the electron spin–orbit coupling term $\hat{H}_{\text{so}}$ has a large effect in some molecules, and in these cases we must include $\hat{H}_{\text{so}}$ in the zero order Hamiltonian. Spin–orbit coupling will be discussed in Chapter 10, and when strong there will not be separation of electron orbital and electron spin terms in a useful zero order Hamiltonian. Other mixing terms can be large in particular cases.

Presuming the symmetry of the rotational, vibrational, and electronic wavefunctions are known (this will be discussed in Chapter 10) we now discuss the symmetry of the nuclear spin and electron spin functions in G_{full}. We will show how electron spin multiplicity restrictions and nuclear spin statistical weights are determined.

Electron Spin Functions

Each electron i in a molecule has a spin s_i with magnitude $\hbar/2$, and the total electron spin function can be written as $|S, m_S\rangle$ in terms of the two quantum numbers S and m_S; these quantum numbers are such that the eigenvalues of $\hat{S}^2$ (the operator for the square of the total electron spin angular momentum) and $\hat{S}_Z$ (the operator for the Z component of the electron spin angular momentum) are $S(S+1)\hbar^2$ and $m_S\hbar$, respectively, i.e.,

$$\hat{S}^2|S,m_S\rangle = S(S+1)\hbar^2|S,m_S\rangle \tag{6-57}$$

and

$$\hat{S}_Z|S,m_S\rangle = m_S\hbar|S,m_S\rangle. \tag{6-58}$$

For a single electron we can only have $S = \frac{1}{2}$ and $m_S = \pm\frac{1}{2}$, so that the possible spin functions are $|\frac{1}{2},\frac{1}{2}\rangle$ and $|\frac{1}{2},-\frac{1}{2}\rangle$, and these single electron functions are usually called α and β, respectively. For a many electron system an appropriate basis set of electron spin functions consists of all possible products of α and β functions for the individual electrons. For example in a two electron system (where we label the electrons 1 and 2) this basis set would consist of the four functions

$$\alpha_1\alpha_2, \alpha_1\beta_2, \beta_1\alpha_2, \text{ and } \beta_1\beta_2, \tag{6-59}$$

or, if we assume the order of the electron labeling subscripts, we can write these four functions simply as

$$\alpha\alpha, \alpha\beta, \beta\alpha, \text{ and } \beta\beta. \tag{6-60}$$

For an n electron system there will be 2^n such product functions, each of which involves the product of n single electron spin functions α or β.

The electron spin–spin Hamiltonian commutes with the operations of the electron permutation group $S_n^{(e)}$ and the 2^n product functions of the type shown in Eqs. (6-59) and (6-60) generate a 2^n dimensional representation, $\Gamma_{\text{espin}}^{(e)}$, say, of $S_n^{(e)}$. This representation is easy to determine, as will be shown in an example to follow, and it can be reduced to its irreducible components Γ_i using Eq. (4-43) together with the character table of $S_n^{(e)}$. We can then use appropriate projection operators to determine combinations of the 2^n product functions that transform irreducibly in $S_n^{(e)}$.

As an example let us consider a four electron molecule such as LiH. The character table of the electron permutation group $S_4^{(e)}$ is given in Table 6-4. Numbering the electrons 1, 2, 3, and 4 we can write the $2^4 = 16$ electron spin product functions [from Eq. (6-60)] as follows:

$$\begin{aligned}
(m_S = 2): &\quad \alpha\alpha\alpha\alpha, \\
(m_S = 1): &\quad \alpha\alpha\alpha\beta, \alpha\alpha\beta\alpha, \alpha\beta\alpha\alpha, \beta\alpha\alpha\alpha, \\
(m_S = 0): &\quad \alpha\alpha\beta\beta, \alpha\beta\alpha\beta, \beta\alpha\alpha\beta, \alpha\beta\beta\alpha, \beta\alpha\beta\alpha, \beta\beta\alpha\alpha, \\
(m_S = -1): &\quad \beta\beta\beta\alpha, \beta\beta\alpha\beta, \beta\alpha\beta\beta, \alpha\beta\beta\beta, \\
(m_S = -2): &\quad \beta\beta\beta\beta,
\end{aligned} \tag{6-61}$$

where each function is written in the order 1, 2, 3, 4 so that, for example, $\alpha\beta\beta\alpha$ is $\alpha_1\beta_2\beta_3\alpha_4$. In each row of Eq. (6-61) the functions have the same

Table 6-4

The character table of the permutation group
$S_4^{(e)}$ for a four electron molecule

$S_4^{(e)}$:	E	(12)	(123)	(1234)	(12)(34)
	1	6	8	6	3
Γ_1:	1	1	1	1	1
Γ_2:	1	-1	1	-1	1
Γ_3:	2	0	-1	0	2
Γ_4:	3	1	0	-1	-1
Γ_5:	3	-1	0	1	-1

value of m_S where

$$m_S = \sum_i m_{S_i}. \qquad (6\text{-}62)$$

We now wish to determine the 16-dimensional representation of $S_4^{(e)}$ that
these functions generate. Each operation of $S_4^{(e)}$ permutes the electrons and
it is easy to appreciate that only functions having the same m_S value can be
interconverted. For example,

$$(123)\alpha\beta\alpha\beta = (123)\alpha_1\beta_2\alpha_3\beta_4 = \alpha_2\beta_3\alpha_1\beta_4 = \alpha\alpha\beta\beta \qquad (6\text{-}63)$$

and there has to be the same number of α's and β's in the function after the
electron permutation. Thus we can separately determine the representations
generated by each set having the same m_S value. A further point to notice is
that the set of functions having $m_S = a$, say, and the set having $m_S = -a$
each generate the same representation. In the four electron example the
functions having $m_S = +2$ and $m_S = -2$ each generate the representation Γ_1
of $S_4^{(e)}$. The transformation properties of the four functions having $m_S = +1$
are given in Table 6-5 and the characters in the representation generated are
$[4, 2, 1, 0, 0]$ which reduces to $\Gamma_1 \oplus \Gamma_4$. The four functions having $m_S = -1$

Table 6-5

The transformation properties of the four $m_S = 1$
electron spin functions of a four electron system in $S_4^{(e)}$

R	E	(12)	(123)	(1234)	(12)(34)
	$\alpha\alpha\alpha\beta$	$\alpha\alpha\alpha\beta$	$\alpha\alpha\alpha\beta$	$\beta\alpha\alpha\alpha$	$\alpha\alpha\beta\alpha$
	$\alpha\alpha\beta\alpha$	$\alpha\alpha\beta\alpha$	$\beta\alpha\alpha\alpha$	$\alpha\alpha\alpha\beta$	$\alpha\alpha\alpha\beta$
	$\alpha\beta\alpha\alpha$	$\beta\alpha\alpha\alpha$	$\alpha\alpha\beta\alpha$	$\alpha\alpha\beta\alpha$	$\beta\alpha\alpha\alpha$
	$\beta\alpha\alpha\alpha$	$\alpha\beta\alpha\alpha$	$\alpha\beta\alpha\alpha$	$\alpha\beta\alpha\alpha$	$\alpha\beta\alpha\alpha$
$\chi(R)$:	4	2	1	0	0

transform similarly. The six functions having $m_S = 0$ generate the representation $\Gamma_1 \oplus \Gamma_3 \oplus \Gamma_4$. Thus for a four electron molecule the 16 electron spin functions generate the representation

$$5\Gamma_1 \oplus \Gamma_3 \oplus 3\Gamma_4 \qquad (6\text{-}64)$$

of the group $S_4^{(e)}$. Notice that the five states of species Γ_1 have $m_S = 2, 1, 0, -1$, and -2; the three states of species Γ_4 have $m_S = 1, 0, -1$; the state of species Γ_3 has $m_S = 0$.

We now look at the classification of the electron spin functions in the group K(spatial). A pair of spin functions (α, β) for a single electron have spin angular momentum quantum number $S = \frac{1}{2}$ and they generate the two-dimensional representation $D^{(1/2)}$ of the group K(spatial)^2. [K(spatial)^2 is the spin double group of K(spatial) and this extension of the group K(spatial) is required for classifying states with half integer angular momentum. This will be discussed further in Chapter 10.] The 2^n product spin functions for an n electron system will therefore generate the representation $D^{(1/2)} \otimes D^{(1/2)} \otimes D^{(1/2)} \otimes \cdots$, where the product is taken n times. From Eqs. (6-40) and (6-41) we deduce that this representation contains $D^{(n/2)}$, $D^{(n/2-1)}$, $D^{(n/2-2)}, \ldots$, $D^{(1/2)}$ or $D^{(0)}$, where some of these irreducible representations will occur more than once. The $(2S + 1)$ functions $|S, m_S\rangle$ transform as $D^{(S)}$, and have eigenvalue $S(S + 1)\hbar^2$ for $\hat{S}^2$ and eigenvalues $m_S \hbar = -S\hbar, (-S + 1)\hbar, \ldots, +S\hbar$ for $\hat{S}_Z$. The linear combinations of the product functions that transform irreducibly in K(spatial)^2 can be determined by using vector coupling coefficients but we will not discuss that here. In fact, these linear combinations are often unambiguously obtained using the reduction in the $S_n^{(e)}$ group. For example, in the four electron case previously discussed we have

$$D^{(1/2)} \otimes D^{(1/2)} \otimes D^{(1/2)} \otimes D^{(1/2)} = D^{(2)} \oplus 3D^{(1)} \oplus 2D^{(0)} \qquad (6\text{-}65)$$

and from Eq. (6-64) we see that the singlet spin functions ($S = 0, m_S = 0$) are of species Γ_3, the triplet spin functions ($S = 1, m_S = -1, 0,$ or $+1$) are each of species Γ_4, and the quintet spin functions ($S = 2, m_S = -2, -1, 0, +1,$ or $+2$) are each of species Γ_1; the *multiplicities* are 1, 3, and 5, respectively.

Using the groups K(spatial)^2 and $S_4^{(e)}$ we have been able to classify the 16 electron spin states of a four electron molecule. The procedure used above can be followed for a molecule containing any number of electrons, but when states of different multiplicity have the same symmetry in $S_n^{(e)}$ it is necessary to use vector coupling coefficients to obtain the combinations of product functions that transform irreducibly in the $S_n^{(e)}$ and K(spatial)^2 groups. The electron spin functions do not involve the nuclear coordinates and they therefore generate the totally symmetric irreducible representation of the group G^{CNP}. The spin functions are also invariant to E^* (S is an axial vector) and all have positive parity.

Nuclear Spin Functions

We will use the same technique for determining the symmetry of the nuclear spin functions in a molecule as used previously for the electron spin functions. The nuclear spin function for a single nucleus can be written $|I_\alpha, m_{I_\alpha}\rangle$, where $m_{I_\alpha} = -I_\alpha, -I_\alpha + 1, \ldots, +I_\alpha$, and I_α and m_{I_α} are the quantum numbers indicating the nuclear spin angular momentum and its Z component, respectively. The functions $|I_\alpha, m_{I_\alpha}\rangle$ will generate the representation $D^{(I_\alpha)}$ of the group K(spatial)^2. Suppose that we are interested in a molecule having the chemical formula $A_a B_b C_c D_d \cdots$ where the A nucleus has spin I_A, the B nucleus has spin I_B, etc. The number of possible nuclear spin states for each A type nucleus is $(2I_A + 1)$, as $m_{I_A} = -I_A, -I_A + 1, \ldots, +I_A$, and so the total number of nuclear spin states for all the A nuclei is $(2I_A + 1)^a$. Considering all nuclei the total number of nuclear spin states for the molecule is given by

$$(2I_A + 1)^a(2I_B + 1)^b \cdots. \tag{6-66}$$

We can write each of these functions as

$$|m_{A1}(A_1)m_{A2}(A_2) \cdots m_{Aa}(A_a)m_{B1}(B_1) \cdots m_{Bb}(B_b) \cdots\rangle, \tag{6-67}$$

where

$$m_{\alpha i} = -I_\alpha, -I_\alpha + 1, \ldots, \text{or } +I_\alpha. \tag{6-68}$$

The effect of a nuclear permutation on one of these functions is trivial to determine since the nuclear labels are just permuted. For example, the cyclic permutation of the three nuclei A_1, A_2, and A_3 has the effect

$$(123)|m_{A1}(A_1)m_{A2}(A_2)m_{A3}(A_3) \cdots\rangle = |m_{A1}(A_2)m_{A2}(A_3)m_{A3}(A_1) \cdots\rangle$$
$$= |m_{A3}(A_1)m_{A1}(A_2)m_{A2}(A_3) \cdots\rangle. \tag{6-69}$$

When we write nuclear spin functions we can use an abbreviated ket notation in which the order of the nuclei is maintained; in this notation Eq. (6-69) becomes

$$(123)|m_{A1}m_{A2}m_{A3} \cdots\rangle = |m_{A3}m_{A1}m_{A2} \cdots\rangle. \tag{6-70}$$

Notice that any permutation operation can only convert a nuclear spin function to another having the same value of $m_I = \sum_{\alpha i} m_{\alpha i}$. Knowing the chemical formula of a molecule and the spins of its nuclei it is thus straightforward, although possibly tedious, to determine the representation of the group G^{CNP} that the nuclear spin wavefunctions generate. Using projection operators and Schmidt orthogonalization [see Eq. (5-83)] the combinations of the nuclear spin functions that transform irreducibly can be determined. It is best to determine the symmetry species of the nuclear spin functions of each set of identical nuclei separately and then to multiply these together to obtain the species of the total nuclear spin function in G^{CNP}.

In the molecule $A_aB_bC_cD_d \cdots$ the species of the nuclear spin wavefunctions of the A nuclei in K(spatial)^2 is determined by forming the product of $D^{(I_A)}$ with itself a times, that of the B nuclei by forming the product of $D^{(I_B)}$ with itself b times, and so on. The species of the complete nuclear spin wavefunction of the molecule is obtained by multiplying all these products together. A given nuclear spin function Φ_{nspin} will transform as $D^{(I)}$ where I is the total nuclear spin angular momentum quantum number of the state.

The nuclear spin wavefunctions are invariant to any electron permutation and to E^* (I is an axial vector), and so they generate the totally symmetric representation of $S_n^{(e)}$ and have positive parity.

Problem 6-1. Determine the species of the nuclear spin states of NH_3 and ND_3 in the CNP group S_3 of the ammonia molecule (see Table 4-2), and in the group K(spatial)^2.

Answer. For NH_3 we can construct the following proton spin functions:

$(m_I = \frac{3}{2})$: $\alpha(1)\alpha(2)\alpha(3) = \Phi_{ns}^{(1)}$,

$(m_I = \frac{1}{2})$: $\alpha(1)\alpha(2)\beta(3) = \Phi_{ns}^{(2)}$, $\alpha(1)\beta(2)\alpha(3) = \Phi_{ns}^{(3)}$, $\beta(1)\alpha(2)\alpha(3) = \Phi_{ns}^{(4)}$,

$(m_I = -\frac{1}{2})$: $\alpha(1)\beta(2)\beta(3) = \Phi_{ns}^{(5)}$, $\beta(1)\alpha(2)\beta(3) = \Phi_{ns}^{(6)}$, $\beta(1)\beta(2)\alpha(3) = \Phi_{ns}^{(7)}$,

$(m_I = -\frac{3}{2})$: $\beta(1)\beta(2)\beta(3) = \Phi_{ns}^{(8)}$,

$$(6\text{-}71)$$

where the nuclei are labeled 1, 2, and 3, α is an $m_I = \frac{1}{2}$ function, and β is an $m_I = -\frac{1}{2}$ function. The representation of S_3 generated by these eight functions is

$$
\begin{array}{ccc}
E & (12) & (123) \\
8 & 4 & 2
\end{array}
$$

where the operation (12) leaves the functions $\Phi_{ns}^{(1)}$, $\Phi_{ns}^{(2)}$, $\Phi_{ns}^{(7)}$, and $\Phi_{ns}^{(8)}$ alone but interchanges the other functions, and (123) leaves the function $\Phi_{ns}^{(1)}$ and $\Phi_{ns}^{(8)}$ alone but interchanges all the other functions. This representation reduces to the following irreducible representations of S_3:

$$4\Gamma_1 \oplus 2\Gamma_3. \qquad (6\text{-}72)$$

The representation of K(spatial)^2 generated by the proton spin functions is

$$D^{(1/2)} \otimes D^{(1/2)} \otimes D^{(1/2)} = D^{(3/2)} \oplus 2D^{(1/2)}, \qquad (6\text{-}73)$$

and so there is a quartet and two doublet proton spin states. The quartet states are each of symmetry Γ_1 and the doublets of species Γ_3. Combinations that transform irreducibly are:

$I = \frac{3}{2}$:

$$
\begin{array}{lll}
m_I = \frac{3}{2}: & \alpha\alpha\alpha & \Gamma_1 \\
m_I = \frac{1}{2}: & (\alpha\alpha\beta + \alpha\beta\alpha + \beta\alpha\alpha)/\sqrt{3} & \Gamma_1 \\
m_I = -\frac{1}{2}: & (\beta\beta\alpha + \beta\alpha\beta + \alpha\beta\beta)/\sqrt{3} & \Gamma_1 \\
m_I = -\frac{3}{2}: & \beta\beta\beta & \Gamma_1
\end{array}
\tag{6-74}
$$

$I = \frac{1}{2}$:

$$
\begin{array}{ll}
m_I = \frac{1}{2}: & [(2\alpha\alpha\beta - \alpha\beta\alpha - \beta\alpha\alpha)/\sqrt{6}, (\alpha\beta\alpha - \beta\alpha\alpha)/\sqrt{2}] \quad \Gamma_3 \\
m_I = -\frac{1}{2}: & [(2\beta\beta\alpha - \beta\alpha\beta - \alpha\beta\beta)/\sqrt{6}, (\beta\alpha\beta - \alpha\beta\beta)/\sqrt{2}] \quad \Gamma_3.
\end{array}
\tag{6-75}
$$

For ND_3 each deuterium nucleus can have $m_I = -1, 0,$ or $+1$ and we label such functions $\nu, \mu,$ and λ, respectively. We have deuterium nuclear spin functions [the nuclear labels (1), (2), and (3) are omitted for brevity] as follows:

$$
\begin{array}{l}
\lambda\lambda\lambda; \; \lambda\lambda\mu, \lambda\mu\lambda, \mu\lambda\lambda; \; \lambda\lambda\nu, \lambda\nu\lambda, \nu\lambda\lambda, \lambda\mu\mu, \mu\lambda\mu, \mu\mu\lambda; \\
\lambda\mu\nu, \lambda\nu\mu, \mu\lambda\nu, \nu\lambda\mu, \mu\nu\lambda, \nu\mu\lambda, \mu\mu\mu; \\
\lambda\nu\nu, \nu\lambda\nu, \nu\nu\lambda, \mu\mu\nu, \mu\nu\mu, \nu\mu\mu; \mu\nu\nu, \nu\mu\nu, \nu\nu\mu; \nu\nu\nu.
\end{array}
\tag{6-76}
$$

In the G^{CNP} group S_3 these 27 functions generate the representation

$$
\begin{array}{ccc}
E & (12) & (123) \\
27 & 9 & 3
\end{array}
$$

which reduces to the following irreducible representations:

$$
\begin{array}{ll}
(m_I = 3 \text{ or } -3): & \Gamma_1, \\
(m_I = 2 \text{ or } -2): & \Gamma_1 \oplus \Gamma_3, \\
(m_I = 1 \text{ or } -1): & \Gamma_1 \oplus \Gamma_3 \oplus \Gamma_1 \oplus \Gamma_3, \\
(m_I = 0): & \Gamma_1 \oplus \Gamma_3 \oplus \Gamma_1 \oplus \Gamma_3 \oplus \Gamma_2.
\end{array}
\tag{6-77}
$$

The representation of $K(\text{spatial})^2$ generated by the deuteron spin functions in ND_3 is given by

$$
D^{(1)} \otimes D^{(1)} \otimes D^{(1)} = D^{(3)} \oplus 2D^{(2)} \oplus 3D^{(1)} \oplus D^{(0)}.
\tag{6-78}
$$

We see that each of the septet states is of species Γ_1, each of the quintet states is of species Γ_3, the components of the triplet states are of species Γ_1 and Γ_3, and the singlet state is of species Γ_2.

The single nitrogen nucleus clearly has a spin function that is totally symmetric in the CNP group. A ^{14}N nucleus has spin 1 and a ^{15}N nucleus has spin $\frac{1}{2}$, so that the species of the total nuclear spin function of the molecule in $K(\text{spatial})^2$ is obtained by multiplying the spin species just given by either

$D^{(1)}$ or $D^{(1/2)}$ depending on which isotope of nitrogen is present; the species in G^{CNP} is obtained by multiplying by $3\Gamma_1$ or $2\Gamma_1$ for ^{14}N or ^{15}N, respectively.

The Complete Basis Functions

We have now shown how to classify the electron spin and nuclear spin functions in the subgroups of G_{full}. We have also stated that Φ_{rot}, Φ_{vib}, and Φ_{elec} each generate representations of G^{CNP}, and we let the symmetry of the product rovibronic functions

$$\Phi^0_{rve} = \Phi_{rot}\Phi_{vib}\Phi_{elec} \qquad (6\text{-}79)$$

in the group G^{CNP} be Γ^{CNP}_{rve}. The functions Φ_{rot} and Φ_{vib} are totally symmetric in $S^{(e)}_n$, and we let the species of Φ_{elec} in $S^{(e)}_n$ be $\Gamma^{(e)}_{elec}$. The functions Φ^0_{rve} are of positive or negative parity, and they generate the representation $D^{(N)}$ of $K(\text{spatial})$ where N is the quantum number for the total rovibronic angular momentum of all particles in the molecule. The total angular momentum F is the sum of the rovibronic angular momentum N, the electron spin angular momentum S, and the nuclear spin angular momentum I. We also introduce $J = N + S$ and for singlet electron spin states $J \equiv N$. We let m_N be the Z component of the rovibronic angular momentum ($m_N = -N$, $-N + 1, \ldots, +N$) and so

$$\hat{N}^2\Phi^0_{rve} = N(N + 1)\hbar^2\Phi^0_{rve} \qquad (6\text{-}80a)$$

and

$$\hat{N}_Z\Phi^0_{rve} = m_N\hbar\Phi^0_{rve}. \qquad (6\text{-}80b)$$

We will now discuss how these basis symmetries are combined to give the symmetry of the functions Φ^0_{int} and Φ_{int}.

Since Φ_{espin} and Φ_{nspin} both have positive parity the parity of Φ^0_{int} is determined by the parity of the Φ^0_{rve} functions. Only states Φ^0_{int} of the same parity are mixed by $\hat{H}'$, and the parity of the final states is given by that of the Φ^0_{int} (i.e., by the parity of Φ^0_{rve}).

The basis functions Φ^0_{rve}, Φ_{espin}, and Φ_{nspin} generate the representations $D^{(N)}$, $D^{(S)}$, and $D^{(I)}$, respectively, of the group $K(\text{spatial})^2$, and hence their product Φ^0_{int} will generate the representation [see Eq. (6-41)]

$$D^{(N)} \otimes D^{(S)} \otimes D^{(I)} = D^{(N+S+I)} \oplus D^{(N+S+I-1)} \oplus \cdots \oplus D^{(||N-S|-I|)}. \qquad (6\text{-}81)$$

If, as is usually the case, N is greater than $(I + S)$ there will be a multiplicity of $(2I + 1)(2S + 1)$ irreducible representations in this product. The representations of $K(\text{spatial})^2$ generated by Φ^0_{int} give the various possible values of the total angular momentum quantum number F, i.e., for the Φ^0_{int} we have

$$F = N + S + I, N + S + I - 1, \ldots, ||N - S| - I|. \qquad (6\text{-}82)$$

total moment

There is, therefore, a multiplicity of different F values for each level E_{int}^0. Components having different F values can be separated by the effect of $\hat{H}'$ since this causes states having the same F value in different E_{int}^0 levels to interact. We see that if we know N, S, and I, and if we know the parity of Φ_{rve}^0, then we can determine the parity and F values of Φ_{int}^0. The term $\hat{H}'$ in the Hamiltonian can mix Φ_{int}^0 states that have the same parity and F value (they need not have the same N, S, or I values).

We can classify Φ_{rve}^0, Φ_{espin}, and Φ_{nspin} in the groups G^{CNP} and $S_n^{(e)}$, but we know that the species of Φ_{int} in these groups is restricted to $\Gamma^{CNP}(A)$ and $\Gamma^{(e)}(A)$, respectively, by the spin statistical formulas. Thus we only need construct Φ_{int}^0 of these species in G^{CNP} and $S_n^{(e)}$ in order to obtain a complete basis set for diagonalizing $\hat{H}_{int}$. To construct such Φ_{int}^0 we only combine Φ_{rve}^0 with those Φ_{espin} and Φ_{nspin} states whose symmetries in G^{CNP} and $S_n^{(e)}$ are such that

$$\Gamma_{rve}^{(e)} \otimes \Gamma_{espin}^{(e)} \otimes \Gamma_{nspin}^{(e)} \supset \Gamma^{(e)}(A) \tag{6-83}$$

and

$$\Gamma_{rve}^{CNP} \otimes \Gamma_{espin}^{CNP} \otimes \Gamma_{nspin}^{CNP} \supset \Gamma^{CNP}(A), \tag{6-84}$$

where the notation used for the basis symmetries is obvious. Since we know that Φ_{espin} is totally symmetric in the group G^{CNP} and that Φ_{rot}, Φ_{vib}, and Φ_{nspin} are all totally symmetric in the group $S_n^{(e)}$, these equations can be rewritten as

$$\Gamma_{elec}^{(e)} \otimes \Gamma_{espin}^{(e)} \supset \Gamma^{(e)}(A) \tag{6-85}$$

and

$$\Gamma_{rve}^{CNP} \otimes \Gamma_{nspin}^{CNP} \supset \Gamma^{CNP}(A). \tag{6-86}$$

Equation (6-85) restricts the electron spin states that a given state Φ_{elec} will be combined with in forming Φ_{int}^0, and Eq. (6-86) restricts the nuclear spin states that a given state Φ_{rve}^0 will be combined with. The former leads to electron spin multiplicity restrictions and the latter to nuclear spin statistical weights.

We illustrate the application of these restrictions by considering the hydrogen molecule, and close the chapter with a worked problem showing the application of Eq. (6-86) to the determination of nuclear spin statistical weights in the NH_3 and ND_3 molecules.

In the hydrogen molecule H_2 we label the two electrons a and b, and the two protons 1 and 2. The electron permutation group is $S_2^{(e)} = \{E, (ab)\}$ and the complete nuclear permutation group is $G^{CNP} = \{E, (12)\}$. These groups are isomorphic to each other and each has two irreducible representations (see Table 5-1) which we call $\Gamma_1^{(e)}$ and $\Gamma_2^{(e)}$ for $S_2^{(e)}$, and Γ_1^{CNP} and

Γ_2^{CNP} for G^{CNP}. Electrons are fermions and so we only construct basis functions Φ_{int}^0 that transform as $\Gamma^{(e)}(A) = \Gamma_2^{(e)}$ in $S_2^{(e)}$. Thus electronic states of species $\Gamma_1^{(e)}$ will only be combined with electron spin functions of species $\Gamma_2^{(e)}$, and electronic states of species $\Gamma_2^{(e)}$ will only be combined with electron spin functions of species $\Gamma_1^{(e)}$, in forming basis functions Φ_{int}^0 for diagonalizing $\hat{H}_{int}$. Similarly, since protons are also fermions we have $\Gamma^{CNP}(A) = \Gamma_2^{CNP}$, and rovibronic functions Φ_{rve}^0 of species Γ_1^{CNP} (or Γ_2^{CNP}) will only be combined with nuclear spin functions Φ_{nspin} of species Γ_2^{CNP} (or Γ_1^{CNP}). These restrictions follow from Eqs. (6-85) and (6-86) by making use of the multiplication rule in S_2 that

$$\Gamma_1 \otimes \Gamma_2 = \Gamma_2. \tag{6-87}$$

To understand the implications of these restrictions for the hydrogen molecule we examine the spin functions in more detail. From Eqs. (6-59) and (6-60) we see that for a two electron molecule such as H_2 the simple product electron spin functions are

$$\alpha\alpha, \ \alpha\beta, \ \beta\alpha, \ \text{and} \ \beta\beta. \tag{6-88}$$

It is easy to determine that these four functions generate the representation

$$3\Gamma_1^{(e)} \oplus \Gamma_2^{(e)} \tag{6-89}$$

of $S_2^{(e)}$. The three combinations transforming irreducibly as $\Gamma_1^{(e)}$ are

$$\begin{aligned} &\alpha\alpha &m_S &= 1, \\ &(\alpha\beta + \beta\alpha)/\sqrt{2} &m_S &= 0, \end{aligned} \tag{6-90}$$

and

$$\beta\beta \qquad\qquad m_S = -1,$$

and the function transforming as $\Gamma_2^{(e)}$ is

$$(\alpha\beta - \beta\alpha)/\sqrt{2} \qquad m_S = 0. \tag{6-91}$$

In $K(\text{spatial})^2$ two-electron spin functions transform as

$$D^{(1/2)} \otimes D^{(1/2)} = D^{(1)} \oplus D^{(0)}. \tag{6-92}$$

Clearly the three functions in Eq. (6-90) are the three components of the $(S = 1)$ triplet state and the function in Eq. (6-91) is the $(S = 0)$ singlet state function. As a result of the spin restrictions discussed before Eq. (6-87) we see that electron orbital functions Φ_{elec} of species $\Gamma_1^{(e)}$ will only be combined with the singlet electron spin function, and electron orbital functions of species $\Gamma_2^{(e)}$ will only be combined with the triplet electron spin functions. The lowest electronic orbital state of the hydrogen molecule has species $\Gamma_1^{(e)}$ and hence will give rise to a singlet electronic state, whereas the first excited electronic state (it is an unbound state) has species $\Gamma_2^{(e)}$ and gives rise to a triplet electronic state. Terms in $\hat{H}'$ (principally electron spin–orbit coupling) can mix states Φ_{int}^0 having different electron spin multiplicities but these coupling effects are usually small and the electron spin multiplicity (i.e., S value) is, therefore, usually well defined.

A similar series of arguments using G^{CNP} for H_2 shows that rovibronic states Φ_{rve}^0 of species Γ_1^{CNP} will only be combined with the single $I = 0$ nuclear spin state, whereas rovibronic states of species Γ_2^{CNP} will only be combined with one of the three $I = 1$ nuclear spin states. Thus rovibronic states of species Γ_1^{CNP} will have a nuclear spin statistical weight of one whereas rovibronic states of species Γ_2^{CNP} will have a nuclear spin statistical weight of three. Again terms in $\hat{H}'$ can mix states having different I values but these effects are usually very small. For the hydrogen molecule in its ground electronic state the mixing of states having different S or I values is very small indeed.

It is instructive to consider the nuclear spin statistical weights of the deuterium molecule D_2. Deuterons have a spin of 1 and are bosons. So for D_2 the wavefunctions Φ_{int} must transform as Γ_1^{CNP} in G^{CNP}, and we only construct Φ_{int}^0 functions of this symmetry. The characters are real, and from Eq. (5-118) we only combine rovibronic functions and nuclear spin functions that have the same symmetry in G^{CNP}. Following the same notation as used in Eq. (6-76) for ND_3 we can write the deuteron spin functions for D_2 as

$$
\begin{aligned}
(m_I = 2): & \quad \lambda\lambda, \\
(m_I = 1): & \quad \lambda\mu, \mu\lambda, \\
(m_I = 0): & \quad \lambda v, v\lambda, \mu\mu, \\
(m_I = -1): & \quad v\mu, \mu v, \\
(m_I = -2): & \quad vv.
\end{aligned}
\tag{6-93}
$$

The species of these functions in G^{CNP} is readily determined to be

$$
\begin{aligned}
(m_I = 2): & \quad \Gamma_1^{CNP}, \\
(m_I = 1): & \quad \Gamma_1^{CNP} \oplus \Gamma_2^{CNP}, \\
(m_I = 0): & \quad \Gamma_1^{CNP} \oplus \Gamma_2^{CNP} \oplus \Gamma_1^{CNP}, \\
(m_I = -1): & \quad \Gamma_1^{CNP} \oplus \Gamma_2^{CNP}, \\
(m_I = -2): & \quad \Gamma_1^{CNP}.
\end{aligned}
\tag{6-94}
$$

The species of these nuclear spin functions in K(spatial) is given by

$$
D^{(1)} \otimes D^{(1)} = D^{(2)} \oplus D^{(1)} \oplus D^{(0)}.
\tag{6-95}
$$

We see that the quintet ($I = 2$) functions are each of species Γ_1^{CNP}, the triplet ($I = 1$) functions are each of species Γ_2^{CNP} and the singlet ($I = 0$) function is of species Γ_1^{CNP}. However, for the purpose of determining the nuclear spin statistical weights we do not need to concern ourselves with anything except the species of the nuclear spin functions in G^{CNP}, and we write this total nuclear spin species as

$$\Gamma_{nspin}^{tot} = 6\Gamma_1^{CNP} \oplus 3\Gamma_2^{CNP}. \qquad (6\text{-}96)$$

Thus, for D_2, rovibronic states of species Γ_1^{CNP} have nuclear spin statistical weight of six and rovibronic states of species Γ_2^{CNP} have nuclear spin statistical weight of three.

Problem 6-2. Determine the nuclear spin statistical weights of the rovibronic states Φ_{rve}^0 of the $^{14}NH_3$ and $^{14}ND_3$ molecules as a function of the symmetry of these states in the CNP group S_3 (see Table 4-2 and Problem 6-1).

Answer. From the solution to Problem 6-1 we see that the total nuclear spin species for $^{14}NH_3$ is $\text{since } I(^{14}N) = 1$

$$\Gamma_{nspin}^{tot} = (4\Gamma_1 \oplus 2\Gamma_3) \otimes 3\Gamma_1 = 12\Gamma_1 \oplus 6\Gamma_3, \qquad (6\text{-}97)$$

and we have $\Gamma^{CNP}(A) = \Gamma_2$. Thus a rovibronic state Φ_{rve}^0 of the NH_3 molecule having symmetry Γ_{rve} in S_3 will only be combined with a nuclear spin state Φ_{nspin} having symmetry Γ_{nspin} if

$$\Gamma_{rve} \otimes \Gamma_{nspin} \supset \Gamma_2. \qquad (6\text{-}98)$$

We can have $\Gamma_{rve} = \Gamma_1, \Gamma_2$, or Γ_3 in S_3 and we determine the nuclear spin statistical weight of each rovibronic symmetry type in turn. If $\Gamma_{rve} = \Gamma_1$ then we can only combine the rovibronic state with a nuclear spin state of species Γ_2 in order to satisfy Eq. (6-98). From Γ_{nspin}^{tot} in Eq. (6-97) we see that there are no nuclear spin states of symmetry Γ_2 for NH_3. Thus rovibronic states of species Γ_1 have no allowed nuclear spin partner and hence have a nuclear spin statistical weight of zero. In forming Φ_{int}^0 functions to diagonalize H_{int} we therefore do not include any functions that contain totally symmetric Φ_{rve}^0 functions. If $\Gamma_{rve} = \Gamma_2$ we can only combine the rovibronic state with a nuclear spin state of species Γ_1 in order to satisfy Eq. (6-98). From Eq. (6-97) we see that there are 12 such nuclear spin states and hence rovibronic states of symmetry Γ_2 have nuclear spin statistical weight of 12. Finally we come to degenerate rovibronic states of species Γ_3, and in S_3 we have

$$\Gamma_3 \otimes \Gamma_3 = \Gamma_1 \oplus \Gamma_2 \oplus \Gamma_3. \qquad (6\text{-}99)$$

Thus a rovibronic state of species Γ_3 can be combined with a doubly degenerate nuclear spin state of species Γ_3 since the product of their species contains Γ_2 which satisfies Eq. (6-98). In this product [see Eq. (6-99)] the species Γ_2 only occurs once, and so the nuclear spin statistical weight of a rovibronic state of species Γ_3 is therefore given by the number of nuclear spin states of species Γ_3, i.e., six in this case. Thus, in summary for $^{14}NH_3$, the nuclear spin statistical weights of rovibronic states of species Γ_1, Γ_2, and Γ_3 are 0, 12, and 6, respectively.

For $^{14}ND_3$ since deuterons are bosons we have $\Gamma^{CNP}(A) = \Gamma_1$ in S_3, and so we must have

$$\Gamma_{rve} \otimes \Gamma_{nspin} \supset \Gamma_1 \qquad (6\text{-}100)$$

in S_3; i.e., we can only combine rovibronic and nuclear spin functions if Γ_{rve} is the same as Γ_{nspin} [see Eq. (5-118); none of the irreducible representations of the CNP group S_3 has complex characters so we must have $\Gamma_{rve} = \Gamma_{nspin}$ here]. From Eq. (6-77) we see that for ND_3 the total nuclear spin representation is

$$\Gamma^{tot}_{nspin} = 30\Gamma_1 \oplus 3\Gamma_2 \oplus 24\Gamma_3 \qquad (6\text{-}101)$$

so that rovibronic states of species Γ_1, Γ_2, and Γ_3 in S_3 have nuclear spin statistical weights of 30, 3, and 24, respectively.

DISCUSSION

The true symmetry labels discussed in this chapter are obtained by using the full Hamiltonian group G_{full}, or its subgroups K(spatial), $\mathscr{E}$, G^{CNP}, and $S_n^{(e)}$. The group G_{full} is the full symmetry group of the exact molecular Hamiltonian and to set up the group for a molecule we only need to know the chemical formula for the molecule. The group is set up without any detailed knowledge of the Hamiltonian, and this is both a strength and a weakness. It is a strength in that the group can be rather easily determined and the symmetry results it leads to are always correct (for the isolated molecule). It is a weakness in that, because none of the details of the Hamiltonian are considered, full allowance is not made for any special properties the Hamiltonian might have. The special properties of the Hamiltonian for a particular molecule may lead in a systematic fashion to what, according to G_{full}, are accidental degeneracies and accidentally vanishing off-diagonal matrix elements of $\hat{H}'$. In such a case if we had a precise enough experimental technique we could show that these degeneracies are not exact and that these off-diagonal matrix elements are not precisely zero. However, for understanding and interpreting molecular properties it is clearly useful to allow for the systematics of such accidents. To allow for the degeneracies (or in reality to allow ourselves to neglect them) we use a subgroup of G_{full} involving the molecular symmetry group, and to allow for the near vanishing of interaction matrix elements we use a near symmetry group to obtain near symmetry labels. The molecular symmetry group will be discussed in Chapter 9 and its application to molecules in Chapter 10.

Near symmetry groups will be discussed in Chapter 11 and the molecular point group is a familiar near symmetry group. Before this we must unfortunately do a bit of hard work in Chapters 7 and 8 in order to understand the molecular wavefunctions Φ_{rve}^0. We must define the coordinates that the wavefunctions involve and determine the explicit expressions for the wavefunctions in terms of the coordinates. Once we know the expressions for the wavefunctions in terms of the coordinates and have determined the effects of the symmetry operations on the coordinates, we can determine the symmetry Γ_{rve} of the rovibronic wavefunctions.

BIBLIOGRAPHICAL NOTES

The Hamiltonian

Moss (1973). Chapters 9 and 10 give a particularly useful account of the molecular Hamiltonian.

Van Vleck (1951). Equation (37) and footnote 35.

Gunther-Mohr, Townes, and Van Vleck (1954). Equation (1).

The K(spatial) Group

Hamermesh (1962). Chapter 9. The group is called $O^+(3)$.

Tinkham (1964). Chapter 5.

Wigner (1959). Chapter 15. Also time reversal symmetry is discussed in Chapter 26.

Parity Violation in Atoms and Molecules

Close (1976).

Baird et al. (1976).

Bouchiat and Pottier (1977).

Sandars (1977).

Polar and Axial Vectors

Margenau and Murphy (1956). Pages 164 and 165.

Spin and Statistical Formulas

Pauli (1940). In this paper a discussion is given of how the statistical formulas follow from the spin of a particle by using relativistic arguments.

7

The Coordinates
in the Rovibronic
Schrödinger Equation

In this chapter the rovibronic Schrödinger equation is set up in coordinates that facilitate its solution and these coordinates are studied in detail. Two techniques are used in order to change coordinates in a Schrödinger equation and each of these techniques is described; the first is used in the rovibronic Schrödinger equation of a diatomic molecule and the second in that of a rigid nonlinear polyatomic molecule. Linear polyatomic molecules and nonrigid molecules are not discussed in this chapter; they are the subject of Chapter 12. The coordinates used in the rovibronic Schrödinger equation are chosen so that the least approximation is necessary in separating the equation into rotational, vibrational, and electronic parts, and we call the coordinates *the rovibronic coordinates*. A key element in the definition of the rovibronic coordinates is the introduction of the Eckart axes, and these are used in a numerical example with the water molecule. Having defined the coordinates it is possible to determine the effects of nuclear permutations and the inversion on them, and these transformation properties are discussed. The discussion given here and in Chapter 8 of the rotation–vibration problem is meant to complement, rather than duplicate, Chapter 11 of Wilson, Decius, and Cross (1955).

114

THE ROVIBRONIC SCHRÖDINGER EQUATION

The rovibronic Hamiltonian $\hat{H}_{rve}$ is given by $(\hat{H}_0 + \hat{T}')$ from Eq. (6-26), and the rovibronic Schrödinger equation is [see Eqs. (6-14), (6-15), and (6-19)]

$$\left\{ -\frac{\hbar^2}{2} \sum_{r=2}^{l} \frac{\mathbf{V}_r^2}{m_r} + \frac{\hbar^2}{2M} \sum_{r,s=2}^{l} \mathbf{V}_r \cdot \mathbf{V}_s + \sum_{r<s=1}^{l} \frac{C_r C_s e^2}{R_{rs}} - E_{rve} \right\} \Phi_{rve}(X_2, \ldots, Z_l) = 0,$$

$$(7\text{-}1)$$

where there are l particles (nuclei and electrons) in the molecule, and E_{rve} and Φ_{rve} are the rovibronic eigenvalues and eigenfunctions, respectively. This is a $(3l - 3)$ dimensional second order partial differential equation and for l greater than 2 (i.e., any molecule) this equation cannot be solved explicitly.

To solve the rovibronic Schrödinger equation given in Eq. (7-1) there are two methods which we will call the *direct method* and the *indirect method*. The direct method is numerical in nature and it involves setting up and using a complex computer program. Using such a numerical technique we determine each Φ_{rve} wavefunction at many points in $(X_2, Y_2, Z_2, \ldots, Z_l)$ space and each associated energy E_{rve}. To do this with a precision that is comparable to the best experiment is impossibly difficult (except for the very lowest eigenstates of three and four particle systems such as H_2^+ and H_2) and hence for practical reasons we use the indirect method. In the indirect method of solving Eq. (7-1) we first make algebraic approximations and then correct for each of the approximations made. We make approximations so that we can separate the variables in Eq. (7-1), and the approximate and separable wave equation obtained is solved partly explicitly and partly by using numerical methods. We correct for the approximations made by using perturbation and variational techniques.

Although solving the Schrödinger equation by the indirect method is forced on us as a practical necessity it has the desirable consequence of allowing us to understand the solutions. The reason for this is that in the course of making the approximations we introduce many concepts that give us a physical insight into the behavior of the nuclei and electrons in the molecule. Some of the more important concepts introduced are electronic state, molecular orbital, potential energy surface, equilibrium nuclear structure, and dipole moment function. We say we understand the solutions in that we can appreciate the pattern of the energy levels and form of the wavefunctions for a given molecule without solving Eq. (7-1), and we appreciate in a general way how they will vary from one molecule to another.

Of the approximations that we make in solving Eq. (7-1) the main one is the Born–Oppenheimer approximation. As a result of this approximation we can separate the rovibronic Schrödinger equation into two parts: one part (the electronic Schrödinger equation) in which the electronic coordinates are variables, and the other (the rotation–vibration Schrödinger equation) in which the nuclear coordinates are the variables. In solving the electronic Schrödinger equation the molecular orbital approximation can be made; this leads to a separate Schrödinger equation for each electron and to electronic wavefunctions that are the product of one-electron molecular orbital functions. In solving the rotation–vibration Schrödinger equation we make the rigid rotor approximation and the harmonic oscillator approximation. The approximate rotation–vibration Schrödinger equation obtained is separable, and the eigenfunctions are each the product of a rotational wavefunction in three variables and a vibrational wavefunction that is the product of $(3N - 6)$ harmonic oscillator wavefunctions, where there are N nuclei in the molecule [for a linear molecule the rotational wavefunction involves two coordinates and the vibrational one involves $(3N - 5)$ coordinates]. These approximations are all made on the basis of the phenomenology of molecular behavior rather than on the basis of an abstract mathematical study of the partial differential equations involved.

We must be able to correct for the approximations made in the indirect method of solving Eq. (7-1) if we are to achieve a precise solution. We must allow for anharmonicity, centrifugal distortion, and Coriolis coupling in the rotation–vibration problem (usually by using perturbation theory) and for electron correlation in the electronic problem (usually by using variation theory). Finally corrections must be made to allow for the breakdown of the Born–Oppenheimer approximation. For our purposes of symmetry labeling molecular energy levels the form of the approximate wavefunctions will be very important since it is from these that we will obtain the symmetry labels.

Before making the above approximations it is necessary to find the most appropriate $(3l - 3)$ coordinates for the Schrödinger equation. We change coordinates in order to facilitate the separation of variables in the equation that is obtained after making the approximations, and the choice of coordinates is nearly as important as the choice of approximation. We choose the coordinates with an eye to the approximations we plan to make, and the choice of the most appropriate coordinates in the rovibronic Schrödinger equation is the subject of this chapter.

TWO METHODS FOR CHANGING COORDINATES
IN A SCHRÖDINGER EQUATION

In this section we discuss two methods for changing coordinates in a Schrödinger equation, and we use the Schrödinger equation of the hydrogen atom as a convenient example. The coordinate change will be from Cartesian coordinates to polar coordinates:

(I)   The first method for changing coordinates involves (a) setting up the classical Hamiltonian in Cartesian coordinates, (b) using the quantum mechanical postulates to derive the quantum mechanical Hamiltonian, and (c) changing coordinates in the resultant Schrödinger equation.

(II) The second method for changing coordinates involves (a) setting up the classical Hamiltonian in Cartesian coordinates as before, (b) changing coordinates in the classical Hamiltonian expression, and (c) using the quantum mechanical postulates plus the *Podolsky trick* to change this to the quantum mechanical Hamiltonian in the new coordinates.

Both methods lead to the same wave equation but when many curvilinear coordinates are involved it is often much quicker to use Method II. Method I was used in Chapter 6 when we changed from $(X_1, Y_1, Z_1, \ldots, X_l, Y_l, Z_l)$ to $(X_0, Y_0, Z_0, X_2, Y_2, Z_2, \ldots, X_l, Y_l, Z_l)$ coordinates in order to achieve a separation of the translational part of the Hamiltonian [see Eqs. (6-2)–(6-18)].

Method I

Using a space fixed (X, Y, Z) axis system for the proton (1) and electron (2) of a hydrogen atom, the classical energy is given by

$$E = \frac{m_1}{2}(\dot{X}_1{}^2 + \dot{Y}_1{}^2 + \dot{Z}_1{}^2) + \frac{m_2}{2}(\dot{X}_2{}^2 + \dot{Y}_2{}^2 + \dot{Z}_2{}^2)$$

$$-\frac{e^2}{[(X_1 - X_2)^2 + (Y_1 - Y_2)^2 + (Z_1 - Z_2)^2]^{1/2}}. \tag{7-2}$$

In order to use the quantum mechanical postulates for obtaining the quantum mechanical Hamiltonian operator we must put E into Hamiltonian form, i.e., express it in terms of coordinates and momenta, and this expression is

$$H = \frac{1}{2m_1}(P_{X_1}^2 + P_{Y_1}^2 + P_{Z_1}^2) + \frac{1}{2m_2}(P_{X_2}^2 + P_{Y_2}^2 + P_{Z_2}^2)$$

$$-\frac{e^2}{[(X_1 - X_2)^2 + (Y_1 - Y_2)^2 + (Z_1 - Z_2)^2]^{1/2}}, \tag{7-3}$$

where $P_{X_1} = m_1\dot{X}_1$, etc. Using the quantum mechanical postulates we replace P_{X_1} by $-i\hbar\,\partial/\partial X_1$ etc. to obtain

$$\hat{H} = -\frac{\hbar^2}{2m_1}\left(\frac{\partial^2}{\partial X_1{}^2} + \frac{\partial^2}{\partial Y_1{}^2} + \frac{\partial^2}{\partial Z_1{}^2}\right) - \frac{\hbar^2}{2m_2}\left(\frac{\partial^2}{\partial X_2{}^2} + \frac{\partial^2}{\partial Y_2{}^2} + \frac{\partial^2}{\partial Z_2{}^2}\right)$$

$$-\frac{e^2}{[(X_1 - X_2)^2 + (Y_1 - Y_2)^2 + (Z_1 - Z_2)^2]^{1/2}} \tag{7-4}$$

and the Schrödinger equation is

$$\hat{H}\Phi_n(X_1, Y_1, Z_1, X_2, Y_2, Z_2) = E_n\Phi_n(X_1, Y_1, Z_1, X_2, Y_2, Z_2). \tag{7-5}$$

Having obtained the Schrödinger equation, Eq. (7-5), we now change coordinates in it in a manner that allows a separation of the equation into explicitly soluble parts. The first coordinate change we make is from $(X_1, Y_1, Z_1, X_2, Y_2, Z_2)$ to (X_0, Y_0, Z_0, X, Y, Z), where the coordinates of the center of mass are (X_0, Y_0, Z_0) in the (X, Y, Z) axis system, and the coordinates of the electron are (X, Y, Z) in an (X, Y, Z) axis system parallel to the (X, Y, Z) axis system but with origin at (X_0, Y_0, Z_0). We make this coordinate change since we know that the potential function is independent of the location of the center of mass so that in these coordinates we are able to separate the translational motion [see Eq. (6-29)] leaving the Schrödinger equation for the internal motion [see Eq. (6-32)] as

$$\left(-\frac{\hbar^2}{2m_2}\left[\frac{\partial^2}{\partial X^2} + \frac{\partial^2}{\partial Y^2} + \frac{\partial^2}{\partial Z^2} \right] - \frac{e^2}{(X^2 + Y^2 + Z^2)^{1/2}} \right)\Phi_{\text{int}}^{(n)}(X, Y, Z)$$
$$= E_{\text{int}}^{(n)}\Phi_{\text{int}}^{(n)}(X, Y, Z), \tag{7-6}$$

where $n = 1, 2, 3, \ldots$ labels the different solutions, and we are neglecting (m_2/m_1) relative to unity. Since the potential energy only depends on R, the separation of the proton and electron, we can separate the variables further by changing from Cartesian coordinates (X, Y, Z) to polar coordinates (R, θ, ϕ), where (θ, ϕ) are the polar angles in the (X, Y, Z) axis system of the vector pointing from the proton to the electron, i.e.,

$$X = R \sin \theta \cos \phi, \tag{7-7}$$
$$Y = R \sin \theta \sin \phi, \tag{7-8}$$

and

$$Z = R \cos \theta. \tag{7-9}$$

This coordinate change in a partial differential equation presents some problems since it is to a *curvilinear* coordinate system but this is well understood [see pages 172–178 in Margenau and Murphy (1956)]. The Schrödinger equation now becomes

$$\left\{ -\frac{\hbar^2}{2m_2}\left[\frac{1}{R^2}\frac{\partial}{\partial R}\left(R^2 \frac{\partial}{\partial R} \right) + \frac{1}{R^2 \sin \theta}\frac{\partial}{\partial \theta}\left(\sin \theta \frac{\partial}{\partial \theta} \right) + \frac{1}{R^2 \sin^2 \theta}\frac{\partial^2}{\partial \phi^2} \right] - \frac{e^2}{R} \right\}$$
$$\times \Phi_{\text{int}}^{(n)}(R, \theta, \phi) = E_{\text{int}}^{(n)}\Phi_{\text{int}}^{(n)}(R, \theta, \phi), \tag{7-10}$$

where the $\Phi_{\text{int}}^{(n)}$ are normalized with

$$\int_0^{2\pi}\int_0^{\pi}\int_0^{\infty} \overline{\Phi_{\text{int}}^{(n)}}\Phi_{\text{int}}^{(m)}R^2 \sin \theta \, dR \, d\theta \, d\phi = \delta_{nm}. \tag{7-11}$$

The two important facts about such a coordinate change are the transformation properties of the differential operator [i.e., the identity of the

terms within the square brackets from Eqs. (7-6) and (7-10)], and the volume
element for the integration ($d\tau = dX\,dY\,dZ = R^2 \sin\theta\,dR\,d\theta\,d\phi$). The chain
rule [see Eqs. (6-8) and (6-9)] is used in determining the transformations
that occur in the differential operators as a result of a coordinate change,
and standard formulas are used to determine the volume element in the
new coordinates [see, for example, Eq. (5-7) in Margenau and Murphy
(1956)].

Equation (7-10) can be separated into a radial (R) part and an angular
(θ, ϕ) part [see, for example, Chapter VI of Eyring, Walter, and Kimball
(1944)], and each of these separate Schrödinger equations can be solved
explicitly. In the rovibronic Schrödinger equation of a molecule it is necessary to make some approximations, after making the coordinate changes,
in order to separate the variables.

Method II

The above technique for obtaining the wave equation in polar coordinates
for a hydrogen atom involves changing coordinates in a partial differential
equation. We now discuss the alternative procedure (II) in which we make
all the coordinate changes in the classical Hamiltonian expression. This
alternative procedure is used in the derivation of the rotation–vibration
Hamiltonian of a polyatomic molecule [see Chapter 11 of Wilson, Decius,
and Cross (1955)].

Changing coordinates from $(X_1, Y_1, Z_1, X_2, Y_2, Z_2)$ to (X_0, Y_0, Z_0, X, Y, Z)
in the classical energy expression of Eq. (7-2) gives

$$E = \frac{m_1 + m_2}{2}(\dot{X}_0{}^2 + \dot{Y}_0{}^2 + \dot{Z}_0{}^2) + \frac{m_2}{2}(\dot{X}^2 + \dot{Y}^2 + \dot{Z}^2)$$
$$- \frac{e^2}{(X^2 + Y^2 + Z^2)^{1/2}}, \tag{7-12}$$

where we again neglect (m_2/m_1) relative to unity. After writing $\dot{X}_0 = P_{X_0}/(m_2 + m_1)$ and replacing P_{X_0} by $\hat{P}_{X_0} = -i\hbar\,\partial/\partial X_0$, etc., the first term
in Eq. (7-12) leads to the translational Hamiltonian operator; the remainder
is the internal energy E_{int}. Changing to polar coordinates [see Eqs. (7-7)–
(7-9)] we obtain

$$E_{\text{int}} = \frac{m_2}{2}(\dot{R}^2 + R^2\dot{\theta}^2 + R^2\sin^2\theta\dot{\phi}^2) - \frac{e^2}{R}. \tag{7-13}$$

We must now express this in Hamiltonian form, i.e., in terms of the coordinates and their *conjugate momenta*, before using the quantum mechanical
postulates. The momentum p_i conjugate to the coordinate $\dot{q}_i$, in a system
having kinetic energy T and potential energy V, is defined by the equation

[see, for example, Chapter II in Eyring, Walter, and Kimball (1944)],

$$p_i = \partial(T - V)/\partial \dot{q}_i. \tag{7-14}$$

If the system is such that V is independent of the velocities $\dot{q}_i$ then V can be omitted in this equation. In the hydrogen atom the conjugate momenta are

$$P_R = \frac{\partial T}{\partial \dot{R}} = m_2 \dot{R}, \tag{7-15}$$

$$P_\theta = \frac{\partial T}{\partial \dot{\theta}} = m_2 R^2 \dot{\theta}, \tag{7-16}$$

and

$$P_\phi = \frac{\partial T}{\partial \dot{\phi}} = m_2 R^2 \sin^2 \theta \dot{\phi}. \tag{7-17}$$

Replacing the velocities in Eq. (7-13) by the momenta, using Eqs. (7-15)–(7-17), we obtain the classical energy in Hamiltonian form as

$$H_{int} = \frac{P_R{}^2}{2m_2} + \frac{P_\theta{}^2}{2m_2 R^2} + \frac{P_\phi{}^2}{2m_2 R^2 \sin^2 \theta} - \frac{e^2}{R}. \tag{7-18}$$

To convert to the quantum mechanical Hamiltonian operator we might think that we could make the replacements

$$P_R \to \hat{P}_R = -i\hbar \frac{\partial}{\partial R}, \tag{7-19}$$

$$P_\theta \to \hat{P}_\theta = -i\hbar \frac{\partial}{\partial \theta}, \tag{7-20}$$

and

$$P_\phi \to \hat{P}_\phi = -i\hbar \frac{\partial}{\partial \phi}, \tag{7-21}$$

which would give

$$\hat{H}_{int} = -\frac{\hbar^2}{2m_2} \frac{\partial^2}{\partial R^2} - \frac{\hbar^2}{2m_2 R^2} \frac{\partial^2}{\partial \theta^2} - \frac{\hbar^2}{2m_2 R^2 \sin^2 \theta} \frac{\partial^2}{\partial \phi^2} - \frac{e^2}{R}. \tag{7-22}$$

Equation (7-22) is not the same as the Hamiltonian operator in Eq. (7-10), since in Eq. (7-10) R^2 occurs between the $\partial/\partial R$ factors and $\sin \theta$ between the $\partial/\partial \theta$ factors. This method for changing coordinates has not led to the correct Schrödinger equation.

We see from the above that although the translational Hamiltonian operator is obtained correctly from the classical Hamiltonian expression by making the replacements $P_{X_0} = -ih\,\partial/\partial X_0$, etc., the internal Hamiltonian operator is not obtained correctly from the classical expression in Eq. (7-18) by making the replacement of P_R by $-ih\,\partial/\partial R$, etc. The reason for this is that to obtain the Hamiltonian operator from the classical Hamiltonian expression by using the quantum mechanical postulates we can only replace the momentum p_i by $-ih\,\partial/\partial q_i$ (where p_i is conjugate to q_i) if q_i is a Cartesian coordinate. In our case R, θ, and ϕ are not Cartesian coordinates.

All is not lost, however, since Podolsky (1928) has shown that we can still set up the classical Hamiltonian in terms of general coordinates q_i and conjugate momenta p_i and replace the p_i by $-ih\,\partial/\partial q_i$ to obtain the Hamiltonian operator, as long as we are careful about how we arrange the classical expression. Working backwards from the correct Hamiltonian operator in q_i and $-ih\,\partial/\partial q_i$ we can see how it is necessary to arrange the way we write the classical Hamiltonian in terms of p_i and q_i so that replacement of p_i by $-ih\,\partial/\partial q_i$ will yield the correct Hamiltonian operator. This recipe, or *Podolsky trick*, will now be explained [see also Smith (1934), Section 11-3 in Wilson, Decius, and Cross (1955), and the Appendix of Watson (1970)].

We can write the classical energy of a system of particles in general coordinates q and velocities $\dot{q}$ as

$$E = \frac{1}{2}\sum_{i,j} g_{ij}\dot{q}_i\dot{q}_j + V(q_i), \qquad (7\text{-}23)$$

where the coefficients g_{ij} can be functions of the coordinates q_i. Expressing this in terms of momenta p_i conjugate to the coordinates [see Eq. (7-14)] we obtain the energy in Hamiltonian form as

$$H = \frac{1}{2}\sum_{i,j} g^{ij}p_ip_j + V(q_i), \qquad (7\text{-}24)$$

where the matrix of g^{ij} is the inverse of the matrix of g_{ij}. At this stage we cannot say that the Hamiltonian operator is obtained from this by replacing p_i by $-ih\,\partial/\partial q_i$, etc. unless all the coordinates are Cartesian. Podolsky (1928) showed that by rewriting H in the form

$$H = \frac{1}{2}g^{1/4}\sum_{i,j} p_i g^{-1/2}g^{ij}p_j g^{1/4} + V(q_i), \qquad (7\text{-}25)$$

where g is the determinant of the matrix of g^{ij}, the Hamiltonian operator *is* correctly obtained by replacing p_i by $-ih\,\partial/\partial q_i$, etc. Since the momenta and coordinates all commute in the classical expressions we see that Eqs. (7-24) and (7-25) are identical. However, in the derived Hamiltonian operator,

$$\hat{H} = \frac{1}{2}g^{1/4}\sum_{i,j} \left(-ih\frac{\partial}{\partial q_i}\right)g^{-1/2}g^{ij}\left(-ih\frac{\partial}{\partial q_j}\right)g^{1/4} + V(q_i), \qquad (7\text{-}26)$$

the operators $\partial/\partial q_i$ and $\partial/\partial q_j$ will not commute with the coordinates, and so Eq. (7-26) is not the same as would be obtained by replacing p_i by $-i\hbar\,\partial/\partial q_i$, etc., in Eq. (7-24).

One final point concerns the volume element in the integration of the eigenfunctions of the Hamiltonian operator in Eq. (7-26). The volume element using $\hat{H}$ of Eq. (7-26) will be $dq_1\,dq_2\cdots dq_n$, but we want to obtain eigenfunctions normalized with volume element $d\tau = dX_1\,dY_1\,dZ_1\cdots$ in general. We write this volume element as $d\tau = s\,dq_1\,dq_2\cdots dq_n$, where s is the *weight factor* involved in the coordinate transformation from Cartesian to general coordinates. To obtain eigenfunctions normalized with the appropriate volume element we must write the classical Hamiltonian in the form

$$H = \frac{1}{2}\,s^{-1/2}g^{1/4}\left\{\sum_{i,j} p_i g^{-1/2}g^{ij}p_j\right\}g^{1/4}s^{1/2} + V(q_i), \qquad (7\text{-}27)$$

before making the replacements $p_i = -i\hbar\,\partial/\partial q_i$, etc., to give $\hat{H}$. The weight factor is easy to determine [see Eq. (5-7) in Margenau and Murphy (1956)], and for the coordinates used in the Hamiltonian for the hydrogen atom

$$s = R^2 \sin\theta. \qquad (7\text{-}28)$$

Problem 7-1. Check that the Podolsky trick works by using it to obtain the Hamiltonian operator for the hydrogen atom from the classical expression of Eq. (7-18) and comparing the result with the Hamiltonian obtained in Eq. (7-10) by the Method I derivation.

Answer. To use the Podolsky trick it is necessary to determine g and to rewrite the Hamiltonian of Eq. (7-18) in the form of Eq. (7-27). From Eq. (7-18) we see that

$$g^{RR} = m_2^{-1}, \qquad (7\text{-}29)$$

$$g^{\theta\theta} = m_2^{-1}R^{-2}, \qquad (7\text{-}30)$$

and

$$g^{\phi\phi} = m_2^{-1}(R \sin\theta)^{-2}; \qquad (7\text{-}31)$$

all other elements g^{ij} vanish. Using these results, and Eq. (7-28) for the weight factor s, we deduce that

$$g = m_2^{-3}R^{-4}(\sin\theta)^{-2}, \qquad (7\text{-}32)$$

and that Eq. (7-18) written in the form of Eq. (7-27) is

$$H = \tfrac{1}{2}[m_2^{-3/4}R^{-2}(\sin\theta)^{-1}]\{P_R[m_2^{3/2}R^2 \sin\theta]m_2^{-1}P_R$$
$$+ P_\theta[m_2^{3/2}R^2 \sin\theta]m_2^{-1}R^{-2}P_\theta$$
$$+ P_\phi[m_2^{3/2}R^2 \sin\theta]m_2^{-1}(R \sin\theta)^{-2}P_\phi\}m_2^{-3/4} - e^2/R. \qquad (7\text{-}33)$$

Since this is in the form prescribed by Podolsky we can obtain the Hamiltonian operator by making the replacements of Eqs. (7-19)–(7-21), which gives

$$
\hat{H} = -\frac{\hbar^2}{2m_2} R^{-2}(\sin\theta)^{-1}\left[\frac{\partial}{\partial R}R^2\sin\theta\frac{\partial}{\partial R}+\frac{\partial}{\partial\theta}\sin\theta\frac{\partial}{\partial\theta}\right.
$$

$$
\left.+\frac{\partial}{\partial\phi}(\sin\theta)^{-1}\frac{\partial}{\partial\phi}\right]-\frac{e^2}{R}
\tag{7-34}
$$

$$
= -\frac{\hbar^2}{2m_2}\left[\frac{1}{R^2}\frac{\partial}{\partial R}\left(R^2\frac{\partial}{\partial R}\right)+\frac{1}{R^2\sin\theta}\frac{\partial}{\partial\theta}\left(\sin\theta\frac{\partial}{\partial\theta}\right)\right.
$$

$$
\left.+\frac{1}{R^2\sin^2\theta}\frac{\partial^2}{\partial\phi^2}\right]-\frac{e^2}{R}.
\tag{7-35}
$$

This is the same as the Hamiltonian operator in Eq. (7-10) obtained by Method I.

Problem 7-2. Change coordinates in Eq. (7-1) so that the coordinates of all particles are referred to an (ξ, η, ζ) axis system parallel to the (X, Y, Z) axis system but with origin at the nuclear center of mass.

Answer. The coordinate change we are making is from the $(3l-3)$ coordinates $(X_2, Y_2, Z_2, \ldots, Z_l)$ to the $(3l-3)$ coordinates $(\xi_2, \eta_2, \zeta_2, \ldots, \zeta_l)$, and since this involves only Cartesian coordinates we will make the coordinate change directly in the Schrödinger equation (i.e., Method I).

In the molecule, particles labeled $1, 2, \ldots, N$ are nuclei and particles labeled $N+1, N+2, \ldots, N+n$ are electrons; there are N nuclei, n electrons, and $N + n = l$. The (X_i, Y_i, Z_i) and (ξ_i, η_i, ζ_i) coordinates of a particle in the molecule are related by

$$
\xi_i = X_i - X_{\text{NCM}}, \quad \eta_i = Y_i - Y_{\text{NCM}}, \quad \text{and} \quad \zeta_i = Z_i - Z_{\text{NCM}}, \tag{7-36}
$$

where $(X_{\text{NCM}}, Y_{\text{NCM}}, Z_{\text{NCM}})$ are the coordinates of the nuclear center of mass in the (X, Y, Z) axis system. If we let M_N be the total mass of all the nuclei in the molecule then we have

$$
\xi_i = X_i + \frac{m_e}{M_N}\sum_{r=N+1}^{l} X_r, \tag{7-37}
$$

with similar equations for η_i and ζ_i (in the following equations we will only write down the equations involving ξ_i since those involving η_i and ζ_i are then obvious). From the chain rule, Eq. (6-8), we have

$$
\frac{\partial}{\partial X_r} = \sum_{i=2}^{l}\frac{\partial\xi_i}{\partial X_r}\frac{\partial}{\partial\xi_i}, \tag{7-38}
$$

where $r \neq 1$, and using Eq. (7-37) this gives

$$\frac{\partial}{\partial X_r} = \sum_{i=2}^{l} \left(\delta_{ir} + \frac{m_e}{M_N} \delta_{er} \right) \frac{\partial}{\partial \xi_i}, \qquad (7\text{-}39)$$

where δ_{ir} is the Kronecker delta, and we have introduced δ_{er} which is such that $\delta_{er} = 1$ if r labels an electron (i.e., $r = N + 1, N + 2, \ldots, l$), and $\delta_{er} = 0$ if r labels a nucleus (i.e., $r = 2, \ldots, N$). Further application of the chain rule gives

$$\frac{\partial^2}{\partial X_r^2} = \frac{\partial^2}{\partial \xi_r^2} + \frac{2m_e}{M_N} \delta_{er} \sum_{j=2}^{l} \frac{\partial^2}{\partial \xi_r \, \partial \xi_j} + \left(\frac{m_e}{M_N} \right)^2 \delta_{er} \sum_{i,j=2}^{l} \frac{\partial^2}{\partial \xi_i \, \partial \xi_j} \qquad (7\text{-}40)$$

and

$$\frac{\partial^2}{\partial X_r \, \partial X_s} = \frac{\partial^2}{\partial \xi_r \, \partial \xi_s} + \frac{m_e}{M_N} \left[\delta_{es} \sum_{j=2}^{l} \frac{\partial^2}{\partial \xi_r \, \partial \xi_j} + \delta_{er} \sum_{i=2}^{l} \frac{\partial^2}{\partial \xi_i \, \partial \xi_s} \right]$$
$$+ \left(\frac{m_e}{M_N} \right)^2 \delta_{er} \delta_{es} \sum_{i,j=2}^{l} \frac{\partial^2}{\partial \xi_i \, \partial \xi_j}, \qquad (7\text{-}41)$$

where $r, s \neq 1$. Summing these expressions and making use of the reductions occurring because of δ_{er} and δ_{es} we obtain (where $M = M_N + nm_e$)

$$\sum_{r,s=2}^{l} \frac{\partial^2}{\partial X_r \, \partial X_s} = \left(\frac{M}{M_N} \right)^2 \sum_{i,j=2}^{l} \frac{\partial^2}{\partial \xi_i \, \partial \xi_j} \qquad (7\text{-}42)$$

and

$$\sum_{r=2}^{l} \frac{1}{m_r} \frac{\partial^2}{\partial X_r^2} = \sum_{i=2}^{l} \frac{1}{m_i} \frac{\partial^2}{\partial \xi_i^2} + \frac{2}{M_N} \sum_{i=N+1}^{l} \sum_{j=2}^{l} \frac{\partial^2}{\partial \xi_i \, \partial \xi_j} + \frac{nm_e}{M_N^2} \sum_{i,j=2}^{l} \frac{\partial^2}{\partial \xi_i \, \partial \xi_j}. \qquad (7\text{-}43)$$

Combining these two equations we have

$$-\sum_{r=2}^{l} \frac{1}{m_r} \frac{\partial^2}{\partial X_r^2} + \frac{1}{M} \sum_{r,s=2}^{l} \frac{\partial^2}{\partial X_r \, \partial X_s}$$
$$= -\sum_{i=2}^{l} \frac{1}{m_i} \frac{\partial^2}{\partial \xi_i^2} + \frac{1}{M_N} \left[\sum_{i,j=2}^{N} \frac{\partial^2}{\partial \xi_i \, \partial \xi_j} - \sum_{i,j=N+1}^{l} \frac{\partial^2}{\partial \xi_i \, \partial \xi_j} \right]. \qquad (7\text{-}44)$$

Using this equation together with the similar ones involving the Y, Z, η, and ζ coordinates we obtain the rovibronic wave equation in this new coordinate frame as

$$(\hat{T}_e + \hat{T}_N + V - E_{rve}) \Phi_{rve}(\xi_2, \eta_2, \zeta_2, \ldots, \zeta_l) = 0, \qquad (7\text{-}45)$$

where

$$\hat{T}_e = -\frac{\hbar^2}{2m_e} \sum_{i=N+1}^{l} \mathbf{V}_i^2 - \frac{\hbar^2}{2M_N} \sum_{i,j=N+1}^{l} \mathbf{V}_i \cdot \mathbf{V}_j, \tag{7-46}$$

$$\hat{T}_N = -\frac{\hbar^2}{2} \sum_{i=2}^{N} \frac{\mathbf{V}_i^2}{m_i} + \frac{\hbar^2}{2M_N} \sum_{i,j=2}^{N} \mathbf{V}_i \cdot \mathbf{V}_j, \tag{7-47}$$

$$\mathbf{V}_i^2 = \frac{\partial^2}{\partial \xi_i^2} + \frac{\partial^2}{\partial \eta_i^2} + \frac{\partial^2}{\partial \zeta_i^2}, \tag{7-48}$$

and

$$\mathbf{V}_i \cdot \mathbf{V}_j = \frac{\partial^2}{\partial \xi_i \partial \xi_j} + \frac{\partial^2}{\partial \eta_i \partial \eta_j} + \frac{\partial^2}{\partial \zeta_i \partial \zeta_j}. \tag{7-49}$$

The kinetic energy is completely separable into an electronic part $\hat{T}_e$ and a nuclear part $\hat{T}_N$ in these coordinates.

INTRODUCTION TO THE MOLECULE FIXED AXIS SYSTEM

In making the Born–Oppenheimer approximation we refer the coordinates of the electrons to the nuclear frame. To do this and to introduce nuclear rotational and vibrational coordinates, it is necessary to introduce a set of axes that are attached in some manner to the nuclear framework and that rotate as the nuclear framework rotates. Such axes ought to be called nuclear fixed axes but they have come to be called *molecule fixed axes*. The molecule fixed axis system (x, y, z) is defined as having origin at the nuclear center of mass and an orientation away from the (ξ, η, ζ) axis system that is determined by the nuclear coordinates (i.e., by the values of $\xi_2, \eta_2, \zeta_2, \ldots, \xi_N, \eta_N, \zeta_N$). The (ξ, η, ζ) axis system was introduced in Problem 7-2. In this section the (x, y, z) axis system is discussed in general and the effect on the electron kinetic energy operator $\hat{T}_e$ [see Eq. (7-46)] of changing from (ξ, η, ζ) coordinates to (x, y, z) coordinates is determined. The effect of this coordinate change on the nuclear kinetic energy operator $\hat{T}_N$ [see Eq. (7-47)] is more difficult to determine and is the subject of later sections of this chapter although it is discussed in general terms here.

To relate the nuclear and electron coordinates in the (ξ, η, ζ) axis system to those in the (x, y, z) axis system we use Euler angles. The Euler angles θ, ϕ, and χ are defined in Fig. 7-1 and we follow the convention used in Wilson, Decius, and Cross (1955). We use θ, ϕ, and χ to define the orientation of the (x, y, z) axis system relative to the (ξ, η, ζ) axis system in the same way that β, α, and γ are used in Fig. 6-1 to relate the orientation of one space fixed axis system (X', Y', Z') to another space fixed axis system (X, Y, Z). The Euler angles are restricted to the ranges $0 \le \theta \le \pi$, $0 \le \phi \le 2\pi$, $0 \le \chi \le 2\pi$. For

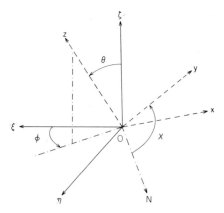

Fig. 7-1. The definition of the Euler angles (θ, ϕ, χ) that relate the orientation of the molecule fixed (x, y, z) axes to the (ξ, η, ζ) axes. The origin of both axis systems is at the nuclear center of mass O, and the node line ON is directed so that a right handed screw is driven along ON in its positive direction by twisting it from ζ to z through θ where $0 \le \theta \le \pi$. ϕ and χ have the ranges 0 to 2π. χ is measured from the node line.

any particle in the molecule we can relate its coordinates in the (ξ, η, ζ) and (x, y, z) axis systems by writing

$$\begin{bmatrix} x_i \\ y_i \\ z_i \end{bmatrix} = \begin{bmatrix} \lambda_{x\xi} & \lambda_{x\eta} & \lambda_{x\zeta} \\ \lambda_{y\xi} & \lambda_{y\eta} & \lambda_{y\zeta} \\ \lambda_{z\xi} & \lambda_{z\eta} & \lambda_{z\zeta} \end{bmatrix} \begin{bmatrix} \xi_i \\ \eta_i \\ \zeta_i \end{bmatrix}, \tag{7-50}$$

where

$$\lambda_{x\xi} = \cos(xO\xi), \quad \text{etc.,} \tag{7-51}$$

and these are elements of the direction cosine matrix. The direction cosines can be expressed in terms of the Euler angles as follows [see Table I-1 in Wilson, Decius, and Cross (1955)]:

$$\lambda_{x\xi} = \cos\theta\cos\phi\cos\chi - \sin\phi\sin\chi, \quad \lambda_{x\eta} = \cos\theta\sin\phi\cos\chi + \cos\phi\sin\chi,$$
$$\lambda_{y\xi} = -\cos\theta\cos\phi\sin\chi - \sin\phi\cos\chi, \quad \lambda_{y\eta} = -\cos\theta\sin\phi\sin\chi + \cos\phi\cos\chi,$$
$$\lambda_{z\xi} = \sin\theta\cos\phi, \qquad\qquad\qquad\qquad \lambda_{z\eta} = \sin\theta\sin\phi,$$

$$\lambda_{x\zeta} = -\sin\theta\cos\chi,$$
$$\lambda_{y\zeta} = \sin\theta\sin\chi, \tag{7-52}$$
$$\lambda_{z\zeta} = \cos\theta.$$

Notice that we have now introduced four Cartesian axis systems:

(a)   An (X, Y, Z) axis system fixed in space. This is the space fixed axis system.

(b)   An (X, Y, Z) axis system parallel to the (X, Y, Z) axis system and with its origin at the molecular center of mass. The molecular center of mass has coordinates (X_0, Y_0, Z_0) in the (X, Y, Z) axis system. This axis system is introduced to facilitate the separation of translation.

(c)   A (ξ, η, ζ) axis system parallel to the (X, Y, Z) and (X, Y, Z) axis systems and with its origin at the nuclear center of mass. The nuclear center of mass has coordinates $(X_{NCM}, Y_{NCM}, Z_{NCM})$ in the (X, Y, Z) axis system. This axis system is introduced to facilitate the separation of the electronic and nuclear coordinates.

(d)   An (x, y, z) axis system with its origin at the nuclear center of mass [as for the (ξ, η, ζ) axis system] and with orientation away from the (ξ, η, ζ) system defined by the nuclear coordinates and given by the Euler angles (θ, ϕ, χ). This is the molecule fixed axis system, and it is introduced to facilitate the separation of the rotational and vibrational coordinates.

We now start on the complicated business of changing coordinates from $(\xi_2, \eta_2, \zeta_2, \ldots, \xi_l, \eta_l, \zeta_l)$ in the Schrödinger equation of Eq. (7-45) to the rovibronic coordinates $(\theta, \phi, \chi, Q_1, \ldots, Q_{3N-6}, x_{N+1}, y_{N+1}, z_{N+1}, \ldots, z_l)$, where the Q_r (vibrational *normal coordinates*) are linear combinations of the (x, y, z) nuclear coordinates; for a linear molecule there are $(3N - 5)$ normal coordinates and two Euler angles. We first look at the effect of this coordinate change on the electronic kinetic energy operator $\hat{T}_e$, and since the Euler angles are independent of the (ξ, η, ζ) coordinates of the electrons we can make the coordinate change in $\hat{T}_e$ rather easily.

Using Eq. (7-50) for an electron we have

$$\frac{\partial x_k}{\partial \xi_j} = \delta_{kj} \lambda_{x\xi} \tag{7-53}$$

and

$$\frac{\partial^2 x_k}{\partial \xi_j^2} = 0, \tag{7-54}$$

where $k, j = N + 1, \ldots, l$, with similar equations for the derivatives involving y, z, η, and ζ. Using Eq. (6-9) we obtain

$$\frac{\partial^2}{\partial \xi_i^2} = \lambda_{x\xi}^2 \frac{\partial^2}{\partial x_i^2} + \lambda_{y\xi}^2 \frac{\partial^2}{\partial y_i^2} + \lambda_{z\xi}^2 \frac{\partial^2}{\partial z_i^2}, \tag{7-55}$$

$$\frac{\partial^2}{\partial \eta_i^2} = \lambda_{x\eta}^2 \frac{\partial^2}{\partial x_i^2} + \lambda_{y\eta}^2 \frac{\partial^2}{\partial y_i^2} + \lambda_{z\eta}^2 \frac{\partial^2}{\partial z_i^2}, \tag{7-56}$$

and

$$\frac{\partial^2}{\partial \zeta_i^2} = \lambda_{x\zeta}^2 \frac{\partial^2}{\partial x_i^2} + \lambda_{y\zeta}^2 \frac{\partial^2}{\partial y_i^2} + \lambda_{z\zeta}^2 \frac{\partial^2}{\partial z_i^2}. \tag{7-57}$$

Adding these three equations we now have [from Eq. (7-48)],

$$\mathbf{\nabla}_i^2 = \frac{\partial^2}{\partial x_i^2} + \frac{\partial^2}{\partial y_i^2} + \frac{\partial^2}{\partial z_i^2}. \tag{7-58}$$

Also it is easy to derive

$$\frac{\partial^2}{\partial \xi_i \, \partial \xi_j} = \sum_{\alpha,\beta} \lambda_{\alpha\xi} \lambda_{\beta\xi} \frac{\partial^2}{\partial \alpha_i \, \partial \beta_j}, \tag{7-59}$$

with similar equations for the η and ζ derivatives; adding these we obtain from Eq. (7-49),

$$\mathbf{V}_i \cdot \mathbf{V}_j = \frac{\partial^2}{\partial x_i \, \partial x_j} + \frac{\partial^2}{\partial y_i \, \partial y_j} + \frac{\partial^2}{\partial z_i \, \partial z_j}. \tag{7-60}$$

The simple results of Eqs. (7-58) and (7-60) follow because the direction cosine matrix is orthogonal.

We see that in the molecule fixed (x, y, z) axis system the electronic kinetic energy operator $\hat{T}_e$ for any molecule is given by Eq. (7-46) with substitutions from Eqs. (7-58) and (7-60). A rotation of the axis system in a manner defined purely by the nuclear coordinates has not changed the form of $\hat{T}_e$ and the use of the Method I coordinate transformation is very simple.

Let us attempt to follow the same procedure (Method I) for the nuclear kinetic energy operator $\hat{T}_N$. From Eq. (7-50) for the nuclear coordinates we have

$$\frac{\partial x_k}{\partial \xi_j} = \delta_{kj} \lambda_{x\xi} + \frac{\partial \lambda_{x\xi}}{\partial \xi_j} \xi_k + \frac{\partial \lambda_{x\eta}}{\partial \xi_j} \eta_k + \frac{\partial \lambda_{x\zeta}}{\partial \xi_j} \zeta_k \tag{7-61}$$

$$= \delta_{kj} \lambda_{x\xi} + (xk), \tag{7-62}$$

where $j = 2, \ldots, N$ and $k = 2, \ldots, l$; we have introduced (xk) as an abbreviation for the last three terms in Eq. (7-61). Similar expressions for $(\partial y_k/\partial \xi_j)$ and $(\partial z_k/\partial \xi_j)$ will involve (yk) and (zk) terms. Using Eq. (6-8) we can write (for a nucleus j)

$$\frac{\partial}{\partial \xi_j} = \sum_{k=2}^{l} \left[\left(\frac{\partial x_k}{\partial \xi_j} \right) \frac{\partial}{\partial x_k} + \left(\frac{\partial y_k}{\partial \xi_j} \right) \frac{\partial}{\partial y_k} + \left(\frac{\partial z_k}{\partial \xi_j} \right) \frac{\partial}{\partial z_k} \right] \tag{7-63}$$

$$= \lambda_{x\xi} \frac{\partial}{\partial x_j} + \lambda_{y\xi} \frac{\partial}{\partial y_j} + \lambda_{z\xi} \frac{\partial}{\partial z_j}$$

$$+ \sum_{k=2}^{l} \left[(xk) \frac{\partial}{\partial x_k} + (yk) \frac{\partial}{\partial y_k} + (zk) \frac{\partial}{\partial z_k} \right]. \tag{7-64}$$

Without writing the second derivatives it is clear that the terms involving (xk), (yk), and (zk) result in derivatives with respect to the electron coordinates being introduced into the expression for $\hat{T}_N$. Hence although by using (ξ, η, ζ) coordinates we achieve a complete separation of the electronic and nuclear coordinates in the kinetic energy operator $(\hat{T}_e + \hat{T}_N)$ when we change to (x, y, z) coordinates (in order to separate the rotational and vibrational coordinates) we introduce the electronic coordinates back into $\hat{T}_N$. The effect of the nuclear–electronic coupling terms introduced into $\hat{T}_N$ is, however, generally small and a price well worth paying for the simplification in the rotation–vibration Hamiltonian that is obtained by using the (x, y, z) co-ordinates. From the mathematics of the coordinate change we see that the derivatives with respect to electron coordinates occur in $\hat{T}_N$ because the (x, y, z) coordinates of the electrons depend on the (ξ, η, ζ) coordinates of the nuclei through the dependence of the direction cosine matrix elements on the nuclear coordinates. Physically we are now referring the electrons to the noninertial (x, y, z) axis system which rotates with the nuclei, and the electrons are subject to Coriolis forces in this axis system; this makes itself felt by the resultant introduction of terms into $\hat{T}_N$ coupling the electronic angular momentum with the rovibronic angular momentum [the terms involving the electron derivatives in $\hat{T}_N$ eventually reduce to this; see Eqs. (7-71)–(7-83) which follow].

In the next section we will derive $\hat{T}_N$ in the (x, y, z) coordinate system for a diatomic molecule in which the derivatives involved in (xk), (yk), and (zk) can be explicitly written down. For a polyatomic molecule these derivatives are more difficult to obtain and the coordinate change is more simply made by Method II.

THE DIATOMIC MOLECULE

Changing to Rovibronic Coordinates

For a diatomic molecule, in which the nuclei are labeled 1 and 2 and the n electrons $3, 4, \ldots, n + 2$, the molecule fixed (x, y, z) axis system is fixed so that the z axis points from nucleus 1 to nucleus 2. This defines the two rotational variables θ and ϕ; we choose $\chi = 0°$ in order to define the location of the x and y axes. The Euler angle χ is arbitrary since there are only two rotational degrees of freedom for the nuclei in a diatomic molecule. The nuclei and axis system are drawn in Fig. 7-2 where nucleus 1 is assumed

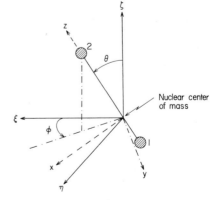

Fig. 7-2. The (x, y, z) axis system for a diatomic molecule where z is defined as pointing from nucleus 1 to nucleus 2, and χ is chosen as zero.

heavier than nucleus 2. We now wish to change coordinates in the wave equation from $(\xi_2, \eta_2, \zeta_2, \ldots, \zeta_{n+2})$ to $(\theta, \phi, R, x_3, y_3, z_3, \ldots, z_{n+2})$, where R is the internuclear distance. This coordinate change is made in the hope of separating the rotational (θ, ϕ), vibrational (R), and electronic (x_i, y_i, z_i) variables in the resultant wave equation, and we use Method I for this coordinate change.

From Fig. 7-2 and Eqs. (7-50) and (7-52) we see that the two sets of coordinates are related by

$$\theta = \arccos[\zeta_2(\xi_2^2 + \eta_2^2 + \zeta_2^2)^{-1/2}], \tag{7-65}$$

$$\phi = \arccos[\xi_2(\xi_2^2 + \eta_2^2)^{-1/2}] = \arcsin[\eta_2(\xi_2^2 + \eta_2^2)^{-1/2}], \tag{7-66}$$

$$R = [(m_1 + m_2)/m_1](\xi_2^2 + \eta_2^2 + \zeta_2^2)^{1/2}, \tag{7-67}$$

$$x_i = \xi_i \cos\theta \cos\phi + \eta_i \cos\theta \sin\phi - \zeta_i \sin\theta, \tag{7-68}$$

$$y_i = -\xi_i \sin\phi + \eta_i \cos\phi, \tag{7-69}$$

and

$$z_i = \xi_i \sin\theta \cos\phi + \eta_i \sin\theta \sin\phi + \zeta_i \cos\theta, \tag{7-70}$$

where $i = 3, 4, \ldots, n + 2$, and the restriction $0 \le \theta \le \pi$ holds. It is important to notice that we can calculate the Euler angles θ and ϕ directly from the nuclear coordinates (ξ_2, η_2, ζ_2) using Eqs. (7-65) and (7-66). Following the Method I coordinate change through for the nuclear kinetic energy operator $\hat{T}_N$ involves a lengthy but straightforward use of the chain rule and we obtain intermediate results such as

$$\frac{\partial}{\partial \xi_2} = \frac{m_1 + m_2}{m_1 R \sin\theta} \left\{ \sum_{k=3}^{n+2} \left[-\cos\theta(\sin\phi \, y_k + \sin\theta \cos\phi z_k) \frac{\partial}{\partial x_k} \right. \right.$$

$$+ \sin\phi(\cos\theta x_k + \sin\theta z_k) \frac{\partial}{\partial y_k} + \sin\theta(\cos\theta \cos\phi x_k - \sin\phi y_k) \frac{\partial}{\partial z_k} \right]$$

$$+ \cos\phi \cos\theta \sin\theta \frac{\partial}{\partial \theta} - \sin\phi \frac{\partial}{\partial \phi} + R \cos\phi \sin^2\theta \frac{\partial}{\partial R} \right\} \tag{7-71}$$

$$= \frac{m_1 + m_2}{m_1 R \sin\theta} \left[-i \sin\theta \sin\phi \left(\frac{\hat{L}_x}{\hbar}\right) - i \cos\theta \sin\theta \cos\phi \left(\frac{\hat{L}_y}{\hbar}\right) \right.$$

$$+ i \cos\theta \sin\phi \left(\frac{\hat{L}_z}{\hbar}\right) + \cos\phi \cos\theta \sin\theta \frac{\partial}{\partial \theta}$$

$$\left. - \sin\phi \frac{\partial}{\partial \phi} + R \cos\phi \sin^2\theta \frac{\partial}{\partial R} \right], \tag{7-72}$$

where the electronic angular momentum operators are defined by

$$\hat{L}_x = -i\hbar \sum_{k=3}^{n+2} \left(y_k \frac{\partial}{\partial z_k} - z_k \frac{\partial}{\partial y_k} \right), \tag{7-73}$$

and cyclically for $\hat{L}_y$ and $\hat{L}_z$. From Eq. (7-47) we deduce that for a diatomic molecule

$$\hat{T}_N = -\frac{\hbar^2}{2m_2} \mathbf{V}_2{}^2 + \frac{\hbar^2}{2(m_1 + m_2)} \mathbf{V}_2{}^2$$

$$= -\frac{\hbar^2 m_1}{2m_2(m_1 + m_2)} \mathbf{V}_2{}^2. \tag{7-74}$$

Changing coordinates using the chain rule [see Eq. (7-72)] gives

$$\hat{T}_N = -\frac{\hbar^2}{2\mu} \mathbf{V}_{R\theta\phi}^2, \tag{7-75}$$

where

$$\mu = \frac{m_1 m_2}{m_1 + m_2} \tag{7-76}$$

and

$$\mathbf{V}_{R\theta\phi}^2 = \frac{1}{R^2} \frac{\partial}{\partial R}\left(R^2 \frac{\partial}{\partial R} \right) + \frac{1}{R^2 \sin^2\theta} \left[\frac{\partial}{\partial \phi} + i\frac{\hat{L}_x}{\hbar} \sin\theta - i\frac{\hat{L}_z}{\hbar} \cos\theta \right]^2$$

$$+ \frac{1}{R^2 \sin\theta} \left[\frac{\partial}{\partial \theta} - i\frac{\hat{L}_y}{\hbar} \right] \sin\theta \left[\frac{\partial}{\partial \theta} - i\frac{\hat{L}_y}{\hbar} \right]. \tag{7-77}$$

The electronic angular momentum has crept into the nuclear kinetic energy operator, and it is instructive to introduce (x, y, z) components of the rovibronic angular momentum operator in place of the operators $\partial/\partial\theta$ and $\partial/\partial\phi$. The rovibronic angular momentum operator[1] $\hat{\boldsymbol{J}}$ has components in the (ξ, η, ζ) axis system given by

$$\hat{J}_\xi = -i\hbar \sum_{j=2}^{n+2} \left(\eta_j \frac{\partial}{\partial \zeta_j} - \zeta_j \frac{\partial}{\partial \eta_j} \right) \tag{7-78}$$

and cyclically for $\hat{J}_\eta$ and $\hat{J}_\zeta$ [see, for example, Eq. (3-97) in Eyring, Walter, and Kimball (1944)]. These three operators do not commute with each other

---

[1] $\hat{\boldsymbol{J}}$ is used here for the rovibronic angular momentum operator instead of the more correct $\hat{N}$ [see the discussion after Eq. (6-79)]. This is in order to follow accepted usage in the rovibronic Hamiltonian. For a singlet electronic state $\hat{\boldsymbol{J}} \equiv \hat{N}$.

and we have

$$[\hat{J}_\xi, \hat{J}_\eta] = \hat{J}_\xi \hat{J}_\eta - \hat{J}_\eta \hat{J}_\xi = i\hbar \hat{J}_\zeta \tag{7-79}$$

with two similar commutators obtained by cyclically permuting $\hat{J}_\xi, \hat{J}_\eta$, and $\hat{J}_\zeta$. Using equations such as Eq. (7-72), the inverses of Eqs. (7-65)–(7-70), and the direction cosine matrix with $\chi = 0°$ to obtain $\hat{J}_x, \hat{J}_y$, and $\hat{J}_z$ from $\hat{J}_\xi, \hat{J}_\eta$, and $\hat{J}_\zeta$, we find that

$$\hat{J}_x = i\hbar \csc\theta \frac{\partial}{\partial\phi} + \cot\theta \hat{L}_z, \tag{7-80}$$

$$\hat{J}_y = -i\hbar \frac{\partial}{\partial\theta}, \tag{7-81}$$

and

$$\hat{J}_z = \hat{L}_z. \tag{7-82}$$

Substituting these equations into Eq. (7-77) we can write Eq. (7-75) as

$$\hat{T}_N = -\frac{1}{2\mu R^2}\left[\hbar^2 \frac{\partial}{\partial R}\left(R^2 \frac{\partial}{\partial R} \right) - (\hat{J}_x - \hat{L}_x)^2 \right.$$
$$\left. - \frac{1}{\sin\theta}(\hat{J}_y - \hat{L}_y)\sin\theta(\hat{J}_y - \hat{L}_y) \right], \tag{7-83}$$

where the terms in $\hat{J}_x\hat{L}_x$ and $\hat{J}_y\hat{L}_y$ are those that give rise to the Coriolis coupling effects on the electrons. The following commutators are important and can be easily derived:

$$[\hat{J}_x, \hat{J}_y] = -i\hbar\cot\theta\hat{J}_x - i\hbar\hat{L}_z, \tag{7-84}$$

$$[\hat{J}_y, \hat{J}_z] = [\hat{J}_z, \hat{J}_x] = 0, \tag{7-85}$$

$$[\hat{J}_x, \hat{L}_\beta] = i\hbar\cot\theta \sum_\gamma \varepsilon_{z\beta\gamma}\hat{L}_\gamma, \tag{7-86}$$

$$[\hat{J}_y, \hat{L}_\beta] = 0, \tag{7-87}$$

$$[\hat{J}_z, \hat{L}_\beta] = i\hbar \sum_\gamma \varepsilon_{z\beta\gamma}\hat{L}_\gamma, \tag{7-88}$$

and

$$[\hat{L}_\alpha, \hat{L}_\beta] = i\hbar \sum_\gamma \varepsilon_{\alpha\beta\gamma}\hat{L}_\gamma, \tag{7-89}$$

where

$$\alpha, \beta, \gamma = x, y, \text{ or } z$$

and

$$\varepsilon_{\alpha\beta\gamma} = +1 \quad \text{if } \alpha\beta\gamma \text{ are cyclic (i.e., } xyz, yzx, \text{ or } zxy),$$
$$= -1 \quad \text{if } \alpha\beta\gamma \text{ are anticyclic (i.e., } zyx, yxz, \text{ or } xzy),$$
$$= 0 \quad \text{otherwise.} \tag{7-90}$$

As a result of this coordinate change we can write the rovibronic Schrödinger equation of a diatomic molecule as

$$[\hat{T}_e + \hat{T}_{vib} + \hat{T}_{rot} + V(R, x_3, y_3, z_3, \ldots, z_{n+2})$$
$$- E_{rve}]\Phi_{rve}(R, \theta, \phi, x_3, \ldots, z_{n+2}) = 0, \tag{7-91}$$

where

$$\hat{T}_e = -\frac{\hbar^2}{2m_e} \sum_{i=3}^{n+2} \mathbf{V}_i^2 - \frac{\hbar^2}{2M_N} \sum_{i,j=3}^{n+2} \mathbf{V}_i \cdot \mathbf{V}_j, \tag{7-92}$$

$$\hat{T}_{vib} = -\frac{\hbar^2}{2\mu R^2} \frac{\partial}{\partial R}\left(R^2 \frac{\partial}{\partial R}\right), \tag{7-93}$$

and

$$\hat{T}_{rot} = \frac{1}{2\mu R^2}\left[(\hat{J}_x - \hat{L}_x)^2 + \frac{1}{\sin\theta}(\hat{J}_y - \hat{L}_y)\sin\theta(\hat{J}_y - \hat{L}_y)\right]. \tag{7-94}$$

This Schrödinger equation does not separate into independent rotational (θ, ϕ), vibrational (R), and electronic $(x_3, y_3, z_3, \ldots, z_{n+2})$ parts because of (a) the occurrence of both R and the electronic coordinates in the potential function, (b) the occurrence of R in $\hat{T}_{rot}$, and (c) the occurrence of $\hat{L}_\alpha$ in $\hat{T}_{rot}$ (which gives rise to Coriolis coupling effects on the electrons). We can achieve a separation of rotation by neglecting the $\hat{L}_\alpha$ terms, and by setting $R = R_e$ (the equilibrium internuclear distance), in $\hat{T}_{rot}$. The approximate (rigid rotor) rotational kinetic energy operator obtained is

$$\hat{T}_{rot}^0 = \frac{1}{2\mu R_e^2}\left[\hat{J}_x^2 + \frac{1}{\sin\theta}\hat{J}_y \sin\theta \hat{J}_y\right]. \tag{7-95}$$

The separation of the vibrational and electronic coordinates is achieved by making the Born–Oppenheimer approximation as will be discussed in the next chapter.

The Transformation Properties of the Rovibronic Coordinates

The transformation properties of the rovibronic coordinates $(\theta, \phi, R, x_3, \ldots, z_{n+2})$ of a diatomic molecule can be determined from those of $(\xi_2, \ldots, \zeta_{n+2})$ by using Eqs. (7-65)–(7-70). We first consider the nuclear permutation (12), which is a symmetry operation only for homonuclear diatomic molecules, and then consider E^*, which is a symmetry operation for all diatomic molecules.

The effect of the nuclear permutation (12) on the nuclear coordinates in the (ξ, η, ζ) axis system can be written as

$$(12)(\xi_1, \eta_1, \zeta_1, \xi_2, \eta_2, \zeta_2) = (\xi_1', \eta_1', \zeta_1', \xi_2', \eta_2', \zeta_2') \qquad (7\text{-}96)$$

$$= (\xi_2, \eta_2, \zeta_2, \xi_1, \eta_1, \zeta_1) \qquad (7\text{-}97)$$

$$= (-\xi_1, -\eta_1, -\zeta_1, -\xi_2, -\eta_2, -\zeta_2). \qquad (7\text{-}98)$$

As a result of (12) the coordinates of the nuclei are interchanged, see Eq. (7-97), and also since $\xi_1 = -\xi_2$, $\eta_1 = -\eta_2$, and $\zeta_1 = -\zeta_2$ for a homonuclear diatomic molecule the result in Eq. (7-98) follows. Using Eqs. (7-65) and (7-66) we have

$$(12)(\theta, \phi) = (\theta', \phi'), \qquad (7\text{-}99)$$

where

$$\theta' = \arccos[\zeta_2'(\xi_2'^2 + \eta_2'^2 + \zeta_2'^2)^{-1/2}], \qquad (7\text{-}100)$$

$$\phi' = \arccos[\xi_2'(\xi_2'^2 + \eta_2'^2)^{-1/2}], \qquad (7\text{-}101)$$

and

$$\phi' = \arcsin[\eta_2'(\xi_2'^2 + \eta_2'^2)^{-1/2}]. \qquad (7\text{-}102)$$

Substituting from Eqs. (7-96) and (7-98) we determine that

$$\cos\theta' = -\cos\theta, \qquad (7\text{-}103)$$

$$\cos\phi' = -\cos\phi, \qquad (7\text{-}104)$$

and

$$\sin\phi' = -\sin\phi, \qquad (7\text{-}105)$$

so that

$$\theta' = \pi - \theta \qquad (7\text{-}106)$$

and

$$\phi' = \phi + \pi. \qquad (7\text{-}107)$$

The result in Eq. (7-106) follows from Eq. (7-103) since we must have $0 \le \theta' \le \pi$.

The result obtained here for the effect of (12) on θ and ϕ for a diatomic molecule can be appreciated by looking at Fig. 7-3. In Fig. 7-3a the molecule fixed axes are attached following the prescription given before Eq. (7-65). In Fig. 7-3b we have *first* permuted the nuclei according to (12) and *then*

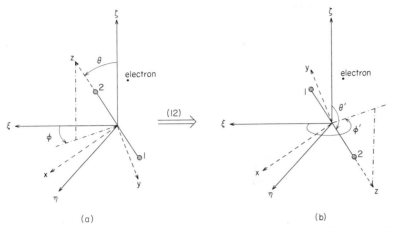

(a)   (b)

Fig. 7-3. The effect of the nuclear permutation (12) on the nuclei, an electron, and the molecule fixed (x, y, z) axes of a homonuclear diatomic molecule. The new values of the Euler angles (θ', ϕ') are related to the original values (θ, ϕ) by $\theta' = \pi - \theta$ and $\phi' = \phi + \pi$. $\chi = 0°$ in both figures by definition of the (x, y, z) axes.

attached the molecule fixed axes according to the prescription given before Eq. (7-65). The z axis is reversed by (12) and the results in Eqs. (7-106) and (7-107) immediately follow. Part of the prescription for attaching the (x, y, z) axes involves choosing $\chi = 0°$ and using this in Fig. 7-3b we find that the y axis is also reversed by (12). Equations (7-68) to (7-70) also lead to this result as we will see below when the transformation properties of the electronic coordinates are considered.

Combining Eqs. (7-98) and (7-67) we see that the vibrational coordinate R is unaffected by (12) and we can write

$$(12)R = R' = R. \tag{7-108}$$

This can be seen by inspection of Fig. 7-3.

The electronic coordinates (ξ_i, η_i, ζ_i) are unaffected by the nuclear permutation (12), but the Euler angles are changed according to Eqs. (7-106) and (7-107) so the (x_i, y_i, z_i) electronic coordinates are affected by (12). We can write

$$(12)x_i = x_i' = \xi_i' \cos \theta' \cos \phi' + \eta_i' \cos \theta' \sin \phi' - \zeta_i' \sin \theta'$$
$$= \xi_i \cos \theta \cos \phi + \eta_i \cos \theta \sin \phi - \zeta_i \sin \theta$$
$$= x_i. \tag{7-109}$$

Similarly

$$(12)y_i = y_i' = -y_i \tag{7-110}$$

and

$$(12)z_i = z_i' = -z_i. \tag{7-111}$$

These results follow from Fig. 7-3 since the y and z axes are reversed by (12) whereas the x axis is unaffected.

Summarizing the effect of (12) for a diatomic molecule we can write:

$$(12)(\theta, \phi, R, x_i, y_i, z_i) = (\pi - \theta, \phi + \pi, R, x_i, -y_i, -z_i). \tag{7-112}$$

The operation E^* is the inversion of the coordinates of all particles in the molecular center of mass and by definition we have

$$E^*(X_i, Y_i, Z_i) = (-X_i, -Y_i, -Z_i). \tag{7-113}$$

Using Eq. (7-37) we see that

$$E^*(\xi_i, \eta_i, \zeta_i) = (-\xi_i, -\eta_i, -\zeta_i) \tag{7-114}$$

for all particles in any molecule. Using this result together with Eqs. (7-65)–(7-70) we deduce that for a diatomic molecule

$$E^*(\theta, \phi, R, x_i, y_i, z_i) = (\pi - \theta, \phi + \pi, R, -x_i, y_i, z_i). \tag{7-115}$$

In Fig. 7-4 the effect of E^* for a diatomic molecule is drawn.

The operation (12)* is also a symmetry operation for homonuclear diatomic molecules, being the product of (12) and E^*. The effect of this operation on the rovibronic coordinates of a diatomic molecule is given by

$$(12)^*(\theta, \phi, R, x_i, y_i, z_i) = (\theta, \phi, R, -x_i, -y_i, -z_i). \tag{7-116}$$

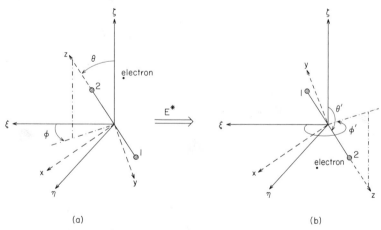

(a)   (b)

Fig. 7-4. The effect of E^* for a diatomic molecule. The new values of the Euler angles θ' and ϕ' are related to the original values by $\theta' = \pi - \theta$ and $\phi' = \phi + \pi$. Note the inversion of the electron coordinates in space at the origin of the (ξ, η, ζ) axis system.

This operation inverts the electronic coordinates through the origin of the molecule fixed axis system and does not alter the orientation of the molecule fixed axes in space. The operation i from the molecular point group has the same effect as $(12)^*$ on the rovibronic coordinates of a homonuclear diatomic molecule. This is discussed further in Chapter 11 [see Eq. (11-15)].

Equation (7-115) applies to all diatomic molecules and Eqs. (7-112) and (7-116) apply to all homonuclear diatomic molecules. The effect of E^* on the rovibronic wavefunctions of a diatomic molecule leads to the $\pm$ label on the rovibronic levels, and (for a homonuclear diatomic molecule) the effect of (12) gives the a or s label. The effect of $(12)^*$ on the electronic wavefunctions gives the g or u label for the electronic states.

RIGID NONLINEAR POLYATOMIC MOLECULES

A large class of polyatomic molecules are nonlinear and rigid, and we focus attention on such molecules in the remainder of the chapter; in Chapter 12 linear molecules and nonrigid molecules are discussed. In this book we always mean by the term *rigid molecule* a molecule that is in an electronic state that has a unique equilibrium configuration and for which the potential energy barriers separating this from other minima in the potential surface are insuperable. A nonrigid molecule is a molecule, such as the ammonia molecule, for which such potential energy barriers are not insuperable and for which observed splittings and/or shifts, due to tunneling, occur in the rotation–vibration energy level pattern.

Changing to Rovibronic Coordinates

We wish to change coordinates in the Schrödinger equation for a rigid nonlinear polyatomic molecule from $(\xi_2, \eta_2, \zeta_2, \ldots, \zeta_l)$ to the rovibronic coordinates $(\theta, \phi, \chi, Q_1, \ldots, Q_{3N-6}, x_{N+1}, \ldots, z_l)$, where the three Euler angles (θ, ϕ, χ) define the orientation of the molecule fixed (x, y, z) axis system relative to the (ξ, η, ζ) axis system, and the $(3N - 6)$ normal coordinates Q_r are particular linear combinations of the nuclear coordinates (x_i, y_i, z_i). As a result of this $\hat{T}_N$ will be expressed in terms of $(\hat{J}_\alpha, \hat{L}_\alpha, Q_1, \ldots, Q_{3N-6}, \hat{P}_1, \ldots, \hat{P}_{3N-6})$ where $\hat{J}_\alpha$ are components of the rovibronic angular momentum and $\hat{P}_r = -i\hbar\, \partial/\partial Q_r$. This coordinate change is made so that the sum of $\hat{T}_N$ and the internuclear potential function V_N (this arises after making the Born–Oppenheimer approximation and will be discussed in the next chapter) is separable with least approximation into a part depending only on $\hat{J}_x$, $\hat{J}_y$, and $\hat{J}_z$ (the rotational kinetic energy) and into $(3N - 6)$ parts each involving one Q_r and associated $\hat{P}_r$. The coordinate change now involves three Euler angles instead of the two in Eqs. (7-65) and (7-66) for a diatomic molecule, and $(3N - 6)$ vibrational coordinates (Q_r) instead of the one (R) in Eq. (7-67) for a diatomic molecule. The effect of this coordinate change on $\hat{T}_e$ [see Eq. (7-46)] is to leave its form unchanged as shown by Eqs. (7-58) and (7-60).

A Method II derivation of $\hat{T}_N$ in the rovibronic coordinates begins with the following expression for the classical nuclear kinetic energy:

$$T_N = \frac{1}{2} \sum_{i=1}^{N} m_i(\dot{\xi}_i^2 + \dot{\eta}_i^2 + \dot{\zeta}_i^2), \qquad (7\text{-}117)$$

where we have the restrictions

$$\sum_{i=1}^{N} m_i\xi_i = \sum_{i=1}^{N} m_i\eta_i = \sum_{i=1}^{N} m_i\zeta_i = 0, \qquad (7\text{-}118)$$

since the (ξ,η,ζ) axis system has origin at the nuclear center of mass. This classical expression for T_N leads to the quantum mechanical expression $\hat{T}_N$ given in Eq. (7-47); see Eqs. (6-2)–(6-18). The derivation of T_N and $\hat{T}_N$ in rovibronic coordinates is given in Chapter 11 of Wilson, Decius, and Cross (1955) and a simplification of this kinetic energy operator is discussed by Watson (1968). We will not duplicate this derivation or simplification here. To determine the effect of symmetry operations on the rovibronic wavefunctions we must, however, understand the rovibronic coordinates and these will be discussed. The two central parts of the coordinate change to the rovibronic coordinates are (a) the Eckart equations [Eckart (1935)] which define the Euler angles, and (b) the l matrix [see Nielsen (1959)] which defines the normal coordinates.

The Eckart equations enable us to determine the Euler angles [i.e., the orientation of the molecule fixed (x, y, z) axes] from the values of the coordinates of the nuclei $(\xi_2,\eta_2,\zeta_2,\ldots,\xi_N,\eta_N,\zeta_N)$. To determine the Euler angles from the nuclear coordinates we need three equations which we might write [by analogy with Eqs. (7-65) and (7-66)] as

$$\theta = f_\theta(\xi_2,\ldots,\zeta_N), \qquad (7\text{-}119)$$

$$\phi = f_\phi(\xi_2,\ldots,\zeta_N), \qquad (7\text{-}120)$$

and

$$\chi = f_\chi(\xi_2,\ldots,\zeta_N). \qquad (7\text{-}121)$$

Having oriented the (x, y, z) axes in the molecule by using the equations represented above we can determine the (x, y, z) coordinates of each nucleus in the molecule. Such a calculation could be made when the nuclei are in their equilibrium configuration and we would obtain the *equilibrium nuclear coordinates* (x_i^e, y_i^e, z_i^e) for each nucleus i. For a given distorted nuclear configuration the displacements $\Delta x_i = (x_i - x_i^e)$, $\Delta y_i = (y_i - y_i^e)$, and $\Delta z_i =$

$(z_i - z_i^e)$ would be the vibrational displacements. We would like to choose f_θ, f_ϕ, and f_χ in Eqs. (7-119)–(7-121) so that if we were to express the nuclear kinetic energy in terms of the Euler angles and vibrational displacements there would be a complete separation into a rotational part (involving only the Euler angles) and a vibrational part (involving only the vibrational displacement coordinates). To do this we would have to eliminate the vibrational angular momentum $\boldsymbol{J}_{\text{vib}}$ in the (x, y, z) axis system; if $\boldsymbol{J}_{\text{vib}}$ were nonvanishing there would be a Coriolis coupling of it to the rovibronic angular momentum $\boldsymbol{J}$ which would spoil the separation of variables. The vibrational angular momentum is given by

$$\boldsymbol{J}_{\text{vib}} = \sum_i m_i \boldsymbol{r}_i \times \dot{\boldsymbol{r}}_i, \qquad (7\text{-}122)$$

where m_i is the mass of the ith nucleus, $\boldsymbol{r}_i$ has components (x_i, y_i, z_i) and the velocity $\dot{\boldsymbol{r}}_i$ has components $(\dot{x}_i, \dot{y}_i, \dot{z}_i)$ in the (x, y, z) axis system. If f_θ, f_ϕ, and f_χ could be chosen so that the $\boldsymbol{r}_i$ and $\dot{\boldsymbol{r}}_i$ satisfy

$$\sum_i m_i \boldsymbol{r}_i \times \dot{\boldsymbol{r}}_i = 0, \qquad (7\text{-}123)$$

then there would be no vibrational angular momentum in the (x, y, z) axis system and no Coriolis coupling terms to spoil the separation of rotation from vibration. Unfortunately, with only three equations [Eqs. (7-119)–(7-121)] at our disposal we cannot define θ, ϕ, and χ so that the resultant (x_i, y_i, z_i) *and* $(\dot{x}_i, \dot{y}_i, \dot{z}_i)$ satisfy Eq. (7-123). However, for a rigid molecule the nuclei do not depart far from the nuclear equilibrium configuration and to a good approximation we have

$$\boldsymbol{J}_{\text{vib}} \cong \sum_i m_i \boldsymbol{r}_i^e \times \dot{\boldsymbol{r}}_i, \qquad (7\text{-}124)$$

where $\boldsymbol{r}_i^e$ has components (x_i^e, y_i^e, z_i^e). We can choose f_θ, f_ϕ, and f_χ so that the $\dot{x}_i, \dot{y}_i$, and $\dot{z}_i$ satisfy

$$\sum_i m_i \boldsymbol{r}_i^e \times \dot{\boldsymbol{r}}_i = 0, \qquad (7\text{-}125)$$

and in the (x, y, z) axis system so defined, $\boldsymbol{J}_{\text{vib}}$ and the Coriolis coupling will be small. Equation (7-125) follows by differentiating with respect to time the *Eckart equation*, which is

$$\sum_i m_i \boldsymbol{r}_i^e \times \boldsymbol{r}_i = 0. \qquad (7\text{-}126)$$

The three components of this equation are

$$\sum_i m_i(x_i^e y_i - y_i^e x_i) = 0, \qquad (7\text{-}127)$$

$$\sum_i m_i(y_i^e z_i - z_i^e y_i) = 0, \qquad (7\text{-}128)$$

and

$$\sum_i m_i(z_i^e x_i - x_i^e z_i) = 0. \qquad (7\text{-}129)$$

Choosing the orientation of the (x, y, z) axes so that Eqs. (7-127)–(7-129) are satisfied results in Eq. (7-125) being satisfied so that $\boldsymbol{J}_{\text{vib}}$ is small. The origin of the (x, y, z) axis system is the nuclear center of mass so the nuclear coordinates must satisfy

$$\sum_i m_i \boldsymbol{r}_i^e = \sum_i m_i \boldsymbol{r}_i = 0. \qquad (7\text{-}130)$$

For the equilibrium nuclear configuration we choose the orientation of the (x, y, z) axes so that they are the principal axes of inertia in order to simplify the rotational kinetic energy. Using principal inertial axes the off-diagonal elements of the inertia matrix (called *products of inertia*) vanish where these elements are given by $I_{\alpha\beta} = -\sum_i m_i \alpha_i \beta_i$, where $\alpha \neq \beta$. Thus the (x, y, z) axes are located in the equilibrium nuclear configuration so that

$$\sum_i m_i x_i^e y_i^e = \sum_i m_i y_i^e z_i^e = \sum_i m_i z_i^e x_i^e = 0. \qquad (7\text{-}131)$$

To use the Eckart equations [Eqs. (7-127)–(7-129)] in order to determine the Euler angles $\theta, \phi,$ and χ from the nuclear coordinates $(\xi_2, \ldots, \zeta_N)$ we proceed as follows. Expressing the (x_i, y_i, z_i) coordinates in terms of the (ξ_i, η_i, ζ_i) coordinates by using the direction cosine matrix elements $\lambda_{\alpha\tau}$ [where $\alpha = x, y,$ or z and $\tau = \xi, \eta,$ or ζ; see Eq. (7-51)] we can write the three Eckart equations as

$$[x\xi]\lambda_{y\xi} + [x\eta]\lambda_{y\eta} + [x\zeta]\lambda_{y\zeta} - [y\xi]\lambda_{x\xi} - [y\eta]\lambda_{x\eta} - [y\zeta]\lambda_{x\zeta} = 0, \quad (7\text{-}132)$$

$$[y\xi]\lambda_{z\xi} + [y\eta]\lambda_{z\eta} + [y\zeta]\lambda_{z\zeta} - [z\xi]\lambda_{y\xi} - [z\eta]\lambda_{y\eta} - [z\zeta]\lambda_{y\zeta} = 0, \quad (7\text{-}133)$$

and

$$[z\xi]\lambda_{x\xi} + [z\eta]\lambda_{x\eta} + [z\zeta]\lambda_{x\zeta} - [x\xi]\lambda_{z\xi} - [x\eta]\lambda_{z\eta} - [x\zeta]\lambda_{z\zeta} = 0, \quad (7\text{-}134)$$

where

$$[\alpha\tau] = \sum_{i=1}^{N} m_i \alpha_i^e \tau_i, \qquad (7\text{-}135)$$

with $\alpha = x, y,$ or z and $\tau = \xi, \eta,$ or ζ. Knowing the equilibrium coordinates of the nuclei in the (x, y, z) axis system [having initially used Eq. (7-131) to determine $x_i^e, y_i^e,$ and z_i^e] we can use the values of the (ξ_i, η_i, ζ_i) coordinates to determine the values of the $[\alpha\tau]$. We can then set up the three transcendental equations, Eqs. (7-132)–(7-134), in the Euler angles using Eq. (7-52). Solving these equations simultaneously gives the values of the Euler angles as we show in a numerical example in the next section. The

Eckart equations lead to these three simultaneous transcendental equations in the Euler angles and not to three simple equations as suggested by Eqs. (7-119)–(7-121).

The equation defining the normal coordinates Q_r in terms of the Cartesian displacement coordinates $\Delta\alpha_i (= \alpha_i - \alpha_i^e)$ involves the elements of the $3N \times (3N - 6)$ l matrix:

$$\Delta\alpha_i = \sum_r m_i^{-1/2} l_{\alpha i,r} Q_r, \tag{7-136}$$

where $\alpha = x, y,$ or z; $r = 1$ to $(3N - 6)$; $i = 1$ to N. As a result each normal coordinate is a linear combination of the $\Delta\alpha_i$. The l matrix is chosen in such a manner that the zero order internuclear potential function V_N (discussed in Chapter 8), when expressed in terms of the Q_r, separates into $(3N - 6)$ harmonic oscillator potential functions. The l matrix will be discussed in the next chapter, and the elements of the l matrix depend on the equilibrium nuclear geometry, the nuclear masses, and the potential function V_N.

Using the Eckart equations and the l matrix in changing coordinates in the classical kinetic energy of Eq. (7-117), using the Podolsky trick to convert this to the quantum mechanical kinetic energy operator, and making Watson's simplification of this expression [Watson (1968)] we obtain

$$\hat{T}_N = \frac{1}{2} \sum_{\alpha,\beta} \mu_{\alpha\beta} (\hat{J}_\alpha - \hat{p}_\alpha - \hat{L}_\alpha)(\hat{J}_\beta - \hat{p}_\beta - \hat{L}_\beta) + \frac{1}{2} \sum_r \hat{P}_r^2 + U, \tag{7-137}$$

where the μ matrix is the inverse of the I' matrix given in Eq. 10 on page 278 of Wilson, Decius, and Cross (1955) (the I' matrix is almost, but not quite, equal to the instantaneous inertia matrix I), $\hat{J}_\alpha$ and $\hat{J}_\beta$ are components of the rovibronic angular momentum operator along the molecule fixed axes, and $\hat{p}_\alpha$ and $\hat{p}_\beta$ are referred to as components of the vibrational angular momentum operator [although they are not quite equal to the quantum mechanical counterparts of the components of J_{vib} given in Eq. (7-122)]. The $\hat{p}_\alpha$ are given by

$$\hat{p}_\alpha = \sum_{r,s=1}^{3N-6} \zeta_{r,s}^\alpha Q_r \hat{P}_s, \tag{7-138}$$

where the $\zeta_{r,s}^\alpha$ (Coriolis coupling constants) depend on the l matrix according to

$$\zeta_{r,s}^{x} = -\zeta_{s,r}^{x} = \sum_{i=1}^{N} (l_{yi,r} l_{zi,s} - l_{zi,r} l_{yi,s}) \tag{7-139}$$

and cyclically for the y and z coefficients. The term U is given by

$$U = -\frac{\hbar^2}{8} \sum_\alpha \mu_{\alpha\alpha}, \tag{7-140}$$

and it can be considered as a mass dependent contribution to V_N. The components of the electronic angular momentum operator $\hat{L}_\alpha$ and $\hat{L}_\beta$ occur in a similar way as they do in Eq. (7-83) for $\hat{T}_N$ of a diatomic molecule. In making the Born–Oppenheimer approximation (discussed in Chapter 8) the $\hat{L}_\alpha$ and $\hat{L}_\beta$ in $\hat{T}_N$ are neglected and we will ignore them in the rest of this chapter.

The $\hat{J}_x, \hat{J}_y$, and $\hat{J}_z$ occurring in $\hat{T}_N$ are components of the rovibronic angular momentum operator $\hat{\boldsymbol{J}}$ [see Eqs. (7-78) and (7-79)]. The expressions for the (x, y, z) components of the angular momentum operator in terms of the Euler angles will be of use to us. The classical expressions for these components are given by [see Wilson, Decius, and Cross (1955), Eq. (6) on page 282],

$$J_x = \sin\chi\, p_\theta - \csc\theta \cos\chi\, p_\phi + \cot\theta \cos\chi\, p_\chi, \tag{7-141}$$

$$J_y = \cos\chi\, p_\theta + \csc\theta \sin\chi\, p_\phi - \cot\theta \sin\chi\, p_\chi, \tag{7-142}$$

and

$$J_z = p_\chi, \tag{7-143}$$

where p_θ, p_ϕ, and p_χ are the momenta conjugate to θ, ϕ, and χ. Converting this to the quantum mechanical expression [see Watson (1970), Eqs. (A27)–(A29)] we obtain

$$\hat{J}_x = -i\hbar\left(\sin\chi\, \frac{\partial}{\partial\theta} - \csc\theta \cos\chi\, \frac{\partial}{\partial\phi} + \cot\theta \cos\chi\, \frac{\partial}{\partial\chi}\right), \tag{7-144}$$

and

$$\hat{J}_y = -i\hbar\left(\cos\chi\, \frac{\partial}{\partial\theta} + \csc\theta \sin\chi\, \frac{\partial}{\partial\phi} - \cot\theta \sin\chi\, \frac{\partial}{\partial\chi}\right), \tag{7-145}$$

$$\hat{J}_z = -i\hbar\, \frac{\partial}{\partial\chi}. \tag{7-146}$$

These operators obey the commutation relations

$$[\hat{J}_\alpha, \hat{J}_\beta] = -i\hbar \sum_\gamma \varepsilon_{\alpha\beta\gamma}\hat{J}_\gamma, \tag{7-147}$$

where $\alpha, \beta, \gamma = x, y$, or z [$\varepsilon_{\alpha\beta\gamma}$ is defined in Eq. (7-90)]. This equation should be compared to Eq. (7-79) for the commutation relations of the components of $\boldsymbol{J}$ along the ξ, η, and ζ axes, which we can write as

$$[\hat{J}_\sigma, \hat{J}_\tau] = +i\hbar \sum_\nu \varepsilon_{\sigma\tau\nu}\hat{J}_\nu, \tag{7-148}$$

where $\sigma, \tau, \nu = \xi, \eta$, or ζ. The reader should also compare Eq. (7-147) to Eqs. (7-84) and (7-85) for the commutation relations of the components of $\hat{\boldsymbol{J}}$ for a diatomic molecule. For nonlinear polyatomic molecules $\hat{J}_x, \hat{J}_y$, and $\hat{J}_z$ each commute with $\hat{L}_x, \hat{L}_y$, and $\hat{L}_z$, unlike the situation for a diatomic molecule [see Eqs. (7-86)–(7-88)]. The components of the electronic angular momentum operator $\hat{L}$ have commutation relations given by Eq. (7-89) for all molecules.

In $\hat{T}_N$ of Eq. (7-137) the $\mu_{\alpha\beta}$ elements can be expressed as a Taylor series in the Q_r [Watson (1968)] as

$$\mu_{\alpha\beta} = \mu_{\alpha\beta}^e - \sum_r \mu_{\alpha\alpha}^e a_r^{\alpha\beta} \mu_{\beta\beta}^e Q_r + \frac{3}{4} \sum_{r,s,\gamma} \mu_{\alpha\alpha}^e a_r^{\alpha\gamma} \mu_{\gamma\gamma}^e a_s^{\gamma\beta} \mu_{\beta\beta}^e Q_r Q_s + \cdots, \quad (7\text{-}149)$$

where $\mu_{\alpha\beta}^e = \{[I^e]^{-1}\}_{\alpha\beta}$ is an element of the inverse of the moment of inertia matrix for the molecule in its equilibrium configuration, and the coefficients $a_r^{\alpha\beta}$ depend on the equilibrium nuclear geometry, on the nuclear masses, and on the potential V_N. If we neglect the dependence of $\mu_{\alpha\beta}$ on the Q_r, neglect the vibrational angular momenta, and neglect U, we obtain

$$\hat{T}_N^{\ 0} = \frac{1}{2} \sum_{\alpha,\beta} \mu_{\alpha\beta}^e \hat{J}_\alpha \hat{J}_\beta + \frac{1}{2} \sum_{r=1}^{3N-6} \hat{P}_r^{\ 2}, \quad (7\text{-}150)$$

which separates into a rotational kinetic energy (the first term) and $3N - 6$ vibrational kinetic energy terms $\frac{1}{2}\hat{P}_1^{\ 2}, \frac{1}{2}\hat{P}_2^{\ 2}, \ldots, \frac{1}{2}\hat{P}_{3N-6}^2$. The dependence of the $\mu_{\alpha\beta}$ on Q_r, and the occurrence of the vibrational angular momentum, give rise to terms in $\hat{T}_N$ such as

$$-\tfrac{1}{2}\mu_{\alpha\alpha}^e a_r^{\alpha\beta} \mu_{\beta\beta}^e Q_r \hat{J}_\alpha \hat{J}_\beta \quad (7\text{-}151)$$

and

$$-\tfrac{1}{2}\mu_{\alpha\beta}^e \hat{J}_\alpha \hat{P}_\beta, \quad (7\text{-}152)$$

which spoil the separation of the rotational and vibrational coordinates in $\hat{T}_N$ [each term involves both $\hat{J}_\alpha$ and Q_r (and/or $\hat{P}_r$)]. The term in Eq. (7-151) is one of the centrifugal distortion terms and the term in Eq. (7-152) is one of the Coriolis coupling terms. We initially approximate $\hat{T}_N$ by $\hat{T}_N^{\ 0}$ and usually correct for centrifugal distortion and Coriolis coupling by perturbation theory.

A Numerical Example Involving the Eckart Equations

We will use a numerical example involving the water molecule to demonstrate the coordinate change $(\xi_2, \ldots, \zeta_l)$ to $(\theta, \phi, \chi, Q_r, x_{N+1}, \ldots, z_l)$, and to show the central place of the Eckart equations in this coordinate change.

In Fig. 7-5 we show a water molecule in its equilibrium configuration [we will take $r(OH) = 0.957\,\text{Å}$ and $H\hat{O}H = 105°$ for the equilibrium configuration of the water molecule]. The molecule fixed axes of the equilibrium configuration of a molecule are located as the principal axes of inertia since this diagonalizes $\mu_{\alpha\beta}^e$ and simplifies the zero order rotational kinetic energy

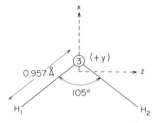

Fig. 7-5. A water molecule in its equilibrium configuration with molecule fixed (x, y, z) axes attached. The oxygen nucleus is labeled 3.

in Eq. (7-150). Since each of the three x, y, and z axes can be identified with any of the three principal axes and the signs of any of the x, y, and z axes can be reversed all without introducing off-diagonal elements into μ^e we have many possible ways of locating the (x, y, z) axes as principal axes. The three principal axes of a molecule are by convention labeled a, b, and c in such a manner that the moments of inertia about the axes are in the order $I_{aa} < I_{bb} < I_{cc}$. We can identify the z axis with any of the three principal axes and depending on whether it is identified with the a, b, or c axis we name the convention adopted as type I, II, or III. We add a superscript r or l depending on whether a right or left handed (x, y, z) axis system is used. For the water molecule we adopt a I^r convention (see Fig. 7-5) in which the z axis is located so that H_2 has a positive z coordinate, the x axis is located so that the oxygen nucleus has a positive x coordinate, and the y axis is located so that the axis system is right handed. Using 1 u and 16 u as the hydrogen and oxygen nuclear masses we determine that (in Å)

$$x_1^e = x_2^e = -0.5178, \qquad x_3^e = 0.06473,$$
$$y_1^e = y_2^e = y_3^e = 0, \tag{7-153}$$
$$z_1^e = -z_2^e = -0.7592, \qquad z_3^e = 0,$$

where we label the oxygen nucleus 3.

As a water molecule moves in space in its equilibrium configuration it is always easy [using the results in Eq. (7-153)] to determine the orientation of the (x, y, z) axes and hence the Euler angles. For example, suppose that the (X_i, Y_i, Z_i) coordinates of the nuclei in the space fixed (X, Y, Z) axis system are (in Å)

$$(0.3733, -0.3626, -0.8031), \quad (0.7529, 0.2949, 0.5118), \quad (0, 0, 0) \tag{7-154}$$

for nuclei H_1, H_2, and O_3, respectively. Moving to the (ξ, η, ζ) axis system with origin at the nuclear center of mass [which we determine to have coordinates $(0.0626, -0.00376, -0.0162)$ in the (X, Y, Z) axis system] we find the (ξ_i, η_i, ζ_i) coordinates to be (in Å)

$$(0.3107, -0.3588, -0.7869), \quad (0.6903, 0.2987, 0.5280),$$
$$\text{and} \quad (-0.0626, 0.00376, 0.0162). \tag{7-155}$$

Knowing the equilibrium nuclear coordinates in the (x, y, z) axis system [Eq. (7-153)] we can determine the Euler angles by using the direction cosine matrix elements [Eq. (7-52)] as follows.

To determine θ we use the fact that all $y_i^e = 0$ and from Eq. (7-52) we obtain

$$\zeta_i = x_i^e(-\sin\theta\cos\chi) + z_i^e\cos\theta, \qquad (7\text{-}156)$$

from which, since $x_1^e = x_2^e$, we obtain

$$(\zeta_2 - \zeta_1) = (z_2^e - z_1^e)\cos\theta, \qquad (7\text{-}157)$$

and thus

$$\theta = \arccos[(\zeta_2 - \zeta_1)/(z_2^e - z_1^e)]. \qquad (7\text{-}158)$$

Since $0 \le \theta \le \pi$ we can unambiguously determine θ from Eq. (7-158) by substituting in the appropriate values for z_i^e and ζ_i from Eqs. (7-153) and (7-155); we obtain

$$\theta = 30°. \qquad (7\text{-}159)$$

To determine ϕ we use equations similar to Eq. (7-157) for ξ_i and η_i, together with Eq. (7-158) for θ, to derive that

$$\phi = \arccos\{(\xi_2 - \xi_1)/[(z_2^e - z_1^e)^2 - (\zeta_2 - \zeta_1)^2]^{1/2}\} \qquad (7\text{-}160)$$

and

$$\phi = \arcsin\{(\eta_2 - \eta_1)/[(z_2^e - z_1^e)^2 - (\zeta_2 - \zeta_1)^2]^{1/2}\}. \qquad (7\text{-}161)$$

Since $0 \le \phi \le 2\pi$ we need both these equations for an unambiguous determination of ϕ. From these equations we determine

$$\phi = 60°. \qquad (7\text{-}162)$$

For the angle χ we obtain

$$\chi = \arccos\{-\zeta_3(z_2^e - z_1^e)[(z_2^e - z_1^e)^2 - (\zeta_2 - \zeta_1)^2]^{-1/2}/x_3^e\} \qquad (7\text{-}163)$$

and to make an unambiguous determination $(0 \le \chi \le 2\pi)$ we also use

$$\cos\theta\cos\phi\cos\chi - \sin\phi\sin\chi = (\xi_1 + \xi_2)/(x_1^e + x_2^e) \qquad (7\text{-}164)$$

to give

$$\chi = 120°. \qquad (7\text{-}165)$$

A water molecule in its equilibrium configuration with this orientation is drawn in Fig. 7-6.

To determine the Euler angles of a molecule that is not in its equilibrium configuration we would proceed as in the preceding example if we defined the (x, y, z) axes as instantaneous principal axes. However for a deformed

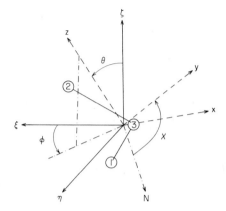

Fig. 7-6. A water molecule in its equilibrium configuration oriented with coordinates given by Eq. (7-155). The Euler angles (θ, ϕ, χ) are shown to be $(30°, 60°, 120°)$ in Eqs. (7-158)–(7-165).

molecule it is better [as we have indicated in Eqs. (7-122)–(7-126)] to choose the (x, y, z) axes to be Eckart axes rather than principal axes. Principal axes are located so that the three equations $I_{xy} = I_{yz} = I_{zx} = 0$ are satisfied, whereas Eckart axes are located so that the three Eckart equations Eqs. (7-127)–(7-129) are satisfied. We see that we need to know the equilibrium (x, y, z) coordinates of the nuclei in order to be able to locate the Eckart axes when the molecule is not at equilibrium. We will use the water molecule as a numerical example of this.

We take a deformed water molecule having $r_{13} = 2.4193\,\text{Å}, r_{23} = 0.9452\,\text{Å}$, and $H\hat{O}H = 103.4°$, and place it so that the (X, Y, Z) coordinates of the nuclei are

$$(0.9066, -0.9251, -2.0434), \quad (0.7690, 0.2778, 0.4742), \quad (0,0,0) \quad (7\text{-}166)$$

for H_1, H_2, and O_3 respectively. The nuclear center of mass is determined to be located at $(0.0930, -0.0360, -0.0872)$ in the (X, Y, Z) axis system, so that the (ξ, η, ζ) coordinates of nuclei 1, 2, and 3 are, respectively

$$(0.8136, -0.8891, -1.9562), \quad (0.6760, 0.3138, 0.5614),$$
$$\text{and} \quad (-0.0930, 0.0360, 0.0872). \quad (7\text{-}167)$$

We now wish to use the Eckart equations and the equilibrium coordinates to determine the Euler angles of this deformed water molecule. The equilibrium coordinates are known and are given in Eq. (7-153). We use the form of the Eckart equations given in Eqs. (7-132)–(7-134) in this determination. The $[\alpha\tau]$ can be calculated from the α_i^{e} in Eq. (7-153) and from the τ_i in Eq. (7-167); the $\lambda_{\alpha\tau}$ are given in terms of the Euler angles in Eq. (7-52). Using these $[\alpha\tau]$ and $\lambda_{\alpha\tau}$ Eq. (7-132) becomes

$$0.8676(\cos\theta\cos\phi\sin\chi + \sin\phi\cos\chi) + 0.3352(-\cos\theta\sin\phi\sin\chi + \cos\phi\cos\chi)$$
$$+ 0.8125(\sin\theta\sin\chi) = 0, \quad (7\text{-}168)$$

Eq. (7-133) becomes

$$-0.1045(\cos\theta\cos\phi\sin\chi + \sin\phi\cos\chi) - 0.9132(-\cos\theta\sin\phi\sin\chi + \cos\phi\cos\chi)$$
$$-1.9114(\sin\theta\sin\chi) = 0, \tag{7-169}$$

and Eq. (7-134) becomes

$$-0.1045(\cos\theta\cos\phi\cos\chi - \sin\phi\sin\chi) + 0.9132(\cos\theta\sin\phi\cos\chi + \cos\phi\sin\chi)$$
$$-1.9114\sin\theta\cos\chi + 0.8676\sin\theta\cos\phi - 0.3352\sin\theta\sin\phi$$
$$-0.8125\cos\theta = 0. \tag{7-170}$$

These three simultaneous transcendental equations can be solved numerically using, for example, the Newton–Raphson method [see Margenau and Murphy (1956), page 493], and one solution is

$$(\theta = 30°, \phi = 60°, \chi = 120°). \tag{7-171}$$

This is depicted in Fig. 7-7. Other solutions, corresponding to other possible ways of locating a right handed (x, y, z) axis system as principal inertial axes on the equilibrium molecule, can also be obtained from the simultaneous solutions of Eqs. (7-168)–(7-170), but since the orientation of the axes must smoothly correlate with those of the equilibrium molecule as the distortion is removed we discard these alternate solutions. We will always discard such solutions in any later similar determinations, such as in the solution of Eq. (7-182) or in obtaining the result in Eq. (7-195). From the Euler angles we can determine the direction cosine matrix elements, and from these and the (ξ, η, ζ) coordinates of the nuclei we can determine the (x, y, z) coordinates of the nuclei [given in Eq. (7-183), which follows].

The above method for determining the Euler angles and (x, y, z) nuclear coordinates from the Eckart equations is applicable to any deformed polyatomic molecule having a unique equilibrium configuration. However, for

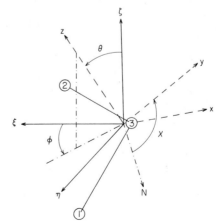

Fig. 7-7. A deformed water molecule with (ξ, η, ζ) coordinates given by Eq. (7-167). The Euler angles are shown to be $(30°, 60°, 120°)$ in Eqs. (7-168)–(7-171); this results from using the Eckart equations.

a triatomic molecule there is a rather simple way of using the Eckart equations. Equations (7-127) and (7-128) can be written (since all $y_i^e = 0$) as

$$\sum_{i=1}^{3} m_i x_i^e y_i = 0 \qquad (7\text{-}172)$$

and

$$\sum_{i=1}^{3} m_i z_i^e y_i = 0. \qquad (7\text{-}173)$$

The center of mass condition Eq. (7-130) gives a third equation

$$\sum_{i=1}^{3} m_i y_i = 0. \qquad (7\text{-}174)$$

These three equations lead directly to the result that

$$y_1 = y_2 = y_3 = 0. \qquad (7\text{-}175)$$

Thus regardless of the extent of the deformation all three atoms lie in the xz plane of the molecule fixed axes if all $y_i^e = 0$. We use the third Eckart equation to determine the orientation of the x and z axes relative to the nuclear frame. To do this it is convenient to introduce the angle ε between the x axis direction and the line joining the nuclear center of mass to the oxygen nucleus as in Fig. 7-8. Using this angle, the cosine rule, and a fair amount of trigonometry we obtain

$$x_1 = 0.9330 \sin \varepsilon - 2.100 \cos \varepsilon, \qquad (7\text{-}176)$$

$$z_1 = -0.9330 \cos \varepsilon - 2.100 \sin \varepsilon, \qquad (7\text{-}177)$$

$$x_2 = -0.9330 \sin \varepsilon - 0.0203 \cos \varepsilon, \qquad (7\text{-}178)$$

$$z_2 = 0.9330 \cos \varepsilon - 0.0203 \sin \varepsilon, \qquad (7\text{-}179)$$

$$x_3 = 0.1325 \cos \varepsilon, \qquad (7\text{-}180)$$

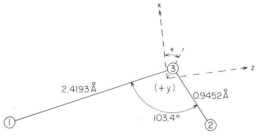

Fig. 7-8. The definition of the angle ε that shows how (x, y, z) axes are oriented on a deformed water molecule. The origin of the (x, y, z) axes is at the nuclear center of mass.

and

$$z_3 = 0.1325 \sin \varepsilon. \tag{7-181}$$

Substituting these values into Eq. (7-129), and using the equilibrium coordinates from Eq. (7-153), we obtain

$$-2.652 \sin \varepsilon + 1.579 \cos \varepsilon = 0, \tag{7-182}$$

from which we determine $\varepsilon = 30.77°$ and that the (x, y, z) nuclear coordinates are (in Å)

$$(-1.3270, 0, -1.8757), \quad (-0.4948, 0, 0.7911), \quad (0.1138, 0, 0.0678). \tag{7-183}$$

Relating these coordinates to the (ξ, η, ζ) coordinates of Eq. (7-167) we can determine the direction cosine matrix elements and thus the Euler angles [given in Eq. (7-171)].

Using the above (x, y, z) coordinates and subtracting the equilibrium coordinates [Eq. (7-153)] we determine the Cartesian displacement coordinates $(\Delta x_i, \Delta y_i, \Delta z_i)$ to be (in Å)

$$(-0.8092, 0, -1.1165), \quad (0.0230, 0, 0.0319), \quad (0.0491, 0, 0.0678). \tag{7-184}$$

It is instructive at this stage to imagine that we can pick up the equilibrium configuration water molecule, with its molecule fixed axes attached, from out of Fig. 7-5 and insert it into Fig. 7-7 in such a way that its (x, y, z) axes are coincident with those of the deformed molecule. In this composite figure the displacements of the nuclei of the deformed molecule away from those of the equilibrium molecule will be as given in Eq. (7-184). Figure 2-4 can be viewed as such a composite figure for ethylene. Writing $\alpha_i = \alpha_i^e + \Delta \alpha_i$ in Eqs. (7-127)–(7-129) we deduce three alternative Eckart equations:

$$\sum_{i=1}^{N} m_i(x_i^e \Delta y_i - y_i^e \Delta x_i) = 0, \tag{7-185}$$

$$\sum_{i=1}^{N} m_i(y_i^e \Delta z_i - z_i^e \Delta y_i) = 0, \tag{7-186}$$

and

$$\sum_{i=1}^{N} m_i(z_i^e \Delta x_i - x_i^e \Delta z_i) = 0. \tag{7-187}$$

The Cartesian displacement coordinates must satisfy these three equations (if we use Eckart axes) as well as the three center of mass equations [from

Eq. (7-130)]

$$\sum_{i=1}^{N} m_i \, \Delta \alpha_i = 0, \tag{7-188}$$

where $\alpha = x$, y, or z.

To determine the values of the normal coordinates from the Cartesian displacement coordinates we use Eq. (7-136) and to do this we need the l matrix. We can obtain the l matrix for the water molecule from the results in Hoy, Mills, and Strey (1972)[2], and the l matrix is such that we can write Eq. (7-136) for a deformed water molecule as

$$
\begin{bmatrix}
\Delta x_1 \\
\Delta y_1 \\
\Delta z_1 \\
\Delta x_2 \\
\Delta y_2 \\
\Delta z_2 \\
\Delta x_3 \\
\Delta y_3 \\
\Delta z_3
\end{bmatrix}
=
\begin{bmatrix}
-0.3931 & 0.5414 & -0.4161 \\
0 & 0 & 0 \\
-0.5742 & -0.4175 & -0.5423 \\
-0.3931 & 0.5414 & 0.4161 \\
0 & 0 & 0 \\
0.5742 & 0.4175 & -0.5423 \\
0.0491 & -0.0677 & 0 \\
0 & 0 & 0 \\
0 & 0 & 0.0678
\end{bmatrix}
\begin{bmatrix}
Q_1 \\
Q_2 \\
Q_3
\end{bmatrix}
\tag{7-189}
$$

where the $\Delta \alpha_i$ are in Å and the Q_r in $u^{1/2} Å$. We can determine the values of the normal coordinates from this equation; for example, combining Eqs. (7-184) and (7-189) we have

$$\Delta x_1 = -0.3931 Q_1 + 0.5414 Q_2 - 0.4161 Q_3 = -0.8092,$$
$$\Delta x_3 = \quad 0.0491 Q_1 - 0.0677 Q_2 \qquad\qquad = \quad 0.0491,$$

and $\tag{7-190}$

$$\Delta z_3 = \qquad\qquad\qquad\qquad 0.0678 Q_3 = \quad 0.0678.$$

Inverting these three equations we determine that (in $u^{1/2}$ Å)

$$(Q_1, Q_2, Q_3) = (1, 0, 1). \tag{7-191}$$

Setting each $Q_i = 1 \, u^{1/2} Å$ in turn in Eq. (7-189) the reader can determine the Cartesian displacement coordinates and picture the molecular deformation that each normal coordinate describes.

---

[2] Specifically from this reference we use the elements of the 3×3 $\mathscr{L}$ matrix [see Eqs. (38) and (39a)] from Table 2, and the U matrix [see Eq. (37)] as implied by the expressions for the S_i in Table 1, to calculate the so-called L matrix using Eq. (39a). The l matrix is calculated from the L matrix using Eq. (18); the so-called B matrix required for this determination is discussed in general terms on pages 54–61 in Wilson, Decius, and Cross (1955). See also Eqs. (8-21)–(8-23) in the next chapter here.

In summary for the deformed water molecule of Fig. 7-8 we have made the coordinate transformation from

$$(\xi_2, \eta_2, \zeta_2, \xi_3, \eta_3, \zeta_3) = (0.6760, 0.3138, 0.5614, -0.0930, 0.0360, 0.0872) \text{ in Å}$$
(7-192)

to

$$(\theta, \phi, \chi, Q_1, Q_2, Q_3) = (30°, 60°, 120°, 1 \text{ u}^{1/2}\text{Å}, 0, 1 \text{ u}^{1/2}\text{Å}) \quad (7\text{-}193)$$

by using the Eckart equations and the l matrix. This coordinate transformation applied generally leads to the expression for $\hat{T}_N$ given in Eq. (7-137).

Problem 7-3. Make the coordinate transformation from the (ξ, η, ζ) coordinates of Eq. (7-192) to $(\theta, \phi, \chi, Q_1, Q_2, Q_3)$ using molecule fixed (x, y, z) axes that are principal inertial axes rather than Eckart axes.

Answer. For a triatomic molecule one principal inertial axis is perpendicular to the plane of the three nuclei. To follow the axis labeling convention adopted for the molecule in its equilibrium configuration we must choose this axis to be the y axis; as a result $I_{xy} = I_{yz} = 0$. To determine the directions of the other principal axes we use the angle ε as introduced in Fig. 7-8, and the principal axis equation that defines ε is

$$\sum_{i=1}^{3} m_i x_i z_i = 0. \quad (7\text{-}194)$$

Substituting Eqs. (7-176)–(7-181) into this equation we determine that for principal axes

$$\varepsilon = 63.63°. \quad (7\text{-}195)$$

Using this value of ε in Eqs. (7-176)–(7-181) we determine the (x, y, z) coordinates of the three nuclei to be (in Å)

$$(-0.0967, 0, -2.296), \quad (-0.8448, 0, 0.3965), \quad (0.0588, 0, 0.1187). \quad (7\text{-}196)$$

Using the direction cosine matrix elements to relate these to the (ξ, η, ζ) coordinates we determine the Euler angles to be

$$(\theta, \phi, \chi) = (30.3°, 128.3°, 59.1°). \quad (7\text{-}197)$$

The $(\Delta x_i, \Delta y_i, \Delta z_i)$ displacements from equilibrium are determined to be (in Å)

$$(0.4211, 0, -1.5368), \quad (-0.3270, 0, -0.3627), \quad (-0.0059, 0, 0.1187). \quad (7\text{-}198)$$

The three principal axis equations are

$$\sum_i m_i x_i y_i = \sum_i m_i y_i z_i = \sum_i m_i z_i x_i = 0, \quad (7\text{-}199)$$

which lead to three nonlinear equations in the $\Delta\alpha_i$. Using Eckart axes we obtain three linear equations [see Eqs. (7-185)–(7-187)] and we relate the normal coordinates to the $\Delta\alpha_i$ by $3N$ linear equations; in this process Eq. (7-189) is obtained [but see Eqs. (7-233)–(7-238), which follow]. Using principal axes the formalism cannot be completely linear. We do not discuss the relationship of the $\Delta\alpha_i$ to the Q_r in these circumstances [the $\Delta\alpha_i$ of Eq. (7-198) are not consistent with Eq. (7-189)].

Comparing the solution to Problem 7-3 with the results in Eq. (7-193) we see that the Euler angles are changed drastically when we use principal axes rather than Eckart axes. In Fig. 7-9 the location of the Eckart axes and principal axes on this deformed water molecule are compared; in each case an equilibrium configuration water molecule having the same Euler angles, so that its (x, y, z) axes are coincident with those of the deformed molecule, is superimposed. We can appreciate from these figures that this deformation (a stretching of the OH_1 bond) causes a significant rotation of principal axes (see Fig. 7-9b) but very little rotation of Eckart axes (see Fig. 7-9a). As we move H_1 towards and away from O_3 the Eckart axes will hardly move

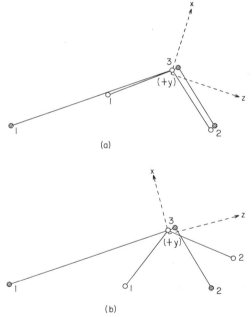

Fig. 7-9. (a) The deformed water molecule of Eq. (7-167) (shaded circles) with molecule fixed *Eckart axes* (x, y, z) and an appropriately oriented water molecule in its equilibrium configuration (open circles). (b) The same with molecule fixed *principal axes* (x, y, z) and an appropriately oriented water molecule in its equilibrium configuration (open circles).

whereas the principal axes will twist back and forth about the y axis and much rotational motion is "mixed into" this vibrational motion. It can be shown that if we use principal axes the rotation–vibration mixing caused by the Coriolis coupling terms in the Hamiltonian is much larger than if we use Eckart axes; in fact the choice of Eckart axes generally minimizes these terms as discussed from Eq. (7-122) to Eq. (7-126). As a result, the separation of $\hat{T}_N$ into rotational and vibrational parts is achieved with minimum approximation using the Eckart axes. This is why Eckart axes are used.

We can view the choice of axes in mathematical or physical terms. Mathematically we are changing the $(3N - 3)$ coordinates from $(\xi_2, \ldots, \zeta_N)$ to $(\theta, \phi, \chi, Q_r)$ where each of the coordinates $(\theta, \phi, \chi, Q_r)$ is a function of the coordinates $(\xi_2, \ldots, \zeta_N)$. We make this coordinate change in such a way that the rotation–vibration Hamiltonian can be separated into $\hat{H}(\hat{J}_x, \hat{J}_y, \hat{J}_z) + \hat{H}(Q_1, \hat{P}_1) + \cdots + \hat{H}(Q_{3N-6}, \hat{P}_{3N-6})$ with minimum approximation. The Eckart conditions and use of the l matrix generally ensure this. Physically we consider the (θ, ϕ, χ) coordinates to define the orientation of the molecule fixed axes, and the Q_r to define vibrational normal coordinate displacements. Our coordinate change then *defines* what we mean by "rotation" and "normal coordinate." For example, if we stretch an OH bond in the water molecule it is obvious by how much the molecular center of mass has been translated (and therefore how much translational motion has occurred), but it is a matter of axis system definition as to how much we consider the molecule (or, equivalently, the molecule fixed axes) to have rotated. In our numerical example the deformation of a water molecule from the equilibrium configuration described by the (ξ, η, ζ) nuclear coordinates of Eq. (7-155) to those of Eq. (7-167) is *defined* as being a purely vibrational deformation using Eckart axes, since the Euler angles of the deformed configuration are the same as those of the equilibrium configuration. Using principal axes this is not the case.

The Transformation Properties of the Rovibronic Coordinates

For a rigid nonlinear polyatomic molecule we want to consider the transformation properties of three Euler angles (θ, ϕ, χ), $3N - 6$ normal coordinates $(Q_1, \ldots, Q_{3N-6})$, and $3n$ electronic coordinates $(x_{N+1}, \ldots, z_{N+n})$ under the effect of nuclear permutations and the inversion. Because of the variety of equilibrium geometries of polyatomic molecules, and varieties of numbers of identical nuclei, we cannot summarize these transformation properties with a few general equations such as Eqs. (7-112), (7-115), and (7-116) for the diatomic molecule. In this section we will consider the transformation properties of the rovibronic coordinates of the water molecule and the methyl fluoride molecule. The results for other molecules will be considered as and when necessary later in the book.

The Euler Angles   To determine the transformation properties of the Euler angles it is simplest to consider the situation in which all Q_r are zero, i.e., when the molecule is in its equilibrium configuration. We can determine the transformation properties of the Euler angles either by studying drawings such as Fig. 7-3 or by studying the equations that define the Euler angles. We will use both approaches here.

The effect of the permutation (12) on a water molecule in its equilibrium configuration is shown in Fig. 7-10. In this figure we *first* permute the nuclei and *then* use the prescription discussed before Eq. (7-153) to attach the (x, y, z) axes. The permutation of identical nuclei in a molecule does not change the orientation of the principal axes in space since identical nuclei have the same mass. However, the choice of which of the molecule fixed axes is to be identified with which principal axis and the choice of sign for the x, y, and z axes depends on the labels on the nuclei. We see that in the water molecule the molecule fixed (x, y, z) axes are rotated by π radians about the x axis by (12), and the change in the Euler angles caused by this is

$$(12)(\theta, \phi, \chi) = (\pi - \theta, \phi + \pi, 2\pi - \chi). \tag{7-200}$$

It is useful to determine the transformations that the Euler angles undergo when the (x, y, z) axes are rotated about various axes. The rotation R_α^π of the (x, y, z) axes through π radians about an axis in the xy plane making an angle α with the x axis (α is measured in a right handed sense about the z axis) changes the Euler angles to $(\pi - \theta, \phi + \pi, 2\pi - 2\alpha - \chi)$. This result is obtained from a careful study of Fig. 7-1 and of the effect of this rotation on the (x, y, z) axes. Similarly the rotation R_z^β by β radians about the z axis (measured in the right handed sense about z) changes the Euler angles to $(\theta, \phi, \chi + \beta)$. These results are given in Table 7-1 for easy reference. Knowing the transformation properties of the Euler angles, those of $\hat{J}_x, \hat{J}_y,$ and $\hat{J}_z$ are easily deduced [see Eqs. (7-144)–(7-146)] and are included in Table 7-1.

The transformation of the Euler angles caused by (12) in the water molecule can equally well be derived by using Eqs. (7-158), (7-160), (7-161), (7-163), and (7-164) together with the result [see Eqs. (1-10) and (1-18)]

$$(12)(\xi_1, \eta_1, \zeta_1, \xi_2, \eta_2, \zeta_2, \xi_3, \eta_3, \zeta_3) = (\xi_1', \eta_1', \zeta_1', \xi_2', \eta_2', \zeta_2', \xi_3', \eta_3', \zeta_3')$$

$$= (\xi_2, \eta_2, \zeta_2, \xi_1, \eta_1, \zeta_1, \xi_3, \eta_3, \zeta_3). \tag{7-201}$$

From Eq. (7-201) we deduce that

$$(12)(\xi_2 - \xi_1) = -(\xi_2 - \xi_1),$$

$$(12)(\eta_2 - \eta_1) = -(\eta_2 - \eta_1),$$

and $(7\text{-}202)$

$$(12)(\zeta_2 - \zeta_1) = -(\zeta_2 - \zeta_1).$$

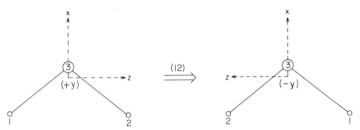

Fig. 7-10. The effect of the nuclear permutation (12) on the nuclei of a water molecule in its equilibrium configuration. The molecule fixed axes are attached afterwards using the convention of Fig. 7-5.

Table 7-1

The transformation properties[a] of the Euler angles and of the components of the rovibronic angular momentum $\hat{J}$

	$R_\alpha{}^\pi$	$R_z{}^\beta$
θ	$\pi - \theta$	θ
ϕ	$\phi + \pi$	ϕ
χ	$2\pi - 2\alpha - \chi$	$\chi + \beta$
$\hat{J}_x$	$\hat{J}_x \cos 2\alpha + \hat{J}_y \sin 2\alpha$	$\hat{J}_x \cos \beta + \hat{J}_y \sin \beta$
$\hat{J}_y$	$\hat{J}_x \sin 2\alpha - \hat{J}_y \cos 2\alpha$	$-\hat{J}_x \sin \beta + \hat{J}_y \cos \beta$
$\hat{J}_z$	$-\hat{J}_z$	$\hat{J}_z$

[a] $R_\alpha{}^\pi$ is a rotation of the molecule fixed (x, y, z) axes through π radians about an axis in the xy plane making an angle α with the x axis (α is measured in the right handed sense about the z axis), and $R_z{}^\beta$ is a rotation of the molecule fixed (x, y, z) axes through β radians about the z axis (β is measured in the right handed sense about the z axis). The expressions for the $\hat{J}_\alpha$ are given in Eqs. (7-144)–(7-146).

The α_i^e are constants [see Eq. (7-153)] unaffected by any symmetry operation. Hence, from Eqs. (7-158), (7-160), (7-161), (7-163), and (7-164),

$$\cos \theta' = -[(\zeta_2 - \zeta_1)/(z_2^e - z_1^e)] = -\cos \theta,$$

$$\cos \phi' = -\cos \phi, \qquad \sin \phi' = -\sin \phi, \qquad (7\text{-}203)$$

$$\cos \chi' = \cos \chi, \qquad \text{and} \qquad \sin \chi' = -\sin \chi.$$

From these equations the result in Eq. (7-200) follows.

The effect of E^* on a water molecule is shown in Fig. 7-11 and the (x, y, z) axes have been attached after performing E^* by using the prescription for

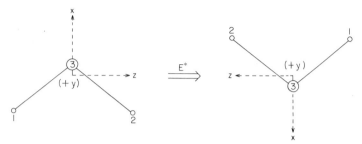

Fig. 7-11.   The effect of the inversion E^* on the nuclei of a water molecule in its equilibrium configuration. The molecule fixed axes are attached afterwards using the convention of Fig. 7-5.

attaching the axes given before Eq. (7-153). This operation for a water molecule has the effect of rotating the axes by π radians about the y axis, and using $R^\pi_{\pi/2}$ from Table 7-1 we deduce that

$$E^*(\theta,\phi,\chi) = (\pi - \theta, \phi + \pi, \pi - \chi). \tag{7-204}$$

Alternatively using

$$E^*(\xi_i,\eta_i,\zeta_i) = (-\xi_i, -\eta_i, -\zeta_i) \tag{7-205}$$

and the expressions for the Euler angles given in Eqs. (7-158)–(7-164) the same result can be obtained. The reader should check this. The successive application of E^* and (12) for a water molecule leads to a rotation of the molecule by π radians about the z axis. Using R_z^π from Table 7-1 we have

$$(12)^*(\theta,\phi,\chi) = (\theta,\phi,\chi + \pi). \tag{7-206}$$

Each element of the MS group of a molecule will, among other things, cause a certain change in the Euler angles of the molecule. This change in the Euler angles can be represented as an *equivalent rotation*. For example, for the H_2O molecule discussed above the equivalent rotations of the operations E, (12), E^*, and (12)* are R^0, $R_0{}^\pi$, $R_{\pi/2}^\pi$, and $R_z{}^\pi$, respectively, where we use the notation of Table 7-1, and R^0 is the identity. We can alternatively write these equivalent rotations as R^0, $R_x{}^\pi$, $R_y{}^\pi$, and $R_z{}^\pi$, respectively.

We will now consider the CH_3F molecule and determine the equivalent rotations of the MS group elements (123) and (23)*. In Fig. 7-12 we show two views of a methyl fluoride molecule in its equilibrium configuration with the molecule fixed (x, y, z) axes attached. In Fig. 7-13 we show the effect of the operations (123) and (23)* on the molecule, using view (b) of Fig. 7-12 as the starting position. After having performed the MS group operations the (x, y, z) axes are attached following the convention of Fig. 7-12. It is obvious that (123) has resulted in the molecule fixed axes being rotated in a right handed sense through $2\pi/3$ radians about the z axis, and that (23)* has resulted in the axes being rotated through π radians about the y axis.

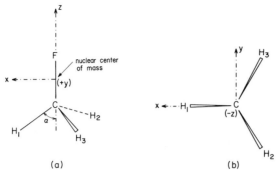

Fig. 7-12. The nuclear labeling convention and (x, y, z) axis definition for a CH_3F molecule in its equilibrium configuration. In (b) the three protons are above the plane of the page and the z axis points down into the page.

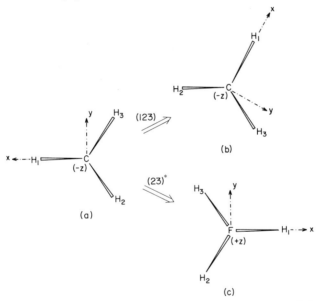

Fig. 7-13. The effects of the operations (123) and (23)* on a CH_3F molecule in its equilibrium configuration. In (c) the three protons are below the plane of the page and the z axis points up out of the page.

The equivalent rotations of (123) and (23)* are $R_z^{2\pi/3}$ and $R_{\pi/2}^\pi$, respectively, and we can write

$$(123)(\theta, \phi, \chi) = R_z^{2\pi/3}(\theta, \phi, \chi) = (\theta, \phi, \chi + 2\pi/3) \tag{7-207}$$

and

$$(23)^*(\theta, \phi, \chi) = R_{\pi/2}^\pi(\theta, \phi, \chi) = (\pi - \theta, \phi + \pi, \pi - \chi). \tag{7-208}$$

The transformation properties of the Euler angles are rather easily obtained from Fig. 7-13, but if you prefer an algebraic proof the same results can be obtained algebraically as follows:

For a CH_3F molecule in its equilibrium configuration we let the distances of the fluorine and carbon nuclei from the nuclear center of mass be r_F and r_C respectively, the CH bond length be r_H, and the angle α be as defined in Fig. 7-12a; the (x, y, z) coordinates of the nuclei can be written in terms of these four parameters. Using the direction cosine matrix [see Eqs. (7-50)–(7-52)] we can write the (ξ, η, ζ) coordinates of each nucleus in terms of the (x, y, z) coordinates (i.e., in terms of r_F, r_C, r_H, and α) and the Euler angles, i.e.,

$$\xi_i = \lambda_{x\xi} x_i + \lambda_{y\xi} y_i + \lambda_{z\xi} z_i,$$
$$\eta_i = \lambda_{x\eta} x_i + \lambda_{y\eta} y_i + \lambda_{z\eta} z_i, \tag{7-209}$$
$$\zeta_i = \lambda_{x\zeta} x_i + \lambda_{y\zeta} y_i + \lambda_{z\zeta} z_i,$$

where

$$x_1 = r_H \sin \alpha, \qquad y_1 = 0, \qquad z_1 = z_2 = z_3 = -r_C - r_H \cos \alpha,$$
$$x_2 = x_3 = -(1/2)r_H \sin \alpha, \qquad y_3 = -y_2 = (\sqrt{3}/2)r_H \sin \alpha, \tag{7-210}$$
$$x_C = y_C = x_F = y_F = 0, \qquad z_C = -r_C, \qquad z_F = r_F.$$

The effect of the operations (123) and (23)* on the Euler angles can now be determined. By definition

$$(123)(\xi_1, \xi_F, \eta_F, \zeta_F) = (\xi_1', \xi_F', \eta_F', \zeta_F') = (\xi_3, \xi_F, \eta_F, \zeta_F) \tag{7-211}$$

and we write

$$(123)(\theta, \phi, \chi) = (\theta', \phi', \chi'). \tag{7-212}$$

From Eq. (7-211) and Eq. (7-52)

$$\zeta_F' = \zeta_F, \tag{7-213}$$

i.e.,

$$r_F \cos \theta' = r_F \cos \theta. \tag{7-214}$$

Thus

$$\cos \theta' = \cos \theta \tag{7-215}$$

and since $0 \le \theta \le \pi$ this implies that

$$\theta' = \theta. \tag{7-216}$$

Similarly

$$\xi_F' = \xi_F \qquad \text{and} \qquad \eta_F' = \eta_F, \tag{7-217}$$

so that

$$r_F \sin \theta' \cos \phi' = r_F \sin \theta \cos \phi \qquad (7\text{-}218)$$

and

$$r_F \sin \theta' \sin \phi' = r_F \sin \theta \sin \phi. \qquad (7\text{-}219)$$

Since $\theta' = \theta$ these equations imply

$$\cos \phi' = \cos \phi \qquad \text{and} \qquad \sin \phi' = \sin \phi \qquad (7\text{-}220)$$

and we deduce the result

$$\phi' = \phi. \qquad (7\text{-}221)$$

We finally determine χ', and to do that we use the equation

$$\xi_1' = \xi_3, \qquad (7\text{-}222)$$

i.e.,

$$(\cos \theta' \cos \phi' \cos \chi' - \sin \phi' \sin \chi')r_H \sin \alpha - \sin \theta' \cos \phi'(r_C + r_H \cos \alpha)$$
$$= (\cos \theta \cos \phi \cos \chi - \sin \phi \sin \chi)(-1/2)r_H \sin \alpha$$
$$- (\cos \theta \cos \phi \sin \chi + \sin \phi \cos \chi)(\sqrt{3}/2)r_H \sin \alpha$$
$$- \sin \theta \cos \phi(r_C + r_H \cos \alpha). \qquad (7\text{-}223)$$

Since $\phi' = \phi$, $\theta' = \theta$, and Eq. (7-223) must be true for all θ and ϕ it follows that

$$\sin \chi' = (\sqrt{3}/2)\cos \chi + (-1/2)\sin \chi \qquad (7\text{-}224a)$$

and

$$\cos \chi' = (-1/2)\cos \chi - (\sqrt{3}/2)\sin \chi \qquad (7\text{-}224b)$$

from which we deduce that

$$\chi' = (\chi + 2\pi/3). \qquad (7\text{-}225)$$

The effect of the operation (23)* can be similarly determined from the equations

$$(23)^*(\xi_1, \xi_F, \eta_F, \zeta_F) = (\xi_1', \xi_F', \eta_F', \zeta_F') = (-\xi_1, -\xi_F, -\eta_F, -\zeta_F) \quad (7\text{-}226)$$

and

$$(23)^*(\theta, \phi, \chi) = (\theta', \phi', \chi'). \qquad (7\text{-}227)$$

From $\xi_F' = -\xi_F$, $\eta_F' = -\eta_F$, and $\zeta_F' = -\zeta_F$ we deduce that

$$\sin \theta' \cos \phi' = -\sin \theta \cos \phi, \qquad (7\text{-}228)$$

$$\sin \theta' \sin \phi' = -\sin \theta \sin \phi, \qquad (7\text{-}229)$$

and

$$\cos\theta' = -\cos\theta, \qquad (7\text{-}230)$$

respectively. From $\xi_1' = -\xi_1$ we deduce that

$$(\cos\theta'\cos\phi'\cos\chi' - \sin\phi'\sin\chi')r_H\sin\alpha - \sin\theta'\cos\phi'(r_C + r_H\cos\alpha)$$
$$= -(\cos\theta\cos\phi\cos\chi - \sin\phi\sin\chi)r_H\sin\alpha + \sin\theta\cos\phi(r_C + r_H\cos\alpha). \qquad (7\text{-}231)$$

These four equations lead to the solution

$$(\theta',\phi',\chi') = (\pi - \theta, \phi + \pi, \pi - \chi). \qquad (7\text{-}232)$$

The Normal Coordinates To determine the transformation properties of the normal coordinates it is useful to extend Eq. (7-136) by adding rotational and translational coordinates. We define rotational coordinates R_x, R_y, and R_z as essentially those combinations of Cartesian displacement coordinates $\Delta\alpha_i$, given in Eqs. (7-185)–(7-187), that vanish using Eckart conditions, and the translational coordinates T_x, T_y, and T_z as essentially those three combinations of $\Delta\alpha_i$ given in Eq. (7-188) that vanish as a result of the center of mass condition. We say "essentially" because we weight these coordinates appropriately so that the extended $3N \times 3N$ l matrix obtained by writing

$$\begin{bmatrix} m_1^{1/2}\,\Delta x_1 \\ \vdots \\ m_N^{1/2}\,\Delta z_N \end{bmatrix} = \begin{bmatrix} \\ l_{\alpha i, r} \\ \\ \end{bmatrix} \begin{bmatrix} Q_1 \\ \vdots \\ Q_{3N-6} \\ R_x \\ \vdots \\ T_z \end{bmatrix}, \qquad (7\text{-}233)$$

is orthogonal, satisfying

$$\sum_{\alpha, i} l_{\alpha i, r} l_{\alpha i, s} = \delta_{rs}, \qquad (7\text{-}234)$$

where $\alpha = x, y$, or z; $i = 1$ to N; $r, s = 1$ to $3N$. This requires that

$$T_\alpha = M_N^{-1/2} \sum_{i=1}^{N} m_i^{1/2}(m_i^{1/2}\,\Delta\alpha_i), \qquad (7\text{-}235)$$

where M_N is the total mass of all the nuclei in the molecule, and (using $\mu_{xx}^e = (I_{xx}^e)^{-1} = \{\sum_i m_i[(y_i^e)^2 + (z_i^e)^2]\}^{-1}$)

$$R_x = (\mu_{xx}^e)^{1/2} \sum_{i=1}^{N} m_i^{1/2}[y_i^e(m_i^{1/2}\,\Delta z_i) - z_i^e(m_i^{1/2}\,\Delta y_i)] \qquad (7\text{-}236)$$

with R_y and R_z given by cyclic permutation of x, y, and z in R_x. We call the six R_α and T_α *zero frequency normal coordinates* and write

$$[R_x, R_y, R_z, T_x, T_y, T_z] = [Q_{3N-5}, Q_{3N-4}, \ldots, Q_{3N}]. \qquad (7\text{-}237)$$

We can then write

$$\Delta\alpha_i = \sum_{r=1}^{3N} m_i^{-1/2} l_{\alpha i, r} Q_r. \qquad (7\text{-}238)$$

The center of mass and Eckart conditions [Eqs. (7-185)–(7-188)] demand that the zero frequency normal coordinates vanish.

The transformation from the $3N$ $\Delta\alpha_i$ to the $3N$ Q_r is orthogonal, and thus the Q_r will generate the same representation of the symmetry group as do the $\Delta\alpha_i$ [see Eqs. (5-54) and remarks after it]. To determine the representation generated by the $3N - 6$ vibrational Q_r we must determine the representation generated by the $3N$ $\Delta\alpha_i$ and then subtract the representation generated by the three R_α and three T_α. We will use the water molecule as an example of this. An alternative procedure for determining the species of the normal coordinates will be discussed in the next chapter, and this technique involves using internal coordinate displacements (bond stretches, angle bends, etc.) rather than Cartesian displacement coordinates.

To determine the symmetry species of the normal coordinates of water in the $C_{2v}(M)$ group (see Table 5-2) we must begin by determining the transformation properties of the Cartesian displacement coordinates under the effect of the symmetry operations of $C_{2v}(M)$. From these transformation properties we can determine the representation of the $C_{2v}(M)$ group generated by the Cartesian displacement coordinates; we will call this representation Γ_{Car}. Once again we have the option of obtaining the results by studying figures or by studying equations. We do not need to do either to see that the character under the identity E in Γ_{Car} using the nine $(\Delta x_1, \ldots, \Delta z_3)$ is

$$\chi[E] = 9. \qquad (7\text{-}239)$$

The effect of the operation (12) is depicted in Fig. 7-14 where Cartesian displacements of the nuclei have been added. We see that

$$(12)(\Delta x_1, \Delta x_2, \Delta x_3) = (\Delta x_1', \Delta x_2', \Delta x_3') = (\Delta x_2, \Delta x_1, \Delta x_3). \qquad (7\text{-}240)$$

Since the y and z axes are reversed by (12) and the nuclei 1 and 2 permuted we have

$$\begin{aligned}
(12)&(\Delta y_1, \Delta z_1, \Delta y_2, \Delta z_2, \Delta y_3, \Delta z_3) \\
&= (\Delta y_1', \Delta z_1', \Delta y_2', \Delta z_2', \Delta y_3', \Delta z_3') \qquad (7\text{-}241) \\
&= (-\Delta y_2, -\Delta z_2, -\Delta y_1, -\Delta z_1, -\Delta y_3, -\Delta z_3).
\end{aligned}$$

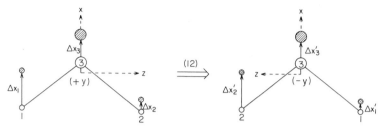

Fig. 7-14. The effect of the nuclear permutation (12) on the Cartesian displacements Δx_i in a water molecule. The new values $(\Delta x_i')$ are related to the old values (Δx_i) by $\Delta x_1' = \Delta x_2$, $\Delta x_2' = \Delta x_1$, and $\Delta x_3' = \Delta x_3$.

As a result we obtain the character in Γ_{Car} under (12) as

$$\chi[(12)] = -1. \tag{7-242}$$

To use equations instead of figures we start by writing

$$\Delta \alpha_i = \alpha_i - \alpha_i^e = (\lambda_{\alpha\xi}\xi_i + \lambda_{\alpha\eta}\eta_i + \lambda_{\alpha\zeta}\zeta_i) - \alpha_i^e. \tag{7-243}$$

The direction cosines are functions of the Euler angles [see Eq. (7-52)] and we can determine the effect of (12) on them from the results in Eqs. (7-200) or (7-203). The effect of (12) on the (ξ_i, η_i, ζ_i) is obtained from Eq. (7-201). The α_i^e are constants [see Eq. (7-153)] and are unaffected by any symmetry operation. We see that

$$
\begin{aligned}
(12)\Delta z_1 = \Delta z_1' &= \sin\theta'\cos\phi'\xi_1' + \sin\theta'\sin\phi'\eta_1' + \cos\theta'\zeta_1' - z_1^e \\
&= -\sin\theta\cos\phi\xi_2 - \sin\theta\sin\phi\eta_2 - \cos\theta\zeta_2 + 0.7592\,\text{Å} \\
&= -z_2 + 0.7592\,\text{Å} = -\Delta z_2,
\end{aligned} \tag{7-244}
$$

where we have used the result that $-z_1^e = z_2^e = 0.7592\,\text{Å}$.

The transformation properties of the Cartesian displacement coordinates under the effect of E^* and $(12)^*$ can be similarly determined and the results are collected together in Table 7-2. The characters of the representation Γ_{Car} are also given in Table 7-2 and using Eq. (4-43) we express this in terms of irreducible representations as:

$$\Gamma_{Car} = 3A_1 \oplus A_2 \oplus 2B_1 \oplus 3B_2. \tag{7-245}$$

To determine the character in Γ_{Car} under the effect of a permutation or permutation-inversion we only need consider the transformations of the coordinates of unpermuted nuclei since these will give the only nonvanishing contributions on the diagonal of the representation matrix. Thus to determine the character in Γ_{Car} under the effect of (12) or $(12)^*$ we only need consider the transformations of the coordinates $\Delta x_3, \Delta y_3$, and Δz_3 in the water molecule.

Table 7-2

The transformation properties of the coordinates
in the $C_{2v}(M)$ group for the water molecule

E	(12)	E^*	(12)*	
Δx_1	Δx_2	Δx_1	Δx_2	
Δx_2	Δx_1	Δx_2	Δx_1	
Δx_3	Δx_3	Δx_3	Δx_3	
Δy_1	$-\Delta y_2$	$-\Delta y_1$	Δy_2	
Δy_2	$-\Delta y_1$	$-\Delta y_2$	Δy_1	
Δy_3	$-\Delta y_3$	$-\Delta y_3$	Δy_3	
Δz_1	$-\Delta z_2$	Δz_1	$-\Delta z_2$	
Δz_2	$-\Delta z_1$	Δz_2	$-\Delta z_1$	
Δz_3	$-\Delta z_3$	Δz_3	$-\Delta z_3$	
χ_{Car}: 9	-1	3	1	
T_x	T_x	T_x	T_x	A_1
T_y	$-T_y$	$-T_y$	T_y	B_1
T_z	$-T_z$	T_z	$-T_z$	B_2
R_x	R_x	$-R_x$	$-R_x$	A_2
R_y	$-R_y$	R_y	$-R_y$	B_2
R_z	$-R_z$	$-R_z$	R_z	B_1

Using Eqs. (7-235) and (7-236) we obtain (in $u^{1/2}$ Å units)

$$T_x = (\Delta x_1 + \Delta x_2 + 16\,\Delta x_3)/18^{1/2},$$
$$T_y = (\Delta y_1 + \Delta y_2 + 16\,\Delta y_3)/18^{1/2},$$
$$T_z = (\Delta z_1 + \Delta z_2 + 16\,\Delta z_3)/18^{1/2},$$
$$R_x = (\Delta y_1 - \Delta y_2)/2^{1/2},$$
$$R_y = 0.3907[1.466(\Delta x_2 - \Delta x_1) + \Delta z_1 + \Delta z_2 - 2\,\Delta z_3],$$

(7-246)

and

$$R_z = 2(-\Delta y_1 - \Delta y_2 + 2\,\Delta y_3)/3.$$

The transformation properties of these coordinates can be determined from the transformation properties of the $\Delta \alpha_i$ coordinates given in Table 7-2 and the results are given in the bottom half of Table 7-2. We see that for H_2O

$$\Gamma(T_x, T_y, T_z, R_x, R_y, R_z) = A_1 \oplus A_2 \oplus 2B_1 \oplus 2B_2.$$ (7-247)

Subtracting Eq. (7-247) from Eq. (7-245) we determine that the normal coordinate representation is

$$\Gamma(Q_1, Q_2, Q_3) = 2A_1 \oplus B_2.$$ (7-248)

As we will see below Eq. (8-194) in the next chapter the normal coordinates each transform according to an irreducible representation of the symmetry group and we have

$$\Gamma(Q_1) = \Gamma(Q_2) = A_1 \qquad (7\text{-}249)$$

and

$$\Gamma(Q_3) = B_2. \qquad (7\text{-}250)$$

The determination of the representation of the group that the zero frequency normal coordinates T_α and R_α generate does not require that we determine the coordinates as done in Eq. (7-246) for the water molecule. As will be shown in Chapter 11 the R_α transform in the same way as the $\hat{J}_\alpha$, and the representation generated by the $\hat{J}_\alpha$ can be determined from the results in Table 7-1 once the equivalent rotations of the elements of the group have been determined. The species of the $\hat{J}_\alpha$ and T_α (the species of the T_α follow from the species of the $\hat{J}_\alpha$ as will also be shown in Chapter 11) are given for each MS group in the character tables of Appendix A.

The Electronic Coordinates   The transformation properties of the electronic coordinates in the (x, y, z) axis system follow from the transformation properties of the Euler angles. For water we have for any electron i the results

$$(12)(x_i, y_i, z_i) = (x_i, -y_i, -z_i),$$
$$E^*(x_i, y_i, z_i) = (x_i, -y_i, z_i), \qquad (7\text{-}251)$$

and

$$(12)^*(x_i, y_i, z_i) = (x_i, y_i, -z_i).$$

Problem 7-4.   We have determined above that a deformed water molecule having nuclear coordinates given as in Eq. (7-167) has rovibronic coordinates given by (where Q_r is in $u^{1/2}$ Å)

$$(\theta, \phi, \chi, Q_1, Q_2, Q_3, x_1, \ldots, z_n) = (30°, 60°, 120°, 1, 0, 1, x_1, \ldots, z_n). \qquad (7\text{-}252)$$

If we permute the two protons the (ξ, η, ζ) coordinates of nuclei H_1, H_2, and O_3 become

$$(0.6760, 0.3138, 0.5614), \quad (0.8136, -0.8891, -1.9562),$$
$$\text{and} \quad (-0.0930, 0.0360, 0.0872), \qquad (7\text{-}253)$$

respectively, and the (ξ, η, ζ) coordinates of the electrons are unaffected. Use the methods of Eqs (7-167)–(7-191) to determine the values of the rovibronic coordinates for the deformed water molecule represented in Eq. (7-253). Check the answer against the results obtained from Eqs. (7-200), (7-249), (7-250), and (7-251) for the effect of (12).

Answer. We first use the Eckart equations, Eqs. (7-132)–(7-134), to determine the Euler angles. The $[\alpha\tau]$ factors are evaluated from the equilibrium nuclear coordinates of Eq. (7-153) and the (ξ,η,ζ) coordinates of Eq. (7-253). Doing this we obtain Eqs. (7-168) and (7-169) as before, but Eq. (7-134) does not give Eq. (7-170) now and instead we obtain

$$+0.1045(\cos\theta\cos\phi\cos\chi - \sin\phi\sin\chi) - 0.9132(\cos\theta\sin\phi\cos\chi + \cos\phi\sin\chi)$$
$$+1.9114\sin\theta\cos\chi + 0.8676\sin\theta\cos\phi - 0.3352\sin\theta\sin\phi$$
$$-0.8125\cos\theta = 0. \tag{7-254}$$

Solving Eqs. (7-168), (7-169), and (7-254) simultaneously we obtain

$$(\theta,\phi,\chi) = (150°, 240°, 240°), \tag{7-255}$$

which differs from the Euler angles of Eq. (7-252) according to the result of Eq. (7-200); hence the molecule fixed axes of the distorted molecule are rotated through π radians about the x axis by (12).

From these Euler angles we can determine the (x, y, z) nuclear coordinates from the (ξ,η,ζ) coordinates in Eq. (7-253) by using the direction cosines, i.e., Eq. (7-50). The (x, y, z) coordinates of nuclei 1, 2, and 3, respectively, are found to be

$$(-0.4948, 0, -0.7911), \quad (-1.3270, 0, 1.8757), \quad (0.1138, 0, -0.0678). \tag{7-256}$$

These results can be compared with the results in Eq. (7-183). From Eqs. (7-256) and (7-153) we determine the Cartesian displacement coordinates, and using the inverse of Eq. (7-190) we determine the values of the normal coordinates to be (where Q_r is in $u^{1/2}$ Å)

$$(Q_1, Q_2, Q_3) = (1, 0, -1), \tag{7-257}$$

which agrees with Eqs. (7-249) and (7-250). From the Euler angles in Eq. (7-255) we determine the (x, y, z) coordinates of the electrons to be changed according to Eq. (7-251) for (12).

BIBLIOGRAPHICAL NOTES

The Rotation–Vibration Hamiltonian

Wilson and Howard (1936). The original derivation of the rotation–vibration Hamiltonian.

Darling and Dennison (1940). This paper gives a modification of the original derivation in order to ensure that the Hamiltonian is hermitian.

Watson (1968). The rotation–vibration Hamiltonian is simplified for a nonlinear rigid polyatomic molecule.

Louck (1976). The rotation–vibration Hamiltonian of a polyatomic molecule is derived by Method I.

8

The Rovibronic Wavefunctions

This chapter is concerned with the determination of approximate rovibronic wavefunctions, and these are functions of the $(3l - 3)$ rovibronic coordinates discussed in Chapter 7. To obtain these wavefunctions we make the following approximations: (i) the Born–Oppenheimer approximation, (ii) the molecular orbital approximation to the electronic wavefunction, (iii) the use of a linear combination of atomic orbitals for each molecular orbital, (iv) the harmonic oscillator approximation for the vibrational Hamiltonian, and (v) the rigid rotor approximation for the rotational Hamiltonian. The approximate rovibronic wavefunctions Φ_{rve}^0 that we obtain are products of rotational, vibrational, and electronic wavefunctions.

THE BORN–OPPENHEIMER APPROXIMATION

The rovibronic Schrödinger equation in the rovibronic coordinates can be written as

$$[\hat{T}_e + \hat{T}_{vib} + \hat{T}_{rot} + V - E_{rve}]\Phi_{rve} = 0. \qquad (8\text{-}1)$$

The equation for a diatomic molecule is given in Eq. (7-91) and that for a polyatomic molecule can be obtained from Eqs. (7-45), (7-46), (7-58), (7-60),

166

and (7-137). In making the Born–Oppenheimer approximation we consider that the motions of the electrons are unaffected by the motions of the nuclei and, on the periodic time scale of their motion, only depend on the nuclear positions. Thus, in this approximation the wavefunctions Φ_{elec} and energies V_{elec} that describe the electronic motion are obtained from the Schrödinger equation that results after neglecting the nuclear kinetic energy term and nuclear–nuclear electrostatic repulsion term V_{nn} in Eq. (8-1), i.e., from

$$[\hat{T}_e + (V - V_{nn})]\Phi_{elec} = V_{elec}\Phi_{elec}. \tag{8-2}$$

In this equation the nuclear coordinates occur (in the nuclear–electron attraction part of V) but these coordinates are not dynamical variables (no integrations or differentiations are to be performed over them), and as a result V_{elec} and Φ_{elec} involve the nuclear coordinates as parameters; at each nuclear geometry we solve Eq. (8-2) to obtain V_{elec} and Φ_{elec}. Using this approximation for the electronic motion we see that from the point of view of the nuclei the electronic energy V_{elec} (together with V_{nn}) acts as a potential energy; to move the nuclei from a configuration having $(V_{elec} + V_{nn})$ low to one in which $(V_{elec} + V_{nn})$ is high requires work. The Schrödinger equation for the nuclear (rotation–vibration) motion is, therefore, given by

$$[\hat{T}_{vib} + \hat{T}_{rot} + (V_{elec} + V_{nn})]\Phi_{rv} = E^0_{rve}\Phi_{rv}, \tag{8-3}$$

where the value of E^0_{rve} obtained is the rovibronic energy in the Born–Oppenheimer approximation. The approximate and separable rovibronic wavefunction Φ^0_{rve} is given by

$$\Phi^0_{rve} = \Phi_{elec}\Phi_{rv} \tag{8-4}$$

from Eqs. (8-2) and (8-3).

We usually concern ourselves with those bound electronic states for which $(V_{elec} + V_{nn})$ has a deep minimum at the equilibrium nuclear geometry (deep relative to kT, where k is Boltzmann's constant and T is the absolute temperature so that at room temperature $kT \cong 200$ cm^{-1}). The problems arising when there is more than one accessible minimum (i.e., for a nonrigid molecule) will be discussed in Chapter 12. The rotation–vibration Schrödinger equation for a bound electronic state is written so that the zero of energy is the minimum value of $(V_{elec} + V_{nn})$, which we call E_{elec}, the electronic energy, and we have

$$[\hat{T}_{vib} + \hat{T}_{rot} + V_N]\Phi_{rv} = E_{rv}\Phi_{rv}, \tag{8-5}$$

where

$$V_N = (V_{elec} + V_{nn}) - E_{elec} \quad \text{and} \quad E_{rv} = E^0_{rve} - E_{elec}.$$

To allow for the breakdown of the Born–Oppenheimer approximation we use the approximate wavefunctions Φ_{rve}^0 as a basis set for diagonalizing the complete rovibronic Hamiltonian $\hat{H}_{rve} = \hat{T}_e + \hat{T}_{vib} + \hat{T}_{rot} + V$.

As a result of making the Born–Oppenheimer approximation we have reduced the problem of solving the $(3l - 3)$-dimensional rovibronic Schrödinger equation, Eq. (8-1), to one of solving two differential equations: the electronic Schrödinger equation, Eq. (8-2), involving $3n$ electronic coordinates, and the rotation–vibration Schrödinger equation, Eq. (8-5), involving $3N - 3$ nuclear coordinates. We now proceed to approximate each of these equations so that they separate into soluble partial differential equations, and we obtain approximate electronic and rovibrational wavefunctions, Φ_{elec}^0 (or Φ_{eo}) and Φ_{rv}^0.

THE ELECTRONIC WAVEFUNCTIONS

From Eqs. (8-2), (7-46), and (6-19) the electronic Schrödinger equation is obtained as

$$\hat{H}_{elec}\Phi_{elec} = V_{elec}\Phi_{elec}, \tag{8-6}$$

where Φ_{elec} and V_{elec} are the eigenfunctions and eigenvalues of

$$\hat{H}_{elec} = -\frac{\hbar^2}{2m_e}\sum_i \mathbf{V}_i^2 - \frac{\hbar^2}{2M_N}\sum_{i,j} \mathbf{V}_i \cdot \mathbf{V}_j + \sum_{i<j}\frac{e^2}{R_{ij}} - \sum_{\alpha,i}\frac{C_\alpha e^2}{R_{i\alpha}}, \tag{8-7}$$

where i and j run over the n electrons and α runs over the N nuclei. To solve Eq. (8-6) we make approximations in order to separate the variables. If we neglect the second and third terms in Eq. (8-7) we reduce $\hat{H}_{elec}$ to an approximate electronic Hamiltonian $\hat{H}_{elec}^0$ that is separable into the sum of n one-electron Hamiltonians, i.e.,

$$\hat{H}_{elec}^0 = \sum_i \left\{ -\frac{\hbar^2}{2m_e}\mathbf{V}_i^2 - \sum_\alpha \frac{C_\alpha e^2}{R_{i\alpha}} \right\} = \sum_i \hat{h}_i. \tag{8-8}$$

As a result of this separation of variables the eigenfunctions and eigenvalues of $\hat{H}_{elec}^0$ are

$$\Phi_{elec}^0 = \phi_a(\mathbf{r}_1)\phi_b(\mathbf{r}_2)\cdots\phi_\lambda(\mathbf{r}_n) \tag{8-9}$$

and

$$V_{elec}^0 = \varepsilon_a + \varepsilon_b + \cdots + \varepsilon_\lambda, \tag{8-10}$$

where

$$\hat{h}\phi_k(\mathbf{r}) = \varepsilon_k\phi_k(\mathbf{r}), \tag{8-11}$$

and r are the electronic coordinates in the (x, y, z) axis system. The $\phi_k(r)$ and the ε_k depend on the nuclear coordinates but this is not put in explicitly to avoid complicating the expressions. The eigenfunctions $\phi_k(r)$ and eigenvalues ε_k of $\hat{h}$ are called *molecular orbitals* (*MO's*) and *molecular orbital energies*, respectively. To determine them we solve Eq. (8-11) for the motion of an electron in the electrostatic field of a given arrangement of the nuclei. We describe the wavefunction in Eq. (8-9) as the electron *configuration* having electron 1 in orbital ϕ_a, electron 2 in orbital ϕ_b, and so on. Because of the indistinguishability of electrons and the Pauli exclusion principle a more complicated determinantal form of the wavefunction is really required, but this aspect of the electronic wavefunction will not be discussed here [see, for example, Section 9c in Eyring, Walter, and Kimball (1944)].

The approximation made in obtaining Eqs. (8-9) and (8-10) as solutions to Eq. (8-6) is drastic. The third term in Eq. (8-7) represents the mutual electrostatic repulsion of the electrons, and this has a large modifying effect on the electronic energies and wavefunctions. As a result of it the motions of the electrons are not independent of each other, as they would be if described by Φ_{elec}^0 in Eq. (8-9); the motions of the electrons are *correlated* with each other. Despite the approximation the eigenfunctions Φ_{elec}^0 are very useful for making a symmetry classification of the eigenfunctions of $\hat{H}_{\text{elec}}$ and for describing them. By adding the "averaged out" repulsive effect of the other electrons to $\hat{h}$ we can still retain an orbital product description of the electronic eigenfunctions but achieve a closer approximation to the exact solution. This improved method is called the self-consistent field approximation (SCF) and the improved one-electron Hamiltonian is called $\hat{h}^{\text{SCF}}$ [see, for example, Eq. (9-99) in Eyring, Walter, and Kimball (1944)]. The second term in Eq. (8-7) also gives rise to a coupling of the electronic motions but this kinetic energy correlation term is a nuclear mass dependent contribution that is of the same order of magnitude as terms neglected in making the Born–Oppenheimer approximation. This kinetic energy coupling term can therefore be neglected in all but the most precise work.

An approximate solution of Eq. (8-11), or an approximate determination of the SCF molecular orbital eigenfunctions of $\hat{h}^{\text{SCF}}$, can be obtained by expressing the molecular orbitals as a linear combination of atomic orbitals (LCAO). Each atomic orbital is an atomic wavefunction for an electron centered on one of the nuclei in the molecule. The coefficients of the atomic orbitals in the molecular orbitals are adjusted in a variational manner to minimize the total energy [the variation method is described, for example, in Section 7c of Eyring, Walter, and Kimball (1944)]. If we use the SCF Hamiltonian then the approximation is referred to as the LCAOSCFMO approximation.

To indicate the appearance of the functions involved we can use the water molecule. With the nuclei of a water molecule at, say, their equilibrium positions we let a single electron loose on the nuclear arrangement and solve Eq. (8-11) for the molecular orbital wavefunctions and energies. To do this we expand the orbitals as a linear combination of atomic orbitals that are centered on the nuclei. If we limit the atomic orbital basis to 1s orbitals at each nucleus together with 2s and 2p orbitals at the oxygen nucleus then we can write each molecular orbital as

$$\phi_k(\mathbf{r}) = c_{k1}1s(H_1) + c_{k2}1s(H_2) + c_{k3}1s(O) + c_{k4}2s(O) + c_{k5}2p_x(O)$$
$$+ c_{k6}2p_y(O) + c_{k7}2p_z(O), \tag{8-12}$$

where the c_{kj} are the parameters to be variationally adjusted, and for example we have

$$1s(H_1) = \pi^{-1/2}a_0^{-3/2}e^{-(r_1/a_0)}, \tag{8-13}$$

where a_0 is the Bohr radius and r_1 is the distance of the electron from nucleus H_1. Slater type orbitals (STO's) are often used in LCAO molecular orbitals and these have the general form

$$\chi_{n\zeta}(r_i, \theta_i, \phi_i) = Nr_i^{n-1}e^{-\zeta r_i} \times \text{(angular part in } \theta_i \text{ and } \phi_i), \tag{8-14}$$

where N is a normalization constant and (r_i, θ_i, ϕ_i) are the polar coordinates of the electron in an (x_i, y_i, z_i) axis system parallel to the (x, y, z) axis system with origin at nucleus i. Using such functions we can write the MO's as

$$\phi_k(\mathbf{r}) = \sum_j c_{kj}\chi_j(\mathbf{r}), \tag{8-15}$$

where $j = n\zeta$ and n and ζ are additional variational parameters.

Starting with a given number of atomic orbitals we obtain a self-consistent field solution for the electronic energies and wave functions if we make an iterative solution of the $\hat{h}^{SCF}$ equation, with variational adjustment of c_{kj}, n, and ζ. We can improve the solution by increasing the number of atomic orbitals used in the self-consistent field equations; using a complete set of atomic orbitals leads to the so-called *Hartree–Fock limit*. Clearly the number of atomic orbital basis functions used is limited to a number that can be conveniently handled computationally. Having obtained a number of self-consistent field solutions for the electronic (orbital) wavefunctions Φ_{eo} such as

$$\Phi_{eo}^{(1)} = \phi_a(\mathbf{r}_1)\phi_b(\mathbf{r}_2) \cdots \phi_\lambda(\mathbf{r}_n),$$
$$\Phi_{eo}^{(2)} = \phi_a(\mathbf{r}_1)\phi_c(\mathbf{r}_2) \cdots \phi_\mu(\mathbf{r}_n), \tag{8-16}$$

and

$$\Phi_{eo}^{(3)} = \phi_d(\mathbf{r}_1)\phi_e(\mathbf{r}_2) \cdots \phi_\nu(\mathbf{r}_n),$$

which correspond to ground and excited electronic (orbital) states, we can obtain an improvement by writing Φ_{elec} as a linear combination of such functions, i.e.,

$$\Phi_{\text{elec}}^{(n)} = \sum_{k=1}^{p} a_{nk}\Phi_{\text{eo}}^{(k)}, \tag{8-17}$$

where the coefficients a_{nk}, and the energies V_{elec}, are obtained from the diagonalization of the matrix of $\hat{H}_{\text{elec}}$ in the basis of functions Φ_{eo}, and the more SCF functions used the better will be the solution. Such a technique for improving the solution is called the configuration interaction (CI) technique. The LCAO(STO)SCFMOCI technique yields results [i.e., potential energy curves $(V_{\text{elec}} + V_{\text{nn}})$] that are very close to the experimentally derived quantities for small molecules.

THE ROTATION–VIBRATION SCHRÖDINGER EQUATION

From Eqs. (7-137) and (8-5) the rotation-vibration equation for a given electronic state is

$$\left[\frac{1}{2}\sum_{\alpha,\beta}\mu_{\alpha\beta}(\hat{J}_\alpha - \hat{p}_\alpha)(\hat{J}_\beta - \hat{p}_\beta) + \frac{1}{2}\sum_r \hat{P}_r^2 + U + V_N\right]\Phi_{\text{rv}} = E_{\text{rv}}\Phi_{\text{rv}}. \tag{8-18}$$

The terms

$$-\frac{1}{2}\sum_{\alpha,\beta}\mu_{\alpha\beta}[(\hat{J}_\alpha - \hat{p}_\alpha)\hat{L}_\beta + \hat{L}_\alpha(\hat{J}_\beta - \hat{p}_\beta)] + \frac{1}{2}\sum_{\alpha,\beta}\mu_{\alpha\beta}\hat{L}_\alpha\hat{L}_\beta \tag{8-19}$$

have been neglected as part of the Born–Oppenheimer separation of the electronic and nuclear (rotation–vibration) parts of the problem [see Eqs. (11-79) and (11-80)]. Before discussing the way we solve Eq. (8-18) we look at the normal coordinates and potential function V_N in more detail.

V_N is a function of the internal coordinates in the molecule (bond lengths and bond angles) and it has the value zero if the nuclei are in their equilibrium configuration; we choose to expand V_N as a Taylor series in the displacements of these internal coordinates from their equilibrium values. We use the symbol $\mathfrak{R}_i$ for the $(3N - 6)$ independent internal coordinate displacements and write

$$V_N = \tfrac{1}{2}f_{ij}\mathfrak{R}_i\mathfrak{R}_j + \tfrac{1}{6}f_{ijk}\mathfrak{R}_i\mathfrak{R}_j\mathfrak{R}_k + \tfrac{1}{24}f_{ijkl}\mathfrak{R}_i\mathfrak{R}_j\mathfrak{R}_k\mathfrak{R}_l + \cdots, \tag{8-20}$$

where summation over repeated indices [such as i, j, k, and l in Eq. (8-20)] is assumed in all the equations of this section. The *force constants* f_{ij}, f_{ijk}, and f_{ijkl} are second, third, and fourth derivatives (at equilibrium) of V_N with respect to the coordinates $\mathfrak{R}_i$. Generally Eq. (8-20) is not taken beyond quartic terms. The equilibrium geometry and the force constants are the molecular parameters that enter into the rotation–vibration Schrödinger equation, Eq. (8-18), and these can come from the solution of the electronic Schrödinger equation, Eq. (8-6); the nuclear masses also occur in the rotation–vibration equation.[1]

The addition of V_N from Eq. (8-20) to T_N from Eq. (7-117) gives the classical rotation–vibration energy. The rotation–vibration Hamiltonian that we use is obtained from this (as discussed in Chapter 7) by changing coordinates to the Euler angles (θ, ϕ, χ) and normal coordinates Q_r, and by converting the expression obtained to quantum mechanical form [see Chapter 11 in Wilson, Decius, and Cross (1955)].

To express V_N in normal coordinates we must express the $\mathfrak{R}_i$ in terms of normal coordinates. In general the $\mathfrak{R}_i$ are nonlinear functions of the Cartesian displacement coordinates $\Delta\alpha_k$, $\Delta\beta_k$, etc., and they are written as

$$\mathfrak{R}_i = B_i^{\alpha k} \Delta\alpha_k + \tfrac{1}{2} B_i^{\alpha k,\beta n} \Delta\alpha_k \Delta\beta_n + \tfrac{1}{6} B_i^{\alpha k,\beta n,\gamma p} \Delta\alpha_k \Delta\beta_n \Delta\gamma_p + \cdots \quad (8\text{-}21)$$

(we are summing over α, β, γ and k, n, p here). The coefficients $B_i^{\alpha k}$, $B_i^{\alpha k,\beta n}$, and $B_i^{\alpha k,\beta n,\gamma p}$ are elements of the so-called $\boldsymbol{B}$ tensor and they are given by the first, second, and third derivatives (at equilibrium) of $\mathfrak{R}_i$ with respect to the Cartesian displacement coordinates. These elements only depend on the equilibrium molecular geometry; the first derivatives $B_i^{\alpha k}$ are the elements of the so-called B matrix [it is a $(3N-6) \times 3N$-dimensional matrix]. Equation (7-136) can be substituted into Eq. (8-21) to give

$$\mathfrak{R}_i = L_{ir}Q_r + \tfrac{1}{2}L_{irs}Q_rQ_s + \tfrac{1}{6}L_{irst}Q_rQ_sQ_t + \cdots, \quad (8\text{-}22)$$

where we introduce L_{ir}, L_{irs}, and L_{irst}, which are elements of the $\boldsymbol{L}$ tensor. The elements L_{ir} are the elements of the so-called L matrix [it is a $(3N-6) \times (3N-6)$-dimensional matrix], and from the result of substituting Eq. (7-136) into the first term of Eq. (8-21) we see that the L matrix is expressed in terms of the B matrix and the $3N \times (3N-6)$-dimensional l matrix according to

$$L = BM^{-1/2}l, \quad (8\text{-}23)$$

where $M^{-1/2}$ is a diagonal $(3N \times 3N)$ matrix containing the reciprocals of the square roots of the nuclear masses. Given the equilibrium nuclear configuration the choice of the elements in the L (or l) matrix defines the normal coordinates since the higher L tensor elements L_{irs}, L_{irst}, ..., can all be expressed in terms of the L_{ir} [see the discussion above Eq. (20) in Hoy, Mills, and Strey (1972)].

Since the $(3N-6)Q_r$ are linearly independent, as are the $(3N-6)\mathfrak{R}_i$, and since each nonvanishing term in the sum in Eq. (8-22) must have the same

---

[1] But see the last paragraph before the Summary in this chapter.

symmetry species in a symmetry group of the vibrational Hamiltonian, we see that the symmetry species of the Q_r is the same as that of the $\mathfrak{R}_i$. As a result it is possible to determine the symmetry of the normal coordinates in a molecule by determining the symmetry of the $(3N - 6)$ independent internal coordinate displacements (bond stretches, angle bends, etc.). This is often simpler than determining the symmetry of the Q_r from the symmetry of the Cartesian displacement coordinates. However, one must use $(3N - 6)$ independent $\mathfrak{R}_i$ coordinates and avoid combinations of internal coordinates that are redundant. An example of this way of determining the symmetry of the normal coordinates is given for the methane molecule in Chapter 10 [see Eqs. (10-36)–(10-38)].

The L (or l) matrix is chosen in order to allow us to separate the variables (the Q_r) in the vibrational part of the Hamiltonian with minimum approximation. To see how this choice should be made we look at V_N. Substituting Eq. (8-22) into V_N of Eq. (8-20) we obtain

$$V_N = \tfrac{1}{2}\Phi_{rs}Q_rQ_s + \tfrac{1}{6}\Phi_{rst}Q_rQ_sQ_t + \tfrac{1}{24}\Phi_{rstu}Q_rQ_sQ_tQ_u + \cdots, \qquad (8\text{-}24)$$

where

$$\Phi_{rs} = L_{ir}f_{ij}L_{js}$$

and the expressions for Φ_{rst} and Φ_{rstu} are more complicated expressions involving the elements of the f and L tensors. The important point is that the L matrix elements are chosen so that Φ_{rs} is diagonal, i.e.,

$$\Phi_{rs} = L_{ir}f_{ij}L_{js} = \delta_{rs}\lambda_r \qquad (8\text{-}25)$$

(we do not sum over r here), and this gives

$$V_N = \tfrac{1}{2}\lambda_r Q_r{}^2 + \tfrac{1}{6}\Phi_{rst}Q_rQ_sQ_t + \tfrac{1}{24}\Phi_{rstu}Q_rQ_sQ_tQ_u + \cdots. \qquad (8\text{-}26)$$

The vibrational kinetic energy operator [see Eq. (8-18)] is

$$\hat{T}_{\text{vib}} = \frac{1}{2}\sum_r \hat{P}_r{}^2, \qquad (8\text{-}27)$$

where $\hat{P}_r = -i\hbar\,\partial/\partial Q_r$. This expression arises from $\tfrac{1}{2}\sum_{\alpha,i} m_i(\Delta\dot{\alpha}_i)^2$ and results from the facts that (i) the $\Delta\alpha_i$ and Q_r are linearly related [see Eq. (7-136)], and (ii) the L matrix is "normalized" so that $\tilde{L}G^{-1}L = E$ [see Appendix VIII of Wilson, Decius, and Cross (1955)]. With this choice for the L matrix we achieve a separation of the variables (the Q_r) in $\hat{T}_{\text{vib}} + V_N$ if we neglect the terms involving cubic and higher powers of the Q_r in V_N; these higher terms give rise to anharmonicity in V_N and their effect is usually small. This is the *harmonic oscillator approximation.*

Using normal coordinates the rotation–vibration Hamiltonian can now be written as

$$\hat{H}_{\rm rv} = \frac{1}{2}\sum_\alpha \mu^e_{\alpha\alpha}\hat{J}_\alpha{}^2 + \frac{1}{2}\sum_r (\hat{P}_r{}^2 + \lambda_r Q_r{}^2) \tag{8-28a}$$

$$+ \frac{1}{2}\sum_{\alpha,\beta}(\mu_{\alpha\beta} - \mu^e_{\alpha\beta})(\hat{J}_\alpha - \hat{p}_\alpha)(\hat{J}_\beta - \hat{p}_\beta) \tag{8-28b}$$

$$- \sum_\alpha \mu^e_{\alpha\alpha}\hat{J}_\alpha\hat{p}_\alpha + \frac{1}{2}\sum_\alpha \mu^e_{\alpha\alpha}\hat{p}_\alpha{}^2 \tag{8-28c}$$

$$+ U + \tfrac{1}{6}\Phi_{rst}Q_r Q_s Q_t + \tfrac{1}{24}\Phi_{rstu}Q_r Q_s Q_t Q_u + \cdots. \tag{8-28d}$$

Equation (8-28a) is the sum of a rigid rotor Hamiltonian and of harmonic oscillator Hamiltonians; the rest of the terms in Eq. (8-28) give rise to the effects of centrifugal distortion, vibrational Coriolis coupling, and anharmonicity. We determine the exact eigenfunctions of the separable Hamiltonian of Eq. (8-28a) and in Chapter 10 we use these eigenfunctions to determine the symmetry labels on the energy levels. We write

$$\hat{H}^0_{\rm rv} = \frac{1}{2}\sum_\alpha \mu^e_{\alpha\alpha}\hat{J}_\alpha{}^2 + \frac{1}{2}\sum_r (\hat{P}_r{}^2 + \lambda_r Q_r{}^2) \tag{8-29}$$

and

$$\hat{H}^0_{\rm rv}\Phi^0_{\rm rv} = E^0_{\rm rv}\Phi^0_{\rm rv}. \tag{8-30}$$

Because of the separation of the variables in $\hat{H}^0_{\rm rv}$, we have

$$\Phi^0_{\rm rv} = \Phi_{\rm rot}(\theta,\phi,\chi)\underbrace{\Phi_{v_1}(Q_1)\Phi_{v_2}(Q_2)\cdots}_{\Phi_{\rm vib}(Q_1,Q_2,\ldots)} \tag{8-31}$$

and

$$E^0_{\rm rv} = E_{\rm rot} + \underbrace{E_{v_1} + E_{v_2} + \cdots}_{E_{\rm vib}}, \tag{8-32}$$

where

$$\left[\frac{1}{2}\sum_\alpha \mu^e_{\alpha\alpha}\hat{J}_\alpha{}^2\right]\Phi_{\rm rot}(\theta,\phi,\chi) = E_{\rm rot}\Phi_{\rm rot}(\theta,\phi,\chi) \tag{8-33}$$

and

$$\tfrac{1}{2}(\hat{P}_r + \lambda_r Q_r{}^2)\Phi_{v_r}(Q_r) = E_{v_r}\Phi_{v_r}(Q_r). \tag{8-34}$$

We now look at the problem of solving Eqs. (8-33) and (8-34) to obtain the wavefunctions $\Phi_{\rm rot}(\theta,\phi,\chi)$ and $\Phi_{v_r}(Q_r)$.

THE RIGID ROTOR SCHRÖDINGER EQUATION

We will first determine the eigenfunctions and eigenvalues of the rigid rotor Schrödinger equation given in Eq. (8-33). It is necessary to consider three cases (linear polyatomic molecules are discussed in Chapter 12):

(i)   The symmetric top molecule, in which two of the $\mu_{\alpha\alpha}^e$ are equal to each other,
(ii)   The spherical top molecule, in which all three $\mu_{\alpha\alpha}^e$ are equal to each other, and
(iii)   The asymmetric top molecule, in which all three $\mu_{\alpha\alpha}^e$ are different.

We write the rigid rotor Hamiltonian (in cm^{-1} units) in the principal axis system as

$$\hat{H}_{rot} = \hbar^{-2}(A_e \hat{J}_a^2 + B_e \hat{J}_b^2 + C_e \hat{J}_c^2), \tag{8-35}$$

where the rotational constants (in cm^{-1}) have been introduced and they are defined by

$$A_e = \hbar^2 \mu_{aa}^e / (2hc), \quad \text{etc.} \tag{8-36}$$

The principal inertial axes of the equilibrium configuration are labeled a, b, and c so that the rotational constants are in the order $A_e \geq B_e \geq C_e$.

The Symmetric Top Molecule

For a symmetric top molecule we can either have $A_e > B_e = C_e$ for a prolate symmetric top (such as CH_3F) or $A_e = B_e > C_e$ for an oblate symmetric top (such as BF_3). We write the rigid rotor Schrödinger equation for a prolate top as

$$\hbar^{-2}[A_e \hat{J}_a^2 + B_e(\hat{J}_b^2 + \hat{J}_c^2)]\Phi_{rot}(\theta,\phi,\chi) = E_{rot}\Phi_{rot}(\theta,\phi,\chi). \tag{8-37}$$

For a prolate top the a axis is chosen as the z axis (I^r convention) and using the Euler angle definitions from Fig. 7-1, and the expressions for $\hat{J}_\alpha$ given in Eqs. (7-144)–(7-146), this Schrödinger equation becomes

$$\left\{ \frac{1}{\sin\theta} \frac{\partial}{\partial\theta}\left(\sin\theta \frac{\partial}{\partial\theta}\right) + \frac{1}{\sin^2\theta} \frac{\partial^2}{\partial\phi^2} + \left(\frac{\cos^2\theta}{\sin^2\theta} + \frac{A_e}{B_e}\right)\frac{\partial^2}{\partial\chi^2} - \frac{2\cos\theta}{\sin^2\theta}\frac{\partial^2}{\partial\phi\,\partial\chi} + \frac{E_{rot}}{B_e} \right\}$$

$$\times \Phi_{rot}(\theta,\phi,\chi) = 0. \tag{8-38}$$

The angles ϕ and χ only occur in derivatives, and so they are cyclic coordinates and we can write

$$\Phi_{rot}(\theta,\phi,\chi) = \Theta(\theta)e^{im\phi}e^{ik\chi}, \tag{8-39}$$

where m and k have the integral values $0, \pm 1, \pm 2, \ldots$, in order that Φ_{rot} be single valued with $\Phi_{\mathrm{rot}}(\theta, \phi, \chi + 2\pi) = \Phi_{\mathrm{rot}}(\theta, \phi + 2\pi, \chi) = \Phi_{\mathrm{rot}}(\theta, \phi, \chi)$; [note that $\exp(2\pi i) = 1$]. Substituting this wavefunction into the wave equation we obtain the following equation for $\Theta(\theta)$

$$\left\{ \frac{1}{\sin\theta} \frac{\partial}{\partial\theta} \left(\sin\theta \frac{\partial}{\partial\theta} \right) + \left[\Delta - \frac{m^2 - 2mk\cos\theta + k^2}{\sin^2\theta} \right] \right\} \Theta(\theta) = 0, \quad (8\text{-}40)$$

where

$$\Delta = \left[E_{\mathrm{rot}} - (A_{\mathrm{e}} - B_{\mathrm{e}})k^2 \right] / B_{\mathrm{e}}. \quad (8\text{-}41)$$

This equation can be solved analytically by writing

$$\Theta(\theta) = x^{|k-m|/2}(1 - x)^{|k+m|/2} F(x), \quad (8\text{-}42)$$

where

$$x = (1 - \cos\theta)/2, \quad (8\text{-}43)$$

and Eq. (8-40) reduces to

$$x(1 - x)\frac{d^2 F}{dx^2} + (\alpha - \beta x)\frac{dF}{dx} + \gamma F = 0, \quad (8\text{-}44)$$

where

$$\alpha = 1 + |k - m|, \quad (8\text{-}45)$$

$$\beta = \alpha + 1 + |k + m|, \quad (8\text{-}46)$$

and

$$\gamma = \Delta - \beta(\beta - 2)/4. \quad (8\text{-}47)$$

Writing $F(x)$ as a power series in x we have

$$F(x) = \sum_{n=0}^{\infty} a_n x^n. \quad (8\text{-}48)$$

Substituting this into Eq. (8-44) we obtain

$$(\gamma a_0 + \alpha a_1) + \left[(\gamma - \beta)a_1 + 2(1 + \alpha)a_2 \right] x$$

$$+ \sum_{n=2}^{\infty} \left\{ [\gamma - n(n - 1) - \beta n]a_n + (n + 1)(n + \alpha)a_{n+1} \right\} x^n = 0. \quad (8\text{-}49)$$

Equating the coefficient of each power of x to zero we obtain

$$a_1 = -(\gamma/\alpha)a_0, \quad (8\text{-}50)$$

$$a_2 = (\beta - \gamma)a_1/(2 + 2\alpha), \quad (8\text{-}51)$$

and the general recursion relation

$$a_{n+1} = \frac{\beta n + n(n-1) - \gamma}{(n+1)(n+\alpha)} a_n. \qquad (8\text{-}52)$$

The coefficient a_0 is chosen so that the wavefunction is normalized. The series must terminate for $\Theta(\theta)$ to be an acceptable wavefunction, and if the highest nonvanishing term has $n = n_{max}$ then

$$a_{n_{max}+1} = 0, \qquad (8\text{-}53)$$

which means that

$$\beta n_{max} + n_{max}(n_{max} - 1) - \gamma = 0. \qquad (8\text{-}54)$$

Substituting Eqs. (8-47) and (8-41) for γ, Eq. (8-54) gives (in cm^{-1})

$$E_{rot} = B_e J(J+1) + (A_e - B_e)k^2, \qquad \cdot \qquad (8\text{-}55)$$

where

$$J = n_{max} + \tfrac{1}{2}|k+m| + \tfrac{1}{2}|k-m|; \qquad (8\text{-}56)$$

that is J is an integer equal to n_{max} plus the larger of the two quantities $K = |k|$ or $M = |m|$. The three quantum numbers introduced can thus have the values

$$J = 0, 1, 2, \ldots; \quad k = 0, \pm 1, \pm 2, \ldots, \pm J; \quad \text{and} \quad m = 0, \pm 1, \pm 2, \ldots, \pm J; \qquad (8\text{-}57)$$

and the energy only depends on the values of J and $K = |k|$. The eigenfunctions depend on all three quantum numbers and can be written down explicitly for any set of (J, k, m) values by using Eqs. (8-39), (8-42), and (8-48) with the recursion relations of Eq. (8-52); the normalization factor a_0 remains to be determined. With the expression for E_{rot} given in Eq. (8-55) we have

$$\Delta = J(J+1) \qquad (8\text{-}58)$$

and

$$\gamma = J(J+1) - \beta(\beta - 2)/4. \qquad (8\text{-}59)$$

Problem 8-1. Write down the unnormalized rigid rotor wavefunctions for a prolate symmetric top molecule for the states (a) $J = 3$, $k = m = 0$; (b) $J = 3$, $k = -1$, $m = -2$; and (c) $J = 3$, $k = 1$, $m = -2$.

Answer. (a) For $J = 3$ and $k = m = 0$, we have $|k+m| = |k-m| = 0$ so that $n_{max} = J = 3$ and $\alpha = 1$, $\beta = 2$, $\gamma = 12$, $a_1 = -12a_0$, $a_2 = 30a_0$, and $a_3 = -20a_0$. Thus [using the notation $F_{J,k,m}(x)$]

$$F_{3,0,0}(x) = (1 - 12x + 30x^2 - 20x^3)a_0 = (-\tfrac{3}{2}\cos\theta + \tfrac{5}{2}\cos^3\theta)a_0,$$

and we can write the unnormalized wavefunction (using the notation $\Phi_{J,k,m}$)

$$\Phi_{3,0,0} = (\tfrac{5}{3}\cos^3\theta - \cos\theta). \tag{8-60}$$

(b) For $J = 3$, $k = -1$, $m = -2$, we have $|k + m| = 3$, $|k - m| = 1$, $\alpha = 2$, $\beta = 6$, $\gamma = 6$, $n_{max} = 1$, and hence

$$F_{3,-1,-2} = [-\tfrac{1}{2} + \tfrac{3}{2}\cos\theta]a_0$$

so that the unnormalized wavefunction is

$$\Phi_{3,-1,-2} = e^{-i\chi}e^{-2i\phi}(1 - \cos\theta)^{1/2}(1 + \cos\theta)^{3/2}(3\cos\theta - 1). \tag{8-61}$$

(c) For $J = 3$, $k = 1$, $m = -2$, we have $|k + m| = 1$, $|k - m| = 3$, $\alpha = 4$, $\beta = 6$, $\gamma = 6$, $n_{max} = 1$, and hence

$$F_{3,1,-2} = (\tfrac{1}{4} + \tfrac{3}{4}\cos\theta)a_0$$

so that the unnormalized function is

$$\Phi_{3,1,-2} = e^{i\chi}e^{-2i\phi}(1 - \cos\theta)^{3/2}(1 + \cos\theta)^{1/2}(1 + 3\cos\theta). \tag{8-62}$$

If $F(x)$ were such that all $a_n = 1$ then $F(x)$ would be a geometric series. By analogy the $F(x)$ introduced above is called a *hypergeometric* series (or hypergeometric function). The standard notation for such a function is such that $F(a, b; c; x)$ is the solution of

$$x(1 - x)\frac{d^2F}{dx^2} + [c - (1 + a + b)x]\frac{dF}{dx} - abF = 0 \tag{8-63}$$

and using this notation we can write the complete rotational wavefunction as

$$\Phi_{Jkm}(\theta, \phi, \chi) = N_{Jkm}x^{|k-m|/2}(1 - x)^{|k+m|/2}F(\tfrac{1}{2}\beta - J - 1, \tfrac{1}{2}\beta + J; \alpha; x)e^{im\phi}e^{ik\chi}, \tag{8-64}$$

where

$$x = (1 - \cos\theta)/2, \qquad \alpha = 1 + |k - m|, \qquad \beta = \alpha + 1 + |k + m|,$$

and N_{Jkm} is the normalization constant a_0 (it also involves a phase factor) and it is chosen so that

$$\int_0^{2\pi}\int_0^{2\pi}\int_0^{\pi} \Phi_{Jkm}^*\Phi_{Jkm} \sin\theta \, d\theta \, d\phi \, d\chi = 1. \tag{8-65}$$

Choosing a real and positive phase factor the normalization constant is given by [see Pauling and Wilson (1935), Eq. (36-14)]

$$N_{Jkm} = \left\{ \frac{(2J + 1)(J + \tfrac{1}{2}|k + m| + \tfrac{1}{2}|k - m|)!(J - \tfrac{1}{2}|k + m| + \tfrac{1}{2}|k - m|)!}{8\pi^2(J - \tfrac{1}{2}|k + m| - \tfrac{1}{2}|k - m|)!(|k - m|!)^2(J + \tfrac{1}{2}|k + m| - \tfrac{1}{2}|k - m|)!} \right\}^{1/2}. \tag{8-66}$$

We will sometimes use the notation $|J, k, m\rangle$ for the rotational wavefunction. An alternative way of writing this function, but with a different phase factor [see Eqs. (4.1.12), (4.1.15), and (4.8.6) in Edmonds (1960)], is

$$|J, k, m\rangle = X_{Jkm} \left\{ \sum_{\sigma} (-1)^{J-m-\sigma} \frac{(\cos \frac{1}{2}\theta)^{m+k+2\sigma}(\sin \frac{1}{2}\theta)^{2J-m-k-2\sigma}}{\sigma!(J-m-\sigma)!(m+k+\sigma)!(J-k-\sigma)!} \right\} e^{im\phi} e^{ik\chi},$$

(8-67)

where

$$X_{Jkm} = [(J + m)!(J - m)!(J + k)!(J - k)!(2J + 1)/(8\pi^2)]^{1/2}.$$

The index σ in the sum runs from 0 or $[-(k + m)]$, whichever is the larger, up to $(J - m)$ or $(J - k)$, whichever is the smaller.

For an oblate rotor the rotational Schrödinger equation is

$$\hbar^{-2}[B_e(\hat{J}_a^2 + \hat{J}_b^2) + C_e\hat{J}_c^2]\Phi_{rot} = E_{rot}\Phi_{rot}.$$

(8-68)

Choosing the c axis as the z axis (a type III^r convention) we obtain exactly the same wave equation as for the prolate rotor except that A_e is replaced by C_e. Thus for an oblate rotor (in cm^{-1})

$$E_{rot} = B_e J(J + 1) - (B_e - C_e)k^2,$$

(8-69)

and Φ_{rot} is as given in Eq. (8-64) or (8-67).

The Spherical Top Molecule

For a spherical top molecule we have $A_e = B_e = C_e$ and so the rigid rotor Schrödinger equation is

$$\hbar^{-2}B_e\hat{J}^2\Phi_{rot} = E_{rot}\Phi_{rot},$$

(8-70)

where $\hat{J}^2 = \hat{J}_x^2 + \hat{J}_y^2 + \hat{J}_z^2$. From the expressions for the $\hat{J}_\alpha^2$ given in Eqs. (7-144)–(7-146) this equation can be reduced to

$$\left\{ \frac{1}{\sin\theta} \frac{\partial}{\partial\theta} \left(\sin\theta \frac{\partial}{\partial\theta} \right) + \frac{1}{\sin^2\theta} \frac{\partial^2}{\partial\phi^2} + \frac{1}{\sin^2\theta} \frac{\partial^2}{\partial\chi^2} - \frac{2\cos\theta}{\sin^2\theta} \frac{\partial^2}{\partial\phi\,\partial\chi} + \frac{E_{rot}}{B_e} \right\} \Phi_{rot} = 0.$$

(8-71)

Following the arguments of Eqs. (8-39)–(8-57) we obtain the result for a spherical top molecule that

$$E_{rot} = B_e \Delta = B_e J(J + 1),$$

(8-72)

and the wavefunctions are as for a symmetric top molecule given in Eqs. (8-64) and (8-67).

The rigid rotor rotational eigenfunctions of all symmetric top and spherical top molecules are the same function [given in Eq. (8-64) or (8-67)] of the quantum numbers J, k, and m and we see that this function does not involve the rotational constants of the molecule; we call this wavefunction the *symmetric top wavefunction*. We can write the symmetric top wavefunction as

$$|J, k, m\rangle = [(2J + 1)/(8\pi^2)]^{1/2} D_{km}^{(J)}([\theta, \phi, \chi]) \qquad (8\text{-}73a)$$

$$= [1/(2\pi)]^{1/2} S_{Jkm}(\theta, \phi) e^{ik\chi}, \qquad (8\text{-}73b)$$

and we can relate the function $S_{Jkm}(\theta, \phi)$ to other standard functions as follows:

$$S_{J0m}(\theta, \phi) = Y_{Jm}(\theta, \phi) \qquad (8\text{-}74)$$

$$= [1/(2\pi)]^{1/2} \Theta_{Jm}(\theta) e^{im\phi}, \qquad (8\text{-}75)$$

$$\Theta_{Jm}(\theta) = \{[(2J + 1)(J - |m|)!]/[2(J + |m|)!]\}^{1/2} P_J^{|m|}(\cos\theta), \qquad (8\text{-}76)$$

$$P_J^{|m|}(\cos\theta) = \sin^{|m|}\theta \, \frac{d^{|m|}}{(d\cos\theta)^{|m|}} \, P_J(\cos\theta), \qquad (8\text{-}77)$$

and

$$P_J(\cos\theta) = \frac{1}{2^J J!} \frac{d^J}{(d\cos\theta)^J} (\cos^2\theta - 1)^J. \qquad (8\text{-}78)$$

$Y_{Jm}(\theta, \phi)$ is a spherical harmonic function, $\Theta_{Jm}(\theta)$ is a normalized associated Legendre polynomial, $P_J^{|m|}(\cos\theta)$ is an associated Legendre polynomial and $P_J(\cos\theta)$ is a Legendre polynomial [see, for example, Chapter IV of Eyring, Walter, and Kimball (1944)]. As discussed in Chapter 4 of Edmonds (1960) the function $D_{km}^{(J)}([\theta, \phi, \chi])$ in Eq. (8-73a) is the (k, m) element in the matrix representation $D^{(J)}$ of the group $\mathbf{K}$ for the rotation operation $[\theta, \phi, \chi]$ [see also the remarks before Eq. (6-40) here and Eq. (15-27) in Wigner (1959)].

The Angular Momentum Ladder Operators

From the form of the symmetric top wavefunction written in Eq. (8-64), and combining Eqs. (8-70) and (8-72), we see that

$$\hat{J}^2 |J, k, m\rangle = J(J + 1)\hbar^2 |J, k, m\rangle \qquad (8\text{-}79)$$

and

$$\hat{J}_z |J, k, m\rangle = -i\hbar \frac{\partial}{\partial\chi} |J, k, m\rangle = k\hbar |J, k, m\rangle. \qquad (8\text{-}80)$$

Thus J is the total rovibronic angular momentum quantum number and k is the quantum number specifying the value of the rovibronic angular momentum about the z axis. We can introduce the operator for the angular momentum about the space fixed ζ axis (see Fig. 7-1) by writing

$$\hat{J}_\zeta = \lambda_{x\zeta}\hat{J}_x + \lambda_{y\zeta}\hat{J}_y + \lambda_{z\zeta}\hat{J}_z = -i\hbar\,\frac{\partial}{\partial\phi} \tag{8-81}$$

and [from Eq. (8-64)] we see that

$$\hat{J}_\zeta|J,k,m\rangle = m\hbar|J,k,m\rangle. \tag{8-82}$$

Thus m is the quantum number specifying the value of the rovibronic angular momentum about the space fixed ζ axis. It is left as an exercise for the reader to show that $\hat{J}^2$, $\hat{J}_z$, and $\hat{J}_\zeta$ commute with each other.

Problem 8-2. Suppose $\hat{A}$ is an operator with eigenfunctions ψ_k and eigenvalues a_k, i.e.,

$$\hat{A}\psi_k = a_k\psi_k. \tag{8-83}$$

Suppose further that we find an operator, $\hat{O}$, say, that is such that its commutator with $\hat{A}$ is a constant multiple of itself, i.e.,

$$[\hat{A},\hat{O}] = b\hat{O}, \tag{8-84}$$

where b is a constant. Prove that in these circumstances $\hat{O}\psi_k$ is an eigenfunction of $\hat{A}$ with eigenvalue $(a_k + b)$, i.e., that

$$\hat{A}[\hat{O}\psi_k] = (a_k + b)[\hat{O}\psi_k]. \tag{8-85}$$

Answer. Starting with Eq. (8-84) and operating both sides on an eigenfunction ψ_k of $\hat{A}$ we have

$$\hat{A}\hat{O}\psi_k - \hat{O}\hat{A}\psi_k = b\hat{O}\psi_k, \tag{8-86}$$

i.e.,

$$\hat{A}[\hat{O}\psi_k] - \hat{O}a_k\psi_k = b[\hat{O}\psi_k]$$

and thus

$$\hat{A}[\hat{O}\psi_k] = a_k[\hat{O}\psi_k] + b(\hat{O}\psi_k),$$

from which Eq. (8-85) directly follows. If b is zero, i.e., if $\hat{A}$ and $\hat{O}$ commute, then $\hat{O}\psi_k$ and ψ_k are both eigenfunctions of $\hat{A}$ with the same eigenvalue.

An operator such as $\hat{O}$, which obeys Eq. (8-84) in Problem 8-2, is called a *ladder operator* for the eigenfunctions of $\hat{A}$ since it changes each eigenfunction of $\hat{A}$ into a new eigenfunction in which the eigenvalue is laddered up or down as b is positive or negative.

We can introduce angular momentum ladder operators as follows. From Eq. (7-147) we deduce that

$$[\hat{J}_z, (\hat{J}_x + i\hat{J}_y)] = -\hbar(\hat{J}_x + i\hat{J}_y) \tag{8-87}$$

and

$$[\hat{J}_z, (\hat{J}_x - i\hat{J}_y)] = +\hbar(\hat{J}_x - i\hat{J}_y), \tag{8-88}$$

and introducing the notation

$$\hat{J}_m{}^{\pm} = (\hat{J}_x \pm i\hat{J}_y), \tag{8-89}$$

where m stands for molecule fixed, we see that $\hat{J}_m{}^+$ and $\hat{J}_m{}^-$ are ladder operators for the eigenfunctions of $\hat{J}_z$; $\hat{J}_m{}^+$ ladders down by $\hbar$ and $\hat{J}_m{}^-$ ladders up by $\hbar$, i.e., if

$$\hat{J}_z\Phi = k\hbar\Phi \tag{8-90}$$

then [unless $k = +J$ (or $-J$) when $\hat{J}_m{}^-\Phi$ (or $\hat{J}_m{}^+\Phi$) will vanish since $|k| \not> J$]

$$\hat{J}_z[\hat{J}_m{}^{\pm}\Phi] = (k \mp 1)\hbar[\hat{J}_m{}^{\pm}\Phi], \tag{8-91}$$

where the $\pm$ and $\mp$ signs are correlated. Since $\hat{J}_x$ and $\hat{J}_y$ commute with $\hat{J}^2$, the ladder operators $\hat{J}_m{}^+$ and $\hat{J}_m{}^-$ each commute with $\hat{J}^2$ so that if

$$\hat{J}^2\Phi = J(J + 1)\hbar^2\Phi \tag{8-92}$$

then

$$\hat{J}^2[\hat{J}_m{}^{\pm}\Phi] = J(J + 1)\hbar^2[\hat{J}_m{}^{\pm}\Phi]. \tag{8-93}$$

Similarly the operators (where s stands for space fixed)

$$\hat{J}_s{}^{\pm} = (\hat{J}_\xi \pm i\hat{J}_\eta) \tag{8-94}$$

can be formed from the space fixed components of $\hat{J}$, and from Eq. (7-148) we see that

$$[\hat{J}_\zeta, \hat{J}_s{}^{\pm}] = \pm\hbar\hat{J}_s{}^{\pm}. \tag{8-95}$$

Thus $\hat{J}_s{}^+$ and $\hat{J}_s{}^-$ are ladder operators for the eigenfunctions of $\hat{J}_\zeta$ and if

$$\hat{J}_\zeta\Phi = m\hbar\Phi, \tag{8-96}$$

then [unless $m = +J$ (or $-J$) when $\hat{J}_s{}^+\Phi$ (or $\hat{J}_s{}^-\Phi$) will vanish since $|m| \not> J$]

$$\hat{J}_\zeta[\hat{J}_s{}^{\pm}\Phi] = (m \pm 1)\hbar[\hat{J}_s{}^{\pm}\Phi]. \tag{8-97}$$

Note that there is a change in sign so that $\hat{J}_s{}^+$ ladders up whereas $\hat{J}_m{}^+$ ladders down (and vice versa for $\hat{J}_s{}^-$ and $\hat{J}_m{}^-$). This change in sign is a result of the different signs in the commutators for the molecule fixed and space fixed components of $\hat{J}$ [see Eqs. (7-147) and (7-148)]. $\hat{J}_s{}^\pm$ commutes with $\hat{J}^2$ and with $\hat{J}_m{}^\pm$.

These ladder operators can be used to determine the symmetric top wavefunctions $|J, k, m\rangle$ and we can write

$$|J, \pm|k|, \pm|m|\rangle = N_\pm{}'(\hat{J}_m{}^\mp)^{|k|}(\hat{J}_s{}^\pm)^{|m|}|J,0,0\rangle \qquad (8\text{-}98a)$$

and

$$|J, \mp|k|, \pm|m|\rangle = N_\pm''(\hat{J}_m{}^\pm)^{|k|}(\hat{J}_s{}^\pm)^{|m|}|J,0,0\rangle, \qquad (8\text{-}98b)$$

with

$$|J,0,0\rangle = \left\{\frac{[(2J+1)/(8\pi^2)]^{1/2}}{(2^J J!)}\right\}\frac{d^J}{(d\cos\theta)^J}(\cos^2\theta - 1)^J \qquad (8\text{-}99)$$

$$= \frac{1}{2\pi}\Theta_{J0}(\theta), \qquad (8\text{-}100)$$

where Eqs. (8-98a) and (8-98b) each represent two equations: one in which the upper $\pm$ (or $\mp$) signs are used and the other in which the lower signs are used. Equation (8-99) defines the phase factor chosen for $|J,0,0\rangle$ and the normalization factors $N_\pm{}'$ and $N_\pm''$, which involve a phase factor, must be determined. To determine them we write

$$|J,k,m \pm 1\rangle = N_\pm \hat{J}_s{}^\pm|J,k,m\rangle, \qquad (8\text{-}101)$$

where $|J,k,m\rangle$ is normalized and $N_\pm$ is required so that the function $|J,k, m \pm 1\rangle$ is normalized. Multiplying each side of Eq. (8-101) by its complex conjugate and integrating over $d\tau = \sin\theta\, d\theta\, d\phi\, d\chi$ we obtain

$$\langle J,k,m \pm 1|J,k,m \pm 1\rangle = |N_\pm|^2\langle J,k,m|(\hat{J}_s{}^\mp\hat{J}_s{}^\pm)|J,k,m\rangle. \qquad (8\text{-}102)$$

Now

$$\hat{J}_s{}^\mp\hat{J}_s{}^\pm = \hat{J}^2 - \hat{J}_\zeta(\hat{J}_\zeta \pm \hbar), \qquad (8\text{-}103)$$

$$\langle J,k,m|[\hat{J}^2 - \hat{J}_\zeta(\hat{J}_\zeta \pm \hbar)]|J,k,m\rangle = [J(J+1) - m(m \pm 1)]\hbar^2, \qquad (8\text{-}104)$$

and we want $N_\pm$ to be such that the left hand side of Eq. (8-102) is unity; therefore

$$|N_\pm|^2 = \frac{1}{\hbar^2[J(J+1) - m(m \pm 1)]} \qquad (8\text{-}105)$$

and

$$N_\pm = \frac{e^{\pm i\delta}}{\hbar[J(J+1) - m(m \pm 1)]^{1/2}}, \qquad (8\text{-}106)$$

where δ is an arbitrary phase factor. By convention we choose $\delta = 0$ so that $N_{\pm}$ is real and positive, and we obtain

$$|J, k, m \pm 1\rangle = \frac{1}{\hbar[J(J+1) - m(m \pm 1)]^{1/2}} \hat{J}_s^{\pm} |J, k, m\rangle, \qquad (8\text{-}107)$$

which leads to the result

$$\langle J, k, m \pm 1 | \hat{J}_s^{\pm} | J, k, m\rangle = \hbar[J(J+1) - m(m \pm 1)]^{1/2}. \qquad (8\text{-}108)$$

Using $\hat{J}_m^{\pm}$, and a similar phase factor choice, we obtain

$$|J, k \mp 1, m\rangle = \frac{1}{\hbar[J(J+1) - k(k \mp 1)]^{1/2}} \hat{J}_m^{\pm} |J, k, m\rangle, \qquad (8\text{-}109)$$

so that

$$\langle J, k \mp 1, m | \hat{J}_m^{\pm} | J, k, m\rangle = \hbar[J(J+1) - k(k \mp 1)]^{1/2}. \qquad (8\text{-}110)$$

From these results we determine that $N_+' = N_+'' = N_-' = N_-'' (= N$, say) and this constant is real and positive. We can thus write Eqs. (8-98a) and (8-98b) as

$$|J, (\pm)|k|, \pm|m|\rangle = N(\hat{J}_m^{(\mp)})^{|k|}(\hat{J}_s^{\pm})^{|m|}|J, 0, 0\rangle, \qquad (8\text{-}111)$$

where J, k, and m are integral and

$$N = \{(J - |m|)!(J - |k|)!/[(J + |m|)!(J + |k|)!]\}^{1/2} \hbar^{-(|k| + |m|)};$$

the $(\pm)$ and $(\mp)$ are correlated and the two $\pm$ are correlated so that Eq. (8-111) is four equations in all. The phase factor choices of Eqs. (8-66) and (8-67) are not the same as that in Eq. (8-111). The nonvanishing matrix elements of $\hat{J}^2$, $\hat{J}_z$, $\hat{J}_\zeta$, $\hat{J}_m^{\pm}$, and $\hat{J}_s^{\pm}$ in the symmetric top wavefunctions $|J, k, m\rangle$, with the phase factor choice of Eq. (8-111), are collected together in Table 8-1. We will use the phase factor choice of Eq. (8-111) in the remainder of this book.

Table 8-1

Nonvanishing matrix elements[a] of components of the rovibronic angular momentum $\hat{J}$

$$\langle J, k, m | \hat{J}^2 | J, k, m\rangle = J(J+1)\hbar^2$$
$$\langle J, k, m | \hat{J}_z | J, k, m\rangle = k\hbar$$
$$\langle J, k, m | \hat{J}_\zeta | J, k, m\rangle = m\hbar$$
$$\langle J, k, m \pm 1 | \hat{J}_s^{\pm} | J, k, m\rangle = \hbar[J(J+1) - m(m \pm 1)]^{1/2}$$
$$\langle J, k \mp 1, m | \hat{J}_m^{\pm} | J, k, m\rangle = \hbar[J(J+1) - k(k \mp 1)]^{1/2}$$

[a] In a basis of symmetric top wavefunctions, where $\hat{J}_m^{\pm} = \hat{J}_x \pm i\hat{J}_y$ and $\hat{J}_s^{\pm} = \hat{J}_\xi \pm i\hat{J}_\eta$.

The Asymmetric Top Molecule

For an asymmetric top molecule the rigid rotor rotational Hamiltonian is [from Eq. (8-35)]

$$\hat{H}_{rot} = \hbar^{-2}(A_e \hat{J}_a^2 + B_e \hat{J}_b^2 + C_e \hat{J}_c^2). \tag{8-112}$$

The expressions for $\hat{J}_a^2$, $\hat{J}_b^2$, and $\hat{J}_c^2$ in terms of the Euler angles depend on the convention used to identify the a, b, and c axes with the x, y, and z axes in Fig. 7-1. Regardless of which convention is used the way we solve this rotational Schrödinger equation is to set up the Hamiltonian matrix in a basis of symmetric top wavefunctions and diagonalize it to obtain the energies and wavefunctions. The wavefunctions are obtained as a linear combination of symmetric top wavefunctions with coefficients that are functions of A_e, B_e, and C_e. We will demonstrate the technique using a type I^r convention and briefly discuss the results of using a type III^r convention at the end of the section.

In a type I^r convention the Hamiltonian for an asymmetric top is

$$\hat{H}_{rot} = \hbar^{-2}(A_e \hat{J}_z^2 + B_e \hat{J}_x^2 + C_e \hat{J}_y^2). \tag{8-113}$$

We set up the Hamiltonian matrix in $|J, k, m\rangle$ symmetric top functions. To do this it is convenient to rewrite the Hamiltonian as

$$\hat{H}_{rot} = \hbar^{-2}\{[(B_e + C_e)/2]\hat{J}^2 + [A_e - (B_e + C_e)/2]\hat{J}_z^2$$
$$+ [(B_e - C_e)/4][(\hat{J}_m^+)^2 + (\hat{J}_m^-)^2]\}. \tag{8-114}$$

We use a symmetric top basis set $|J, k, m\rangle$ with a type I^r convention and so the k quantum number refers to rotational angular momentum about the a axis. As a result in a type I^r convention we sometimes write the basis set as $|J, k_a, m\rangle$ where we introduce k_a as the a axis rotational quantum number. To determine matrix elements of the Hamiltonian we need the matrix elements of $\hat{J}^2, \hat{J}_z^2, (\hat{J}_m^+)^2$, and $(\hat{J}_m^-)^2$. From the results in Table 8-1 the only nonvanishing matrix elements of these operators are

$$\langle J, k, m|\hat{J}^2|J, k, m\rangle = J(J + 1)\hbar^2, \tag{8-115}$$

$$\langle J, k, m|\hat{J}_z^2|J, k, m\rangle = k^2\hbar^2, \tag{8-116}$$

$$\langle J, k - 2, m|(\hat{J}_m^+)^2|J, k, m\rangle = \{[J(J + 1) - (k - 1)(k - 2)][J(J + 1)$$
$$- k(k - 1)]\}^{1/2}\hbar^2, \tag{8-117}$$

and

$$\langle J, k + 2, m|(\hat{J}_m^-)^2|J, k, m\rangle = \{[J(J + 1) - (k + 1)(k + 2)][J(J + 1)$$
$$- k(k + 1)]\}^{1/2}\hbar^2. \tag{8-118}$$

We see that $\hat{H}_{\text{rot}}$ for an asymmetric top only has nonvanishing matrix elements between states of the same J and m and between states having the same k values or k values differing by two. As a result the Hamiltonian matrix factors into blocks, one for each J value, and each of these blocks consists of $2J + 1$ identical blocks, one for each m value. In the absence of external fields this m degeneracy only affects intensities; we neglect it and focus attention only on $m = 0$ states. Each $(m = 0)$ J block can be block diagonalized into four blocks by forming sum $(+)$ and difference $(-)$ combinations of $|J, K, 0\rangle$ and $|J, -K, 0\rangle$ functions where $K = |k|$. This is because $\hat{H}_{\text{rot}}$ has no matrix elements between k-even and k-odd functions or between $+$ and $-$ functions. The four blocks are called E^+, E^-, O^+, and O^- depending on whether k is even or odd, and on whether they are $+$ or $-$ functions. The solution to the next problem will demonstrate this point. A general result for the asymmetric top Hamiltonian matrix of a given J is that for J even the E^+ block has dimension $(J + 2)/2$ and the other three blocks have dimension $J/2$, whereas for J odd the E^- block has dimension $(J - 1)/2$ and the other three blocks have dimension $(J + 1)/2$.

Problem 8-3. Determine the rigid rotor energies and wavefunctions for an asymmetric top molecule with rotational constants A_{e}, B_{e}, and C_{e} for the states with $J = 0, 1,$ and 2.

Answer. We consider $J = 0, 1,$ and 2 separately and for each J block of the Hamiltonian matrix we neglect the m degeneracy and suppress the m label so that we label the basis functions $|J, k\rangle$.

The $J = 0$ block of the Hamiltonian matrix is a 1×1 block involving the $J = k = 0$ basis set wavefunction. This matrix element is zero and so $E_{\text{rot}}(J = 0) = 0$ and $\Phi_{\text{rot}}(J = 0) = (8\pi^2)^{-1/2}$.

The $J = 1$ block is a 3×3 block involving $|J, k\rangle$ functions with $k = -1, 0,$ and $+1$; i.e., $|1, -1\rangle$, $|1, 0\rangle$, and $|1, +1\rangle$. We form sum and difference combinations and we label them $|J, K, A^{\pm}\rangle$ where $A = O$ or E (for odd or even)

$$|1, 1, O^+\rangle = [|1, -1\rangle + |1, +1\rangle]/\sqrt{2}, \tag{8-119}$$

$$|1, 1, O^-\rangle = [|1, -1\rangle - |1, +1\rangle]/\sqrt{2}, \tag{8-120}$$

and

$$|1, 0, E^+\rangle = |1, 0\rangle. \tag{8-121}$$

These functions are O^+, O^-, and E^+ type functions, respectively, and $\hat{H}_{\text{rot}}$ has no off-diagonal matrix elements between them [the reader can test this by using the results in Eqs. (8-115)–(8-118)]. Thus these are eigenfunctions of $\hat{H}_{\text{rot}}$ and the eigenvalues are given by the diagonal matrix elements of of $\hat{H}_{rot}$ in these functions. For the O^+ level the energy, using the notation $E_{rot}(J, A^{\pm})$, is given by

$$E_{rot}(1, O^+) = \tfrac{1}{2}[\langle 1, -1| + \langle 1, +1|]\hat{H}_{rot}[|1, -1\rangle + |1, +1\rangle]. \quad (8\text{-}122)$$

From Eqs. (8-115) and (8-116) with Eq. (8-114) we see that

$$\begin{aligned}
\langle 1, -1|\hat{H}_{rot}|1, -1\rangle &= \langle 1, +1|\hat{H}_{rot}|1, +1\rangle \\
&= [(B_e + C_e)/2]2 + [A_e - (B_e + C_e)/2] \\
&= A_e + (B_e + C_e)/2, \quad\quad\quad\quad\quad (8\text{-}123)
\end{aligned}$$

and from Eqs. (8-117) and (8-118) with Eq. (8-114) we have

$$\langle 1, -1|\hat{H}_{rot}|1, +1\rangle = \langle 1, +1|\hat{H}_{rot}|1, -1\rangle = [(B_e - C_e)/4]2 = (B_e - C_e)/2. \quad (8\text{-}124)$$

Substituting Eqs. (8-123) and (8-124) into Eq. (8-122) we obtain

$$E_{rot}(1, O^+) = A_e + B_e. \quad (8\text{-}125)$$

It is straightforward to determine in a similar manner that

$$E_{rot}(1, O^-) = A_e + C_e \quad (8\text{-}126)$$

and

$$E_{rot}(1, E^+) = B_e + C_e. \quad (8\text{-}127)$$

For $J = 2$ the appropriate basis functions are

$$|2, 2, E^+\rangle = [|2, -2\rangle + |2, +2\rangle]/\sqrt{2}, \quad (8\text{-}128)$$

$$|2, 2, E^-\rangle = [|2, -2\rangle - |2, +2\rangle]/\sqrt{2}, \quad (8\text{-}129)$$

$$|2, 1, O^+\rangle = [|2, -1\rangle + |2, +1\rangle]/\sqrt{2}, \quad (8\text{-}130)$$

$$|2, 1, O^-\rangle = [|2, -1\rangle - |2, +1\rangle]/\sqrt{2}, \quad (8\text{-}131)$$

and

$$|2, 0, E^+\rangle = |2, 0\rangle. \quad (8\text{-}132)$$

The only nonvanishing off-diagonal matrix element of $\hat{H}_{rot}$ between these five functions is that between the two E^+ functions. From the three 1×1 blocks we obtain energies $E_{rot}(J, A^{\pm})$ given by

$$E_{rot}(2, E^-) = 4A_e + B_e + C_e, \quad (8\text{-}133)$$

$$E_{rot}(2, O^+) = A_e + 4B_e + C_e, \quad (8\text{-}134)$$

and

$$E_{rot}(2, O^-) = A_e + B_e + 4C_e. \tag{8-135}$$

The 2×2 block of $\hat{H}_{rot}$ for the $J = 2 E^+$ functions is

	$|2,0,E^+\rangle$	$|2,2,E^+\rangle$
$\langle 2,0,E^+|$	$3(B_e + C_e)$	$\sqrt{3}(B_e - C_e)$
$\langle 2,2,E^+|$	$\sqrt{3}(B_e - C_e)$	$4A_e + B_e + C_e$

The eigenvalues of this matrix are

$$E^{\pm}_{rot}(2, E^+) = 2(A_e + B_e + C_e) \pm [3(B_e - C_e)^2 + (2A_e - B_e - C_e)^2]^{1/2} \tag{8-136}$$

and the eigenfunctions $\Phi^{\pm}_{rot}(J, E^+)$ are

$$\Phi^-_{rot}(2, E^+) = [c^+|2,0,E^+\rangle - c^-|2,2,E^+\rangle]/\sqrt{2}, \tag{8-137}$$

and

$$\Phi^+_{rot}(2, E^+) = [c^-|2,0,E^+\rangle + c^+|2,2,E^+\rangle]/\sqrt{2}, \tag{8-138}$$

where

$$c^{\pm} = \{1 \pm (2A_e - B_e - C_e)/[3(B_e - C_e)^2 + (2A_e - B_e - C_e)^2]^{1/2}\}^{1/2}. \tag{8-139}$$

These two functions are a mixture of $K = 0$ and $K = 2$ functions with co-efficients depending on the rotational constants. As a result these functions are not eigenfunctions of $\hat{J}_z$ and we say that K (i.e., K_a) is not a *good* quantum number for them. The idea of a good quantum number will be discussed further in Chapter 11.

In the above discussion of the asymmetric top wavefunctions and energy levels a I^r convention was used and consequently the basis set consisted of $|J, k_a, m\rangle$ symmetric top functions. It is of interest to consider briefly what happens when we use a III^r convention and $|J, k_c, m\rangle$ basis functions. In a III^r convention the rigid rotor Hamiltonian becomes

$$\hat{H}_{rot} = \hbar^{-2}(A_e\hat{J}_x^2 + B_e\hat{J}_y^2 + C_e\hat{J}_z^2), \tag{8-140}$$

which can be rewritten as

$$\hbar^{-2}\{[(A_e + B_e)/2]\hat{J}^2 + [C_e - (A_e + B_e)/2]\hat{J}_z^2 + [(A_e - B_e)/4][(\hat{J}_m^+)^2 + (\hat{J}_m^-)^2]\}. \tag{8-141}$$

We set up the matrix of this Hamiltonian in a $|J,k,m\rangle = |J,k_c,m\rangle$ basis and diagonalize to get the energies and wavefunctions. The energies will be the same as obtained in the I^r basis but the wavefunctions will look different since the Euler angles are defined differently; also the $E^\pm$ and $O^\pm$ designations are different since K_c rather that K_a is used. In a I^r basis the Euler angles $\theta(I^r)$ and $\phi(I^r)$ are the polar angles of the a axis in the (ξ,η,ζ) axis system, and $\chi(I^r)$ is the angle between the c axis and the line of intersection of the bc and $\xi\eta$ planes. In a III^r basis the angles $\theta(III^r)$ and $\phi(III^r)$ are the polar angles of the c axis in the (ξ,η,ζ) axis system, and $\chi(III^r)$ is the angle between the b axis and the line of intersection of the ab and $\xi\eta$ planes.[2] The relationship between these two sets of Euler angles is complicated; for example,

$$\theta(I^r) = \arccos[-\sin\theta(III^r)\cos\chi(III^r)]. \quad (8\text{-}142)$$

Using relationships such as this it is possible to show that the asymmetric top wavefunctions obtained in the I^r basis are identical to those obtained in the III^r basis for a given molecule.

If the asymmetric top molecule under study is a near prolate top, i.e., $B_e \cong C_e$, then in a type I^r basis the off-diagonal matrix elements will be small. If the molecule under study is a near oblate top; i.e., $A_e \cong B_e$, then in a III^r basis the off-diagonal matrix elements will be small. The degree of asymmetry in an asymmetric top is conveniently given by the value of

$$\kappa = (2B_e - A_e - C_e)/(A_e - C_e). \quad (8\text{-}143)$$

For a prolate top $\kappa = -1$, for an oblate top $\kappa = +1$, and for an asymmetric top $-1 < \kappa < +1$, with $\kappa = 0$ being the "most" asymmetric when B_e is halfway between A_e and C_e.

In Fig. 8-1 we show the correlation of the $J = 0, 1$, and 2 energy levels of a $\kappa = 0$ asymmetric top molecule with the $\kappa = \pm1$ symmetric top limits. The $E^\pm$ and $O^\pm$ designations of the levels are shown for the asymmetric rotor levels in the I^r basis. It is customary to label the asymmetric top levels $J_{K_aK_c}$ where the labels K_a and K_c indicate the prolate and oblate levels respectively with which the level correlates ($K_a = |k_a|$ and $K_c = |k_c|$). These labels are very useful and we will also often refer to an asymmetric rotor level as being ee, eo, oe, or oo depending on whether K_a and K_c are even (e) or odd (o), respectively; e.g., the level 2_{21} is an eo level. The asymmetric top wavefunctions are not in general eigenfunctions of $\hat{J}_a$ or $\hat{J}_c$ and neither K_a or K_c is a good quantum number.

[2] See Figs. 10-3 and 10-4.

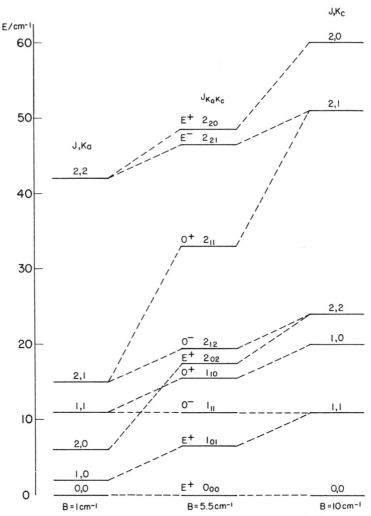

Fig. 8-1. The correlation of the $J = 0$, 1, and 2 energy levels of a rigid asymmetric top molecule having $A = 10$ cm^{-1}, $B = 5.5$ cm^{-1}, and $C = 1$ cm^{-1} (i.e., $\kappa = 0$) with those for the prolate top having $(A, B, C) = (10, 1, 1)$ cm^{-1} on the left, and with those for the oblate top having $(A, B, C) = (10, 10, 1)$ cm^{-1} on the right. The $E^{\pm}$ and $O^{\pm}$ designations depend on whether K_a is even (E) or odd (O), and on whether sum of difference functions $(|J, K_a\rangle \pm |J, -K_a\rangle)/\sqrt{2}$ occur in the basis set for the level considered.

THE HARMONIC OSCILLATOR SCHRÖDINGER EQUATION

Before looking at the solution of the harmonic oscillator Schrödinger equation the following problem should be worked.

Problem 8-4. In the harmonic oscillator equation the normal coordinate Q and its conjugate momentum $\hat{P} = -i\hbar\, \partial/\partial Q$ occur. Determine the values of the commutators $[\hat{P}^2, Q]$ and $[Q^2, \hat{P}]$.

Answer. Straightforward commutator manipulation gives

$$[\hat{P}^2, Q] = \hat{P}[\hat{P}, Q] + [\hat{P}, Q]\hat{P} = \hat{P}(-i\hbar) + (-i\hbar)\hat{P} = -2i\hbar\hat{P} \quad (8\text{-}144)$$

and

$$[Q^2, \hat{P}] = Q[Q, \hat{P}] + [Q, \hat{P}]Q = Q(i\hbar) + (i\hbar)Q = 2i\hbar Q. \quad (8\text{-}145)$$

The One-Dimensional Harmonic Oscillator

The one-dimensional harmonic oscillator wave equation can be written as

$$\hat{H}_{\mathrm{ho}}\Phi_v = \tfrac{1}{2}(\hat{P}^2 + \lambda Q^2)\Phi_v = E_v\Phi_v. \quad (8\text{-}146)$$

To determine the eigenfunctions and eigenvalues of this Hamiltonian we introduce the operators

$$\hat{R}^+ = (1/\sqrt{2})(\hat{P} + i\lambda^{1/2}Q) \quad (8\text{-}147)$$

and

$$\hat{R}^- = (1/\sqrt{2})(\hat{P} - i\lambda^{1/2}Q). \quad (8\text{-}148)$$

The commutators of these with the harmonic oscillator Hamiltonian are [using Eqs. (8-144) and (8-145)]

$$[\hat{H}_{\mathrm{ho}}, \hat{R}^+] = (1/2\sqrt{2})(i\lambda^{1/2}[\hat{P}^2, Q] + \lambda[Q^2, \hat{P}]) = \hbar\lambda^{1/2}\hat{R}^+ \quad (8\text{-}149)$$

and

$$[\hat{H}_{\mathrm{ho}}, \hat{R}^-] = -\hbar\lambda^{1/2}\hat{R}^-. \quad (8\text{-}150)$$

Thus from Eqs. (8-84) and (8-85) in Problem 8-2 we see that $\hat{R}^+$ is a "ladder up" operator and $\hat{R}^-$ is a "ladder down" operator so that if

$$\hat{H}_{\mathrm{ho}}\Phi_v = E_v\Phi_v, \quad (8\text{-}151)$$

then

$$\hat{H}_{\mathrm{ho}}[\hat{R}^\pm\Phi_v] = (E_v \pm \hbar\lambda^{1/2})[\hat{R}^\pm\Phi_v]. \quad (8\text{-}152)$$

It is easy to see that

$$\hat{H}_{\mathrm{ho}} = \hat{R}^+\hat{R}^- + \tfrac{1}{2}\hbar\lambda^{1/2} \quad (8\text{-}153)$$

and

$$\hat{H}_{\text{ho}} = \hat{R}^- \hat{R}^+ - \tfrac{1}{2}\hbar\lambda^{1/2}. \tag{8-154}$$

Thus, for example, eigenfunctions of $\hat{R}^+\hat{R}^-$ are eigenfunctions of $\hat{H}_{\text{ho}}$ and vice versa. The eigenvalues of $\hat{R}^+\hat{R}^-$ must be real and nonnegative as we now show. If Φ_v is a normalized eigenfunction of $\hat{R}^+\hat{R}^-$ with eigenvalue a we have

$$a = \int \Phi_v^*(\hat{R}^+\hat{R}^-)\Phi_v\,dQ. \tag{8-155}$$

This equation can be rewritten as

$$a = \int (\hat{R}^-\Phi_v)^*(\hat{R}^-\Phi_v)\,dQ \tag{8-156}$$

$$= \int |\hat{R}^-\Phi_v|^2\,dQ \tag{8-157}$$

and hence a (an eigenvalue of $\hat{R}^+\hat{R}^-$) must be real and nonnegative. Since by definition $\lambda^{1/2}$ is real and positive the eigenvalues of $\hat{H}_{\text{ho}}$ must be real and nonnegative as a result of Eqs. (8-153) and (8-157). Since the eigenvalues of $\hat{H}_{\text{ho}}$ are real and nonnegative there must be a lowest one which we call E_0. We call the eigenfunction having this eigenvalue Φ_0. Applying $\hat{R}^-$ to Φ_0 must give zero since there is no eigenfunction having a lower eigenvalue. Thus we can write

$$\hat{R}^-\Phi_0 = [(-i\hbar/\sqrt{2})\,\partial/\partial Q - i(\lambda/2)^{1/2}Q]\Phi_0 = 0, \tag{8-158}$$

which is a first order differential equation for Φ_0 that yields as solution the normalized function

$$\Phi_0 = (2\lambda^{1/2}/h)^{1/4}\exp[-(\lambda^{1/2}/2\hbar)Q^2], \tag{8-159}$$

where a real and positive phase factor choice has been made. Applying $\hat{R}^+$ to Eq. (8-158) we obtain

$$\hat{R}^+\hat{R}^-\Phi_0 = 0, \tag{8-160}$$

and from Eq. (8-153) we see that we can rewrite Eq. (8-160) as

$$(\hat{H}_{\text{ho}} - \hbar\lambda^{1/2}/2)\Phi_0 = 0, \tag{8-161}$$

i.e.,

$$E_0 = \hbar\lambda^{1/2}/2. \tag{8-162}$$

We have thus determined the lowest eigenvalue E_0 and associated eigenfunction Φ_0 of the harmonic oscillator Hamiltonian.

The higher eigenvalues E_v and eigenfunctions Φ_v (where $v = 1, 2, 3, \ldots$) are obtained by operating with $(\hat{R}^+)^v$ on Φ_0. Clearly from the fact that $\hat{R}^+$

ladders up by $\hbar\lambda^{1/2}$ we must have

$$E_v = E_0 + v(\hbar\lambda^{1/2}) = (v + \tfrac{1}{2})\hbar\lambda^{1/2}. \tag{8-163}$$

The constant λ is not always used and other constants defined from it are given by

$$\lambda^{1/2} = \omega = \hbar\gamma = 2\pi c\omega_e = 2\pi v, \tag{8-164}$$

where the units of λ, ω, γ, ω_e, and v are $\sec^{-2}$, radians $\sec^{-1}$, $\mathrm{erg}^{-1}\ \sec^{-2}$, cm^{-1}, and cycles $\sec^{-1}$ (Hz), respectively. In the same manner as Eq. (8-98) we can write

$$\Phi_v = N_v'(\hat{R}^+)^v\Phi_0, \tag{8-165}$$

where N_v' ensures normalization of Φ_v and it involves a phase factor. To follow arguments very similar to those used in Eqs. (8-101)–(8-106) we write

$$\Phi_{v\pm 1} = N_\pm \hat{R}^\pm\Phi_v, \tag{8-166}$$

where Φ_v is normalized and $N_\pm$ is required for $\Phi_{v\pm 1}$ to be normalized. Multiplying each side on the left by its complex conjugate and integrating, and noting that [from Eqs. (8-153) and (8-154) and using $\hbar^2\gamma = \hbar\lambda^{1/2}$]

$$\langle\Phi_v|\hat{R}^-\hat{R}^+|\Phi_v\rangle = (v + 1)\hbar^2\gamma \tag{8-167}$$

and that

$$\langle\Phi_v|\hat{R}^+\hat{R}^-|\Phi_v\rangle = v\hbar^2\gamma, \tag{8-168}$$

we determine that

$$N_\pm = e^{\pm i\delta}/[(v + \tfrac{1}{2} \pm \tfrac{1}{2})\hbar^2\gamma]^{1/2}. \tag{8-169}$$

We choose the phase factor δ so that

$$e^{\pm i\delta} = \mp i, \tag{8-170}$$

and we obtain

$$N_\pm = \mp i/[(v + \tfrac{1}{2} \pm \tfrac{1}{2})\hbar^2\gamma]^{1/2}. \tag{8-171}$$

For example, for $v = 1$ we have

$$\Phi_1 = [-i/(\hbar^2\gamma)^{1/2}]\hat{R}^+\Phi_0 \tag{8-172}$$

$$= (\gamma/4\pi)^{1/4}(2\gamma^{1/2}Q)\exp(-\gamma Q^2/2), \tag{8-173}$$

In general we can write

$$\Phi_v = N_v H_v(\gamma^{1/2}Q)\exp(-\gamma Q^2/2), \tag{8-174}$$

where

$$N_v = \gamma^{1/4}/(\pi^{1/2}2^v v!)^{1/2} \tag{8-175}$$

and $H_v(\gamma^{1/2}Q)$ is a Hermite polynomial for which the first four values are [see Appendix III of Wilson, Decius, and Cross (1955)]

$$H_0(\gamma^{1/2}Q) = 1, \tag{8-176}$$

$$H_1(\gamma^{1/2}Q) = 2\gamma^{1/2}Q, \tag{8-177}$$

$$H_2(\gamma^{1/2}Q) = 4\gamma Q^2 - 2, \tag{8-178}$$

and

$$H_3(\gamma^{1/2}Q) = 8\gamma^{3/2}Q^3 - 12\gamma^{1/2}Q. \tag{8-179}$$

In general $H_v(\gamma^{1/2}Q)$ contains $(\gamma^{1/2}Q)$ to the powers $v, v - 2, v - 4, \ldots, 1$ or 0, i.e., either all even powers or all odd powers as v is even or odd, respectively.

Matrix elements that are particularly useful when working with harmonic oscillator functions are those of $\hat{P}$ and Q. From Eqs. (8-166) and (8-171) we can write

$$\langle \Phi_{v\pm1}|\hat{R}^\pm|\Phi_v\rangle = \pm i[(v + \tfrac{1}{2} \pm \tfrac{1}{2})\hbar^2\gamma]^{1/2}, \tag{8-180}$$

and these are the only nonvanishing matrix elements of the operators $\hat{R}^+$ and $\hat{R}^-$. From Eqs. (8-147) and (8-148) we see that

$$\hat{P} = (\hat{R}^+ + \hat{R}^-)/\sqrt{2} \tag{8-181}$$

and

$$Q = (\hat{R}^+ - \hat{R}^-)/(i\hbar\gamma\sqrt{2}), \tag{8-182}$$

so that the nonvanishing matrix elements of $\hat{P}$ and Q can be deduced from Eq. (8-180); they are given in Table 8-2. The phase factor choice given in Eq. (8-170) has resulted in the matrix elements of Q being real and those of $\hat{P}$ being imaginary. We see that the only nonvanishing matrix elements of

Table 8-2

Nonvanishing matrix elements of normal coordinate Q and momentum $\hat{P}$ for the one-dimensional harmonic oscillator[a]

$\langle v + 1|Q|v\rangle = \sqrt{(v + 1)/(2\gamma)}$	$\langle v + 1|\hat{P}|v\rangle = i\hbar\sqrt{(v + 1)\gamma/2}$
$\langle v - 1|Q|v\rangle = \sqrt{v/(2\gamma)}$	$\langle v - 1|\hat{P}|v\rangle = -i\hbar\sqrt{v\gamma/2}$

[a] $\hat{P} = -i\hbar\,\partial/\partial Q$ and the one-dimensional harmonic oscillator functions $\Phi_v = |v\rangle$. We use $\gamma = \lambda^{1/2}/\hbar$ where the one-dimensional harmonic oscillator Hamiltonian, is $\hat{H}_{\text{ho}} = \tfrac{1}{2}(\hat{P}^2 + \lambda Q^2)$.

$\hat{P}$ or Q have v values differing by ± 1. Matrix elements of $\hat{P}^r Q^s$ where r and s are integers can be evaluated from the results in Table 8-2 (see Problem 8-5); the nonvanishing matrix elements $\langle v'|\hat{P}^r Q^s|v''\rangle$ are those for which

$$v' = v'' + r + s, v'' + r + s - 2, v'' + r + s - 4, \ldots, v'' - r - s.^3 \quad (8\text{-}183)$$

Problem 8-5. Determine the matrix element $\langle \Phi_4|\hat{P}Q^2|\Phi_3\rangle$ for the harmonic oscillator basis functions Φ_3 and Φ_4 (with $v = 3$ and 4, respectively) from the results in Table 8-2.

Answer. We can write the matrix element as follows:

$$\langle \Phi_4|\hat{P}Q^2|\Phi_3\rangle = \sum_{v,v'} \langle \Phi_4|\hat{P}|\Phi_v\rangle\langle \Phi_v|Q|\Phi_{v'}\rangle\langle \Phi_{v'}|Q|\Phi_3\rangle, \quad (8\text{-}184)$$

where we have used the law of matrix multiplication [see Eq. (4-3)]. Nonvanishing matrix elements of $\hat{P}$ and Q have $\Delta v = \pm 1$ so that there are only three nonvanishing terms in this sum and

$$
\begin{aligned}
\langle \Phi_4|\hat{P}Q^2|\Phi_3\rangle &= \langle \Phi_4|\hat{P}|\Phi_5\rangle\langle \Phi_5|Q|\Phi_4\rangle\langle \Phi_4|Q|\Phi_3\rangle \\
&\quad + \langle \Phi_4|\hat{P}|\Phi_3\rangle\langle \Phi_3|Q|\Phi_4\rangle\langle \Phi_4|Q|\Phi_3\rangle \\
&\quad + \langle \Phi_4|\hat{P}|\Phi_3\rangle\langle \Phi_3|Q|\Phi_2\rangle\langle \Phi_2|Q|\Phi_3\rangle \\
&= [(-ih\sqrt{5\gamma/2})\sqrt{5/2\gamma}\sqrt{4/2\gamma}] + [(ih\sqrt{4\gamma/2})\sqrt{4/2\gamma}\sqrt{4/2\gamma}] \\
&\quad + [(ih\sqrt{4\gamma/2})\sqrt{3/2\gamma}\sqrt{3/2\gamma}] \\
&= ih(2/\gamma)^{1/2}. \quad (8\text{-}185)
\end{aligned}
$$

The harmonic oscillator Hamiltonian for a molecule is

$$\frac{1}{2}\sum_r (\hat{P}_r{}^2 + \lambda_r Q_r{}^2) \quad (8\text{-}186)$$

from Eq. (8-28a). From the results of this section we see that the eigenfunctions of this Hamiltonian can be written

$$
\begin{aligned}
\Phi_{\text{vib}} &= \Phi_{v_1}(Q_1)\Phi_{v_2}(Q_2)\cdots\Phi_{v_{3N-6}}(Q_{3N-6}) \\
&= \exp\left[-\frac{1}{2}\sum_r \gamma_r Q_r{}^2\right] \prod_r N_{v_r} H_{v_r}(\gamma_r^{1/2} Q_r) \quad (8\text{-}187)
\end{aligned}
$$

and the eigenvalues are

$$E_{\text{vib}} = \sum_r (v_r + \tfrac{1}{2})h^2\gamma_r, \quad (8\text{-}188)$$

where $\gamma_r = \lambda_r^{1/2}/h$.

---

3 The lower limit is one or zero if $v'' < (r + s)$.

The Two-Dimensional Isotropic Harmonic Oscillator

If the normal coordinate representation of a molecule contains a doubly degenerate irreducible representation then the degenerate pair of normal coordinates, Q_a and Q_b, say, must have the same values for λ_a and λ_b, as we now show. Since (Q_a, Q_b) form the basis for a degenerate representation they must be "mixed" by at least one operation, R, say, of the symmetry group of the molecule, i.e.,

$$RQ_a = c_{aa}Q_a + c_{ab}Q_b \tag{8-189}$$

and

$$RQ_b = c_{ba}Q_a + c_{bb}Q_b, \tag{8-190}$$

where $c_{\alpha\beta}$ are transformation coefficients and the matrix of the $c_{\alpha\beta}$ is the unitary matrix representing R in the irreducible representation generated by Q_a and Q_b [i.e., $R(Q_a^2 + Q_b^2) = (Q_a^2 + Q_b^2)$]. However, since R is a symmetry operation it commutes with the Hamiltonian and the following operator equation must be true

$$R\hat{H}_{\text{tdho}} = \hat{H}_{\text{tdho}}R, \tag{8-191}$$

where tdho stands for two-dimensional harmonic oscillator, but

$$\begin{aligned}R\hat{H}_{\text{tdho}} &= R[\tfrac{1}{2}(\hat{P}_a^2 + \hat{P}_b^2 + \lambda_a Q_a^2 + \lambda_b Q_b^2)] \\ &= \{\tfrac{1}{2}[\hat{P}_a^2 + \hat{P}_b^2 + \lambda_a(c_{aa}^2 Q_a^2 + 2c_{aa}c_{ab}Q_aQ_b + c_{ab}^2 Q_b^2) \\ &\quad + \lambda_b(c_{ba}^2 Q_a^2 + 2c_{ba}c_{bb}Q_aQ_b + c_{bb}^2 Q_b^2)]\}R, \end{aligned} \tag{8-192}$$

and the only way that the right hand sides of Eqs. (8-191) and (8-192) can be equal is if either

$$c_{ab} = c_{ba} = 0 \tag{8-193}$$

or

$$\lambda_a = \lambda_b. \tag{8-194}$$

This means that either [from Eq. (8-193)] Q_a and Q_b are not mixed by any symmetry operations if $\lambda_a \neq \lambda_b$, or if they are mixed then [from Eq. (8-194)] they must have the same frequency. In other words if $\lambda_a \neq \lambda_b$ then Q_a and Q_b cannot be the basis for a doubly degenerate irreducible representation of a symmetry group of the Hamiltonian, but if Q_a and Q_b are the basis for such a representation then it is necessary that $\lambda_a = \lambda_b$.

As a result of the above argument [see Eq. (8-193)] the normal coordinates Q_r that have nondegenerate frequencies λ_r will form the basis for one-dimensional (and hence irreducible) representations of a symmetry group of the Hamiltonian. A set of l normal coordinates $Q_{s1}, Q_{s2}, \ldots, Q_{sl}$ that all have the same normal frequency λ_s will form the basis for an l-dimensional representation of the group. This l-dimensional representation can be reducible or irreducible. If the representation is reducible then it is an *accident* requiring a fortuitous relationship to occur between the force constants and the nuclear masses; this is rare. Even if an accidental degeneracy occurs it is still possible to construct normal coordinates that transform irreducibly.

From Eq. (8-188) the energy of a pair of degenerate vibrations described by the normal coordinates Q_a and Q_b is given by

$$E(v_a, v_b) = [(v_a + \tfrac{1}{2}) + (v_b + \tfrac{1}{2})]\hbar^2\gamma, \tag{8-195}$$

where $\hbar\gamma = \lambda_a^{1/2} = \lambda_b^{1/2}$, and there are systematic degeneracies, i.e.,

$$
\begin{aligned}
E(0,0) &= \hbar^2\gamma, \qquad E(1,0) = E(0,1) = 2\hbar^2\gamma, \\
E(2,0) &= E(1,1) = E(0,2) = 3\hbar^2\gamma, \qquad \text{etc.}
\end{aligned} \tag{8-196}
$$

Thus the energy of a level depends only on $(v_a + v_b)$ and each level has a degeneracy of $(v_a + v_b + 1)$. The wavefunction for this pair of vibrations is [from Eq. (8-187)] given by

$$\Phi_{v_a v_b}(Q_a, Q_b) = N_{v_a} N_{v_b} \exp[-\gamma(Q_a^2 + Q_b^2)/2] H_{v_a}(\gamma^{1/2} Q_a) H_{v_b}(\gamma^{1/2} Q_b), \tag{8-197}$$

i.e.,

$$\Phi_{1,0} = (\gamma/2\pi)^{1/2} \exp[-\gamma(Q_a^2 + Q_b^2)/2] 2\gamma^{1/2} Q_a \tag{8-198}$$

and

$$\Phi_{0,1} = (\gamma/2\pi)^{1/2} \exp[-\gamma(Q_a^2 + Q_b^2)/2] 2\gamma^{1/2} Q_b \tag{8-199}$$

(both with energy $E = 2\hbar^2\gamma$) and

$$\Phi_{2,0} = (\gamma/8\pi)^{1/2} \exp[-\gamma(Q_a^2 + Q_b^2)/2](4\gamma Q_a^2 - 2), \tag{8-200}$$

$$\Phi_{0,2} = (\gamma/8\pi)^{1/2} \exp[-\gamma(Q_a^2 + Q_b^2)/2](4\gamma Q_b^2 - 2), \tag{8-201}$$

and

$$\Phi_{1,1} = (\gamma/4\pi)^{1/2} \exp[-\gamma(Q_a^2 + Q_b^2)/2] 4\gamma Q_a Q_b \tag{8-202}$$

(all with energy $E = 3\hbar^2\gamma$), etc. Equations (8-195) and (8-197) give the energy levels and wavefunctions of the *two-dimensional (isotropic) harmonic oscillator Hamiltonian*

$$\hat{H}_{\text{tdho}} = \frac{1}{2}[\hat{P}_a^2 + \hat{P}_b^2 + \lambda(Q_a^2 + Q_b^2)]. \tag{8-203}$$

For the two-dimensional harmonic oscillator we could use the wavefunctions as given in Eq. (8-197), but it proves more convenient to express

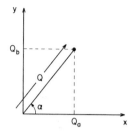

Fig. 8-2.   A diagrammatic representation of the relationship between the coordinates (Q_a, Q_b) and (Q, α) given in Eqs.
(8-204)–(8-207) for a two-dimensional isotropic harmonic
oscillator.

them in terms of new coordinates Q and α, instead of Q_a and Q_b, and to
introduce new quantum numbers v and l instead of v_a and v_b. The coordinates
Q and α are defined by

$$Q_a = Q \cos \alpha, \tag{8-204}$$

and

$$Q_b = Q \sin \alpha, \tag{8-205}$$

i.e.,

$$Q = |(Q_a{}^2 + Q_b{}^2)^{1/2}|, \tag{8-206}$$

and

$$\alpha = \arctan(Q_b/Q_a) = \tfrac{1}{2}\arccos[(Q_a{}^2 - Q_b{}^2)/(Q_a{}^2 + Q_b{}^2)], \tag{8-207}$$

where $Q > 0$ and $0 \leq \alpha \leq 2\pi$. The relationship between these coordinates is
represented in Fig. 8-2. Changing coordinates in Eq. (8-203) [using the chain
rule, Eqs. (6-8) and (6-9)]

$$\hat{H}_{\text{tdho}} = \frac{1}{2}\left[-\hbar^2\left(\frac{\partial^2}{\partial Q^2} + \frac{1}{Q}\frac{\partial}{\partial Q} + \frac{1}{Q^2}\frac{\partial^2}{\partial \alpha^2} \right) + \lambda Q^2 \right]. \tag{8-208}$$

This operator commutes with the *vibrational angular momentum operator* $\hat{M}$
which is given by

$$\hat{M} = (Q_a\hat{P}_b - Q_b\hat{P}_a) = -i\hbar\frac{\partial}{\partial \alpha}, \tag{8-209}$$

and the eigenfunctions of $\hat{H}_{\text{tdho}}$ can be written

$$\Psi_{v,l} = F_{v,l}(Q)e^{il\alpha}, \tag{8-210}$$

where

$$\hat{M}\Psi_{v,l} = l\hbar\Psi_{v,l}, \tag{8-211}$$

and $F_{v,l}(Q)$ is a function only of Q. As we will see the vibrational angular
momentum quantum number l can have any one of the $(v + 1)$ values v,
$v - 2, v - 4, \ldots, -v$.

The eigenfunctions and eigenvalues of $\hat{H}_{\text{tdho}}$ and the matrix elements of the normal coordinates and momenta can be determined using the ladder operator technique [see Moffitt and Liehr (1957)] and we will outline how this determination proceeds. We can rewrite $\hat{H}_{\text{tdho}}$ as

$$\hat{H}_{\text{tdho}} = \tfrac{1}{2}[\hat{P}^2 - ih(1/Q)\hat{P} + (\hat{M}^2/Q^2) + \lambda Q^2] = \tfrac{1}{2}(\hat{P}^+\hat{P}^- + \lambda Q^+ Q^-), \quad (8\text{-}212)$$

where

$$\hat{P} = -ih\,\partial/\partial Q$$

and we have introduced

$$Q^{\pm} = Q_a \pm iQ_b = Qe^{\pm i\alpha} \tag{8-213}$$

and

$$\hat{P}^{\pm} = \hat{P}_a \pm i\hat{P}_b = e^{\pm i\alpha}[\hat{P} \pm i(\hat{M}/Q)]. \tag{8-214}$$

The ladder operators appropriate now are

$$\hat{R}^{\pm(\pm)} = e^{(\pm)i\alpha}[\hat{P} \pm i\lambda^{1/2}Q(\pm)i(\hat{M}/Q)] = \hat{P}^{(\pm)} \pm i\lambda^{1/2}Q^{(\pm)}, \tag{8-215}$$

where the $\pm$ signs are correlated with each other and the $(\pm)$ signs are also correlated with each other so that there are four ladder operators in all. The commutators of the ladder operators with $\hat{H}_{\text{tdho}}$ and $\hat{M}$ are

$$[\hat{H}_{\text{tdho}}, \hat{R}^{\pm(\pm)}] = \pm h^2\gamma\hat{R}^{\pm(\pm)} \tag{8-216}$$

and

$$[\hat{M}, \hat{R}^{\pm(\pm)}] = (\pm)h\hat{R}^{\pm(\pm)}. \tag{8-217}$$

The eigenvalues of $\hat{H}_{\text{tdho}}$ and $\hat{M}$ are E and lh, respectively, so that

$\hat{R}^{+(+)}$ ladders E up by $h^2\gamma$ and ladders l up by 1,
$\hat{R}^{+(-)}$ ladders E up by $h^2\gamma$ and ladders l down by 1,
$\hat{R}^{-(+)}$ ladders E down by $h^2\gamma$ and ladders l up by 1, and
$\hat{R}^{-(-)}$ ladders E down by $h^2\gamma$ and ladders l down by 1.

In the notation of Moffitt and Liehr (1957) these four ladder operators are called F, $G^{\dagger}$, G, and $F^{\dagger}$, respectively. Equations like Eq. (8-160) can be set up for the lowest eigenfunction $\Psi_{0,l}$ of $\hat{H}_{\text{tdho}}$, i.e.,

$$\hat{R}^{+(+)}\hat{R}^{-(-)}\Psi_{0,l} = (2\hat{H}_{\text{tdho}} + 2\lambda^{1/2}\hat{M} - 2h\lambda^{1/2})\Psi_{0,l} = 0 \tag{8-218}$$

and

$$\hat{R}^{+(-)}\hat{R}^{-(+)}\Psi_{0,l} = (2\hat{H}_{\text{tdho}} - 2\lambda^{1/2}\hat{M} - 2h\lambda^{1/2})\Psi_{0,l} = 0. \tag{8-219}$$

By adding and subtracting these two equations we determine that the lowest level is nondegenerate with

$$E_0 = h^2\gamma \tag{8-220}$$

Table 8-3

Matrix elements of $Q^{\pm}$ and $\hat{P}^{\pm}$ for the two-dimensional isotropic harmonic oscillator[a]

$\langle v+1,l+1\|Q^+\|v,l\rangle = -i\sqrt{(v+l+2)/(2\gamma)}$	$\langle v+1,l+1\|\hat{P}^+\|v,l\rangle = \hbar\sqrt{(v+l+2)\gamma/2}$
$\langle v+1,l-1\|Q^-\|v,l\rangle = -i\sqrt{(v-l+2)/(2\gamma)}$	$\langle v+1,l-1\|\hat{P}^-\|v,l\rangle = \hbar\sqrt{(v-l+2)\gamma/2}$
$\langle v-1,l+1\|Q^+\|v,l\rangle = i\sqrt{(v-l)/(2\gamma)}$	$\langle v-1,l+1\|\hat{P}^+\|v,l\rangle = \hbar\sqrt{(v-l)\gamma/2}$
$\langle v-1,l-1\|Q^-\|v,l\rangle = i\sqrt{(v+l)/(2\gamma)}$	$\langle v-1,l-1\|\hat{P}^-\|v,l\rangle = \hbar\sqrt{(v+l)\gamma/2}$

[a] $Q^{\pm} = Q_a \pm iQ_b = Qe^{\pm i\alpha}$ and $\hat{P}^{\pm} = \hat{P}_a \pm i\hat{P}_b = e^{\pm i\alpha}(\hat{P} \pm i\hat{M}/Q)$ and the two-dimensional isotropic harmonic oscillator wavefunctions $\Psi_{v,l} = |v,l\rangle$.

and

$$l = 0. \tag{8-221}$$

Solving

$$\hat{R}^{-(-)}\Psi_{0,0} = 0 \tag{8-222}$$

gives

$$\Psi_{0,0} = (\gamma/\pi)^{1/2}\exp[-(\gamma Q^2/2)], \tag{8-223}$$

where this function is normalized and a real and positive phase factor choice has been adopted. We can write

$$\Psi_{v,l} = N'_{v,l}[\hat{R}^{+(-)}]^{(v-l)/2}[\hat{R}^{+(+)}]^{(v+l)/2}\Psi_{0,0}, \tag{8-224}$$

where $l = v, v-2, \ldots, -v$, and

$$E_v = (v+1)\hbar^2\gamma. \tag{8-225}$$

We can determine the matrix elements of $\hat{R}^{\pm(\pm)}$ in the functions $\Psi_{v,l}$, and from these (we choose the phase factor to be real and positive) the matrix elements of $Q^{\pm}$ and $\hat{P}^{\pm}$ (or Q_a, Q_b, $\hat{P}_a$, and $\hat{P}_b$) follow by using relations such as

$$\hat{P}^+ = [\hat{R}^{+(+)} + \hat{R}^{-(+)}]/2. \tag{8-226}$$

The nonvanishing matrix elements of $Q^{\pm}$ and $\hat{P}^{\pm}$ in the functions $\Psi_{v,l}$ are given in Table 8-3. The same phase factor choice has been made here as that adopted by Moffitt and Liehr (1957) in their Eq. (28).

For a molecule having triply degenerate normal coordinates the eigenfunctions are written $\Psi_{v,l,n}(Q,\alpha,\beta)$, in a similar way to the preceding,[4] where l and n are vibrational angular momentum quantum numbers and α and β are vibrational angular coordinates. The complete vibrational wavefunctions of a molecule in the harmonic oscillator approximation is written as a

[4] See Eqs. (20)–(22) on page 355 in Wilson, Decius, and Cross (1955), where α, β are called θ, ϕ.

product of one-, two-, and three-dimensional harmonic oscillator functions according to the symmetry degeneracies of the normal coordinates. As an afterthought it should be mentioned that although nuclear masses occur in the vibration–rotation Schrödinger equation, it is better to use atomic masses in calculations in order to allow for the mass of the electrons; in fact, this partly allows for the breakdown of the Born–Oppenheimer approximation [see, for example, Oka and Morino (1961), Eqs. (29)–(37)].

SUMMARY

As a result of making the Born–Oppenheimer approximation, using LCAOSCFMO electronic orbital functions, and making the rigid rotor and harmonic oscillator approximations to the rotation–vibration Hamiltonian, we have obtained useful approximate rovibronic wavefunctions. Such functions are the product of rotational, vibrational, and electronic orbital wavefunctions Φ_{rot}, Φ_{vib}, and Φ_{eo}, respectively. In Eq. (8-111) Φ_{rot} is given for a symmetric or spherical top molecule, and Φ_{rot} is a linear combination of such functions for an asymmetric top molecule. Φ_{vib} is a product of harmonic oscillator functions and Φ_{eo} is a product of LCAO molecular orbital functions. In Chapter 10 we show how these functions can be classified in the molecular symmetry group and in Chapter 11 we show how the effects of the breakdown of the various approximations made here are treated.

BIBLIOGRAPHICAL NOTES

The Born–Oppenheimer Approximation

Born and Oppenheimer (1927).
Born (1951).
Born and Huang (1956).

Molecular Orbital Theory

Herzberg (1966).   Chapter III.
Mulliken (1975).   Part VII.

The Rotational Schrödinger Equation

Herzberg (1945).   Pages 22–26 and 42–50.
King, Hainer, and Cross (1943).
Pauling and Wilson (1935).   Section 36.
Van Vleck (1951).

The Vibrational Schrödinger Equation

Herzberg (1945).   Pages 76–82 and 204–219.
Pauling and Wilson (1935).   Section 11.
Wilson, Decius, and Cross (1955).   Pages 34–38, 193–197, and 352–358.

9

The Definition of the Molecular Symmetry Group

In this chapter the drawbacks of using the complete nuclear permutation inversion group to label molecular rovibronic energy levels are discussed. The molecular symmetry group is introduced and it is shown how its use overcomes these drawbacks. It is also shown how the symmetry labels obtained from the use of these two groups are related, and how the definition of the MS group can be extended to situations in which two or more electronic states of a molecule are considered simultaneously.

THE COMPLETE NUCLEAR PERMUTATION INVERSION GROUP

Having read Chapters 1 and 2 it is presumed that the reader would have no trouble in determining the elements of the complete nuclear permutation inversion (CNPI) group of a molecule. This group is the direct product of the complete nuclear permutation (CNP) group G^{CNP} [see Eq. (1-55)] and the inversion group $\mathscr{E} = \{E, E^*\}$. The CNPI group can be set up for any molecule once we know its chemical formula. As discussed in Chapter 6 the Hamiltonian of an isolated molecule in field free space is invariant to the operations of the CNPI group and we can, in principle, classify the rovibronic wavefunctions and label the energy levels, according to the irreducible representations of the group. However, this is often not a sensible thing to do.

Let us consider the number of elements in the CNPI group. For a series of simple molecules the order of the CNPI group (twice the order of the CNP group) is as follows:

H_2 $2! \times 2 = 4$, SF_6 $6! \times 2 = 1440$,
H_2O $2! \times 2 = 4$, C_2H_6 $2! \times 6! \times 2 = 2880$,
BF_3 $3! \times 2 = 12$, C_2H_5OH $2! \times 6! \times 2 = 2880$,
CH_3F $3! \times 2 = 12$, C_6H_6 $6! \times 6! \times 2 = 1036800$,
CH_4 $4! \times 2 = 48$, $CH_3COCH_2CH_2OH$ $4! \times 8! \times 2! \times 2 = 3870720$.
C_2H_4 $2! \times 4! \times 2 = 96$,

Clearly the order of the CNPI group can be very large, very much larger than the order of any finite point group. In these circumstances it would often be very laborious to use the CNPI group for classifying the molecular states.

As well as being very large the CNPI group almost invariably produces, as we shall see, a symmetry classification of the levels of a molecule in which there are systematic "accidental" degeneracies. Such systematic degeneracies are not really accidental but we call them accidental here since they are not required by the symmetry of the CNPI group. These degeneracies, which we will now call *structural degeneracies*, are caused by the presence of more than one *symmetrically equivalent nuclear equilibrium structure* in a given electronic state of a molecule. We will first explain what symmetrically equivalent nuclear equilibrium structures are and then show how structural degeneracy arises.

The idea of symmetrically equivalent nuclear equilibrium structures can be explained using the methane molecule as an example. In Fig. 9-1a a methane molecule is shown with the nuclei in the equilibrium configuration for the ground electronic state. If we deform the molecule through the planar

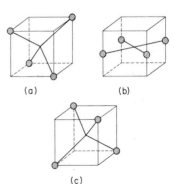

Fig. 9-1. A methane molecule in two different symmetrically equivalent nuclear equilibrium structures, (a) and (c), which can be interconverted by a deformation through the planar configuration shown in (b).

configuration (Fig. 9-1b) we can invert the molecule to obtain the symmetrically equivalent nuclear equilibrium structure shown in Fig. 9-1c. The configurations in Figs. 9-1a and 9-1c are the nuclear configurations at two identically shaped deep minima in the potential energy surface V_N of the ground electronic state, and a configuration such as shown in Fig. 9-1b represents a maximum in V_N in a coordinate connecting the minima. Symmetrically equivalent nuclear equilibrium structures arise not only in molecules, such as methane, in which the configurations can be interchanged by inverting the molecule, but also rather commonly in molecules such as ethane (H_3CCH_3) in which the configurations can be interchanged by the torsion of one part of the molecule. There are other, less common, ways of interchanging symmetrically equivalent forms, as we shall see when we consider the ethylene molecule.

To distinguish between the different symmetrically equivalent nuclear equilibrium structures of a molecule it is convenient to label identical nuclei. Having labeled the nuclei of a molecule in its equilibrium configuration, we determine the number of symmetrically equivalent nuclear structures for the given equilibrium configuration by finding out how many distinct labeled forms of the molecule can be obtained by permuting the labels on the nuclei with and without inverting the molecule. Distinct labeled forms are such that to interconvert them one cannot merely rotate the molecule in space but one must deform the molecule across a potential barrier.

For the methane molecule there are only two symmetrically equivalent equilibrium structures as we can see once we have labeled the protons (see Fig. 9-2). We call the forms in Figs. 9-2a and 9-2b the A (for anticlockwise) form and the C (for clockwise) form, respectively, since in the A (C) form $1 \rightarrow 2 \rightarrow 3$ is anticlockwise (clockwise) looking in the $C \rightarrow H_4$ direction.

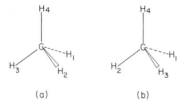

Fig. 9-2. The two symmetrically equivalent nuclear equilibrium structures of a methane molecule. They are distinguishable because we have numbered the protons. The form in (a) is called the anticlockwise (A) form in the text because $1 \rightarrow 2 \rightarrow 3$ is anticlockwise when viewed in the $C \rightarrow H_4$ direction, and (b) is the clockwise (C) form.

The ethylene molecule was considered in Chapters 1 and 2 and it presents a further example of structural degeneracy. Labeling the protons one to four and the carbon nuclei five and six in an ethylene molecule, we see that in its electronic ground state there are altogether 12 distinct numbered forms (shown schematically in Fig. 1-3). To interchange pairs of these forms we either have to twist one CH_2 group relative to the other or we have to break CH bonds. The interconversions involving the breaking of CH bonds are opposed by very high barriers in the potential function V_N (the dissociation energy of a CH bond is about 30,000 cm^{-1}).

To understand structural degeneracy, and how it arises when there are symmetrically equivalent nuclear equilibrium structures, we need only have a qualitative appreciation of the solution of the rotation–vibration Schrödinger equation. For the methane molecule we can choose either structure A or C in Fig. 9-2 as the equilibrium configuration for the purpose of defining the Eckart axes (x, y, z) and hence the Euler angles and vibrational displacements $\Delta\alpha_i$. Depending upon which structure (A or C) we choose we obtain rotation–vibration wavefunctions $\Psi_A^{(n)}$ and energies $E_A^{(n)}$, or $\Psi_C^{(n)}$ and $E_C^{(n)}$, where $n = 1, 2, 3, \ldots$, for the successive eigenstates. If, as is true in methane, the barrier in V_N between the minima A and C is very high then the wavefunctions $\Psi_A^{(n)}$ are localized in the A minimum and the wavefunctions $\Psi_C^{(n)}$ in the C minimum, with no effective penetration of either into the other minimum. In other words the molecular vibrations just occur in the region around each minimum. Since the two minima in V_N for methane have the same shape (they are symmetrically equivalent) the energies $E_A^{(n)}$ and $E_C^{(n)}$ will be identical (and identical to the observed rotation–vibration energies $E^{(n)}$). Thus each observed rotation–vibration energy level will be doubly degenerate, corresponding to the energies $E_A^{(n)}$ and $E_C^{(n)}$, and this is structural degeneracy.

If there are n symmetrically equivalent minima in the potential function V_N for a molecule, with no effective penetration (tunneling) of the barriers between these minima by the local wavefunctions, then each level will be n-fold structurally degenerate. When tunneling occurs the degeneracy is split, and this happens if the barrier is not high relative to the vibrational energy. In actual fact there will always be some tunneling, since potential energy barriers are not infinitely high, but often the experimental resolution is not high enough to detect the splitting. This is the case at the moment for methane. Structural degeneracy is not required by the symmetry of the CNPI group since the symmetry group is the same regardless of the height of the barrier to tunneling, and the possibility of the splitting must therefore be allowed for by the symmetry labels obtained using the CNPI group. A molecule for which no observable tunneling between minima in V_N occurs [whether these minima be symmetrically equivalent or not; see answer (vi) to Problem 9-1] is said to be rigid, and one for which observable tunneling occurs is said to be nonrigid.

No experimental splittings, torsional or otherwise, resulting from tunneling through the barriers that separate the 12 symmetrically equivalent forms of ethylene (see Fig. 1-3) have been observed; the barriers in the potential function that separate these 12 forms are insuperable and the molecule is, therefore, a rigid molecule. To determine the rotation–vibration energy level pattern of ethylene we would only need to consider one numbered form and the shape of the one deep minimum containing that form in the potential energy surface V_N. Each of the twelve forms would have identical rotation–vibration energy levels and these would match the observed rotation–vibration energy levels of the molecule. Thus each of the rotation–vibration energy levels of ethylene in its electronic ground state has a twelve-fold structural degeneracy.

Structural degeneracy is present in nearly all molecules that contain identical nuclei. The lower the structural symmetry of a molecule the more, in general, is the structural degeneracy, and the structural degeneracy increases at a dramatic rate with the size of the molecule. Simple symmetrical molecules such as SO_2 or BF_3 in their ground electronic states have no structural degeneracy since for each there is only one distinct numbered form. However, it is possible that SO_2 has unequal bond lengths at equilibrium in an excited electronic state; if true this means that in this excited electronic state there would be two distinct numbered forms of the molecule at equilibrium, as shown in Fig. 9-3, and each level would be structurally doubly degenerate if there was no tunneling between the forms. Interestingly enough for unsymmetrical $S^{16}O_2$ one of each of the pairs that would be split apart by tunneling has symmetry such that its nuclear spin statistical weight is zero (^{16}O nuclei being bosons), and hence it cannot occur. Thus the tunneling produces a shift but no splitting of the levels [see Fig. 5 of Redding (1971)], and although each rotation–vibration level has a structural double degeneracy each complete state Φ_{int} does not. A similar statistical weight effect occurs for the A_1 and A_2 [in $C_{3v}(M)$] levels of NH_3, the B_{1g} and B_{1u} [in $D_{2h}(M)$] levels of C_2H_4, and the A_1, A_2, F_1, and F_2 [in $T_d(M)$] levels of CH_4, when inversion or torsional tunneling is allowed for [see Tables B-4(i), (vi), and (viii) in Appendix B]. If there were an excited electronic state of BF_3 in which the molecule were planar with three unequal bond lengths at equilibrium, with no observable tunneling splittings, then each level would have a structural degeneracy of six. In methane the rotation–vibration levels of a nonplanar electronic state with all four bond lengths unequal at equilibrium, and no tunneling splittings, would have a structural degeneracy of 48.

We see that, although mathematically correct, the use of the CNPI group has the drawbacks of often being inconveniently large and of often producing a symmetry labeling in which accidental degeneracies occur in a systematic fashion. We only need the chemical formula of a molecule in order to set

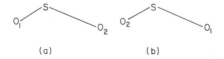

(a)   (b)

Fig. 9-3. The two symmetrically equivalent nuclear equilibrium structures of an unsymmetrical bent SO_2 molecule.

up the CNPI group and although we invariably know, or can guess, the equilibrium structure of a molecule and the tunneling possibilities, this information is not used in setting up the group.

THE MOLECULAR SYMMETRY GROUP

The definition of the molecular symmetry (MS) group is easier to appreciate after we have considered the question "what do we do with a symmetry group"? What we do with a symmetry group is to label the energy levels of the molecule using the irreducible representations of the group in order to identify all the zero order levels that can and cannot interact as a result of considering (a) the effect of previously neglected terms in the complete Hamiltonian, or (b) the effect of applying an external perturbation such as an electric or magnetic field or electromagnetic radiation. We wish to use symmetry labels that enable us to determine all the levels that can interact in these circumstances, but we do not need to do more than this. We can say that if our symmetry group permits this then it is a sufficiently large symmetry group and it provides sufficient symmetry labeling.

We are also pragmatic in a similar way when we solve the rotation–vibration problem. The potential energy surface V_N for the methyl fluoride molecule in its ground electronic state has two deep minima centered at the anticlockwise (A) and clockwise (C) labeled forms shown in Figs. 9-4a and 9-4b, respectively. To determine the rotation–vibration energy levels a mathematician might insist that we should determine the eigenvalues of the rotation–vibration Hamiltonian using the complete double minimum potential energy surface. However, experience tells us that there will be no observable splittings as a result of tunneling between the A and C forms for the rotation–vibration levels that we study. Thus for all practical purposes we can determine the rotation–vibration energy levels of the methyl fluoride molecule by solving the rotation–vibration equation for the energy levels in one minimum, appropriate for the A form, say, with the complete neglect of the potential energy surface in the region of the other minimum. Rotation–vibration interactions or external perturbations will only connect one A form level with another A form level in the absence of inversion tunneling.

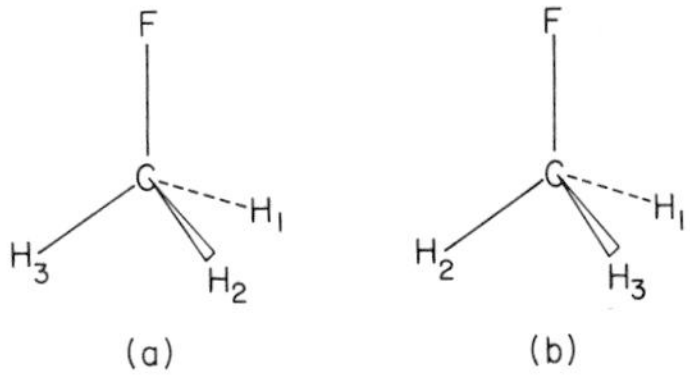

Fig. 9-4. The two symmetrically equivalent nuclear equilibrium structures of a methyl fluoride molecule where (a) is the anticlockwise (A) form and (b) is the clockwise (C) form.

The C form will provide a duplicate set of energy levels and interactions so that the methyl fluoride molecule has the twofold structural degeneracy just discussed.

We can completely understand the rotation–vibration energy levels and all possible interactions within the ground electronic state of methyl fluoride (apart from inversion tunneling) by considering only one of the symmetrically equivalent forms of the molecule. Hence we need only make a symmetry labeling of the levels of one form to obtain a sufficient symmetry labeling of the levels of the molecule. To do this we just need the CNPI group of one form of the molecule, i.e., the group

$$\{E, (123), (132), (12)^*, (23)^*, (13)^*\}; \qquad (9\text{-}1)$$

this group, called $C_{3v}(M)$, was introduced in the answer to Problem 2-2. Elements in the CNPI group of methyl fluoride that interconvert the A and C forms, such as (12) or (123)*, are not present in the CNPI group of one form and constitute what are called *unfeasible* elements of the CNPI group of methyl fluoride; we could equally well call them *useless* elements of the CNPI group. The group in Eq. (9-1) is the molecular symmetry group of methyl fluoride when inversion tunneling splittings are not observed.

The MS group of a molecule is obtained by deleting all elements from the CNPI group of the molecule that are unfeasible. An unfeasible element is one that interconverts numbered equilibrium forms of the molecule when these forms are separated by an insuperable barrier in the potential energy surface; an insuperable barrier is one that does not allow observable tunneling to occur through it on the time scale of the experiment being performed. Tunneling may not be observable if we use a low resolution experiment yet it may occur if we use a high resolution experiment. The MS group that we use to analyze the results will then be different for the two cases since the elements associated with the tunneling are feasible in the latter case but not in the former case. If splittings from inversion tunneling were observed in methyl fluoride (perhaps in highly excited vibrational states), and we wished to symmetry label the split levels, the MS group of the molecule would become equal to its CNPI group since all elements of it would be feasible. In the same way that to set up the point group for a molecule we need to know the equilibrium nuclear geometry, so to set up the MS group we need to know the equilibrium nuclear geometry and the situation with regard to rotation–vibration tunneling.

For the ethylene molecule the CNPI group has 96 elements and is the direct product $S_4^{(H)} \otimes S_2^{(C)} \otimes \mathcal{E}$ (see Problem 2-3). The CNPI group of one of the forms (a), (b), (c), or (d) shown in Fig. 1-3 can be used as the MS group of ethylene, and it consists of the eight elements given in Eq. (2-14). If we wish to use one of the numbered forms (e), (f), (g), or (h) in Fig. 1-3, in the rotation–vibration problem, then the MS group we use is

$$\{E, (13)(24), (12)(34)(56), (14)(23)(56),$$
$$E^*, (13)(24)^*, (12)(34)(56)^*, (14)(23)(56)^*\}, \tag{9-2}$$

whereas if we want to use one of the numbered forms (i), (j), (k), or (l) in Fig. 1-3 the appropriate MS group is

$$\{E, (14)(23), (12)(34)(56), (13)(24)(56),$$
$$E^*, (14)(23)^*, (12)(34)(56)^*, (13)(24)(56)^*\}. \tag{9-3}$$

To solve the rotation–vibration problem we use just one form of the ethylene molecule, and the MS group is the one appropriate for that form. Regardless of which form we chose we obtain an identical energy level pattern and symmetry labeling.

The MS group is thus defined for a particular electronic state (and numbered form) of a molecule with regard to the experimental observation or nonobservation of tunneling splittings. It therefore uses the information of the nuclear structure to obtain a symmetry group that is much smaller than the CNPI group (in most cases), but which provides a sufficient symmetry labeling of the observed levels.

The systematic way to determine the MS group of a molecule is first to label the nuclei and then to write down the elements of the CNPI group of the molecule. The next stage is to draw all the distinct symmetrically equivalent nuclear equilibrium configurations of the molecule (by permuting the labels on identical nuclei with and without inverting the molecule) and to form sets of these forms that one wishes to be connected by observable tunneling effects. To obtain the MS group it is then necessary to consider one particular set and to delete from the CNPI group of the molecule any elements that convert the forms in that set to the forms in other sets. Different sets may have different (but isomorphic) MS groups [for example, see the groups in Eqs. (2-14), (9-2), and (9-3)]. In the determination of some MS groups it is useful to know that they can be written as the product of subgroups [see Woodman (1970)].

Problem 9-1. Set up the MS groups of the following molecules in their ground electronic states: (i) CH_2F_2, (ii) HN_3, (iii) BF_3, (iv) NF_3, (v) CH_4, (vi) trans $C(HF)CHF$, and (vii) C_2H_2. Each of these molecules is such that either it does not possess symmetrically equivalent nuclear equilibrium structures or, if it does, no observable tunneling between them occurs.

Answer. (i) CH_2F_2. We choose to label the protons 1 and 2, and the fluorine nuclei 3 and 4, so that the CNPI group is

$$\{E, (12), (34), (12)(34), E^*, (12)^*, (34)^*, (12)(34)^*\}. \tag{9-4}$$

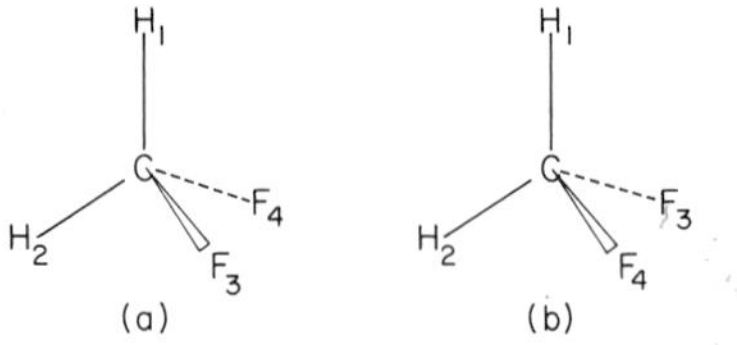

Fig. 9-5. 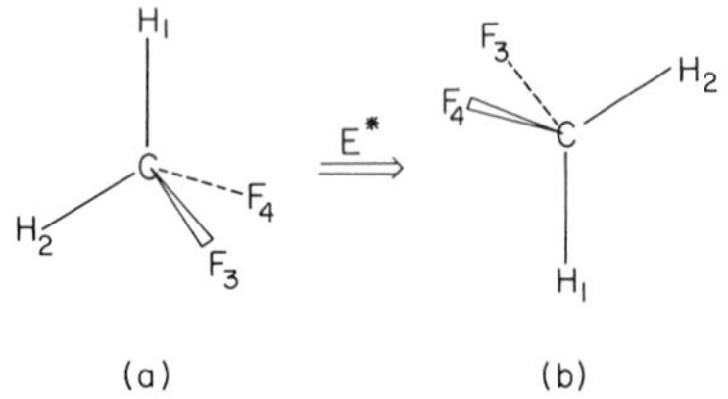   Fig. 9-6.

Fig. 9-5. The two symmetrically equivalent nuclear equilibrium structures of a CH_2F_2 molecule.

Fig. 9-6. The effect of the operation E^* on the form of CH_2F_2 given in Fig. 9-5a.

There are only two symmetrically equivalent nuclear equilibrium structures as shown in Fig. 9-5, and no tunneling between these forms is observed. The form in Fig. 9-5a has 2–3–4 clockwise when looking in the $C \rightarrow H_1$ direction and the form in Fig. 9-5b has them anticlockwise. The permutations (12) and (34) interconvert the forms and are thus unfeasible, whereas (12)(34) is feasible since it does not interconvert the forms. The operation E^* inter-converts the forms (see Fig. 9-6 and note that the form in Fig. 9-6b is identical to the form in Fig. 9-5b; i.e., it is the anticlockwise form) so that E^* is un-feasible. The operations (12)* or (34)* do not interconvert the forms and are both feasible. The operation (12)(34)* does interconvert the forms and is unfeasible. Thus the MS group of either form of CH_2F_2 is the group

$$\{E,(12)(34),(12)^*,(34)^*\}. \tag{9-5}$$

(ii) HN_3. We choose to label the nitrogen nuclei 1, 2, and 3 so that the elements of the CNPI group are

$$\{E,(12),(23),(13),(123),(132),E^*,(12)^*,(23)^*,(13)^*,(123)^*,(132)^*\}. \tag{9-6}$$

The symmetrically equivalent nuclear equilibrium structures are shown in Fig. 9-7. There is no observable tunneling between these forms. It is easy to see that all the permutations interconvert forms. For example, we determine that the form in Fig. 9-7a is transformed as follows:

$$(12)(a) = (c), \qquad (23)(a) = (b), \qquad (13)(a) = (f),$$
$$(123)(a) = (d), \qquad \text{and} \qquad (132)(a) = (e). \tag{9-7}$$

The inversion E^* is feasible since it does not interconvert forms but all permutations accompanied by the inversion are unfeasible. Thus the MS group of HN_3 is

$$\{E,E^*\}. \tag{9-8}$$

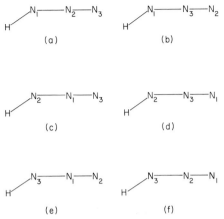

Fig. 9-7.   The six symmetrically equivalent nuclear equilibrium structures of an HN_3 molecule.

(iii) BF_3.   We label the three fluorine nuclei 1, 2, and 3 and hence the CNPI group is the same as that of HN_3 given in Eq. (9-6). However, in its equilibrium configuration BF_3 is planar with three equal BF bonds, and as a result there are no different symmetrically equivalent nuclear equilibrium structures; all operations of the CNPI group are feasible. Thus the MS group of BF_3 is the same as its CNPI group and there is no structural degeneracy. The H_2O molecule is another example of a simple molecule having no structural degeneracy for which the MS group is equal to the CNPI group.

(iv) NF_3.   Labeling the three fluorine nuclei 1, 2, and 3 we obtain the CNPI group of Eq. (9-6) as for HN_3 and BF_3. For this pyramidal molecule there are two symmetrically equivalent nuclear equilibrium structures as shown in Fig. 9-8. To interconvert these forms we must invert the molecule, and inversion splittings are not observed. Deleting elements from the CNPI group that interconvert these forms we obtain the MS group of NF_3 as

$$\{E,(123),(132),(12)^*,(23)^*,(13)^*\}. \qquad (9\text{-}9)$$

(v) CH_4.   Labeling the four protons 1, 2, 3, and 4 we can write down the CNPI group of the molecule as the direct product of the CNP group $S_4^{(H)}$ containing $4! = 24$ elements and the inversion group $\mathscr{E}$; this has 48 elements.

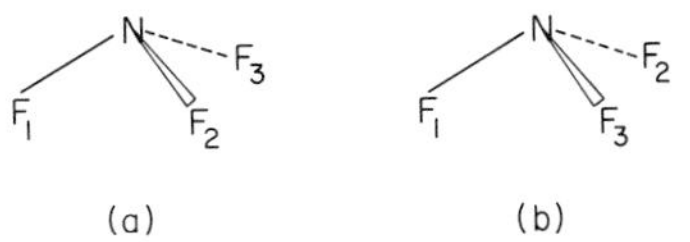

Fig. 9-8.   The two symmetrically equivalent nuclear equilibrium structures of an NF_3 molecule.

As discussed earlier in this chapter there are two distinct numbered forms (the A and C forms of Fig. 9-2) and inversion tunneling is not observed. Any operation, such as (12), in which one pair of protons is permuted interconverts the forms and is hence unfeasible. Any cyclic permutation of three protons is clearly feasible, and the inversion E^* is equally clearly unfeasible. From these three results we can rather quickly deduce the MS group of CH_4 by making use of the following rules:

(a) any operation that is the product of a feasible operation and an unfeasible operation is itself unfeasible, and

(b) any operation that is the product of two feasible operations is itself feasible.

For an example of the use of (a) we can look at the product

$$(123)(34) = (1234). \tag{9-10}$$

The operation (34) converts the form A to form C (since it is an unfeasible operation) and (123) then sends C to C (since it is feasible), so the product of these two operations sends A to C and is unfeasible. For methane any cyclic permutation of all four protons is unfeasible. An example of rule (b) is provided by the product

$$(123)(234) = (12)(34). \tag{9-11}$$

The operation (234) converts A to A and then (123) converts A to A, so that the product converts A to A and is feasible. We can see this latter result another way by looking at this operation as the product (12)(34), and this is the successive application of two unfeasible operations: $(34)A = C$, and $(12)C = A$, so that the product sends A to A and is feasible. We can thus appreciate a third rule:

(c) For molecules with only two distinct numbered forms the product of two unfeasible operations is itself a feasible operation. For molecules having more than two distinct numbered forms the product of two unfeasible operations may, or may not, be feasible.

For methane we deduce that all cyclic permutations of three protons and all permutations consisting of two successive pair transpositions are feasible whereas all pair transpositions, and cyclic permutations of all four protons, are unfeasible. Since E^* is unfeasible we can use rules (a) and (c) to deduce that the product of a pair transposition and E^* or the product of a cyclic permutation of all four protons and E^* is feasible. The MS group of methane thus consists of the following 24 elements which we arrange in classes that follow.

E	(123)	(12)(34)	(1234)*	(12)*	
	(132)	(13)(24)	(1243)*	(13)*	
	(124)	(14)(23)	(1324)*	(14)*	
	(142)		(1342)*	(23)*	(9-12)
	(134)		(1423)*	(24)*	
	(143)		(1432)*	(34)*	
	(234)				
	(243)				

The MS group of methane in its ground electronic state is a subgroup of the CNPI group that consists of whole classes of elements from the CNPI group. Such a subgroup is called an *invariant* subgroup. The MS group is not always an invariant subgroup of the CNPI group (see, for example, the MS group of ethylene).

(vi) trans C(HF)CHF. Numbering the protons 1 and 2, the carbon nuclei 3 and 4, and the fluorine nuclei 5 and 6, we obtain the CNPI group as the direct product

$$\{E, (12), (34), (56), (12)(34), (12)(56), (34)(56), (12)(34)(56)\} \otimes \mathscr{E}. \quad (9\text{-}13)$$

The symmetrically equivalent nuclear equilibrium structures are given in Fig. 9-9. It is important to realize that a structure such as given in Fig. 9-10

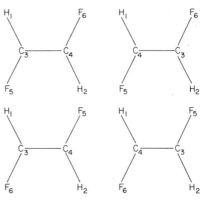

Fig. 9-9. The four symmetrically equivalent nuclear equilibrium structures of a trans C(HF)CHF molecule.

Fig. 9-10. A numbered cis form of the C(HF)CHF molecule.

is not symmetrically equivalent to those in Fig. 9-9 since it is not obtained from any of them by relabeling identical nuclei with or without inverting the molecule. The form in Fig. 9-10 is the cis configuration of the molecule and is related to the trans configuration by twisting the molecule over a potential barrier. Thus in this molecule in the ground electronic state potential energy surface there are four symmetrically equivalent minima for the trans configuration and four other symmetrically equivalent minima for the cis configuration at a different energy from that of the trans configuration. The shape of the potential energy surface around the cis and trans minima will be different and the rotation–vibration energy level patterns of the cis and trans forms will be different. For trans $C_2H_2F_2$ the feasible elements are easy to determine and the MS group consists of

$$\{E, (12)(34)(56), E^*, (12)(34)(56)^*\}. \tag{9-14}$$

The MS group of cis $C_2H_2F_2$ is identical and if cis–trans tunneling occurs the MS group is still the same. This tunneling is not between symmetrically equivalent nuclear equilibrium structures, so although this tunneling makes the molecule nonrigid it does not cause the splitting of any structural degeneracy and it does not enlarge the MS group.

(vii) C_2H_2. In the acetylene molecule HCCH we number the protons 1 and 2 and the carbon nuclei 3 and 4; the CNPI group is the group of eight elements given in Eq. (9-4). The operations $(34), (12), (34)^*$, and $(12)^*$ are unfeasible and the MS group is

$$\{E, (12)(34), E^*, (12)(34)^*\}. \tag{9-15}$$

For any linear molecule of $D_{\infty h}$ structure the MS group consists of the four elements E, (p), E^*, and $(p)^*$, where (p) is the simultaneous transposition of all pairs of identical nuclei symmetrically located about the molecular midpoint. For any linear molecule of $C_{\infty v}$ structure the MS group is $\{E, E^*\}$.

Problem 9-2. Set up the MS groups of the following molecules in their ground electronic states: (i) NH_3 allowing for inversion tunneling; (ii) C_2H_4 and (iii) H_2O_2 allowing for torsional tunneling in each case.

Answer. (i) NH_3 with inversion tunneling. Numbering the protons 1, 2, and 3 the CNPI group is as for HN_3 given in Eq. (9-6), and there are two symmetrically equivalent nuclear equilibrium structures as for NF_3 (see Fig. 9-8). Allowing for the effects of inversion tunneling in setting up the MS group we must include as feasible elements permutations and permutation inversions that interconvert forms across the inversion barrier. Thus the MS group of inverting NH_3 is the same as its CNPI group and this is the same as the MS group of planar BF_3.

(ii) C_2H_4 with torsional tunneling.   The CNPI group of ethylene is the direct product $S_4^{(H)} \otimes S_2^{(C)} \otimes \mathscr{E}$ with 96 elements as discussed before. In Fig. 1-3 the 12 symmetrically equivalent nuclear equilibrium structures of an ethylene molecule in its ground electronic state are drawn schematically. Grouping these into sets within which the forms are connected by torsional tunneling we obtain the six sets $[(a),(b)]$, $[(c),(d)]$, $[(e),(f)]$, $[(g),(h)]$, $[(i),(j)]$, and $[(k),(l)]$ from Fig. 1-3. Let us consider the set $[(a),(b)]$. Examples of elements that convert the forms in this set into forms in other sets are (13), (56), (24), and (234), and these are hence unfeasible elements of the CNPI group for the set $[(a),(b)]$. Elements such as (12) and (34) are feasible, and combining these with the elements of the MS group of nontorsionally tunneling ethylene in the forms (a) and (b) [see Eq. (2-14)] we obtain the MS group of the torsionally tunneling forms in set $[(a),(b)]$ of Fig. 1-3 as

$$\left\{ \begin{array}{llllll} E & (12) & (12)(34) & (13)(24)(56) & (1324)(56) \\ & (34) & & (14)(23)(56) & (1423)(56) \end{array} \right\} \otimes \mathscr{E}, \qquad (9\text{-}16)$$

where the class structure of the group is also indicated. The MS group for the set $[(c),(d)]$ is the same as this. The MS group for the set $[(e),(f)]$ or the set $[(g),(h)]$ is the group

$$\left\{ \begin{array}{llllll} E & (13) & (13)(24) & (12)(34)(56) & (1234)(56) \\ & (24) & & (14)(23)(56) & (1432)(56) \end{array} \right\} \otimes \mathscr{E}, \qquad (9\text{-}17)$$

and the MS group for the set $[(i),(j)]$ or the set $[(k),(l)]$ is the group

$$\left\{ \begin{array}{llllll} E & (14) & (14)(23) & (12)(34)(56) & (1243)(56) \\ & (23) & & (13)(24)(56) & (1342)(56) \end{array} \right\} \otimes \mathscr{E}. \qquad (9\text{-}18)$$

Since we only consider a single set of symmetrically equivalent nuclear equilibrium structures connected by allowed tunneling, and neglect the others, in a rotational–vibrational analysis, the fact that there are different MS groups poses no problems as long as we are careful to use the one appropriate to the particular set considered. These groups are isomorphic and lead to identical energy levels labelings of the appropriate sets.

(iii) H_2O_2 with torsional tunneling.   The equilibrium structure of H_2O_2 in its ground electronic state is nonplanar with a dihedral angle of approximately $120°$, as shown in Fig. 9-11. Labeling the protons 1 and 2, and the

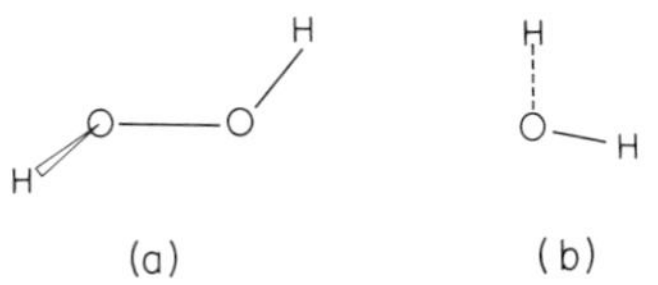

Fig. 9-11. The H_2O_2 molecule in its equilibrium configuration; (a) is a view from the side and (b) is an end on view.

(a)   (b)

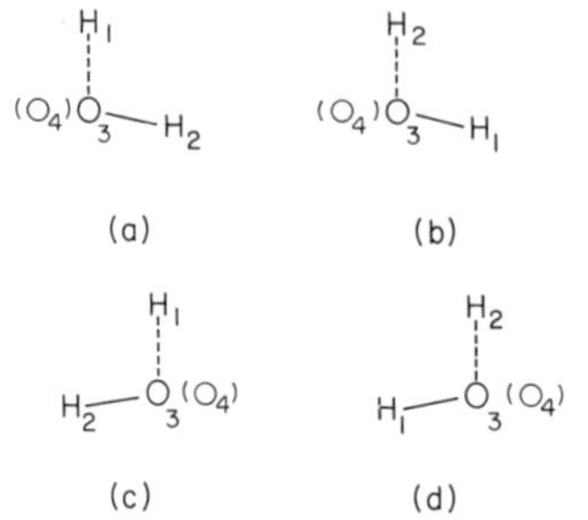

$$\begin{array}{cc}
\overset{\overset{H_1}{|}}{(O_4)O_3{-}H_2} & \overset{\overset{H_2}{|}}{(O_4)O_3{-}H_1}\\
(a) & (b)
\end{array}$$

Fig. 9-12. The four symmetrically equivalent nuclear equilibrium structures of an H_2O_2 molecule shown viewed end on. The oxygen atom O_4 is behind the oxygen atom O_3 in the four figures.

$$\begin{array}{cc}
\overset{\overset{H_1}{|}}{H_2{-}O_3(O_4)} & \overset{\overset{H_2}{|}}{H_1{-}O_3(O_4)}\\
(c) & (d)
\end{array}$$

oxygen nuclei 3 and 4, we can label the nuclei of the configuration of Fig. 9-11 in the two ways shown in (a) and (b) of Fig. 9-12. A nonsuperposable symmetrically equivalent nuclear equilibrium configuration is obtained from that of Fig. 9-11 by inverting it, and the two ways of labeling this configuration are shown in (c) and (d) of Fig. 9-12. There are thus four symmetrically equivalent nuclear equilibrium structures of H_2O_2 in its ground electronic state. If there were no observable tunneling between these structures each rotation–vibration level would have a fourfold structural degeneracy and the MS group would be

$$\{E,(12)(34)\}. \tag{9-19}$$

If we allow for torsional tunneling then the interconversions (a)$\leftrightarrow$(c) and (b)$\leftrightarrow$(d) (of Fig. 9-12) occur, and we have two sets of structures, $[(a),(c)]$ and $[(b),(d)]$. The MS group for either of these sets is

$$\{E,(12)(34), E^*,(12)(34)^*\}. \tag{9-20}$$

In H_2O_2 torsional tunneling is observed and the MS group is as given in Eq. (9-20).

The H_2S_2 molecule has a similar equilibrium structure to that of H_2O_2, with a dihedral angle of about $90°$. Except in ultrahigh resolution Lamb-dip experiments [Winton and Winnewisser (1970)] H_2S_2 has no observable torsional tunneling in its ground state, and thus for normal spectroscopic work on this state its MS group is as given in Eq. (9-19); no permutation inversion elements are present. A molecule having no permutation inversion elements in its MS group has optically active forms and H_2S_2 is a simple example of this.

If tunneling from structure (a) to (b) of Fig. 9-12 [and equivalently (c) to (d)] was observed in addition to torsional tunneling then the MS group of H_2O_2 would be the same as its CNPI group, i.e.,

$$\{E,(12),(34),(12)(34)\} \otimes \mathscr{E}. \tag{9-21}$$

Such a type of tunneling, with bonds breaking and reforming, is observed in certain molecules having weak van der Waals bonds [see, for example, Dyke (1977)].

THE CHARACTER TABLES AND CORRELATION TABLES OF MS GROUPS

Using the definition given above we can set up the MS group for any molecule in a given electronic state of that molecule if we know the equilibrium structure and tunneling properties of that state. As will be discussed in Chapter 11, it turns out that the MS group is isomorphic to the molecular point group for any rigid nonlinear molecule. We have named these MS groups by the name of the appropriate point group followed by (M), e.g. the MS group of CH_2F_2 in its ground electronic state is called $C_{2v}(M)$. Further, because of the isomorphism, the character tables of these MS groups are the same as those of the point groups; we have chosen to name the irreducible representation of these MS groups by the same names as used for the point groups. It is very important to realize that the MS group and molecular point group are not identical; each element of the MS group for a rigid nonlinear molecule involves the product of a molecular point group operation and a molecular rotation group operation as will be discussed in Chapter 11. In Appendix A at the end of the book the character tables of the most common MS groups are listed including those of linear and nonrigid molecules to be discussed in Chapter 12. The MS group of a nonrigid molecule is called G_n, where n is the order of the group. In the rest of this section we discuss the *correlation* of the irreducible representations of MS groups and CNPI groups and the application of the correlation rule to tunneling effects in molecules.

The correlation rule, as it applies to showing how the irreducible representations of a group correlate to those of a subgroup, is well known and easy to understand. Suppose we have labeled the energy levels of a molecule using the irreducible representation labels of a symmetry group G and that we wish to label them in a subgroup H of G. The symmetry labels obtained using the irreducible representations of H, can be deduced from those of G once the correlation of the irreducible representations of G to those of H is known.

To appreciate how this correlation is determined let us consider that the group G, of order g, has elements $\{G_1, G_2, \ldots, G_g\}$ and that its subgroup H of order $h < g$, has elements $\{H_1, H_2, \ldots, H_h\}$. Further suppose that $H_1 = G_1, H_2 = G_2, \ldots, H_h = G_h$. Any irreducible matrix representation, Γ_α, say, of G will provide a matrix representation of H by considering only the matrices corresponding to the elements $G_1, G_2, \ldots, G_h$ of G. This matrix representation of H will in general be reducible and the irreducible representation, Γ_i, say, of H will occur $a_i^{(\alpha)}$ times where [from Eq. (4-43)]

$$a_i^{(\alpha)} = \frac{1}{h} \sum_{r=1}^{h} \chi^{\Gamma_\alpha}[H_r] \chi^{\Gamma_i}[H_r]^*, \tag{9-22}$$

and $\chi^{\Gamma_\alpha}[H_r]$ and $\chi^{\Gamma_i}[H_r]$ are the characters in the representations Γ_α and Γ_i for the operation $H_r = G_r$. We can represent this reduction by writing

$$\Gamma_\alpha = a_1^{(\alpha)} \Gamma_1 \oplus a_2^{(\alpha)} \Gamma_2 \oplus \cdots = \sum_i a_i^{(\alpha)} \Gamma_i. \tag{9-23}$$

Equation (9-23) shows how the irreducible representation Γ_α of G correlates with the irreducible representations (Γ_i) of H; if a molecular wavefunction transforms as Γ_α in G it will transform as the right hand side of Eq. (9-23) in H, where the $a_i^{(\alpha)}$ are obtained from Eq. (9-22).

As an example of this let us consider the correlation of the irreducible representations of the CNPI group of ethylene G_{96}, where

$$G_{96} = S_4^{(H)} \otimes S_2^{(C)} \otimes \mathscr{E}, \tag{9-24}$$

with the irreducible representations of the group $D_{2h}(M)$ of a rigidly planar ethylene molecule in the configuration of Fig. 1-3a. The elements and character table of the group $D_{2h}(M)$ are given in Table A-6 (in Appendix A). The character tables of the groups $S_4^{(H)}$, $S_2^{(C)}$, and $\mathscr{E}$ are given in Table 9-1 and the character table of G_{96} can be deduced from these by multiplication. The

Table 9-1

The groups $S_4^{(H)}$, $S_2^{(C)}$ and $\mathscr{E}$ for ethylene[a]

$S_4^{(H)}$:	E	(12)	(12)(34)	(123)	(1234)
	1	6	3	8	6
A:	1	1	1	1	1
B:	1	-1	1	1	-1
E:	2	0	2	-1	0
F:	3	1	-1	0	-1
G:	3	-1	-1	0	1

$S_2^{(C)}$:	E	(56)	$\mathscr{E}$:	E	E^*
s:	1	1	$+$:	1	1
a:	1	-1	$-$:	1	-1

[a] Here the CNPI group G_{96} is given by $S_4^{(H)} \otimes S_2^{(C)} \otimes \mathscr{E}$.

Table 9-2
Correlation table for the CNPI group of ethylene (G_{96}) to
the MS group $D_{2h}(M)$

G_{96}	$D_{2h}(M)$	G_{96}	$D_{2h}(M)$
$A_s{}^+$	A_g	$A_s{}^-$	A_u
$B_s{}^+$	A_g	$B_s{}^-$	A_u
$E_s{}^+$	$2A_g$	$E_s{}^-$	$2A_u$
$F_s{}^+$	$B_{1u} \oplus B_{2u} \oplus B_{3g}$	$F_s{}^-$	$B_{1g} \oplus B_{2g} \oplus B_{3u}$
$G_s{}^+$	$B_{1u} \oplus B_{2u} \oplus B_{3g}$	$G_s{}^-$	$B_{1g} \oplus B_{2g} \oplus B_{3u}$
$A_a{}^+$	B_{1u}	$A_a{}^-$	B_{1g}
$B_a{}^+$	B_{1u}	$B_a{}^-$	B_{1g}
$E_a{}^+$	$2B_{1u}$	$E_a{}^-$	$2B_{1g}$
$F_a{}^+$	$A_g \oplus B_{2u} \oplus B_{3g}$	$F_a{}^-$	$A_u \oplus B_{2g} \oplus B_{3u}$
$G_a{}^+$	$A_g \oplus B_{2u} \oplus B_{3g}$	$G_a{}^-$	$A_u \oplus B_{2g} \oplus B_{3u}$

20 irreducible representations of G_{96} are labeled $\Gamma_\alpha{}^\beta$ where $\Gamma = A$, B, E, F, or G according to the character of the representation under the elements of $S_4^{(H)}$, $\alpha =$ s or a as the character is positive or negative under (56), and $\beta = +$ or $-$ as the character under E^* is positive or negative. The correlation of the irreducible representations of G_{96} with those of the $D_{2h}(M)$ group can be obtained by using Eqs. (9-22) and (9-23) and the correlation table is given in Table 9-2. Examples of the results of using Eq, (9-22) are

$$a_{B_{1u}}^{(F_s{}^+)} = a_{B_{3g}}^{(F_s{}^+)} = a_{B_{2u}}^{(F_s{}^+)} = 1, \qquad (9\text{-}25)$$

with all other $a_i^{(F_s{}^+)} = 0$, and

$$a_{A_u}^{(E_s{}^-)} = 2, \qquad (9\text{-}26)$$

with all other $a_i^{(E_s{}^-)} = 0$. From these equations we can use Eq. (9-23) to obtain

$$F_s{}^+ = B_{1u} \oplus B_{2u} \oplus B_{3g} \qquad (9\text{-}27)$$

and

$$E_s{}^- = 2A_u. \qquad (9\text{-}28)$$

Problem 9-3. Determine the correlation table of the CNPI group of NF_3, given in Eq. (9-6), with the MS group of the molecule, given in Eq. (9-9). These groups are called $D_{3h}(M)$ and $C_{3v}(M)$, respectively, and their character tables are given in Tables A-8 and A-9.

Answer. The group $D_{3h}(M)$ has six irreducible representations, A_1', A_2', E', A_1'', A_2'', and E'', and the group $C_{3v}(M)$ has three, A_1, A_2, and E. The representation A_1' of $D_{3h}(M)$ has character $+1$ under each operation and hence correlates with the representation A_1 of $C_{3v}(M)$. Similarly A_2'' of $D_{3h}(M)$ has character $+1$ under the operations E, (123), (132), (12)*, (23)*, and (13)* so that it correlates with A_1 of $C_{3v}(M)$. The representations A_2' and A_1'' each have characters 1, 1, 1, -1, -1, and -1 under the operations E, (123), (132), (12)*, (23)*, and (13)*, respectively, and so each correlates with the representation A_2 of $C_{3v}(M)$. Finally E' and E'' of $D_{3h}(M)$ each have characters 2, -1, -1, 0, 0, and 0 under E, (123), (132), (12)*, (23)*, and (13)*, respectively, and so each correlates with the representation E of $C_{3v}(M)$. Thus, for example, a pair of functions transforming as E' or E'' in $D_{3h}(M)$ transform as E in $C_{3v}(M)$.

We can equally well discuss the correlation of the irreducible representations in the other direction, i.e., from the subgroup H to the group G, and this is useful when we want to consider the effect of small tunneling splittings; the group G would then be the MS group allowing for tunneling and H a subgroup in which elements corresponding to tunneling were omitted. Suppose we have symmetry labeled a level as Γ_i in H. We wish to know what the symmetry of the level would be if tunneling splittings were allowed for and a larger MS group G used. It has been shown by Watson (1965) that the symmetry in G, which we call $\Gamma(\Gamma_i)$ is given by

$$\Gamma(\Gamma_i) = \sum_\alpha a_i^{(\alpha)} \Gamma_\alpha \qquad (9\text{-}29)$$

in terms of the irreducible representations Γ_α of G, where the $a_i^{(\alpha)}$ are as given in Eq. (9-22). The representation $\Gamma(\Gamma_i)$ is the representation of G *induced* by the representation Γ_i of the subgroup H. Thus given the symmetry species of a level in the group H and the correlation table relating the irreducible representations of G to H, we can use the correlation table *backwards* to determine the symmetries of the levels in the larger group G.

We can use the correlation table of the $D_{2h}(M)$ and G_{96} groups of ethylene as an example. For ethylene the rotation–vibration levels of the ground electronic state can be labeled using the eight irreducible representation labels of the $D_{2h}(M)$ group since there are no tunneling splittings. If we wished to use the G_{96} group to classify the levels then the species of each level could be obtained from the species in $D_{2h}(M)$ by using Table 9-2 backwards. For example, a level of species A_u in $D_{2h}(M)$ will have species $\Gamma(A_u)$ in G_{96} where

$$\Gamma(A_u) = \sum_\alpha a_{A_u}^{(\alpha)} \Gamma_\alpha \qquad (9\text{-}30)$$

and Γ_α are the irreducible representations of G_{96}. From Table 9-2 we have

$$a_{A_u}^{(E_s^-)} = 2, \qquad a_{A_u}^{(A_s^-)} = a_{A_u}^{(B_s^-)} = a_{A_u}^{(F_a^-)} = a_{A_u}^{(G_a^-)} = 1, \qquad (9\text{-}31)$$

Table 9-3
Reverse correlation table for $D_{2h}(M)$ to the CNPI group of ethylene (G_{96})

$D_{2h}(M)$	G_{96}	$D_{2h}(M)$	G_{96}
A_g	$A_s{}^+ \oplus B_s{}^+ \oplus 2E_s{}^+ \oplus F_a{}^+ \oplus G_a{}^+$	A_u	$A_s{}^- \oplus B_s{}^- \oplus 2E_s{}^- \oplus F_a{}^- \oplus G_a{}^-$
B_{1g}	$A_a{}^- \oplus B_a{}^- \oplus 2E_a{}^- \oplus F_s{}^- \oplus G_s{}^-$	B_{1u}	$A_a{}^+ \oplus B_a{}^+ \oplus 2E_a{}^+ \oplus F_s{}^+ \oplus G_s{}^+$
B_{2g}	$F_s{}^- \oplus G_s{}^- \oplus F_a{}^- \oplus G_a{}^-$	B_{2u}	$F_s{}^+ \oplus G_s{}^+ \oplus F_a{}^+ \oplus G_a{}^+$
B_{3g}	$F_s{}^+ \oplus G_s{}^+ \oplus F_a{}^+ \oplus G_a{}^+$	B_{3u}	$F_s{}^- \oplus G_s{}^- \oplus F_a{}^- \oplus G_a{}^-$

and all other $a_{A_u}^{(\alpha)}$ are zero. Thus

$$\Gamma(A_u) = A_s{}^- \oplus B_s{}^- \oplus 2E_s{}^- \oplus F_a{}^- \oplus G_a{}^-. \tag{9-32}$$

To obtain the representation in G_{96} induced by the irreducible representations of $D_{2h}(M)$ we use the normal correlation table backwards and the *reverse correlation table* obtained for $D_{2h}(M) \to G_{96}$ is given in Table 9-3.

We are now in a position to appreciate the utility of the MS group. In the ethylene molecule no tunneling splittings are observed in the ground electronic state energy levels, and the MS group is $D_{2h}(M)$. Using this group the rotation–vibration energy levels are labeled according to the irreducible representation labels A_g, A_u, B_{1g}, etc. If we used the CNPI group G_{96} to label the levels then a level labeled A_g in $D_{2h}(M)$ would be labeled $(A_s{}^+ \oplus B_s{}^+ \oplus 2E_s{}^+ \oplus F_a{}^+ \oplus G_a{}^+)$, and so on according to Table 9-3 for each of the eight possible irreducible representation labels of $D_{2h}(M)$. We see that according to the CNPI group classification each nondegenerate level of ethylene [in $D_{2h}(M)$] is twelvefold degenerate and there is a lot of "accidental" degeneracy. Since no tunneling splittings are observed we have no use for the symmetry labels of the individual split levels, and a classification that distinguishes between the eight possible symmetry types without allowing for the splitting is enough. This is what the MS group, $D_{2h}(M)$ in this case, provides. Notice from Table 9-3 that the twelvefold structural degeneracy is not completely resolvable by tunneling; the A_g, B_{1g}, A_u, and B_{1u} levels in $D_{2h}(M)$ can only at most be split into six levels by tunneling and the B_{2g}, B_{3g}, B_{2u}, and B_{3u} levels into four. In an excited electronic state of ethylene it may be that torsional tunneling is observable, in which case the MS group would be the group G_{16} (see Table A-23). Part of the structural degeneracy would then be allowed for in the symmetry classification but not all of it; only that part capable of experimental resolution.

Problem 9-4. Using the $D_{2h}(M)$ group (see Table A-6) there are eight symmetry types for the rotation–vibration levels of nontunneling planar ethylene. Determine the symmetry that each of these levels has in the G_{16}

group (see Table A-23) by using the correlation table of G_{16} to $D_{2h}(M)$; the G_{16} group allows for torsional tunneling.

Answer. To determine the reverse correlation $D_{2h}(M) \rightarrow G_{16}$ we must first determine the correlation $G_{16} \rightarrow D_{2h}(M)$. Using the character tables in Tables A-6 and A-23 this is straightforward, and the correlation table obtained is given in the left hand half of Table 9-4. The reverse correlation table, obtained by using the correlation table backwards according to Eq. (9-29), is given in the right hand half of Table 9-4. We see that levels of symmetry B_{2g}, B_{3g}, B_{2u}, and B_{3u} cannot be split by torsional tunneling. The right hand half of Table 9-4 and the reverse correlation $G_{16} \rightarrow G_{96}$ can be combined to show the successive reverse correlations $D_{2h}(M) \rightarrow G_{16} \rightarrow G_{96}$. For example,

$$A_g \rightarrow \begin{cases} A_1^+ \rightarrow A_s^+ \oplus E_s^+ \oplus F_a^+ \\ B_1^+ \rightarrow B_s^+ \oplus E_s^+ \oplus G_a^+. \end{cases} \tag{9-33}$$

Such a connected reverse correlation would be useful if one were interested in the effect of successive tunnelings. The reader can easily construct a similar reverse correlation table for $C_2(M) \rightarrow G_4 \rightarrow G_8$ for the H_2O_2 molecule (see Tables A-3, A-20, A-22, and B-4(v)).

Table 9-4

Correlation and reverse correlation tables for the $D_{2h}(M)$ and G_{16} groups of ethylene

G_{16}	$D_{2h}(M)$	$D_{2h}(M)$	G_{16}
A_1^+	A_g	A_g	$A_1^+ \oplus B_1^+$
A_2^+	B_{1u}	B_{1g}	$A_2^- \oplus B_2^-$
B_1^+	A_g	B_{2g}	E^-
B_2^+	B_{1u}	B_{3g}	E^+
E^+	$B_{2u} \oplus B_{3g}$		
		A_u	$A_1^- \oplus B_1^-$
A_1^-	A_u	B_{1u}	$A_2^+ \oplus B_2^+$
A_2^-	B_{1g}	B_{2u}	E^+
B_1^-	A_u	B_{3u}	E^-
B_2^-	B_{1g}		
E^-	$B_{2g} \oplus B_{3u}$		

In the next chapter the determination of the statistical weights using the MS group is discussed and the addition of statistical weights to these correlation tables will be examined. Some useful reverse correlation tables, with statistical weights, are collected together in Appendix B at the end of the book.

THE MS GROUP FOR LEVELS OF MORE
THAN ONE ELECTRONIC STATE

So far in this chapter we have set up the MS group in order to classify the rotation–vibration levels of a single electronic state of a molecule. The electronic wavefunctions of a molecule depend on the nuclear coordinates, and for a particular electronic state, we can classify the electronic wavefunction, and hence rovibronic wavefunctions, in the MS group of that electronic state. It is sometimes necessary to consider rovibronic levels of more than one electronic state in a given problem; for example, we might want to consider interactions between rovibronic levels belonging to different electronic states or to consider electric dipole transitions between electronic states. In these circumstances we must generalize our definition of the MS group so that it will be appropriate for classifying simultaneously the rovibronic levels of more than one electronic state.

Clearly, just as for the purposes of classifying the levels of a single electronic state, we could use the CNPI group for classifying the levels of all electronic states of a molecule. However, we wish to use the smallest group necessary for a sufficient symmetry labeling of the levels, and this group is the MS group. The definition of the MS group given earlier for one electronic state can be applied when we wish to study simultaneously more than one electronic state by generalizing the concept of feasibility. The generalization that seems useful in practical cases is that an unfeasible element of the CNPI group interconverts symmetrically equivalent equilibrium nuclear structures of one of the electronic states being considered when these structures are separated by an insuperable potential energy barrier and when they cannot be interconverted by transitions to and from any of the other electronic states being considered.

An example of the application of this extended definition is to the determination of the MS group appropriate in the study of the $\tilde{A} \leftarrow \tilde{X}$ (sometimes called $V \leftarrow N$) electronic transition in ethylene [see Merer and Watson (1973)]. The ethylene molecule has D_{2h} point group symmetry in the $\tilde{X}$ state (the ground electronic state) and D_{2d} point group symmetry in the excited $\tilde{A}$ electronic state. The MS group of the numbered ground state form given in Fig. 1-3a is given in Eq. (2-14). Within the ground state this numbered form has negligible tunneling into the other numbered forms. However, by transitions to and from the $\tilde{A}$ state (of D_{2d} symmetry) the form in Fig. 1-3a can be converted into the form in Fig. 1-3b. Thus the elements (12) and (34) are feasible here and are in the MS group for use in simultaneously classifying the levels of the $\tilde{X}$ and $\tilde{A}$ electronic states of ethylene; these elements are unfeasible for either of the electronic states taken by themselves. The MS group for use in simultaneously classifying the levels

Table 9-5
Reverse correlation table
for $D_{2d}(M)$ and G_{16} of
ethylene

$D_{2d}(M)$	G_{16}
A_1	$A_1{}^+ \oplus A_1{}^-$
A_2	$A_2{}^+ \oplus A_2{}^-$
B_1	$B_1{}^+ \oplus B_1{}^-$
B_2	$B_2{}^+ \oplus B_2{}^-$
E	$E^+ \oplus E^-$

of the $\tilde{X}$ and $\tilde{A}$ states is hence the G_{16} group given in Eq. (9-16). Having symmetry classified the levels of the $\tilde{A}$ state in the $D_{2d}(M)$ group and those of the $\tilde{X}$ state in the $D_{2h}(M)$ group, the reverse correlation tables $D_{2h}(M) \rightarrow G_{16}$ and $D_{2d}(M) \rightarrow G_{16}$ can be used to determine the species of the levels in the G_{16} group. These reverse correlation tables are given in Tables 9-4 and 9-5, respectively. The excited $\tilde{R}$ electronic state of ethylene is nonplanar and to study the $\tilde{R}-\tilde{X}$ band system the MS group G_{16} is required [see Merer and Schoonveld (1969)].

SUMMARY

The CNPI group for a molecule can be used to classify the rotational, vibrational, and electronic wavefunctions. The irreducible representation labels obtained are true symmetry labels for the rovibronic states. However, for a molecule in an electronic state in which there are two or more symmetrically equivalent nuclear equilibrium structures, between which tunneling produces no observable splittings, this classification scheme is more detailed than necessary. To symmetry label the energy levels as much as is necessary, that is in order to distinguish between energy levels that are distinguishable experimentally, it is sufficient to use a particular subgroup of the CNPI group. This subgroup, called the molecular symmetry group, is obtained from the CNPI group by deleting all unfeasible elements. If we are only interested in considering a single electronic state of a molecule then an unfeasible element is one that interconverts symmetrically equivalent equilibrium nuclear structures that are separated by an insuperable barrier in the potential energy function (on the time scale of the experiment); feasible elements maintain an equilibrium form as itself or interconvert forms between which tunneling (such as by torsion or inversion) produces observable splittings. If we are interested in considering two or more electronic states at once, because we wish to study interactions or transitions between them, then the definition of a feasible element must be extended to include elements that connect forms that can be interconverted by transitions to and from any of the other electronic states being considered.

The MS group, rather than the CNPI group, is to be used to classify the rovibronic levels E_{rve} of a molecule, and it gives a true symmetry classification of the states. We can also use the MS group to classify the nuclear spin states and electron spin states as we shall see.

BIBLIOGRAPHICAL NOTE

Bunker (1964).   The correlation of the symmetry species from a smaller MS group to a larger MS group is discussed in Section 4 without using the correlation table.

10

The Classification
of Molecular
Wavefunctions in the
Molecular Symmetry
Group

In this chapter the determination of the symmetry species of the nuclear spin, rotational, vibrational, electron orbital, and electron spin wavefunctions of a molecule in its molecular symmetry group is discussed. The determination of nuclear spin statistical weights using the molecular symmetry group is also explained. To classify the rotational and electron spin wavefunctions for molecules that have an odd number of electrons and strong spin–orbit coupling it is necessary to use the spin double group of the molecular symmetry group, and this type of group is introduced and used.

THE CLASSIFICATION OF THE COMPLETE
INTERNAL WAVEFUNCTION

As discussed in Chapter 6 the complete internal wavefunction Φ_{int} can have $+$ or $-$ parity according to the effect of E^*, i.e.,

$$E^*\Phi_{int} = \pm\Phi_{int}, \tag{10-1}$$

and Φ_{int} can also be labeled by the quantum numbers F and m_F using the group K(spatial). From the rules of Fermi–Dirac and Bose–Einstein statistical formulas we have seen that Φ_{int} is changed in sign by any nuclear permuta-

226

tion, $P_{(odd)}$, say, that involves an odd permutation of nuclei that have half integer spin (fermions), i.e.,

$$P_{(odd)}\Phi_{int} = -\Phi_{int}, \tag{10-2}$$

and Φ_{int} is invariant to a nuclear permutation, $P_{(even)}$, say, that involves an even permutation of fermion nuclei regardless of whether the permutation involves an even or odd permutation of nuclei that have integer spin (bosons), i.e.,

$$P_{(even)}\Phi_{int} = +\Phi_{int}. \tag{10-3}$$

Thus Φ_{int} transforms as one of two nondegenerate irreducible representations of the CNPI group which we write as Γ^+ or Γ^- as the character is $+1$ or -1 under E^* (i.e., as the parity[1] is $+$ or $-$); the characters under the permutation operations are fixed by Eqs. (10-2) and (10-3).

Problem 10-1. The CNPI groups of the molecules H_2O, BF_3, and CH_4 are given in Tables A-4, A-9, and A-26, respectively. Determine the allowed species of the complete internal wavefunctions of these molecules using Eqs. (10-1)–(10-3). Also determine the allowed species for the complete internal wavefunctions of D_2O and CD_4, and the allowed species of the complete internal wavefunctions of $^{12}C_2H_4$ and $^{13}C_2H_4$ in the CNPI group G_{96} (see Table 9-1).

Answer. For H_2O in the $C_{2v}(M)$ group the operation (12) is an odd permutation of fermion nuclei and thus changes the sign of Φ_{int}. As a result the species of Φ_{int} is B_2 or B_1 as the parity is $+$ or $-$ (i.e., $\Gamma^+ = B_2$ and $\Gamma^- = B_1$ in this case). Since deuterium nuclei are bosons the operation (12) does not change the sign of Φ_{int} and the species of the overall wavefunction of D_2O in $C_{2v}(M)$ is A_1 or A_2 as the parity is $+$ or $-$ (i.e., $\Gamma^+ = A_1$ and $\Gamma^- = A_2$ in this case). For BF_3 since the fluorine nuclei are fermions, Φ_{int} is changed in sign by (23) but not by (123) [since (123) is an even permutation of the fluorine nuclei; see Eq. (1-25)]. Hence Φ_{int} has species A_2' or A_2'' in $D_{3h}(M)$ as the parity is $+$ or $-$. For CH_4 the operations (23) and (1423) are odd permutations whereas (123) and (14)(23) are even so that Φ_{int} is of species A_2^+ or A_1^- in G_{48} as the parity is $+$ or $-$. For CD_4 the overall species is A_1^+ or A_2^- as the parity is $+$ or $-$. For $^{12}C_2H_4$ the species of the overall wavefunction is B_s^+ or B_s^- in G_{96} as the parity is $+$ or $-$, and for $^{13}C_2H_4$ it is B_a^+ or B_a^- as the parity is $+$ or $-$; ^{12}C nuclei are bosons and ^{13}C nuclei are fermions.

In Chapter 9 we saw that a considerable simplification without loss of useful information is obtained by classifying the rovibronic wavefunctions

---

[1] It is interesting to speculate as to whether there is a purely experimental method of determining the parity of an E_{int} level [Oka (1973)].

in the MS group rather than in the CNPI group. When we do this we use rovibronic wavefunctions that are only evaluated over a small region of vibrational coordinate space; we can denote such functions as $\Phi_{rve}^{(l)}$ where (l) stands for *local*. For a rigid molecule, such as methane or methyl fluoride, the region of vibrational coordinate space considered only contains one of the symmetrically equivalent equilibrium structures, whereas for a molecule such as H_2O_2, in which observable tunneling effects occur, the region contains more than one of the symmetrically equivalent equilibrium structures. For methane we have two parallel sets of rovibronic wavefunctions: $\Phi_{rve}^{(lA)}$ centered on configuration A [see Fig. (9-2a)], and $\Phi_{rve}^{(lC)}$ centered on configuration C[see Fig. (9-2b)]. Classifying one of these in the MS group provides a sufficient symmetry labeling of the energy levels. Similarly we need only consider local functions $\Phi_{int}^{(l)}$ and need only classify them in the MS group of the molecule rather than in the CNPI group.

The classification of $\Phi_{int}^{(l)}$ in the MS group is straightforward. Since the MS group is a subgroup of the CNPI group we can use the correlation table of the groups to determine the MS group species from the species Γ^+ and Γ^- in the CNPI group. For the methane molecule the species of Φ_{int} in the CNPI group G_{48} can be either $\Gamma^+ = A_2^+$ or $\Gamma^- = A_1^-$. The correlation table of G_{48} to the MS group $T_d(M)$ yields the MS species A_2 or A_1, respectively, for levels that are A_2^+ or A_1^- in the CNPI group. Thus $\Phi_{int}^{(l)}$ functions of methane can have species A_1 or A_2 in the MS group. For $^{12}C_2H_4$ we have $\Gamma^+ = B_s^+$ and $\Gamma^- = B_s^-$ in G_{96}, so that from Table 9-2 the $\Phi_{int}^{(l)}$ functions of the planar nontunneling molecule in the $D_{2h}(M)$ group can be A_g or A_u, respectively. Similarly for $^{13}C_2H_4$ we have $\Gamma^+ = B_a^+ \to B_{1u}$ in $D_{2h}(M)$ and $\Gamma^- = B_a^- \to B_{1g}$.

Rather than determining the allowed species of $\Phi_{int}^{(l)}$ in the MS group from the species in the CNPI group using the correlation table, we can determine them directly. To do this we use the rules expressed in Eqs. (10-2) and (10-3). For example in $^{12}CH_4$, using the rules in Eqs. (10-2) and (10-3) with $T_d(M)$ we determine from the characters under the permutation operations (123) and (14)(23), both of which are $P_{(even)}$ operations, that $\Phi_{int}^{(l)}$ can only transform as A_1 or A_2. Similarly for $^{12}C_2H_4$ we see that $\Phi_{int}^{(l)}$ can only be of A_g or A_u symmetry in $D_{2h}(M)$ since these are the only two representations having character $+1$ for all the permutations [all the permutations in $D_{2h}(M)$ are $P_{(even)}$ operations for $^{12}C_2H_4$]. For $^{13}C_2H_4$ the permutations (13)(24)(56) and (14)(23)(56) are odd permutations of fermions (the ^{13}C nuclei) and hence $\Phi_{int}^{(l)}$ can only be B_{1g} or B_{1u}.

For molecules whose MS group contains one or more permutation–inversion operations we always obtain two allowed species for the $\Phi_{int}^{(l)}$ functions. For molecules such as CHIFCl or H_2S_2, whose MS group contains no permutation–inversion operations, the $\Phi_{int}^{(l)}$ functions can only be of one species in the MS group. For example the overall wavefunction of $H_2\,^{32}S_2$ (^{32}S nuclei are bosons) must be changed in sign by the operation (12)(34) of the $C_2(M)$ group (see Table A-3), and hence $\Phi^{(l)}_{int}$ can only have species B in that group. The interpretation of this is appreciated if we look at the correlation of the species Γ^+ and Γ^- of the CNPI group to the MS group species in a general way. The representation Γ^+ has the same character under a given permutation operation of the CNPI group as does Γ^- ($+1$ or -1 as the permutation is $P_{(even)}$ or $P_{(odd)}$), but Γ^+ and Γ^- have characters of opposite sign under a given permutation–inversion operation (for Γ^+ the character under permutation–inversions of the type $P^*_{(even)}$ is $+1$ and under $P^*_{(odd)}$ is -1, whereas for Γ^- the character under $P^*_{(even)}$ is -1 and under $P^*_{(odd)}$ is $+1$). Each of Γ^+ and Γ^- will correlate with an irreducible representation of the MS group. If the MS group contains permutation–inversion operations then Γ^+ and Γ^- will correlate with different nondegenerate irreducible representations of the MS group (having opposite signs for the permutation–inversion operations). If the MS group contains no permutation–inversion operations then Γ^+ and Γ^- will each correlate with the same nondegenerate irreducible representation of the MS group and each overall state of the molecule has a structural double degeneracy. Using Φ_{int} functions that are not local such a structural double degeneracy consists of a pair of states of opposite parity. For molecules with P^* operations in the MS group, and for which two MS group species are obtained for $\Phi^{(l)}_{int}$, we can identify one of the pair of species as correlating with a Φ_{int} state of $+$ parity (this will have character $+1$ under $P^*_{(even)}$ operations) and the other as correlating with a Φ_{int} state of $-$ parity (this will have character -1 under $P^*_{(even)}$ operations). We could then distinguish between the two types of overall state using the parity label rather than the MS group label. However for the purpose of determining the nuclear spin statistical weights we must use the MS group label and nothing is lost if we use the MS group label rather than the parity label for the overall states.

Having determined the allowed species of $\Phi^{(l)}_{int}$ in the MS group as described previously, we can use the result to determine the nuclear spin states with which the $\Phi^{(l)}_{rve}$ functions can allowably combine. As a result we can use the MS group to determine the nuclear spin statistical weights of the levels.

THE CLASSIFICATION OF THE NUCLEAR SPIN WAVEFUNCTIONS AND THE DETERMINATION OF NUCLEAR SPIN STATISTICAL WEIGHTS

In Chapter 6 a complete set of basis nuclear spin wavefunctions for a molecule was introduced, and the transformation properties of these wavefunctions under the effect of nuclear permutations was discussed [see Eqs.

230 10. The Classification of Molecular Wavefunctions

(6-66)–(6-70)]. The classification of the nuclear spin wavefunctions of NH_3 and ND_3 in the CNP group was the subject of Problem 6-1. The classification of the nuclear spin wavefunctions in the MS group is equally straightforward when it is remembered that such wavefunctions are invariant to E^*; hence the permutation–inversion operation $P^* = PE^*$ has the same effect on a nuclear spin wavefunction as does the permutation P.

Problem 10-2. Determine the representations of the MS group $D_{2h}(M)$ (see Table A-6) generated by the nuclear spin wavefunctions of $^{12}C_2H_4$.

Answer. Since ^{12}C nuclei have a spin of zero the carbon nuclear spin function is $|m_{I_5}, m_{I_6}\rangle = |0,0\rangle$ (the carbon nuclei are labeled 5 and 6). The function has symmetry A_g in the $D_{2h}(M)$ group. The proton spin functions ($|m_{I_1}, m_{I_2}, m_{I_3}, m_{I_4}\rangle$) are

$m_I = +2:$ $|\tfrac{1}{2},\tfrac{1}{2},\tfrac{1}{2},\tfrac{1}{2}\rangle;$

$m_I = +1:$ $|\tfrac{1}{2},\tfrac{1}{2},\tfrac{1}{2},-\tfrac{1}{2}\rangle, |\tfrac{1}{2},\tfrac{1}{2},-\tfrac{1}{2},\tfrac{1}{2}\rangle, |\tfrac{1}{2},-\tfrac{1}{2},\tfrac{1}{2},\tfrac{1}{2}\rangle, |-\tfrac{1}{2},\tfrac{1}{2},\tfrac{1}{2},\tfrac{1}{2}\rangle;$

$m_I = 0:$ $|\tfrac{1}{2},\tfrac{1}{2},-\tfrac{1}{2},-\tfrac{1}{2}\rangle, |\tfrac{1}{2},-\tfrac{1}{2},\tfrac{1}{2},-\tfrac{1}{2}\rangle, |-\tfrac{1}{2},\tfrac{1}{2},\tfrac{1}{2},-\tfrac{1}{2}\rangle, |\tfrac{1}{2},-\tfrac{1}{2},-\tfrac{1}{2},\tfrac{1}{2}\rangle,$
$\qquad\qquad |-\tfrac{1}{2},\tfrac{1}{2},-\tfrac{1}{2},\tfrac{1}{2}\rangle, |-\tfrac{1}{2},-\tfrac{1}{2},\tfrac{1}{2},\tfrac{1}{2}\rangle;$

$m_I = -1:$ $|-\tfrac{1}{2},-\tfrac{1}{2},-\tfrac{1}{2},\tfrac{1}{2}\rangle, |-\tfrac{1}{2},-\tfrac{1}{2},\tfrac{1}{2},-\tfrac{1}{2}\rangle, |-\tfrac{1}{2},\tfrac{1}{2},-\tfrac{1}{2},-\tfrac{1}{2}\rangle,$
$\qquad\qquad |\tfrac{1}{2},-\tfrac{1}{2},-\tfrac{1}{2},-\tfrac{1}{2}\rangle;$

$m_I = -2:$ $|-\tfrac{1}{2},-\tfrac{1}{2},-\tfrac{1}{2},-\tfrac{1}{2}\rangle.$

(10-4)

The function having $m_I = +2$ and the function having $m_I = -2$ are each of A_g symmetry. The four $m_I = +1$ functions and the four $m_I = -1$ functions generate the representation

$$A_g \oplus B_{3g} \oplus B_{1u} \oplus B_{2u}, \qquad (10\text{-}5)$$

and the six $m_I = 0$ functions generate the representation

$$3A_g \oplus B_{3g} \oplus B_{1u} \oplus B_{2u}. \qquad (10\text{-}6)$$

The 16 proton spin functions therefore generate the representation

$$\Gamma^{\text{tot}}_{\text{nspin}} = 7A_g \oplus 3B_{3g} \oplus 3B_{1u} \oplus 3B_{2u}, \qquad (10\text{-}7)$$

and since the carbon nuclear spin species is A_g, this is the total nuclear spin species.

In the answer to Problem 10-2 we have determined the symmetry species of the 16 nuclear spin functions of $^{12}C_2H_4$ in the $D_{2h}(M)$ group. We can equally well classify these functions in the G_{16} group appropriate if internal rotation splittings are resolved or even in the CNPI group G_{96} if we wish. This requires us to determine the transformation properties of the spin functions of Eq. (10-4) under the effect of the permutation and permutation–inversion operations of these groups. The results are (see Table A-23)

$$\Gamma_{nspin}^{tot}(G_{16}) = 6A_1^+ \oplus B_1^+ \oplus 3B_2^+ \oplus 3E^+ \qquad (10\text{-}8)$$

and (see Table 9-1)

$$\Gamma_{nspin}^{tot}(G_{96}) = 5A_s^+ \oplus 3F_s^+ \oplus E_s^+. \qquad (10\text{-}9)$$

These results will be used in the following.

In Chapter 6 the use of the CNPI group for the purpose of determining nuclear spin statistical weights was explained. We can equally well determine the statistical weights of molecular energy levels using the MS group. We can only combine a rovibronic state $\Phi_{rve}^{(l)}$ having symmetry Γ_{rve} in the MS group with a nuclear spin state having symmetry Γ_{nspin} in the MS group if the product of these symmetries contains Γ_{int}, where Γ_{int} is the species for $\Phi_{int}^{(l)}$ that is allowed by the statistical formulas, i.e., we must have

$$\Gamma_{rve} \otimes \Gamma_{nspin} \supset \Gamma_{int}. \qquad (10\text{-}10)$$

We will use the $^{12}C_2H_4$ ethylene molecule as an example of the application of this rule. The rovibronic states of ethylene can be classified according to the eight irreducible representations of the $D_{2h}(M)$ group. The nuclear spin states have the symmetries given in Eq. (10-7) and the complete internal wavefunction can only be of either A_g or A_u species. To determine the statistical weights it is convenient to construct a table such as is shown in Table 10-1. In this table we first complete column 1 with all possible Γ_{rve} and then put the pairs of possible Γ_{int} in column 3. We now determine which of the nuclear spin functions [from Eq. (10-7)] can be combined with each rovibronic function so that the product of the nuclear spin symmetry and rovibronic symmetry is A_g or A_u; those that combine with Γ_{rve} to give an overall species of A_g are put on the left of the semicolon in column 2 and those that combine to give A_u overall are put on the right. In this manner column 2 in Table 10-1 under Γ_{nspin} is completed. The statistical weights can now be easily determined

Table 10-1

The statistical weights of the rovibronic states of $^{12}C_2H_4$ in the $D_{2h}(M)$ group

Γ_{rve}	Γ_{nspin}	Γ_{int}	Statistical weight	Γ_{rve}	Γ_{nspin}	Γ_{int}	Statistical weight
A_g	$7A_g;—$	$A_g; A_u$	7	A_u	$—; 7A_g$	$A_g; A_u$	7
B_{1g}	$—; 3B_{1u}$	$A_g; A_u$	3	B_{1u}	$3B_{1u};—$	$A_g; A_u$	3
B_{2g}	$—; 3B_{2u}$	$A_g; A_u$	3	B_{2u}	$3B_{2u};—$	$A_g; A_u$	3
B_{3g}	$3B_{3g};—$	$A_g; A_u$	3	B_{3u}	$—; 3B_{3g}$	$A_g; A_u$	3

Table 10-2

The statistical weights of the rovibronic states of $^{12}C_2H_4$ in the G_{16} group

Γ_{rve}	Γ_{nspin}	Γ_{int}	Statistical weight	Γ_{rve}	Γ_{nspin}	Γ_{int}	Statistical weight
A_1^+	$B_1^+;-$	$B_1^+;A_1^-$	1	A_1^-	$-;6A_1^+$	$B_1^+;A_1^-$	6
A_2^+	$3B_2^+;-$	$B_1^+;A_1^-$	3	A_2^-	$-;-$	$B_1^+;A_1^-$	0
B_1^+	$6A_1^+;-$	$B_1^+;A_1^-$	6	B_1^-	$-;B_1^+$	$B_1^+;A_1^-$	1
B_2^+	$-;-$	$B_1^+;A_1^-$	0	B_2^-	$-;3B_2^+$	$B_1^+;A_1^-$	3
E^+	$3E^+;-$	$B_1^+;A_1^-$	3	E^-	$-;3E^+$	$B_1^+;A_1^-$	3

Table 10-3

The statistical weights of the rovibronic states of $^{12}C_2H_4$ in the G_{96} group

Γ_{rve}	Γ_{nspin}	Γ_{int}	Statistical weight	Γ_{rve}	Γ_{nspin}	Γ_{int}	Statistical weight
A_s^+	$-;-$	$B_s^+;B_s^-$	0	A_s^-	$-;-$	$B_s^+;B_s^-$	0
B_s^+	$5A_s^+;-$	$B_s^+;B_s^-$	5	B_s^-	$-;5A_s^+$	$B_s^+;B_s^-$	5
E_s^+	$E_s^+;-$	$B_s^+;B_s^-$	1	E_s^-	$-;E_s^+$	$B_s^+;B_s^-$	1
F_s^+	$-;-$	$B_s^+;B_s^-$	0	F_s^-	$-;-$	$B_s^+;B_s^-$	0
G_s^+	$3F_s^+;-$	$B_s^+;B_s^-$	3	G_s^-	$-;3F_s^+$	$B_s^+;B_s^-$	3

as the number of possible $\Phi_{int}^{(1)}$ of the acceptable symmetry species that can be obtained from each $\Phi_{rve}^{(1)}$ by combining with nuclear spin functions.

In Tables 10-2 and 10-3 similar statistical weight tables for $^{12}C_2H_4$ using the groups G_{16} and G_{96} are given. In G_{96} all "a" rovibronic states have zero statistical weight, since Γ_{nspin} and Γ_{int} must be "s," and these rovibronic states have been omitted from the table. To obtain these results the direct product tables of the species of the groups are needed, and these are given

Table 10-4

The direct product table of
the irreducible representations of $G_{16}{}^a$

	A_1	A_2	B_1	B_2	E
$A_1:$	A_1	A_2	B_1	B_2	E
$A_2:$	A_2	A_1	B_2	B_1	E
$B_1:$	B_1	B_2	A_1	A_2	E
$B_2:$	B_2	B_1	A_2	A_1	E
$E\ :$	E	E	E	E	$(A_1 \oplus A_2 \oplus B_1 \oplus B_2)$

^{a} We omit the $+$ and $-$ labels since $+ \otimes + = - \otimes - $ $= +$, and $+ \otimes - = -$.

Table 10-5

The direct product table of the irreducible representations of S_4

	A	B	E	F	G
A:	A	B	E	F	G
B:	B	A	E	G	F
E:	E	E	$(A \oplus B \oplus E)$	$(F \oplus G)$	$(F \oplus G)$
F:	F	G	$(F \oplus G)$	$(A \oplus E \oplus F \oplus G)$	$(B \oplus E \oplus F \oplus G)$
G:	G	F	$(F \oplus G)$	$(B \oplus E \oplus F \oplus G)$	$(A \oplus E \oplus F \oplus G)$

in Table 10-4 for G_{16} and in Table 10-5 for $S_4^{(H)}$, from which that for G_{96} is easily derived since the subscripts and superscripts multiply as follows:

$$a \otimes a = s \otimes s = s, \quad a \otimes s = a, \quad + \otimes + = - \otimes - = +, \quad \text{and} \quad + \otimes - = -.$$

The results in Table 10-2 have been used to add statistical weights to the reverse correlation table $D_{2h}(M) \to G_{16}$ given in Table B-4(vi) in Appendix B. Such a table is very useful in the identification of tunneling splittings, and it is particularly important to be aware of the presence of levels having zero statistical weight since these will be *missing* levels.

Problem 10-3. In Table B-4(v) in Appendix B the reverse correlation table of the groups $C_2(M) \to G_4 \to G_8$ for $H_2{}^{16}O_2$ or $H_2{}^{32}S_2$ is given and the statistical weights are added. Check that these statistical weights are correct.

Answer. The species of the complete internal wavefunctions in the groups $C_2(M)$, G_4, and G_8 are $2B$ (allowing for the structural parity double degeneracy), B_1 or B_2, and B_1'' or B_2'', respectively. The proton nuclear spin wavefunctions $|\frac{1}{2}, \frac{1}{2}\rangle, |\frac{1}{2}, -\frac{1}{2}\rangle, |-\frac{1}{2}, \frac{1}{2}\rangle$, and $|-\frac{1}{2}, -\frac{1}{2}\rangle$ generate the representations $3A \oplus B$, $3A_1 \oplus B_2$, and $3A_1{}' \oplus B_2''$ in the three groups, respectively. The statistical weights given in Table B-4(v) follow directly from these results. From Table B-4(v) we see that the final tunneling ($G_4 \to G_8$) will not produce any splittings of the observed energy levels (although it will cause shifts in the levels), since in every case one of the pair of levels split by the tunneling has zero statistical weight.

In Appendix B (Table B-4) reverse correlation tables with statistical weights are given for ammonia, methanol, methylsilane, acetone, hydrogen peroxide, ethylene, ethane, and methane [see, also, Watson (1965)].

As a result of the facts that nuclear spin wavefunctions have positive parity and that the complete internal wavefunction can have positive or negative parity without restriction, we could determine the statistical weights of the levels of any molecule by using the *permutation subgroup* of the MS group. This is the subgroup obtained by deleting all permutation–inversion elements in the MS group. This is in fact the customary way of determining nuclear spin statistical weights although the group is called "the rotational subgroup of the molecular point group." This will be discussed further in the next chapter. Since in studying a molecule we determine the symmetry of the rovibronic levels in the MS group it is expedient to continue using this symmetry in the determination of the statistical weights, and that is why we have discussed the determination of the statistical weights using the MS group rather than its permutation subgroup.

Having determined the allowed species for the wavefunction $\Phi_{\text{int}}^{(l)}$, the nuclear spin symmetry species, and the statistical weights of the rovibronic levels as functions of the symmetry of the rovibronic wavefunctions, the symmetry of the rovibronic wavefunctions $\Phi_{\text{rve}}^{(l)}$ in the MS group remains to be determined. We will now drop the superscript (l) on the wavefunctions although it is to be understood. To classify by symmetry the rovibronic wavefunctions Φ_{rve} we will classify by symmetry the rotational, vibrational, and electronic wavefunctions of which Φ_{rve} is composed in zero order [see Eq. (6-79)]. The product of these symmetries gives the rovibronic symmetry in the MS group as a result of the vanishing integral rule [see Eq. (5-133)].

THE CLASSIFICATION OF THE ROTATIONAL WAVEFUNCTIONS

The rigid rotor wave equation given in Eq. (8-33) leads to the rotational eigenfunctions $|J, k, m\rangle$ as given in Eq. (8-111) for a symmetric top molecule. For an asymmetric top molecule the rotational eigenfunctions are linear combinations of symmetric top functions (see Problem 8-3). The symmetric top functions depend on the Euler angles (θ, ϕ, χ) and to determine the transformation properties of the functions we first must determine the transformation properties of the Euler angles. To determine the effect of an MS group element on a rotational wavefunction we replace each MS group element by its *equivalent rotation* [this describes the effect of the MS group element on the Euler angles and we use the notation R_z^β or R_α^π as defined in Table 7-1; see also the discussion after Eq. (7-206)], and we use the expressions for the symmetric top wavefunctions given in Eq. (8-111).

We will determine the general effect of the rotation operations R_z^β and R_α^π on any symmetric top function $|J, k, m\rangle$. This will allow us to determine the transformation properties of any symmetric top or asymmetric top wavefunction in an MS group once we have identified the equivalent rotation of each MS group operation (this is done in the MS group character tables in Appendix A and R^0 is the identity rotation). The symmetry of spherical top wavefunctions are obtained by reduction of the representations of the group $K(\text{mol})$. In this section we only consider states having J integral. States having J half-integral will be discussed at the end of the chapter.

Using the rotational ladder operators $\hat{J}_m{}^{\pm}$ defined in Eq. (8-89), we obtain the following operator equations (each of which is two equations)

$$R_z{}^{\beta}\hat{J}_m{}^{\pm} = [(\hat{J}_x\cos\beta + \hat{J}_y\sin\beta) \pm i(\hat{J}_y\cos\beta - \hat{J}_x\sin\beta)]R_z{}^{\beta}$$
$$= [\cos\beta\hat{J}_m{}^{\pm} \mp i\sin\beta\hat{J}_m{}^{\pm}]R_z{}^{\beta} = e^{\mp i\beta}\hat{J}_m{}^{\pm}R_z{}^{\beta} \qquad (10\text{-}11)$$

and

$$R_\alpha{}^{\pi}\hat{J}_m{}^{\pm} = [(\hat{J}_x\cos 2\alpha + \hat{J}_y\sin 2\alpha) \pm i(\hat{J}_x\sin 2\alpha - \hat{J}_y\cos 2\alpha)]R_\alpha{}^{\pi}$$
$$= e^{\pm i2\alpha}\hat{J}_m{}^{\mp}R_\alpha{}^{\pi}. \qquad (10\text{-}12)$$

Clearly the space fixed axes are unaffected by $R_z{}^{\beta}$ and $R_\alpha{}^{\pi}$ so that for the ladder operators $\hat{J}_s{}^{\pm}$ [defined in Eq. (8-94)] we have the operator equations

$$R_z{}^{\beta}\hat{J}_s{}^{\pm} = \hat{J}_s{}^{\pm}R_z{}^{\beta} \qquad (10\text{-}13)$$

and

$$R_\alpha{}^{\pi}\hat{J}_s{}^{\pm} = \hat{J}_s{}^{\pm}R_\alpha{}^{\pi}. \qquad (10\text{-}14)$$

Using Eq. (8-99) for the symmetric top wavefunction $|J,0,0\rangle$ and the transformation properties of the Euler angles given in Table 7-1 we deduce that

$$R_z{}^{\beta}|J,0,0\rangle = |J,0,0\rangle \qquad (10\text{-}15)$$

and

$$R_\alpha{}^{\pi}|J,0,0\rangle = (-1)^J|J,0,0\rangle. \qquad (10\text{-}16)$$

Thus from Eqs. (8-111), (10-11), (10-13), and (10-15) we have

$$R_z{}^{\beta}|J,|k|, \pm|m|\rangle = R_z{}^{\beta}N(\hat{J}_m{}^-)^{|k|}(\hat{J}_s{}^{\pm})^{|m|}|J,0,0\rangle$$
$$= e^{+i|k|\beta}|J,|k|, \pm|m|\rangle. \qquad (10\text{-}17)$$

Similarly

$$R_z{}^{\beta}|J, -|k|, \pm|m|\rangle = e^{-i|k|\beta}|J, -|k|, \pm|m|\rangle. \qquad (10\text{-}18)$$

Also from Eqs. (8-111), (10-12), (10-14), and (10-16) we have

$$R_\alpha{}^{\pi}|J,|k|, \pm|m|\rangle = R_\alpha{}^{\pi}N(\hat{J}_m{}^-)^{|k|}(\hat{J}_s{}^{\pm})^{|m|}|J,0,0\rangle$$
$$= (-1)^J e^{-i2|k|\alpha}|J, -|k|, \pm|m|\rangle \qquad (10\text{-}19)$$

and

$$R_\alpha{}^{\pi}|J, -|k|, \pm|m|\rangle = (-1)^J e^{i2|k|\alpha}|J,|k|, \pm|m|\rangle. \qquad (10\text{-}20)$$

Summarizing these results we have

$$R_z{}^{\beta}|J,k,m\rangle = e^{ik\beta}|J,k,m\rangle \qquad (10\text{-}21)$$

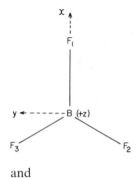

Fig. 10-1. The molecule fixed axes and nuclear labeling convention for the BF_3 molecule in its ground electronic state. The z axis is pointing up out of the page.

and

$$R_\alpha^\pi |J, k, m\rangle = (-1)^J e^{-2ik\alpha} |J, -k, m\rangle. \tag{10-22}$$

The unnormalized functions $|3, -1, -2\rangle$ and $|3, 1, -2\rangle$ are given in Eqs. (8-61) and (8-62); they can be used to test Eq. (10-22). The $(2J + 1)$ m degeneracy of each (J, k) state can be neglected (unless an external field is applied). We will now use the BF_3, CH_3F, and H_2O molecules as examples.

The BF_3 molecule, and the molecule fixed axis system used, is drawn in Fig. 10-1. Its MS group is $D_{3h}(M)$. The character table of $D_{3h}(M)$ and the equivalent rotations are given in Table A-9. Using Eqs. (10-15) and (10-16) we see that the representation generated by the function $|J, 0, 0\rangle$ is [where the operations are given in the same order as in Table A-9 for the group $D_{3h}(M)$]

$$1, 1, (-1)^J, 1, 1, (-1)^J. \tag{10-23}$$

For the pair of functions $|J, K, m\rangle$, $|J, -K, m\rangle$ the representation generated is [using Eqs. (10-21) and (10-22)]

$$2, 2\cos(2\pi K/3), 0, 2(-1)^K, 2\cos(-\pi K/3), 0. \tag{10-24}$$

The representations in Eqs. (10-23) and (10-24) can be written in terms of the irreducible representations of $D_{3h}(M)$ and these symmetry species, as a function of J and $K = |k|$, are collected in Table 10-6. The similar table for CH_3F (obtained using Table A-8) is given in Table 10-7. It is necessary to multiply these species by $(2J + 1)$ if one wishes to allow for the m degeneracy.

The rotational wavefunctions of the water molecule in a I^r basis are linear combinations of symmetric top rotational wavefunctions $|J, k_a, m\rangle$, and in a III^r basis they are linear combinations of $|J, k_c, m\rangle$ functions as discussed in Chapter 8 (see Problem 8-3). It will be instructive to classify both the $|J, k_a, m\rangle$ basis wavefunctions and the $|J, k_c, m\rangle$ basis wavefunctions in the MS group $C_{2v}(M)$ (see Table A-4) and to consider the correlation of these symmetry labels. As a result we will be able to label by symmetry the asymmetric rotor wavefunctions and determine a general set of rules for enabling this to be done for any asymmetric top.

Table 10-6

Symmetry species Γ_{rot} of rotational wavefunctions of BF_3 in the $D_{3h}(M)$ group[a]

K		Γ_{rot}
0	J even	A_1'
	J odd	A_2'
$6n \pm 1$		E''
$6n \pm 2$		E'
$6n \pm 3$		$A_1'' \oplus A_2''$
$6n \pm 6$		$A_1' \oplus A_2'$

[a] The $(2J + 1)$ m degeneracy is ignored. n is integral, and $K = |k|$.

Table 10-7

Symmetry species Γ_{rot} of rotational wavefunctions of CH_3F in the $C_{3v}(M)$ group[a]

K		Γ_{rot}
0	J even	A_1
	J odd	A_2
$3n \pm 1$		E
$3n \pm 3$		$A_1 \oplus A_2$

[a] The $(2J + 1)$ m degeneracy is ignored. n is integral, and $K = |k|$.

The water molecule and its axis system are shown in Fig. 10-2. In a I^r representation the Euler angles are $\theta(I^r)$, $\phi(I^r)$, and $\chi(I^r)$, and in a III^r representation they are $\theta(III^r)$, $\phi(III^r)$, and $\chi(III^r)$, as discussed in Chapter 8 and indicated in Figs. 10-3 and 10-4 here. The transformation properties of $\theta(I^r)$,

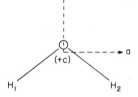

Fig. 10-2. The inertial axes for the water molecule labeled so that the moments of inertia are in the order $I_{aa} < I_{bb} < I_{cc}$.

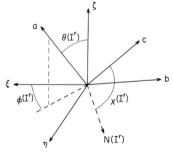

Fig. 10-3.

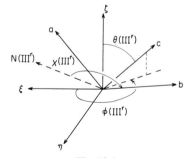

Fig. 10-4.

Fig. 10-3. The definition of the Euler angles using a type I^r convention, i.e., $abc \equiv zxy$.

Fig. 10-4. The definition of the Euler angles using a type III^r convention, i.e., $abc \equiv xyz$.

Table 10-8
Transformation properties of the I^r and III^r Euler angles[a]
for the H_2O molecule of Fig. (10-2) in the $C_{2v}(M)$ group[b]

E	(12)	E^*	(12)*
Equiv. rot.	R_b^π	R_c^π	R_a^π
$\theta(I^r)$	$\pi - \theta(I^r)$	$\pi - \theta(I^r)$	$\theta(I^r)$
$\phi(I^r)$	$\phi(I^r) + \pi$	$\phi(I^r) + \pi$	$\phi(I^r)$
$\chi(I^r)$	$2\pi - \chi(I^r)$	$\pi - \chi(I^r)$	$\chi(I^r) + \pi$
$\theta(III^r)$	$\pi - \theta(III^r)$	$\theta(III^r)$	$\pi - \theta(III^r)$
$\phi(III^r)$	$\phi(III^r) + \pi$	$\phi(III^r)$	$\phi(III^r) + \pi$
$\chi(III^r)$	$\pi - \chi(III^r)$	$\chi(III^r) + \pi$	$2\pi - \chi(III^r)$

[a] See Figs. (10-3) and (10-4).
[b] See Table A-4.

$\phi(I^r)$, and $\chi(I^r)$ were discussed in Chapter 7 [see Eqs. (7-200), (7-204), and
(7-206)], and those of $\theta(III^r)$, $\phi(III^r)$, and $\chi(III^r)$ can be similarly determined.
The results are collected together in Table 10-8. These results can be obtained
from the results in Table 7-1 once we have identified the equivalent rotations
of each MS group element; these identifications are

$$(12): \quad R_b^\pi, \qquad E^*: \quad R_c^\pi, \qquad (12)^*: \quad R_a^\pi. \qquad (10\text{-}25)$$

Thus, for example, (12)* rotates the molecule fixed axes through π radians
about the a axis (R_a^π), and in a III^r basis $(abc \equiv xyz)$ this is R_0^π in the notation
of Table 7-1. From the results in Table 10-8, and using the results in Eqs.
(10-21) and (10-22), we can obtain the representations of $C_{2v}(M)$ generated
by the $|J, k_a, m\rangle$ and $|J, k_c, m\rangle$ functions. The results are given in Table 10-9.
These results must be multiplied by $(2J + 1)$ if we wish to allow for the
m degeneracy. To determine the symmetry species of the asymmetric rotor
functions we can correlate between the K_a and K_c species labels given in

Table 10-9
Representations of the $C_{2v}(M)$ group
for H_2O generated by basis functions
$|J, k_a, m\rangle$ and $|J, k_c, m\rangle$[a]

K_a		Γ_{rot}	K_c		Γ_{rot}
0	J even	A_1	0	J even	A_1
	J odd	B_1		J odd	B_2
Odd		$A_2 \oplus B_2$	Odd		$A_2 \oplus B_1$
Even		$A_1 \oplus B_1$	Even		$A_1 \oplus B_2$

[a] $K_a = |k_a|$, $K_c = |k_c|$.

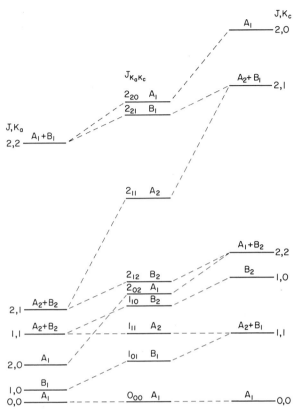

Fig. 10-5. The $C_{2v}(M)$ symmetry labels for the asymmetric rotor energy levels of H_2O (center) and their correlation with the prolate rotor (left) and oblate rotor (right) levels.

Table 10-9, and the result is shown in Fig. 10-5. Alternately we could set up + and − combinations of $|J, K_a, m\rangle$ and $|J, -K_a, m\rangle$ functions and determine their transformation properties. The symmetry species for a rigid rotor state $J_{K_aK_c}$ depends only on the evenness or oddness of K_a and K_c, and the result is summarized in Table 10-10 where ee is any state $J_{K_aK_c}$ for

Table 10-10

Symmetry species of $J_{K_aK_c}$ levels of H_2O in the $C_{2v}(M)$ group

K_aK_c	Γ_{rot}	K_aK_c	Γ_{rot}
ee	A_1	oe	B_2
eo	B_1	oo	A_2

which K_a and K_c are even (e), and eo is any state for which K_a is even (e) and K_c odd (o), etc.

From the correlation between the oblate and prolate rotor species given in Table 10-9 we can appreciate a simple way of determining the species of asymmetric rotor levels in the MS group. We first determine the equivalent rotation of each MS group element (for an asymmetric top this can only be one of E, R_a^π, R_b^π, and R_c^π). We then use *the asymmetric top symmetry rule* which states: The ee functions will transform as the totally symmetric representation, the eo functions as the representation having $+1$ for R_a^π (and -1 for R_b^π and R_c^π), the oe functions as the representation having $+1$ for R_c^π (and -1 for R_a^π and R_b^π), and the oo functions as the representation having $+1$ under R_b^π (and -1 for R_a^π and R_c^π).

Problem 10-4. Classify the asymmetric rotor wavefunctions of formaldehyde (CH_2O), ethylene, and trans $C_2H_2F_2$ in their respective MS groups.

Answer. The formaldehyde molecule and its inertial axes are shown in Fig. 10-6. The MS group is the $C_{2v}(M)$ group as for H_2O but now since the a axis is the twofold symmetry axis of the molecule the identifications of the equivalent rotations with the MS group elements is different from Eq. (10-25) (or as given in Table A-4) and instead we have:

$$(12): \quad R_a^\pi, \qquad E^*: \quad R_c^\pi, \qquad (12)^*: \quad R_b^\pi. \qquad (10\text{-}26)$$

Using the asymmetric top symmetry rule, the symmetry species of the asymmetric rotor states in the $C_{2v}(M)$ group for formaldehyde are thus as given in Table 10-11.

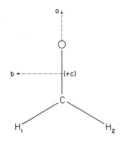

Fig. 10-6.   The inertial axes of the formaldehyde molecule.

Table 10-11

Symmetry species of $J_{K_a K_c}$ levels of CH_2O in the $C_{2v}(M)$ group

$K_a K_c$	Γ_{rot}	$K_a K_c$	Γ_{rot}
ee	A_1	oe	B_2
eo	A_2	oo	B_1

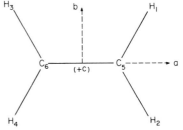

Fig. 10-7.   The inertial axes of the ethylene molecule.

Table 10-12
Symmetry species of $J_{K_aK_c}$ levels of
ethylene in the $D_{2h}(M)$ group

K_aK_c	Γ_{rot}	K_aK_c	Γ_{rot}
ee	A_g	oe	B_{3g}
eo	B_{1g}	oo	B_{2g}

The ethylene molecule and its inertial axes are shown in Fig. 10-7. The MS group, $D_{2h}(M)$ is given in Table A-6. The equivalent rotations of the elements of the $D_{2h}(M)$ group are

$$E \text{ and } (14)(23)(56)^*: \quad R^0,$$
$$(12)(34) \text{ and } (13)(24)(56)^*: \quad R_a^\pi,$$
$$(13)(24)(56) \text{ and } (12)(34)^*: \quad R_b^\pi, \qquad (10\text{-}27)$$
$$(14)(23)(56) \text{ and } E^*: \quad R_c^\pi.$$

The symmetries of the asymmetric top functions of ethylene are thus determined from the asymmetric top symmetry rule to be as given in Table 10-12.

The principal axes of trans $C_2H_2F_2$ are located as shown in Fig. 10-8 and the MS group, $C_{2h}(M)$ is shown in Table A-5. The rotation of the axes

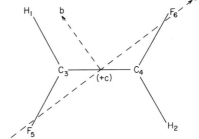

Fig. 10-8.   The inertial axes for trans $C_2H_2F_2$.

242   10.   The Classification of Molecular Wavefunctions

Table 10-13
Symmetry species of $J_{K_aK_c}$ levels of
trans $C_2H_2F_2$ in the $C_{2h}(M)$ group

K_aK_c	Γ_{rot}	K_aK_c	Γ_{rot}
ee	A_g	oe	A_g
eo	B_g	oo	B_g

caused by the elements of the $C_{2h}(M)$ group are

$$E \text{ and } (12)(34)(56)^*: \quad R^0,$$
$$(12)(34)(56) \text{ and } E^*: \quad R_c^{\pi}. \tag{10-28}$$

The symmetries of the asymmetric top functions of trans $C_2H_2F_2$ are deduced from these results to be as given in Table 10-13. Notice that although the $C_{2h}(M)$ group has four irreducible representations there are only two possible species for the rotational wavefunctions. This is because the rotational wavefunctions are invariant to the operation $(12)(34)(56)^*$. Similarly for ethylene the rotational wavefunctions are invariant to the MS group operation $(14)(23)(56)^*$. The reason for this is that these operations do not affect the orientation of the molecule fixed axes and do not change the Euler angles. Each of these operations has the same effect on the rovibronic coordinates as the operation i of the molecular point group, as will be discussed further in the next chapter [see Eq. (11-12) and the discussion following that equation].

Table 10-14
Symmetry species of rotational wavefunctions of methane in $T_d(M)$
as a function of J^a

J	Γ_{rot}
$12n$	$n(A_1 \oplus A_2 \oplus 2E \oplus 3F_1 \oplus 3F_2) \oplus A_1$
$12n+1$	$n(A_1 \oplus A_2 \oplus 2E \oplus 3F_1 \oplus 3F_2) \oplus F_1$
$12n+2$	$n(A_1 \oplus A_2 \oplus 2E \oplus 3F_1 \oplus 3F_2) \oplus E \oplus F_2$
$12n+3$	$n(A_1 \oplus A_2 \oplus 2E \oplus 3F_1 \oplus 3F_2) \oplus A_2 \oplus F_1 \oplus F_2$
$12n+4$	$n(A_1 \oplus A_2 \oplus 2E \oplus 3F_1 \oplus 3F_2) \oplus A_1 \oplus E \oplus F_1 \oplus F_2$
$12n+5$	$n(A_1 \oplus A_2 \oplus 2E \oplus 3F_1 \oplus 3F_2) \oplus E \oplus 2F_1 \oplus F_2$
$12n+6$	$n(A_1 \oplus A_2 \oplus 2E \oplus 3F_1 \oplus 3F_2) \oplus A_1 \oplus A_2 \oplus E \oplus F_1 \oplus 2F_2$
$12n+7$	$n(A_1 \oplus A_2 \oplus 2E \oplus 3F_1 \oplus 3F_2) \oplus A_2 \oplus E \oplus 2F_1 \oplus 2F_2$
$12n+8$	$n(A_1 \oplus A_2 \oplus 2E \oplus 3F_1 \oplus 3F_2) \oplus A_1 \oplus 2E \oplus 2F_1 \oplus 2F_2$
$12n+9$	$n(A_1 \oplus A_2 \oplus 2E \oplus 3F_1 \oplus 3F_2) \oplus A_1 \oplus A_2 \oplus 2E \oplus 3F_1 \oplus 2F_2$
$12n+10$	$n(A_1 \oplus A_2 \oplus 2E \oplus 3F_1 \oplus 3F_2) \oplus A_1 \oplus A_2 \oplus 2E \oplus 2F_1 \oplus 3F_2$
$12n+11$	$n(A_1 \oplus A_2 \oplus 2E \oplus 3F_1 \oplus 3F_2) \oplus A_2 \oplus 2E \oplus 3F_1 \oplus 3F_2$

^{a} n is integral.

The MS group of methane $T_d(M)$ (see Table A-11) involves rotations of the molecule about several axes, and for some of the operations of the group the Euler angle transformations are very complicated. For a given value of the J quantum number there are $(2J + 1)^2$ states having m, $k = -J$, $-J + 1, \ldots, +J$ all with the same energy in the rigid rotor approximation. This set of functions will transform as the representation $D^{(J)}$ of the group $K(\text{mol})$. Using the correlation table of the group $K(\text{mol})$ onto the group T_d (see Appendix B) we can reduce the representation $D^{(J)}$ and determine the symmetry species in $T_d(M)$ of each set of $(2J + 1)^2$ functions $|J, k, m\rangle$ for a given value of J. The result is given in Table 10-14 where we have not included the m degeneracy [to include the m degeneracy it is necessary to multiply the species by $(2J + 1)$]. Because of the complicated Euler angle transformations to which some elements of the $T_d(M)$ group give rise, it is not very easy to determine the transformation properties of individual $|J, k, m\rangle$ functions.

THE CLASSIFICATION OF THE VIBRATIONAL WAVEFUNCTIONS

To determine the symmetry species of the vibrational wavefunctions of a molecule in a particular electronic state we begin by determining the symmetry species of the normal coordinates. To determine the symmetry species of the normal coordinates we proceed, as described in Chapter 7, from the symmetry species of the Cartesian displacement coordinates [see, for example, Eqs. (7-245)–(7-250)]. The vibrational wavefunction of a molecule in the harmonic oscillator approximation is the product of harmonic oscillator functions (one for each normal coordinate), and each harmonic oscillator function can be classified in the MS group, once we know the symmetry of the normal coordinate, by using the expressions given in Chapter 8 for the harmonic oscillator wavefunctions. The symmetry of the complete vibrational wavefunction is the product of the symmetries of the individual harmonic oscillator wavefunctions.

The harmonic oscillator wavefunction for a nondegenerate vibration when the vibrational quantum number v is zero is given by [see Eq. (8-159)]

$$\Phi_0 = (\gamma/\pi)^{1/4} \exp(-\gamma Q^2/2), \tag{10-29}$$

where $\gamma = \lambda^{1/2}/\hbar$. The potential energy in the harmonic oscillator approximation is $(\hbar^2\gamma^2/2)Q^2$ and this must be totally symmetric in the MS group since the MS group is a symmetry group of the Hamiltonian. As a result Q^2, and hence the $v = 0$ function Φ_0, must be totally symmetric in the MS group. For a nondegenerate normal coordinate the excited state harmonic oscillator wavefunction Φ_v, with vibrational quantum number v, is given by [see Eq. (8-165)]

$$\Phi_v = N_v'[(-i\hbar/\sqrt{2})\,\partial/\partial Q + (i\hbar\gamma/\sqrt{2})Q]^v\Phi_0 = N_v'(\hat{R}^+)^v\Phi_0, \tag{10-30}$$

where N_v' is a normalization constant. If the normal coordinate Q transforms according to the one-dimensional representation, $\Gamma^{(i)}$, say, of the MS group, the ladder operator $\hat{R}^+$ will also transform according to the representation $\Gamma^{(i)}$. Since $\hat{R}^+$ transforms as $\Gamma^{(i)}$ then $(\hat{R}^+)^v$ and (since Φ_0 is totally symmetric)$\cdot\Phi_v$ will transform as the product $\Gamma^{(i)} \otimes \Gamma^{(i)} \otimes \Gamma^{(i)} \otimes \cdots$ taken v times; for v even this is the totally symmetric representation and for v odd this is $\Gamma^{(i)}$.

A pair of doubly degenerate normal coordinates Q_a and Q_b (which are such that $\gamma_a = \gamma_b = \gamma$ in the potential function) will form the basis for a two-dimensional irreducible representation, $\Gamma^{(j)}$, say, of the MS group. From the discussion in Chapter 8 [see Eqs. (8-223) and (8-224)] we deduce that the vibrational wavefunction for the lowest level ($v = l = 0$) is given by

$$\Psi_{0,0} = (\gamma/\pi)^{1/2} \exp[-(\gamma Q^2/2)], \tag{10-31}$$

and that the excited state vibrational wavefunctions are given by

$$\Psi_{v,l} = N_{v,l}'(\hat{R}^{+(-)})^{(v-l)/2}(\hat{R}^{+(+)})^{(v+l)/2}\Psi_{0,0}, \tag{10-32}$$

where the ladder operators $\hat{R}^{\pm(\pm)}$ are defined in Eq. (8-215) and $Q^2 = Q_a^2 + Q_b^2$. From the invariance of the potential function term $(\hbar^2\gamma^2/2)Q^2$ we deduce that $\Psi_{0,0}$ is totally symmetric in the MS group. Given that $\Psi_{0,0}$ is totally symmetric then, from Eq. (10-32), we see that the $(v+1)$ degenerate functions $\Psi_{v,l}$, where $l = v, v-2, v-4, \ldots, -v$, transform in the same way as the $(v+1)$ operators

$$(\hat{R}^{+(+)})^v, (\hat{R}^{+(-)})(\hat{R}^{+(+)})^{v-1}, (\hat{R}^{+(-)})^2(\hat{R}^{+(+)})^{v-2}, \ldots, (\hat{R}^{+(-)})^v, \tag{10-33}$$

i.e., as the symmetric vth power of the two-dimensional representation generated by the pair of ladder operators $(\hat{R}^{+(+)}, \hat{R}^{+(-)})$ [see Eq. (5-114)]. From Eqs. (8-213)–(8-215) we see that $(\hat{R}^{+(+)} + \hat{R}^{+(-)})$ transforms as Q_a and that $(\hat{R}^{+(+)} - \hat{R}^{+(-)})$ transforms as Q_b, so that $(\hat{R}^{+(+)}, \hat{R}^{+(-)})$ transforms like (Q_a, Q_b), i.e., as $\Gamma^{(j)}$. Thus the $(v+1)$ degenerate functions $\Psi_{v,l}$ transform as the symmetric vth power of the normal coordinate species $\Gamma^{(j)}$; this is written as $[\Gamma^{(j)}]^v$ [see Eq. (5-114)].

For a triply degenerate vibration (v_a, v_b, v_c) with coordinates $Q_a, Q_b,$ and Q_c, and $\gamma_a = \gamma_b = \gamma_c = \gamma$, each level has an energy of

$$E(v_a, v_b, v_c) = [v + \tfrac{3}{2}]\hbar^2\gamma, \tag{10-34}$$

where $v = (v_a + v_b + v_c)$, and each level has a degeneracy of $(v+1)(v+2)/2$ as

$$(v_a, v_b, v_c) = (v, 0, 0), (v-1, 1, 0), (v-1, 0, 1), \ldots, (0, 0, v).$$

The $v = 0$ function is totally symmetric and the three $v = 1$ functions transform as the normal coordinates (Q_a, Q_b, Q_c), i.e., as the three-dimensional irreducible representation, F, say, of the MS group. For v greater than one, the symmetry of the $(v + 1)(v + 2)/2$ independent functions is the symmetric vth power of the normal coordinate representation and this is written $[F]^v$. For a three-dimensional species F the character under the operation R in the representation $[F]^v$ is given by [see Section 7-3 in Wilson, Decius, and Cross (1955)]

$$\chi^{[F]^v}[R] = \tfrac{1}{3}[2\chi^F[R]\chi^{[F]^{v-1}}[R] + \tfrac{1}{2}\{\chi^F[R^2] - (\chi^F[R])^2\}\chi^{[F]^{v-2}}[R] + \chi^F[R^v]]. \tag{10-35}$$

This is an extension of Eq. (5-114) to a three-dimensional representation.

As an example we will determine the symmetry of a vibrational state of methane. To determine the symmetry species of the normal coordinates we use the internal coordinate displacements $\mathfrak{R}_i$ [see Eqs. (8-21)–(8-27)] as follows

$$\Delta r_1, \Delta r_2, \Delta r_3, \Delta r_4, \qquad \Delta\alpha_{12}, \Delta\alpha_{13}, \Delta\alpha_{14}, \Delta\alpha_{23}, \Delta\alpha_{24}, \Delta\alpha_{34}, \tag{10-36}$$

and the redundant combination of these coordinates is

$$R_{\text{red}} = \sum_{i<j} \Delta\alpha_{ij} = 0, \tag{10-37}$$

where $\Delta\alpha_{ij}$ is an increase in the bond angle H_iCH_j and Δr_i is an increase in the bond length CH_i. The transformation properties of these coordinates are given in Table 10-15. By subtracting the species of R_{red} (i.e., A_1) from that of the ten displacement coordinates in Eq. (10-36) we see that the normal coordinate representation is

$$\Gamma(Q_r) = A_1 \oplus E \oplus 2F_2 \tag{10-38}$$

Table 10-15

Transformation properties of the internal coordinates of methane in $T_d(M)$

E	(123)	(14)(23)	(1423)*	(23)*
Δr_1	Δr_3	Δr_4	Δr_3	Δr_1
Δr_2	Δr_1	Δr_3	Δr_4	Δr_3
Δr_3	Δr_2	Δr_2	Δr_2	Δr_2
Δr_4	Δr_4	Δr_1	Δr_1	Δr_4
$\Delta\alpha_{12}$	$\Delta\alpha_{13}$	$\Delta\alpha_{34}$	$\Delta\alpha_{34}$	$\Delta\alpha_{13}$
$\Delta\alpha_{13}$	$\Delta\alpha_{23}$	$\Delta\alpha_{24}$	$\Delta\alpha_{23}$	$\Delta\alpha_{12}$
$\Delta\alpha_{14}$	$\Delta\alpha_{34}$	$\Delta\alpha_{14}$	$\Delta\alpha_{13}$	$\Delta\alpha_{14}$
$\Delta\alpha_{23}$	$\Delta\alpha_{12}$	$\Delta\alpha_{23}$	$\Delta\alpha_{24}$	$\Delta\alpha_{23}$
$\Delta\alpha_{24}$	$\Delta\alpha_{14}$	$\Delta\alpha_{13}$	$\Delta\alpha_{14}$	$\Delta\alpha_{34}$
$\Delta\alpha_{34}$	$\Delta\alpha_{24}$	$\Delta\alpha_{12}$	$\Delta\alpha_{12}$	$\Delta\alpha_{24}$

and we can write the species of the normal coordinates as

$$\Gamma^{(1)} = A_1, \qquad \Gamma^{(2)} = E, \qquad \Gamma^{(3)} = F_2, \qquad \text{and} \qquad \Gamma^{(4)} = F_2. \quad (10\text{-}39)$$

A vibrational energy level is described by the values of the four quantum numbers v_1, v_2, v_3, and v_4, e.g., $(v_1, v_2, v_3, v_4) = (3, 2, 1, 2)$. Let us determine the symmetry of the vibrational state $(3, 2, 1, 2)$ in the $T_d(M)$ group for methane (see Table A-11). We first determine the symmetry species of the vibrational wavefunctions of each of the four modes and then multiply these species together. For the v_1 mode of species A_1 we have $v_1 = 3$ and

$$\Gamma(\Phi_{v_1 = 3}) = A_1 \otimes A_1 \otimes A_1 = A_1. \quad (10\text{-}40)$$

For the v_2 mode we have [using Eq. (5-112) for $[E]^2 = [E \otimes E]$]

$$\Gamma(\Phi_{v_2 = 2}) = [E]^2 = A_1 \oplus E, \quad (10\text{-}41)$$

and similarly

$$\Gamma(\Phi_{v_3 = 1}) = F_2, \quad (10\text{-}42)$$

and [using Eq. (10-35)]

$$\Gamma(\Phi_{v_4 = 2}) = [F_2]^2 = A_1 \oplus E \oplus F_2. \quad (10\text{-}43)$$

Thus

$$\Gamma(\Phi_{v_1 = 3}\Phi_{v_2 = 2}\Phi_{v_3 = 1}\Phi_{v_4 = 2}) = A_1 \otimes (A_1 \oplus E) \otimes F_2 \otimes (A_1 \oplus E \oplus F_2)$$
$$= 2A_1 \oplus A_2 \oplus 3E \oplus 7F_1 \oplus 8F_2. \quad (10\text{-}44)$$

In the general case of a molecule with normal coordinates of species $\Gamma^{(1)}, \Gamma^{(2)}, \ldots, \Gamma^{(f)}$ in a vibrational state with quantum numbers $v_1, v_2, \ldots, v_f$ the species of the vibrational wavefunctions is

$$\Gamma(v_1, v_2, \ldots, v_f) = [\Gamma^{(1)}]^{v_1} \otimes [\Gamma^{(2)}]^{v_2} \otimes \cdots \otimes [\Gamma^{(f)}]^{v_f}, \quad (10\text{-}45)$$

where $[\ ]^v$ is the symmetric vth product for a degenerate species and the ordinary vth product for a nondegenerate species.

THE CLASSIFICATION OF THE ELECTRONIC
ORBITAL WAVEFUNCTIONS

The electronic (orbital) wavefunctions Φ_{elec} of a molecule can be written as the sum of single configuration wavefunctions Φ_{eo}, each of which is the product of molecular orbital functions ϕ_k [see Eqs. (8-16) and (8-17)]. The molecular orbital wavefunctions can be written as a linear combination of atomic orbital functions [see Eq. (8-15)]. The symmetry of the electronic orbital wavefunctions is given by the symmetries of the molecular orbitals and these, in turn, can be obtained from the transformation properties of the atomic orbitals. Under the effect of an element of the MS group the atomic orbitals transform in a straightforward way and we will use the water molecule as a convenient example.

For the water molecule we consider molecular orbitals constructed from the following limited set of atomic orbitals centered on the oxygen nucleus, $1s(O), 2s(O), 2p_x(O), 2p_y(O), 2p_z(O)$, together with the atomic orbitals $1s(H_1)$ and $1s(H_2)$ centered on the protons. The effect of the MS group operation (12) is shown in Fig. 10-9 and we see that

$$(12)2p_z(O) = -2p_z(O) \qquad (10\text{-}46)$$

since although the electrons are unaffected by (12) the molecule fixed z axis is reversed. We can determine that the following combinations of atomic orbitals transform irreducibly:

$$A_1 \begin{cases} 1s(O) \\ 2s(O) \\ 2p_x(O) \\ [1s(H_2) + 1s(H_1)] \end{cases} , \quad B_1 \quad 2p_y(O), \quad B_2 \begin{cases} 2p_z(O) \\ [1s(H_2) - 1s(H_1)] \end{cases}. \qquad (10\text{-}47)$$

These symmetry adapted combinations of atomic orbitals are called *symmetry orbitals* (SO) and MO's written as LCAO's [see Eq. (8-12)] can only consist of one symmetry type of SO as a result of the vanishing integral rule [Eq. (5-133)].

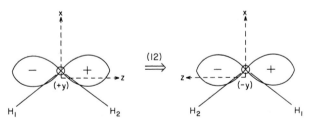

Fig. 10-9. The effect of the operation (12) for the water molecule with the atomic orbital $2p_z(O)$ marked.

To solve the SCF equations for the water molecule we build up Hamiltonian matrices for each of the above three representations and diagonalize them separately. To do this we need to know how many orbitals of a given symmetry type are occupied and to do that we argue as follows. The lowest lying MO will obviously be largely a 1s orbital on the oxygen nucleus, $(1a_1) = 1s(O)$. The two SO's of B_2 symmetry will form a bonding and an antibonding pair of MO's qualitatively given by

$$(1b_2) = 2p_z(O) + [1s(H_2) - 1s(H_1)]$$

and   (10-48)

$$(1b_2{}^*) = 2p_z(O) - [1s(H_2) - 1s(H_1)].$$

The 2s(O) and $2p_x(O)$ orbitals will each similarly form bonding and anti-bonding MO's with the $[1s(H_2) + 1s(H_1)]$ SO; the bonding orbitals are qualitatively given by

$$(2a_1) = 2s(O) + [1s(H_2) + 1s(H_1)]$$

and   (10-49)

$$(3a_1) = 2p_x(O) + [1s(H_2) + 1s(H_1)].$$

Finally the $2p_y(O)$ orbital will give a nonbonding orbital of symmetry B_1:

$$(1b_1) = 2p_y(O).\qquad(10\text{-}50)$$

Filling the bonding and nonbonding MO's we obtain the electronic con-figuration of the ground state of H_2O (called the $\tilde{X}$ state) as

$$(1a_1)^2(2a_1)^2(1b_2)^2(3a_1)^2(1b_1)^2.\qquad(10\text{-}51)$$

The orbitals are written in order of increasing energy and the symmetry of the complete electronic orbital wavefunction is given by

$$(A_1)^2 \otimes (A_1)^2 \otimes (B_2)^2 \otimes (A_1)^2 \otimes (B_1)^2 = A_1.\qquad(10\text{-}52)$$

The symmetry of the complete electronic orbital wavefunction is built up from the MO symmetries and the occupation numbers (0, 1, or 2 from the Pauli exclusion principle) in a way similar to that in which we build up the symmetries of complete vibrational wavefunctions [see Eq. (10-45)]. Excited electronic state wavefunctions are obtained by promoting one or more of the electrons out of these MO's into MO's of higher energy. For example, the excited $\tilde{A}$ state of water is obtained by promoting an electron out of the $(1b_1)$ nonbonding orbital and into the $(3sa_1)$ orbital (largely an atomic 3s(O) orbital) to give the configuration

$$(1a_1)^2(2a_1)^2(1b_2)^2(3a_1)^2(1b_1)^1(3sa_1)^1\qquad(10\text{-}53)$$

which has symmetry B_1.

The symmetry of the lowest electronic state of a molecule is not always clear from simple MO arguments such as those just presented. However, the nuclear spin statistical weights of the rotational levels of a state depend on the electronic symmetry (as well as on the rotational and vibrational symmetry) and the relative intensities of the rotational lines in a spectrum depend on the statistical weights. Thus the experimental determination of these relative intensities can sometimes enable the symmetry of the lowest electronic state to be determined [see, for example, Carlotti, Johns, and Trombetti (1974)].

THE CLASSIFICATION OF THE ELECTRON SPIN WAVEFUNCTIONS

It often happens, as for example in the ground electronic states of the CH and NH_2 molecules, that the magnetic coupling of the electron spin to the electron and nuclear orbital motions is small. In these circumstances it is appropriate to use electron spin basis wavefunctions that are quantized along space fixed axes in the same manner that nuclear spin functions were used at the beginning of this chapter. Such a situation is termed *Hund's case (b)* and basis functions in which the electron spin functions are quantized along space fixed axes are termed Hund's case (b) functions. Sometimes spin–orbit coupling is strong so that the electron spin is "tied" to the molecule fixed axis system (for low values of the rotational quantum numbers at least). In these circumstances the wavefunctions are better represented using electron spin functions that are quantized along molecule fixed axes;[2] this is *Hund's case (a)* and a Hund's case (a) basis. We must be able to classify each type of basis set function in the MS group.

Hund's Case (b) Basis Functions

An electron spin function quantized along space fixed axes is unaffected either by any permutation of identical nuclei or by the inversion operation E^*, so that it is totally symmetric in the MS group. Hence, when spin–orbit coupling is small, so that a Hund's case (b) basis set is appropriate, the symmetry species of the molecular wavefunctions can be determined without considering the spin functions. The effect of including electron spin in the wavefunction, and of constructing electronic wavefunctions that satisfy the Pauli exclusion principle, is merely to add a multiplicity label $2S + 1$ as a superscript to the species of the electronic wavefunction. For example, for the states discussed for the water molecule we have

$$(1a_1)^2(2a_1)^2(1b_2)^2(3a_1)^2(1b_1)^2 \qquad {}^1A_1$$

and   (10-54)

$$(1a_1)^2(2a_1)^2(1b_2)^2(3a_1)^2(1b_1)^1(3sa_1)^1 \qquad {}^1B_1 \quad \text{and} \quad {}^3B_1.$$

[2] In a Hund's case (a) basis $\hat{T}_N$ [see Eq. (7-137)] involves $\hat{L}_\alpha + \hat{S}_\alpha$ and $\hat{L}_\beta + \hat{S}_\beta$ rather than $\hat{L}_\alpha$ and $\hat{L}_\beta$, respectively.

When Hund's case (b) functions are used the quantum numbers J, k, and m associated with the solution of the rotational wave equation describe, respectively, the total angular momentum of the molecule *excluding* electron spin and its projections along the molecule fixed z axis and the space fixed ζ axis direction. For nonsinglet states in case (b) we use N and m_N rather than J and m since J is reserved for total angular momentum *including* electron spin. The quantum numbers F and m_F are reserved for total angular momentum, and its component in the ζ axis direction, including both electron and nuclear spin.

Hund's Case (a) Basis Functions

Molecule fixed electron spin functions can be classified in the group K(mol) and will have the species $D^{(S)}$ where $S(S + 1)\hbar^2$ is the eigenvalue of $\hat{S}^2$. To determine the species of the electron spin functions in the MS group we use the correlation table[3] of K(mol) to the MS group (see Table B-2). For integral values of S this presents no problems. When S is half-integral (i.e., for a molecule with an odd number of electrons) the classification of the spin functions in K(mol) and in the MS group presents problems, but before getting involved in these extra complications we will complete the general argument and apply it to an example in which the molecule has an even number of electrons.

When Hund's case (a) electron spin functions are used the quantum numbers J, p, and m_J associated with the solution of the rotational equation equation describe, respectively, the total angular momentum of the molecule including electron spin, its projection along the molecule fixed z axis, and its projection along the space fixed ζ axis direction. We will now consider an example of a molecule with an even number of electrons and show how the symmetry classifications in the Hund's case (a) and (b) limits are made.

We consider a C_{3v}(M) pyramidal XY_3Z molecule, such as CH_3F, in an electronic state of symmetry $^3A_2$ and in a totally symmetric, A_1, vibrational state. The species of the rovibronic–electron spin wavefunctions Φ_{rves} are obtained by multiplying the species of the rotational, vibrational, electronic orbital, and electron spin wavefunctions.

In the Hund's case (b) limit for the CH_3F example the rotational species Γ_{rot} are obtained from Table 10-7, with the replacement of J by N, and these are given on the left in Fig. 10-10. The vibrational species is A_1 and the electronic orbital species is A_2 so that the rovibronic species Γ_{rve} are obtained by multiplying Γ_{rot} by A_2. The rovibronic species are given down the center of Fig. 10-10. The triplet electron spin functions are each of symmetry A_1 in

---

[3] This involves using the equivalent rotations of the MS group and reducing K(mol) onto the equivalent rotation group.

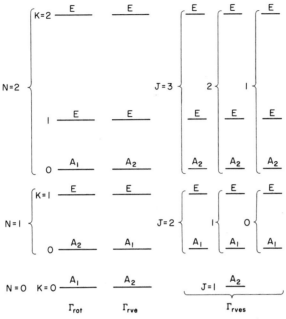

Fig. 10-10. The symmetry labels Γ_{rves} for the rovibronic–electron spin energy levels of the CH_3F molecule in a $^3A_2$ vibronic state using Hund's case (b) electron spin functions and the group $C_{3v}(M)$.

case (b) so that the rovibronic–electron spin species Γ_{rves} are as given on the right in Fig. 10-10, and we have $J = N + 1, N, N - 1$ for a triplet state (except for $N = 0$ when $J = 1$ only).

In a Hund's case (a) basis for the CH_3F molecule the electron spin species is obtained first from the reduction[4] of $D^{(1)}$ of $K(mol)$ onto $C_{3v}(M)$, and this gives $\Gamma_{espin} = A_2 \oplus E$. Multiplying by the electronic orbital species A_2 we obtain the electronic spin–orbit species Γ_{eso} as $A_1 \oplus E$ and these two spin–orbit states are split apart by spin–orbit coupling as indicated on the left in Fig. 10-11; in Hund's case (a) this splitting is much larger than the rotational energy level spacings. Multiplying the species Γ_{eso} by the vibrational species A_1 and then by the rotational species Γ_{rot} (obtained from Table 10-7 with the replacement of K by $P = |p|$) we obtain the Γ_{rves} species as given on the right in Fig. 10-11.

As we see by comparing the right hand sides of Figs. 10-10 and 10-11 the case (a) and case (b) rovibronic–electron spin energy level patterns are quite different. In the case (b) limit (Fig. 10-10) the rovibronic–electron spin energy

---

[4] This involves using the equivalent rotations of the MS group and reducing $K(mol)$ onto the equivalent rotation group.

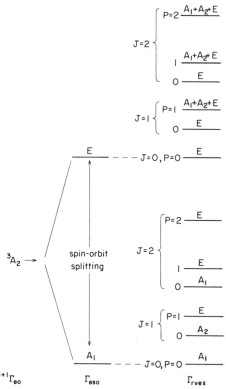

Fig. 10-11. The symmetry labels Γ_{rves} for the rovibronic–electron spin energy levels of the CH_3F molecule in a $^3A_2$ electronic state (and an A_1 vibrational state) using Hund's case (a) electron spin functions and the group $C_{3v}(M)$.

level pattern is like that of a singlet state but every energy level (except that with $N = 0$) is a triplet; a small spin–orbit interaction would make these triplets resolvable. In the case (a) limit the energy level pattern is that of two distinct electronic states (one of which is degenerate). Regardless of the extent of the spin–orbit interaction the rovibronic–electron spin Hamiltonian commutes with the MS group and with K(mol) so the symmetry labels J and Γ_{rves} remain valid. Thus we can correlate the case (a) and case (b) rovibronic–electron spin energy levels, maintaining J and Γ_{rves}. The reader can test for himself that this can be done and, for example, for $J = 1$ we have the case (a) and case (b) symmetries as follows:

$$\Gamma_{\text{rves}}(\text{case (b)}) = A_2 \oplus A_1 \oplus E \oplus A_2 \oplus E \oplus E = A_1 \oplus 2A_2 \oplus 3E \qquad (10\text{-}55)$$

and

$$\Gamma_{\text{rves}}(\text{case (a)}) = A_2 \oplus E \oplus E \oplus (A_1 \oplus A_2 \oplus E) = A_1 \oplus 2A_2 \oplus 3E. \qquad (10\text{-}56)$$

Spin Double Groups

When the number of electrons is odd the value of S is half-integral and a special discussion of the representation $D^{(S)}$ of K is required. It is also necessary to give a special discussion of the MS groups, and of the correlation of $D^{(S)}$ to the species of MS groups, when S is half-integral.

Let us consider the implication of taking the elements of the matrices in the representations $D^{(j)}$ of K, when j is half-integral, to be as given when j is integral by combining Eqs. (8-73a) and (8-67). The character of this $D^{(j)}$ under an operation involving a rotation through an angle ε is given by [see Tinkham (1964), page 109 ; or Hamermesh (1962), page 355 ; or Wigner (1959), Eq. (15-28)]

$$\chi^{(j)}(\varepsilon) = \frac{\sin[(2j + 1)\varepsilon/2]}{\sin(\varepsilon/2)}. \qquad (10\text{-}57)$$

The character for a rotation through an angle $\varepsilon + 2\pi$ radians is thus given by

$$\chi^{(j)}(\varepsilon + 2\pi) = \frac{\sin[(2j + 1)\varepsilon/2 + (2j + 1)\pi]}{\sin[(\varepsilon/2) + \pi]}. \qquad (10\text{-}58)$$

For j integral or half-integral we can write

$$\chi^{(j)}(\varepsilon + 2\pi) = (-1)^{2j}\chi^{(j)}(\varepsilon) \qquad (10\text{-}59)$$

and also

$$\chi^{(j)}(\varepsilon + 4\pi) = \chi^{(j)}(\varepsilon). \qquad (10\text{-}60)$$

From Eq. (10-59) we see that if j is half-integral then

$$\chi^{(j)}(\varepsilon + 2\pi) = -\chi^{(j)}(\varepsilon). \qquad (10\text{-}61)$$

Since a rotation through 2π is equivalent to no rotation at all the character in this $D^{(j)}$, when j is half-integral, under each operation in K is ambiguous with regard to its sign; the character is *double valued*. It would appear that we cannot set up representations for half-integral j using these $D^{(j)}$ since we do not have $D[P_1]D[P_2] = D[P_{12}]$ but only $D[P_1]D[P_2] = \pm D[P_{12}]$ [see Eq. (5-155)]. We deal with this situation by introducing the fictitious operation R which is a rotation through 2π but which is not considered to be the identity. In this way we double the number of operations in K and we call the group K^2, *the spin double group of the three-dimensional rotation group*. In this group the rotation through an angle $\varepsilon + 2\pi$ is considered distinct from a rotation through ε and Eq. (10-61) no longer results in a sign ambiguity since ε and $\varepsilon + 2\pi$ are considered as different operations. The character of a representation matrix in $D^{(j)}$ having j half-integral will have opposite sign under a rotation through $\varepsilon + 2\pi$ from its value under a rotation through ε; this representation is a single valued representation of the spin double group $\boldsymbol{K}^2$ or a (so-called) double valued representation of the group $\boldsymbol{K}$. A representation $D^{(j)}$ for which j is integral will have the same character under a rotation through ε as under a rotation through $\varepsilon + 2\pi$ and constitutes a single valued representation (i.e., a true representation) of the group $\boldsymbol{K}$. In the group $\boldsymbol{K}^2$ a rotation through $\varepsilon + 4\pi$ is equivalent to a rotation of ε about the axis and R^2 is the identity.

We now discuss the *spin double groups* of the MS groups, and these are obtained by doubling the number of elements in each MS group in the same way that we doubled the number of elements in $\boldsymbol{K}$ to obtain $\boldsymbol{K}^2$. To do this we must include the operation R, which is a rotation through 2π, in the MS group. Thus for each permutation or permutation–inversion operation O that occurs in an MS group, we add the operation RO which is defined as causing the same permutation (or permutation–inversion) of the nuclei as O but which causes a rotation of 2π in addition to the rotation caused by O. The effect of RO on the spatial coordinates of the nuclei and electrons is the same as the effect of O.

Let us consider the equivalent rotations (see Table 7-1) of permutation or permutation–inversion elements O, of the operation R, and of the combined operations RO. In the ordinary MS group, in which a rotation through 2π radians is the identity, we can write the effect of the equivalent rotation of each MS group element using the following equations

$$R_z^{\beta}(\theta, \phi, \chi) = (\theta, \phi, \chi + \beta) \tag{10-62}$$

and

$$R_\alpha^{\pi}(\theta, \phi, \chi) = (\pi - \theta, \phi + \pi, 2\pi - 2\alpha - \chi), \tag{10-63}$$

where $0 \le \theta \le \pi$, $0 \le \phi \le 2\pi$, and $0 \le \chi \le 2\pi$. In a spin double group of an MS group the rotation through 2π radians is not the identity. We can allow for this by taking the ranges of the Euler angles as $0 \le \theta \le \pi$, $0 \le \phi \le 4\pi$, and $0 \le \chi \le 4\pi$ (where ϕ and χ are mod 4π), and by taking

$$R^0(\theta, \phi, \chi) = (\theta, \phi, \chi) \text{ or } (\theta, \phi + 2\pi, \chi + 2\pi),$$
$$R_z^{\beta}(\theta, \phi, \chi) = (\theta, \phi, \chi + \beta) \text{ or } (\theta, \phi + 2\pi, \chi + 2\pi + \beta), \tag{10-64}$$
$$R_\alpha^{\pi}(\theta, \phi, \chi) = (\pi - \theta, \phi + \pi, 2\pi - 2\alpha - \chi) \text{ or } (\pi - \theta, \phi + 3\pi, 4\pi - 2\alpha - \chi);$$

$$R^{2\pi}(\theta, \phi, \chi) = (\theta, \phi, \chi + 2\pi) \text{ or } (\theta, \phi + 2\pi, \chi),$$
$$R^{2\pi}R_z^{\beta}(\theta, \phi, \chi) = (\theta, \phi, \chi + 2\pi + \beta) \text{ or } (\theta, \phi + 2\pi, \chi + \beta), \tag{10-65}$$
$$R^{2\pi}R_\alpha^{\pi}(\theta, \phi, \chi) = (\pi - \theta, \phi + \pi, 4\pi - 2\alpha - \chi) \text{ or } (\pi - \theta, \phi + 3\pi, 2\pi - 2\alpha - \chi).$$

In this treatment there are two Euler angle transformations for each equivalent rotation but this does not present any difficulty in applications, as we will now see.

For example, the spin double group of the MS group of CH_3F is called $C_{3v}(M)^2$ and it consists of the following 12 elements (see Table A-8):

$$\{E, (123), (132), (12)^*, (23)^*, (13)^*, R, R(123), R(132), R(12)^*, R(23)^*, R(13)^*\}.$$
(10-66)

The equivalent rotations of these 12 elements are, respectively,

$$R^0, R_z^{2\pi/3}, R_z^{4\pi/3}, R_{\pi/6}^{\pi}, R_{\pi/2}^{\pi}, R_{5\pi/6}^{\pi}, R^{2\pi}, R^{2\pi}R_z^{2\pi/3},$$
$$R^{2\pi}R_z^{4\pi/3}, R^{2\pi}R_{\pi/6}^{\pi}, R^{2\pi}R_{\pi/2}^{\pi}, R^{2\pi}R_{5\pi/6}^{\pi}.$$
(10-67)

For convenience in the character tables of Appendix A we write the equivalent rotations $R^{2\pi}R_z^{\beta}$ and $R^{2\pi}R_\alpha^{\pi}$ as $R_z^{\beta+2\pi}$ and $R_\alpha^{3\pi}$, respectively.

The spin double group of an MS group is a group and its character table can be set up. In Appendix A the character tables of the spin double groups of the MS groups are given; the normal MS group character table is to the left and above the dashed line divisions in each case. To show how the character tables of the spin double groups of the MS groups are determined we consider the groups $C_{2v}(M)^2$ and $C_{3v}(M)^2$ as examples.

$C_{2v}(M)^2$ The operations of the $C_{2v}(M)^2$ group are

$$E, (12), E^*, (12)^*, R, R(12), RE^*, \text{ and } R(12),$$
(10-68)

and the equivalent rotations of these elements for the H_2O molecule are (see Fig. 10-2)

$$R^0, R_b^{\pi}, R_c^{\pi}, R_a^{\pi}, R^{2\pi}, R_b^{3\pi}, R_c^{3\pi}, \text{ and } R_a^{3\pi},$$
(10-69)

respectively. The Euler angle transformations (using the I^r representation) are given in Table 10-16 [only one of the two possibilities from Eqs. (10-64) and (10-65) are given]. The equivalent rotations in Eq. (10-69) (or the results in Table 10-16) are required in the calculation of the multiplication table of the group elements. For example, as far as the effect of the elements on the coordinates of the nuclei and electrons in space is concerned we can write

Table 10-16

Transformation properties of the Euler angles of H_2O^a for the elements of the group $C_{2v}(M)^2$

E	(12)	E^*	$(12)^*$	R	$R(12)$	RE^*	$R(12)^*$
θ	$\pi - \theta$	$\pi - \theta$	θ	θ	$\pi - \theta$	$\pi - \theta$	θ
ϕ	$\phi + \pi$	$\phi + \pi$	ϕ	ϕ	$\phi + \pi$	$\phi + \pi$	ϕ
χ	$2\pi - \chi$	$\pi - \chi$	$\chi + \pi$	$\chi + 2\pi$	$4\pi - \chi$	$3\pi - \chi$	$\chi + 3\pi$

^{a} A I^r convention is used; ϕ and χ are mod 4π and $0 \leq \theta \leq \pi$.

the product E^* times $(12)^*$ as

$$E^*(12)^* = (12) \text{ or } R(12), \qquad (10\text{-}70)$$

but the Euler angle transformations distinguish between these two possibilities. We can write

$$E^*(12)^*(\theta,\phi,\chi) = E^*(\theta',\phi',\chi') = (\pi - \theta',\phi' + \pi, \pi - \chi')$$
$$= (\pi - \theta, \phi + \pi, -\chi), \qquad (10\text{-}71)$$

where

$$(12)^*(\theta,\phi,\chi) = (\theta',\phi',\chi') = (\theta,\phi,\chi + \pi). \qquad (10\text{-}72)$$

From Eq. (10-71) and Table (10-16) we see that

$$E^*(12)^* = R(12) \qquad (10\text{-}73)$$

since $(4\pi - \chi)$ and $-\chi$ are equivalent (χ is defined mod 4π in the spin double group). The multiplication table[5] of the elements of the $C_{2v}(M)^2$ group, derived by considering equations such as the preceding, is shown in Table 10-17. From the results in Table 10-17 we can determine the class structure of the elements of the $C_{2v}(M)^2$ group [see Eq. (4-47)], and this is given in Table 10-18. We are now in a position to use Eqs. (4-61) and (4-62) to determine the characters of the irreducible representations of $C_{2v}(M)^2$. Following Eq. (4-61) the following class products can be determined from the results in Tables 10-17 and 10-18:

$$C_5{}^2 = C_1 \qquad (10\text{-}74)$$
$$C_2{}^2 = 2C_1 + 2C_5, \qquad (10\text{-}75)$$
$$C_3{}^2 = 2C_1 + 2C_5, \qquad (10\text{-}76)$$

and

$$C_2 C_3 = 2C_4. \qquad (10\text{-}77)$$

For a two-dimensional representation of $C_{2v}(M)^2$ the character under E is 2, i.e., $\chi[E] = \chi_1 = 2$, and the preceding equations, in conjunction with Eq. (4-62), lead to the following relationships among the characters of a two-dimensional representation:

$$\chi_5{}^2 = 4, \qquad (10\text{-}78)$$
$$\chi_2{}^2 = 2 + \chi_5, \qquad (10\text{-}79)$$
$$\chi_3{}^2 = 2 + \chi_5, \qquad (10\text{-}80)$$

[5] In a spin double group E^*P and PE^* are not necessarily the same and $(E^*)^2 = R$.

Table 10-17

Multiplication table for the elements of $C_{2v}(M)^2$ obtained using the Euler angle transformations given in Table 10-16

	E	(12)	E^*	$(12)^*$	R	$R(12)$	RE^*	$R(12)^*$
E:	E	(12)	E^*	$(12)^*$	R	$R(12)$	RE^*	$R(12)^*$
(12):	(12)	R	$R(12)^*$	E^*	$R(12)$	E	$(12)^*$	RE^*
E^*:	E^*	$(12)^*$	R	$R(12)$	RE^*	$R(12)^*$	E	(12)
$(12)^*$:	$(12)^*$	RE^*	(12)	R	$R(12)^*$	E^*	$R(12)$	E
R:	R	$R(12)$	RE^*	$R(12)^*$	E	(12)	E^*	$(12)^*$
$R(12)$:	$R(12)$	E	$(12)^*$	RE^*	(12)	R	$R(12)^*$	E^*
RE^*:	RE^*	$R(12)^*$	E	(12)	E^*	$(12)^*$	R	$R(12)$
$R(12)^*$:	$R(12)^*$	E^*	$R(12)$	E	$(12)^*$	RE^*	(12)	R

Table 10-18

The class structure of the group $C_{2v}(M)^2$

C_1	C_2	C_3	C_4	C_5
E	(12) $R(12)$	E^* RE^*	$(12)^*$ $R(12)^*$	R

Table 10-19

The determination of the characters of the two-dimensional representations of $C_{2v}(M)^2$ from Eqs. (10-78)–(10-81).

Equation	Character	Value				
	χ_1	$+2$				
Eq. (10-78)	χ_5	$+2$				-2
Eq. (10-79)	χ_2	$+2$		-2		0
Eq. (10-80)	χ_3	$+2$	-2	$+2$	-2	0
Eq. (10-81)	χ_4	$+2$	-2	-2	$+2$	0

and

$$\chi_2\chi_3 = 2\chi_4. \tag{10-81}$$

The characters for the two-dimensional representations of $C_{2v}(M)^2$ that follow from these equations are given in Table 10-19. The two-dimensional representations thus have characters as given in Table 10-20. From Eq. (4-44)

Table 10-20

Characters of two-dimensional
representations of $C_{2v}(M)^2$ obtained
from Table 10-19

	C_1	C_2	C_3	C_4	C_5
Γ_1:	2	2	2	2	2
Γ_2:	2	2	-2	-2	2
Γ_3:	2	-2	2	-2	2
Γ_4:	2	-2	-2	2	2
Γ_5:	2	0	0	0	-2

we see that the representation Γ_5 in Table 10-20 is irreducible but that Γ_1 to Γ_4 are reducible. The reduction of Γ_1 to Γ_4 is obvious (they are each twice an irreducible representation) and the character table (see Table A-4) follows.

$C_{3v}(M)^2$ From the definitions given in Eq. (10-67) we can draw up Table 10-21 for the effect of each element of $C_{3v}(M)^2$ on the Euler angles. Using these results the multiplication table for the group elements can be determined and this is given in Table 10-22. Since $R^2 = E$, and since R commutes with all the elements of an MS group, we only need to construct part of the multiplication table as done in Table 10-22. Using the multiplication table we can determine the class structure of the group (see Table 10-23) and determine the following class multiplication results:

$$C_4{}^2 = C_1,$$
$$C_2 C_4 = C_5,$$
$$C_2 C_5 = C_2 + 2C_4,$$
$$C_3{}^2 = 3C_4 + 3C_2,$$
$$C_3 C_4 = C_6.$$

(10-82)

For the two-dimensional representations of $C_{3v}(M)^2$ we have $\chi[E] = \chi_1 = 2$ and the preceding equations lead to the following equations involving the characters (using Eq. (4-62))

$$\chi_4{}^2 = 4,$$
$$\chi_2 \chi_4 = 2\chi_5,$$
$$\chi_2 \chi_5 = \chi_2 + \chi_4,$$
$$\chi_3{}^2 = 2(\chi_4 + 2\chi_2)/3,$$
$$\chi_6 = \chi_3 \chi_4/2.$$

(10-83)

From these equations we deduce the characters of the two-dimensional

Table 10-21
The definition of the effect of each operation of the $C_{3v}(M)^2$ group
on the Euler angles[a]

E	(123)	(132)	(12)*	(23)*	(13)*
θ	θ	θ	$\pi - \theta$	$\pi - \theta$	$\pi - \theta$
ϕ	ϕ	ϕ	$\phi + \pi$	$\phi + \pi$	$\phi + \pi$
χ	$\chi + 2\pi/3$	$\chi + 4\pi/3$	$5\pi/3 - \chi$	$\pi - \chi$	$\pi/3 - \chi$
R	$R(123)$	$R(132)$	$R(12)^*$	$R(23)^*$	$R(13)^*$
θ	θ	θ	$\pi - \theta$	$\pi - \theta$	$\pi - \theta$
ϕ	ϕ	ϕ	$\phi + \pi$	$\phi + \pi$	$\phi + \pi$
$\chi + 2\pi$	$\chi + 8\pi/3$	$\chi + 10\pi/3$	$11\pi/3 - \chi$	$3\pi - \chi$	$7\pi/3 - \chi$

[a] ϕ and χ are mod 4π and $0 \le \theta \le \pi$.

Table 10-22
Part of the multiplication table of the
elements of the $C_{3v}(M)^2$ group[a]

	E	(123)	(132)	(12)*	(23)*	(13)*
E:	E	(123)	(132)	(12)*	(23)*	(13)*
(123):	(123)	(132)	R	$R(13)^*$	(12)*	(23)*
(132):	(132)	R	$R(123)$	$R(23)^*$	$R(13)^*$	(12)*
(12)*:	(12)*	(23)*	(13)*	R	$R(123)$	$R(132)$
(23)*:	(23)*	(13)*	$R(12)^*$	(132)	R	$R(123)$
(13)*:	(13)*	$R(12)^*$	$R(23)^*$	(123)	(132)	R

[a] Since $R^2 = E$, and since R commutes with all the elements of the group, the rest of the multiplication table can be easily derived from these results.

Table 10-23
The class structure of the $C_{3v}(M)^2$ group

C_1	C_2	C_3	C_4	C_5	C_6
E	(123)	(12)*	R	$R(123)$	$R(12)^*$
	$R(132)$	$R(23)^*$		(132)	(23)*
		(13)*			$R(13)^*$

representations to be as given in Table 10-24. The characters of the irreducible representations can be deduced from the results in Table 10-24 and these are given in Table 10-25. In the character table of $C_{3v}(M)^2$ given in Table A-8 the pair of separably degenerate irreducible representations are considered together as $E_{3/2}$ (see Tables 5-4 and 6-2).

Table 10-24

Characters of the two-dimensional
representations of $C_{3v}(M)^2$ as obtained
from the results in Eq. (10-83)

C_1	C_2	C_3	C_4	C_5	C_6
2	2	2	2	2	2
2	2	-2	2	2	-2
2	-1	0	2	-1	0
2	-2	$2i$	-2	2	$-2i$
2	-2	$-2i$	-2	2	$2i$
2	1	0	-2	-1	0

Table 10-25

Characters of the irreducible representations of
$C_{3v}(M)^2$ as deduced from the results in
Table 10-24

	C_1	C_2	C_3	C_4	C_5	C_6
A_1:	1	1	1	1	1	1
A_2:	1	1	-1	1	1	-1
E:	2	-1	0	2	-1	0
$E_{3/2}$:	$\begin{cases}1 \\ 1\end{cases}$	$\begin{matrix}-1 \\ -1\end{matrix}$	$\begin{matrix}i \\ -i\end{matrix}$	$\begin{matrix}-1 \\ -1\end{matrix}$	$\begin{matrix}1 \\ 1\end{matrix}$	$\begin{matrix}-i \\ i\end{matrix}$
$E_{1/2}$:	2	1	0	-2	-1	0

We could equally well have defined the Euler angle transformations of
(132) and R(132) to be interchanged from the definition in Table 10-21, and
(23)* and R(23)* to be similarly interchanged. If we did this the class structure
would be as follows:

$$\begin{array}{cccccc}
E & (123) & (12)* & R & R(123) & R(12)* \\
 & (132) & (23)* & & R(132) & R(23)* \\
 & & (13)* & & & R(13)*
\end{array} \qquad (10\text{-}84)$$

By this renaming of some of the elements of $C_{3v}(M)^2$ we have achieved a
neater looking class structure. This is pointed out to show that Euler angle
transformations caused by O and RO (where O is any element of the MS
group) can be interchanged and the choice is simply one of convenience in
the spin double group.

Having set up the spin double groups of the MS groups, having determined
the equivalent rotations for each of their operations, and having obtained
the character tables of the groups, we can use the equivalent rotations of the
MS groups to correlate the irreducible representations with the representa-

tions $D^{(j)}$ of $K(\text{mol})^2$ for j half-integral. These correlations are included in Appendix B. The subscript notation $(\frac{1}{2}, \frac{3}{2}, \frac{5}{2}, \ldots)$ for the extra irreducible representations in the spin double groups of the MS groups is (where possible) that adopted by Herzberg (1966) and is obtained from the lowest j value of the representation $D^{(j)}$ of $K(\text{mol})^2$ with which the irreducible representation correlates (see Table B-2).

As an example of the use of the spin double groups let us consider the NF_2 molecule in an excited $^2B_1$ electronic state and determine the symmetry labels in the case (a) and (b) limits. We will consider the levels in which the vibrational state is totally symmetric. In the case (b) limit the electron spin wavefunctions are totally symmetric and the rovibronic species Γ_{rve} are obtained by multiplying the rotational species Γ_{rot} for the levels $N_{K_aK_c}$ (given in Table 10-10 with $N = J$) by the vibronic species B_1. Since $S = \frac{1}{2}$ we have $J = N \pm \frac{1}{2}$, and the rovibronic–electron spin species Γ_{rves} are as given down the right hand side in Fig. 10-12. In the case (a) limit the electron spin species Γ_{espin} is obtained by reducing $D^{(S)} = D^{(1/2)}$ of $K(\text{mol})^2$ onto the double group $C_{2v}(M)^2$ and this gives $\Gamma_{\text{espin}} = E_{1/2}$. Multiplying by the electron orbital species B_1 we obtain the spin–orbit species $\Gamma_{\text{eso}} = E_{1/2}$. The rotational species Γ_{rot} cannot be obtained from Table 10-10 since now we have to

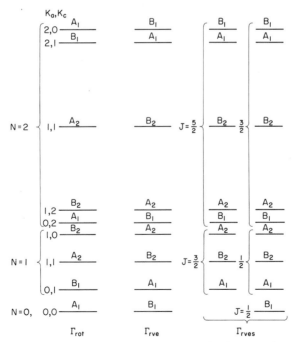

Fig. 10-12. The symmetry labels Γ_{rves} for the rovibronic–electron spin energy levels of NF_2 in a $^2B_1$ vibronic state using Hund's case (b) electron spin functions.

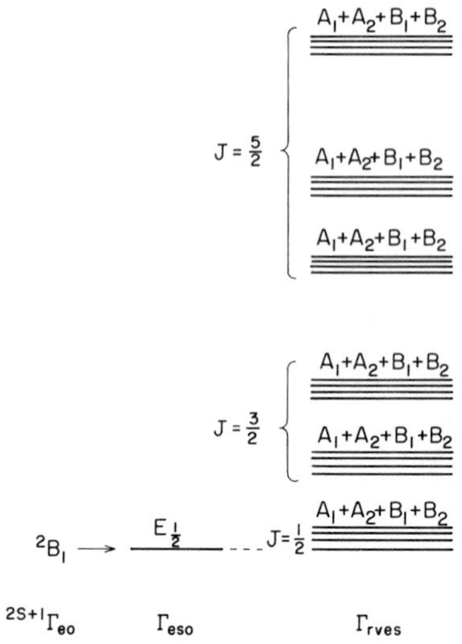

10. The Classification of Molecular Wavefunctions

$A_1 + A_2 + B_1 + B_2$

$J = \frac{5}{2}$ $A_1 + A_2 + B_1 + B_2$

$A_1 + A_2 + B_1 + B_2$

Fig. 10-13. The symmetry labels Γ_{rves} for the rovibronic–electron spin energy levels of NF_2 in a $^2B_1$ electronic state (and an A_1 vibrational state) using Hund's case (a) electron spin functions.

$A_1 + A_2 + B_1 + B_2$

$J = \frac{3}{2}$ $A_1 + A_2 + B_1 + B_2$

$A_1 + A_2 + B_1 + B_2$

$^2B_1 \longrightarrow$ $E_{\frac{1}{2}}$ $\cdots J = \frac{1}{2}$

$^{2S+1}\Gamma_{eo}$ Γ_{eso} Γ_{rves}

classify the functions $J_{P_a P_c}$ where J is half-integral; in case (a), for a molecule with an odd number of electrons, S and hence J, are always half-integral. For NF_2 we can use the correlation table for $K(\text{mol})^2 \rightarrow C_{2v}(M)^2$ to obtain the following results for the species of the rotational wavefunctions as a function of J:

$$\Gamma_{rot}(J = \tfrac{1}{2}) = E_{1/2}, \qquad \Gamma_{rot}(J = \tfrac{3}{2}) = 2E_{1/2},$$

and   (10-85)

$$\Gamma_{rot}(J = \tfrac{5}{2}) = 3E_{1/2}.$$

Using the result that in $C_{2v}(M)^2$

$$E_{1/2} \otimes E_{1/2} = A_1 \oplus A_2 \oplus B_1 \oplus B_2 \qquad (10\text{-}86)$$

we obtain the rovibronic–electron spin species Γ_{rves} as given down the right hand side in Fig. 10-13. For a given value of J we obtain, as we must, the same species as obtained using the case (b) basis. To classify the individual $J_{P_a P_c}$ functions we must determine the transformation properties of the symmetric top basis functions with half-integral J in the group $C_{2v}(M)^2$ and this general problem is discussed in the next section.

THE CLASSIFICATION OF ROTATIONAL WAVEFUNCTIONS HAVING HALF-INTEGRAL J

The determination of the species of the rotational wavefunctions of a molecule for half-integral values of the rotational quantum numbers has not been discussed above except by reduction of the species of the group $K(\text{mol})^2$, and we now do this by using the spin double group of the MS group.

The results in Eq. (10-21) and (10-22) for the transformation properties of the symmetric top functions $|J, k, m\rangle$ (with J integral) under the rotations R_z^β and R_α^π were obtained from the transformation properties of $|J, 0, 0\rangle$ and the transformation properties of the ladder operators $\hat{J}_m^\pm$ and $\hat{J}_s^\pm$. For J half-integral we use the same technique except that we start with the function $|J, \frac{1}{2}, \frac{1}{2}\rangle$ rather than the function $|J, 0, 0\rangle$. Using Eqs. (8-107) and (8-109) we obtain

$$|J, (\pm)|p|, \pm|m|\rangle = N_\pm^{(\pm)} (\hat{J}_m^{(\mp)})^{|p|(\mp)(1/2)} (\hat{J}_s^\pm)^{|m| \mp (1/2)} |J, \tfrac{1}{2}, \tfrac{1}{2}\rangle, \quad (10\text{-}87)$$

$$N_\pm^{(\pm)} = \left[\frac{(J - |p|)!(J - |m|)!}{(J + |p|)!(J + |m|)!} \right]^{1/2} \hbar^{-(|p| + |m|)} \left[\left(J + \tfrac{1}{2} \right) \hbar \right]^{\pm(1/2)(\pm)(1/2)} \quad (10\text{-}88)$$

and J, p and m are half-integral. [Compare Eq. (8-111).]

From Eq. (8-67) we have (with an appropriate choice of phase factor)

$$\left| J, \tfrac{1}{2}, \tfrac{1}{2} \right\rangle = \sqrt{\frac{2J+1}{8\pi^2}} \left(J + \tfrac{1}{2} \right)! \left(J - \tfrac{1}{2} \right)! \, e^{i(\phi + \chi)/2} \sum_{\sigma=0}^{J-(1/2)} (-1)^{J-(1/2)-\sigma}$$

$$\times \frac{(\cos\frac{1}{2}\theta)^{2\sigma+1}(\sin\frac{1}{2}\theta)^{2J-1-2\sigma}}{\sigma!(\sigma+1)![(J - \frac{1}{2} - \sigma)!]^2}. \quad (10\text{-}89)$$

Reversing the order of the terms in the sum and writing $x = (J - \frac{1}{2} - \sigma)$ as the summation index we can rewrite this expression as

$$\left| J, \tfrac{1}{2}, \tfrac{1}{2} \right\rangle = \sqrt{\frac{2J+1}{8\pi^2}} \left(J + \tfrac{1}{2} \right)! \left(J - \tfrac{1}{2} \right)! \, e^{i(\phi + \chi)/2} \sum_{x=0}^{J-(1/2)} (-1)^x$$

$$\times \frac{(\cos\frac{1}{2}\theta)^{2J-2x}(\sin\frac{1}{2}\theta)^{2x}}{(x!)^2(J + \frac{1}{2} - x)!(J - \frac{1}{2} - x)!}. \quad (10\text{-}90)$$

Using Eq. (10-87) and the expression for $|J, \frac{1}{2}, \frac{1}{2}\rangle$ in Eq. (10-90) we have

$$|J, -\tfrac{1}{2}, \tfrac{1}{2}\rangle = [\hbar(J + \tfrac{1}{2})]^{-1} \hat{J}_m^+ |J, \tfrac{1}{2}, \tfrac{1}{2}\rangle \qquad (10\text{-}91)$$

$$= (-1) \sqrt{\frac{2J+1}{8\pi^2}} \left(J + \tfrac{1}{2} \right)! \left(J - \tfrac{1}{2} \right)! \, e^{i(\phi - \chi)/2} \sum_{\sigma=0}^{J-(1/2)} (-1)^{J-(1/2)-\sigma}$$

$$\times \frac{(\sin\frac{1}{2}\theta)^{2J-2\sigma}(\cos\frac{1}{2}\theta)^{2\sigma}}{(\sigma!)^2(J + \frac{1}{2} - \sigma)!(J - \frac{1}{2} - \sigma)!}$$

$$= (-1)^{J+(1/2)} \sqrt{\frac{2J+1}{8\pi^2}} \left(J + \tfrac{1}{2} \right)! \left(J - \tfrac{1}{2} \right)! \, e^{i(\phi - \chi)/2} \sum_{\sigma=0}^{J-(1/2)} (-1)^\sigma$$

$$\times \frac{(\sin\frac{1}{2}\theta)^{2J-2\sigma}(\cos\frac{1}{2}\theta)^{2\sigma}}{(\sigma!)^2(J + \frac{1}{2} - \sigma)!(J - \frac{1}{2} - \sigma)!}. \quad (10\text{-}92)$$

Using Eqs. (10-90)–(10-92) we determine (see Table 7-1 for the definition of R_z^β and R_α^π) that

$$R_z^\beta \left| J, \tfrac{1}{2}, \tfrac{1}{2} \right\rangle = e^{i\beta/2} \left| J, \tfrac{1}{2}, \tfrac{1}{2} \right\rangle \tag{10-93}$$

and

$$R_\alpha^\pi \left| J, \tfrac{1}{2}, \tfrac{1}{2} \right\rangle = \exp\left[\tfrac{1}{2}i(3\pi - 2\alpha)\right](-1)^{J+(1/2)} \left| J, -\tfrac{1}{2}, \tfrac{1}{2} \right\rangle$$
$$= e^{i\pi J} e^{-i\alpha} \left[\hbar(J + \tfrac{1}{2})\right]^{-1} \hat{J}_m^+ \left| J, \tfrac{1}{2}, \tfrac{1}{2} \right\rangle. \tag{10-94}$$

We can use Eqs. (10-93) and (10-94) in conjunction with Eqs. (10-11)–(10-14) for $\hat{J}_m^\pm$ and $\hat{J}_s^\pm$, and Eq. (10-87) for $|J, p, m\rangle$, to determine the transformation properties of the rotational function $|J, p, m\rangle$. The results obtained are

$$R_z^\beta |J, p, m\rangle = e^{ip\beta} |J, p, m\rangle \tag{10-95}$$

and

$$R_\alpha^\pi |J, p, m\rangle = e^{i\pi J} e^{-2i\alpha p} |J, -p, m\rangle, \tag{10-96}$$

where J, p, and m are half-integral.

As an example of an application we consider the CH_3 radical which has a planar $^2A_2''$ electronic ground state; the spin double group of its MS group is the $D_{3h}(M)^2$ group (see Table A-9). Using space fixed electron spin func-

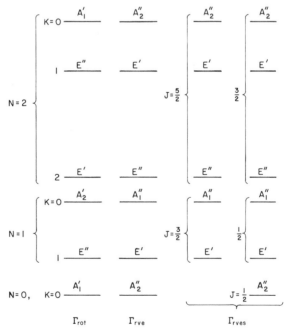

Fig. 10-14. The symmetry labels Γ_{rves} for the rovibronic–electron spin energy levels of CH_3 in a $^2A_2''$ vibronic state using Hund's case (b) electron spin functions.

tions [case (b)] the symmetry Γ_{rves} of the levels of a totally symmetric vibrational state of the $^2A_2''$ electronic state can be determined using the results in Table 10-6 and these are given in Fig. 10-14. This figure is similar to Fig. 10-10 except that now we have $J = N \pm \frac{1}{2}$ and the extra symmetry distinctions (′) and (″) arising from the feasibility of E^*. Using molecule fixed spin functions [case (a)] we determine from the $K(\text{mol})^2 \to D_{3h}(M)^2$ correlation table that the pair of $S = \frac{1}{2}$ spin functions generates the representation $E_{1/2}$ of $D_{3h}(M)^2$. Multiplying this by the electron orbital species A_2'' we obtain $E_{5/2}$ as the electron spin–orbit species Γ_{eso}. The transformation properties of the rotational wavefunctions $|J, p, m\rangle$, with J half-integral, under the effect of the operations of the $D_{3h}(M)^2$ group can be obtained from a knowledge of the equivalent rotations of the elements in conjunction with Eqs. (10-95) and (10-96). The characters of the representation generated by the pair of functions $|J, P, m\rangle$ and $|J, -P, m\rangle$ are [giving the characters in the order the elements are listed in Table A-9 for $D_{3h}(M)^2$]:

$$2, 2\cos(2\pi P/3), 0, 2\cos(P\pi), 2\cos(-\pi P/3),$$

$$0, 2\cos(2P\pi), 2\cos(8P\pi/3), 2\cos(5P\pi/3). \qquad (10\text{-}97)$$

This should be compared to Eq. (10-24). The symmetry species obtained are listed in Table 10-26. In Fig. 10-15 the symmetry Γ_{rves} of the rovibronic–electron spin functions obtained using these results are given in a case (a) basis for CH_3 in the $^2A_2''$ state. The reader can correlate the levels of Figs. 10-14 and 10-15 if it is remembered that J and Γ_{rves} do not change. Note that Γ_{rves} is always single-valued.

For most polyatomic molecules the coupling of the electron spin to the molecular framework is small and a good understanding of the energy level

Table 10-26
Symmetry species of
case (a) rotational
wavefunctions
$|J, p, m\rangle$ of CH_3
in the $D_{3h}(M)^2$
group[a]

P	Γ_{rot}
$6n \pm (\frac{1}{2})$	$E_{1/2}$
$6n \pm (\frac{3}{2})$	$E_{3/2}$
$6n \pm (\frac{5}{2})$	$E_{5/2}$

[a] n is integral; $P = |p|$ is positive half-integral.

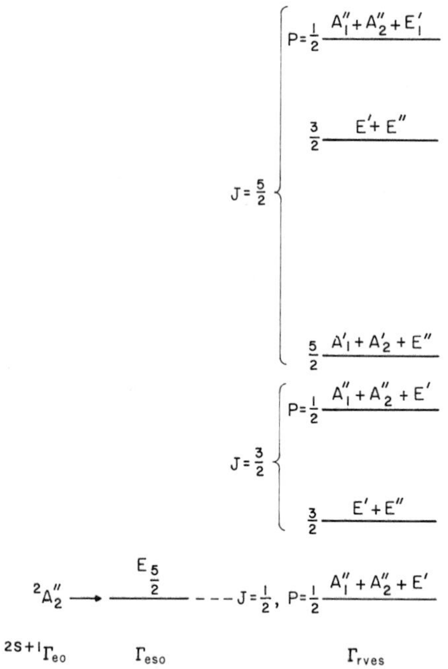

Fig. 10-15. The symmetry labels Γ_{rves} for the rovibronic–electron spin energy levels of CH_3 in a $^2A_2''$ electronic state (and an A_1 vibrational state) using Hund's case (a) electron spin functions.

pattern is obtained using a Hund's case (b) basis. In this case the energy level pattern is like that of a singlet state except for the fact that each level has a multiplicity of $2S + 1$ (or if $N < S$ the multiplicity is $2N + 1$).

SUMMARY

Using the rigid rotor–harmonic oscillator–LCAOMO approximations for the rotation–vibration–electron orbital states we have shown how to classify the rovibronic wavefunctions of a molecule in its MS group. The case (a) or case (b) electron spin function symmetries have also been determined and the spin double groups of the MS groups have been introduced. The classification of the rotational wavefunctions when the rotational quantum numbers are half-integral requires the use of the spin double group and this has been explained. The nuclear spin symmetries, and the application of the statistical formulas in determining the species of the complete internal wavefunction Φ_{int}, and the nuclear spin statistical weights, have been described using the MS group.

BIBLIOGRAPHICAL NOTES

Parity

Oka (1973). In this paper the two possible symmetries for Φ_{int} are shown to correspond to the + and − parity labels, and the use of the parity label on rotational levels (i.e., on rovibronic states E_{rve}) is advocated. However, as discussed above, the rotational levels are better labeled using Γ_{rve} from the MS group since this label provides more information than the parity label concerning inter and intra molecular interactions and interactions with electromagnetic radiation [except for molecules with $C_s(M)$ symmetry for which Γ_{rve} and parity are equivalent]. See pages 89 and 90 of Bunker (1975).

Double Groups

Bethe (1929).
Opechowski (1940).
Hougen (1963).

11

Near Symmetry, Perturbations, and Optical Selection Rules

In this chapter we introduce the concepts of near symmetry and near quantum number, and we explain their use. The molecular point group and molecular rotation group are important near symmetry groups, and they provide useful energy level labels which are called near symmetry labels; true symmetry labels are obtained from the MS group and the group K(spatial). The perturbations that can occur between molecular energy levels are discussed and it is shown how true symmetry labels can be used to determine that some perturbation terms vanish and that some levels cannot interact. Near quantum numbers and near symmetry labels can be used to determine which are the strong perturbations. The selection rules for one-photon electric dipole transitions are derived using both near symmetry labels and near quantum numbers as well as true symmetry labels. Forbidden transitions are discussed and, at the end of the chapter, magnetic dipole transitions, electric quadrupole transitions, multiphoton processes (including the Raman effect), the Zeeman effect, and the Stark effect are briefly considered.

NEAR SYMMETRY

In the absence of external fields, the Hamiltonian $\hat{H}$ of an isolated molecule commutes with the elements of the molecular symmetry group of the mole-

268

cule and with the elements of the group K(spatial). Thus these two groups are true symmetry groups of $\hat{H}$ and we can label the molecular states according to the irreducible representations of these groups; such labels are called true symmetry labels. On the other hand a *near symmetry group* of $\hat{H}$ is such that its elements do not commute with $\hat{H}$ but rather they commute with an approximate Hamiltonian $\hat{H}^0$; we can consider $\hat{H}^0$ as being obtained from $\hat{H}$ by neglecting a small term $\hat{H}'$. Writing

$$\hat{H} = \hat{H}^0 + \hat{H}' \tag{11-1}$$

then for any element G of the near symmetry group

$$[G, \hat{H}^0] = 0 \tag{11-2}$$

but

$$[G, \hat{H}] = [G, (\hat{H}^0 + \hat{H}')] = [G, \hat{H}'] \neq 0. \tag{11-3}$$

We can label the eigenstates of $\hat{H}^0$ according to the irreducible representations of the near symmetry group [because Eq. (11-2) is true] and such labels will be useful *near symmetry labels* on the exact states (the eigenstates of $\hat{H}$) if the symmetry breaking effects of $\hat{H}'$ are small. The term $\hat{H}'$ can mix eigenstates of $\hat{H}^0$ that have different near symmetry labels (and hence break that symmetry) but, of course, it cannot mix states having different true symmetry labels [see the vanishing integral rule, Eq. (5-133)]. A symmetry group, and the symmetry labels obtained with it, are used to determine which of many possible terms in the Hamiltonian vanish and to determine which states are coupled by internally and externally caused perturbations. True symmetry groups give rigorously correct results. Near symmetry groups give rise to many more restrictions which, although only approximately true, usually enable us to understand the major perturbation effects.

The two near symmetry groups that we will consider here are the *molecular rotation group* and the *molecular point group*. We will also discuss the concept of a *near quantum number* since this is closely related to the idea of near symmetry. We will not discuss the dynamical groups that are near symmetry groups of the electronic Hamiltonian; a review article on this subject has been written by Wulfman (1971).

The Molecular Rotation Group

The molecular rotation group of a molecule consists of all Euler angle transformations that leave the rigid rotor Hamiltonian of the molecule invariant, and we can picture each operation of the group as being a bodily rotation of the molecule about an axis that has a definite orientation within the molecule fixed axis system. Each element of the molecular rotation group of a molecule changes the Euler angles but has no effect on the vibronic or spin coordinates. The terms in the complete Hamiltonian in which the rotational coordinates are coupled to the vibronic or spin coordinates [such as the Coriolis coupling term Eq. (8-28c) or the term $\hat{H}_{nsr}$ in Table 6-1] will not necessarily commute with the elements of the molecular rotation group, and they can therefore break this symmetry. However, these effects are often small and in these circumstances the molecular rotation group provides useful near symmetry labels on the molecular energy levels.

For a spherical top molecule the rigid rotor Hamiltonian is [see Eq. (8-70)] in cm^{-1}

$$\hat{H}_{sph} = \hbar^{-2}B\hat{J}^2. \tag{11-4}$$

This Hamiltonian is invariant to the Euler angle transformation caused by rotation of the molecule about any axis that has a definite orientation within the molecule fixed axis system. Thus the molecular rotation group of a spherical top molecule is K(mol); this group provides the label J for the states, and each J level has a $(2J + 1)$-fold k degeneracy. This degeneracy, and the symmetry in K(mol), is removed by interactions such as caused by centrifugal distortion and Coriolis coupling effects.

For a symmetric top molecule the rigid rotor Hamiltonian is [see Eqs. (8-37) and (8-68)] in cm^{-1}

$$\hat{H}_{symm} = \hbar^{-2}[X\hat{J}^2 + (Z - X)\hat{J}_z^2], \tag{11-5}$$

where we use the letters X and Z for the rotational constants ($X = B$ and $Z = A$ or C). This Hamiltonian is invariant to the Euler angle transformations caused by any rotation about the molecule fixed z axis and by any twofold rotation about an axis perpendicular to the z axis. It is left as an exercise for the reader to show that the operations R_z^β and R_α^π (see Table 7-1) commute with $\hat{J}^2$ and $\hat{J}_z^2$ [see Eqs. (7-144)–(7-146)] and hence with $\hat{H}_{symm}$. The molecular rotation group of a symmetric top molecule is thus the group D_∞. This group is a subgroup of the group K(mol) and the character table of the spin double group D_∞^2 is given in Table 11-1. We can classify the symmetric top functions $|J,k,m\rangle$ in D_∞^2, by using the results in Table 7-1, and Eqs. (10-21), (10-22), (10-95), and (10-96); the results of this classification are included in Table 11-1. We see that the distinctions made by classifying the states in D_∞^2 parallel the distinctions made by using the unsigned quantum number K (or P). Rotation–vibration levels of different D_∞^2 species [i.e., of different K (or P) value] can be mixed by Coriolis coupling and centrifugal distortion terms in the Hamiltonian [see, for example, Eqs. (11-105) and (11-108) that follow], and these effects spoil the symmetry of the complete Hamiltonian in D_∞^2. The symmetry in D_∞^2, i.e., the quantum number K, deteriorates in usefulness the more vibrationally excited the molecule be-

Table 11-1

The character table of D_∞^2 and the species of symmetric top functions $|J, \pm K, m\rangle$ in the group

	E	$2R_z^\varepsilon$	$\cdots$	∞R_a^π	$R^{2\pi}$	$2R_z^{2\pi+\varepsilon}$	$\cdots$	Symmetry species as function of K^a
Σ^+:	1	1	$\cdots$	1	1	1	$\cdots$	0 (J even)
Σ^-:	1	1	$\cdots$	-1	1	1	$\cdots$	0 (J odd)
Π:	2	$2\cos\varepsilon$	$\cdots$	0	2	$2\cos\varepsilon$	$\cdots$	1
Δ:	2	$2\cos 2\varepsilon$	$\cdots$	0	2	$2\cos 2\varepsilon$	$\cdots$	2
Φ:	2	$2\cos 3\varepsilon$	$\cdots$	0	2	$2\cos 3\varepsilon$	$\cdots$	3
$\vdots$	$\vdots$	$\vdots$	$\cdots$	$\vdots$	$\vdots$	$\vdots$	$\cdots$	$\vdots$
$E_{1/2}$:	2	$2\cos(\varepsilon/2)$	$\cdots$	0	-2	$-2\cos(\varepsilon/2)$	$\cdots$	$\frac{1}{2}$
$E_{3/2}$:	2	$2\cos(3\varepsilon/2)$	$\cdots$	0	-2	$-2\cos(3\varepsilon/2)$	$\cdots$	$\frac{3}{2}$
$E_{5/2}$:	2	$2\cos(5\varepsilon/2)$	$\cdots$	0	-2	$-2\cos(5\varepsilon/2)$	$\cdots$	$\frac{5}{2}$
$\vdots$	$\vdots$	$\vdots$	$\cdots$	$\vdots$	$\vdots$	$\vdots$	$\cdots$	$\vdots$

[a] P in Hund's case (a).

comes. This is because the density of vibrational states increases rapidly with vibrational excitation and there is, therefore, more opportunity for Coriolis and centrifugal distortion interaction.

For an asymmetric top molecule the rigid rotor Hamiltonian is [see Eq. (8-112)] in cm^{-1}

$$\hat{H}_{\text{asym}} = \hbar^{-2}(A\hat{J}_a^2 + B\hat{J}_b^2 + C\hat{J}_c^2). \tag{11-6}$$

This Hamiltonian is invariant to the Euler angle transformations caused by a twofold rotation about the a, b, or c axis. The molecular rotation group is hence the group $D_2 = \{E, R_a^\pi, R_b^\pi, R_c^\pi\}$, which is a subgroup of D_∞, and the character table of the spin double group D_2^2 is given in Table 11-2. We can

Table 11-2

The character table of the group D_2^2 and the species of asymmetric top functions $|J_{K_aK_c}\rangle$ in the group

	E	R_a^π	R_b^π	R_c^π	$R^{2\pi}$	K_aK_c
	1	2	2	2	1	
A:	1	1	1	1	1	ee
B_a:	1	1	-1	-1	1	eo
B_b:	1	-1	1	-1	1	oo
B_c:	1	-1	-1	1	1	oe
$E_{1/2}$:	2	0	0	0	-2	Half-integral J

classify the asymmetric rotor wavefunctions in this group by following the procedure used in obtaining Table 10-10, and the results are also included in Table 11-2. For molecules for which the $J_{K_a K_c}$ states ee, oe, eo, and oo transform as four different irreducible representations of the molecular symmetry group we achieve no new distinctions by using the molecular rotation group. However, for molecules for which this is not true, such as trans C(HF)CHF (see Table 10-13) or completely unsymmetrical molecules like CHIFCl, we do obtain extra energy level distinctions that are of great use (particularly for the levels of the vibrational ground state for which the symmetry breaking effects are usually negligible). The factorization of the rigid rotor Hamiltonian into four blocks (see Problem 8-3) follows from the symmetry of the states in the group D_2.

The correlation of the species of $K(\text{mol})^2$ to D_∞^2 and to D_2^2 is given in Table B-1 of Appendix B for $J \leq 3$.

The Molecular Point Group

For any molecule the point group symmetry of the equilibrium nuclear arrangement is easily determined (see Chapter 3). In using the point group to transform molecular wavefunctions the elements of the group are interpreted as being rotations and reflections of the vibronic variables (vibrational displacements and electronic coordinates) in the molecule fixed axis system [see Section 5-5 and Fig. 5-7 in Wilson, Decius, and Cross (1955)]. The molecular point group is a symmetry group of the vibronic Hamiltonian since all interparticle distances are maintained by its elements. The Euler angles, angular momentum components $\hat{J}_\alpha$ and nuclear spin coordinates are unaffected by the elements of the molecular point group. If we neglect terms in the Hamiltonian that couple the vibronic coordinates to the other degrees of freedom (particularly the Coriolis coupling and centrifugal distortion terms) we obtain an approximate Hamiltonian that commutes with the elements of the molecular point group. The molecular point group is thus a near symmetry group of the complete molecular Hamiltonian, and the coupling effects due to Coriolis coupling and centrifugal distortion are the principal symmetry breaking terms in the Hamiltonian. However, the molecular point group is generally not used for labeling rovibronic states, but rather for labeling vibrational and electronic states and for studying vibronic interactions. It is a true symmetry group of the vibronic (and electronic) Hamiltonian.

It is important to appreciate exactly how the molecular coordinates transform under the elements of the molecular point group, and to appreciate how the elements are related to the elements of the molecular symmetry group. We will use the water molecule as an example, and then discuss a general rule

Table 11-3

The character table of the
C_{2v} point group

	E	C_{2x}	σ_{xz}	σ_{xy}
A_1:	1	1	1	1
A_2:	1	1	-1	-1
B_1:	1	-1	-1	1
B_2:	1	-1	1	-1

Table 11-4

The transformation properties of the Cartesian displacement coordinates of H_2O in
the C_{2v} group given in Table 11-3[a]

E	C_{2x}	σ_{xz}	σ_{xy}	E	C_{2x}	σ_{xz}	σ_{xy}	
Δx_1	Δx_2	Δx_1	Δx_2	T_x	T_x	T_x	T_x	$: A_1$
Δx_2	Δx_1	Δx_2	Δx_1	T_y	$-T_y$	$-T_y$	T_y	$: B_1$
Δx_3	Δx_3	Δx_3	Δx_3	T_z	$-T_z$	T_z	$-T_z$	$: B_2$
Δy_1	$-\Delta y_2$	$-\Delta y_1$	Δy_2	R_x	R_x	$-R_x$	$-R_x$	$: A_2$
Δy_2	$-\Delta y_1$	$-\Delta y_2$	Δy_1	R_y	$-R_y$	R_y	$-R_y$	$: B_2$
Δy_3	$-\Delta y_3$	$-\Delta y_3$	Δy_3	R_z	$-R_z$	$-R_z$	R_z	$: B_1$
Δz_1	$-\Delta z_2$	Δz_1	$-\Delta z_2$					
Δz_2	$-\Delta z_1$	Δz_2	$-\Delta z_1$					
Δz_3	$-\Delta z_3$	Δz_3	$-\Delta z_3$					
χ_{Car}: 9	-1	3	1					

[a] The protons are numbered 1 and 2 and the oxygen nucleus 3. The (x, y, z) axes are defined in Fig. 7-5.

that relates the molecular point group elements and the molecular symmetry group elements for any nonlinear rigid molecule.

Locating the molecule fixed (x, y, z) axes on a water molecule following the I^r convention as shown in Fig. 7-5 (the xz plane is the molecular plane and the x axis is the twofold symmetry axis) the elements of the molecular point group C_{2v} are $\{E, C_{2x}, \sigma_{xz}, \sigma_{xy}\}$ and the character table of this group is as given in Table 11-3. To determine the symmetry of the normal coordinates in this group we first determine the transformation properties of the Cartesian displacement coordinates. The results of doing this are given in Table 11-4. As an example in Fig. 11-1 the effect of C_{2x} on the Δx coordinates is shown, and we see that

$$C_{2x}(\Delta x_1, \Delta x_2, \Delta x_3) = (\Delta x_1', \Delta x_2', \Delta x_3') = (\Delta x_2, \Delta x_1, \Delta x_3). \quad (11\text{-}7)$$

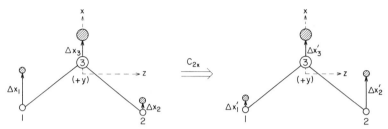

Fig. 11-1. The effect of the molecular point group operation C_{2x} on the x displacement coordinates of the nuclei in H_2O.

The vibrational displacements are rotated about the x axis by C_{2x} but the molecule fixed axes are unaffected. The transformation properties of the translational and rotational combinations of the Cartesian displacement coordinates [see Eq. (7-246)] are also given in Table 11-4, and we deduce the species of the normal coordinates in C_{2v} to be

$$\Gamma(Q_1, Q_2, Q_3) = 2A_1 \oplus B_2. \tag{11-8}$$

The symmetry of the lower molecular orbitals can be determined from those of the minimal basis set of atomic orbitals $1s(O)$, $2s(O)$, $2p_x(O)$, $2p_y(O)$, $2p_z(O)$, $1s(H_1)$, and $1s(H_2)$. The transformation properties of these orbitals in the C_{2v} group are given in Table 11-5. In Fig. 11-2 the effect of C_{2x} on the $2p_z(O)$ orbital is shown. Using the LCAO approximation we deduce that the lowest bonding and nonbonding symmetry adapted molecular orbitals of H_2O in the C_{2v} group, can be written qualitatively as

$$A_1 \begin{cases} (1a_1): & 1s(O), \\ (2a_1): & 2s(O) + [1s(H_2) + 1s(H_1)], \\ (3a_1): & 2p_x(O) + [1s(H_2) + 1s(H_1)]; \end{cases} \tag{11-9}$$
$$B_2 \quad (1b_2): \quad 2p_z(O) + [1s(H_2) - 1s(H_1)];$$
$$B_1 \quad (1b_1): \quad 2p_y(O).$$

Comparison of Tables 11-4 and 11-5 with the results given in Table 7-2 and Eq. (10-47), and of Figs. 11-1 and 11-2 with Figs. 7-14 and 10-9, shows that the point group operations C_{2x}, σ_{xz}, and σ_{xy} have the same effect on the vibronic variables in H_2O as the MS group operations (12), E^*, and (12)*, respectively. By studying the point groups of all rigid nonlinear molecules we find that there is a general rule, which we can state as follows:

For a rigid nonlinear molecule the molecular point group and the molecular symmetry group are isomorphic, and each element in the molecular point group has the same effect on the vibronic variables as its partner in the molecular symmetry group.

Table 11-5

The transformation properties of the atomic orbitals in
the C_{2v} group for H_2O

E	C_{2x}	σ_{xz}	σ_{xy}	
1s(O)	1s(O)	1s(O)	1s(O) :	A_1
2s(O)	2s(O)	2s(O)	2s(O) :	A_1
$2p_x$(O)	$2p_x$(O)	$2p_x$(O)	$2p_x$(O) :	A_1
$2p_y$(O)	$-2p_y$(O)	$-2p_y$(O)	$2p_y$(O) :	B_1
$2p_z$(O)	$-2p_z$(O)	$2p_z$(O)	$-2p_z$(O) :	B_2
1s(H$_1$)	1s(H$_2$)	1s(H$_1$)	1s(H$_2$) $\Big\}$	$A_1 \oplus B_2$
1s(H$_2$)	1s(H$_1$)	1s(H$_2$)	1s(H$_1$) $\Big\}$:	

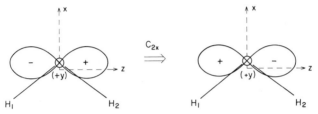

Fig. 11-2. The effect of the molecular point group operation C_{2x} on the $2p_z$(O) orbital in H_2O.

Since the molecular point group and the molecular symmetry group are isomorphic for rigid nonlinear molecules, we choose to use common character tables and irreducible representation labels for them and this is done in Appendix A. The results of a classification of vibronic states would then be the same using either group, but we should remember that for the complete Hamiltonian the molecular point group is a near symmetry group whereas the molecular symmetry group is a true symmetry group.

We will now look in more detail at the relationship between the molecular point group and the molecular symmetry group. Each operation O of the molecular symmetry group of a molecule will, in general, transform the vibronic variables, the Euler angles, and the nuclear spins [and the electron spins in Hund's case (a)]. We can therefore write each O as the product of three commuting operations O_a, O_b, and O_c, where O_a is an operation that produces the change in vibronic variables [and Hund's case (a) electron spin functions] caused by O, O_b is an operation that produces the change in Euler angles caused by O, and O_c is an operation that produces the nuclear spin permutation caused by O. Any of these operations could be an identity operation for which we use the notation E, R^0, and p_0, respectively. We can, therefore, write each operation of an MS group as

$$O = O_a O_b O_c. \tag{11-10}$$

For rigid nonlinear molecules the group of all operations O_a is the molecular point group. The operations O_b will occur in the molecular rotation group of the molecule. but the group of all operations O_b will sometimes be only a subgroup of the molecular rotation group. The operations O_c will be in the near symmetry group whose elements only permute the spins (but not the coordinates) of the nuclei; we do not discuss this near symmetry group (this *nuclear spin permutation group* could be used for classifying nuclear spin states). For the water molecule we obtain

$$E = ER^0 p_0, \tag{11-11a}$$

$$(12) = C_{2x} R_x^\pi p_{12}, \tag{11-11b}$$

$$E^* = \sigma_{xz} R_y^\pi p_0, \tag{11-11c}$$

and

$$(12)^* = \sigma_{xy} R_z^\pi p_{12}, \tag{11-11d}$$

where p_{12} permutes the spins of H_1 and H_2. For the water molecule it so happens that the group of all operations O_b (i.e., the group $\{R^0, R_x^\pi, R_y^\pi, R_z^\pi\}$) is the molecular rotation group.

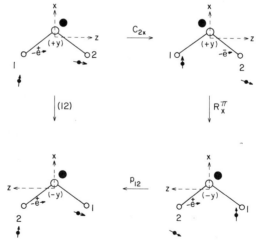

Fig. 11-3. The effect of the successive operations C_{2x}, R_x^π, and p_{12} on a distorted water molecule and the equivalence of this to the permutation (12). The molecule fixed (x, y, z) axes are right handed in all figures.

In Figs. 11-3, 11-4, and 11-5 we represent the effect of the operations O, O_a, O_b, and O_c on the water molecule for $O = (12)$, E^*, and $(12)^*$, respectively [see also Bunker and Papousek (1969)]. In each molecular figure the three black dots represent the instantaneous position of the oxygen nucleus and of the hydrogen nuclei in a rotating and vibrating water molecule. The solid arrows passing through the protons represent the space quantized nuclear

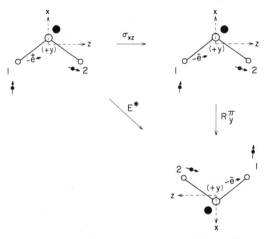

Fig. 11-4. The effect of the successive operations σ_{xz} and R_y^{π} on a distorted water molecule, and the equivalence of this to the inversion E^*. The molecule fixed (x, y, z) axes are right handed in all figures.

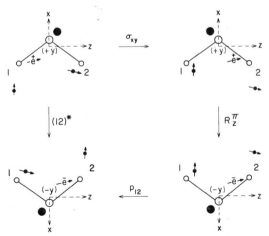

Fig. 11-5. The effect of the successive operations σ_{xy}, R_z^{π}, and p_{12} on a distorted water molecule, and the equivalence of this to the permutation–inversion operation $(12)^*$. The molecule fixed (x, y, z) axes are right handed in all figures.

spins. Applying the center of mass and Eckart conditions we can locate the molecule fixed (x, y, z) axes and an appropriately oriented water molecule in its equilibrium configuration (open circles). The e is an electron with its space quantized spin marked by an arrow through it, and the $+$ or $-$ above the e indicates whether the electron is above or below the plane of the paper.

Figure 11-3 shows the point group operation C_{2x}; it rotates the vibrational displacements and electronic coordinates about the x axis through π radians [Hund's case (a) electron spin functions would also be rotated by this operation]. The orientation of the molecule fixed axes and the nuclear spins are unaffected by any point group operation. The rotation operation R_x^{π} is a bodily rotation of the molecule about the molecule fixed x axis through π radians, and p_{12} is the permutation of the nuclear spins. Doing the operations C_{2x}, R_x^{π} and p_{12} successively is equivalent to the nuclear permutation (12). Figures 11-4 and 11-5 are similar representations of Eqs. (11-11c) and (11-11d).

The point group of a centrosymmetric molecule, such as ethylene, contains the point group operation i, and this operation merits special attention. In the molecular symmetry group of a centrosymmetric molecule there is always present an operation that we will call $\hat{O}_i$; the operation $\hat{O}_i$ is a permutation inversion operation that can be written as

$$\hat{O}_i = (AA')(BB')(CC') \cdots (NN')^* \qquad (11\text{-}12)$$

where AA', BB', CC', ..., NN' are all pairs of identical nuclei that, in the equilibrium configuration, are located symmetrically about the nuclear center of mass. For the ethylene molecule labeled as in Fig. 10-7 we have

$$\hat{O}_i = (14)(23)(56)^*, \qquad (11\text{-}13)$$

for the ethane molecule labeled as in Table A-10 we have

$$\hat{O}_i = (14)(26)(35)(78)^*, \qquad (11\text{-}14)$$

and for a homonuclear diatomic molecule [see Eq. (7-116)]

$$\hat{O}_i = (12)^*. \qquad (11\text{-}15)$$

The operation $\hat{O}_i$ does not change the positions in space of the nuclei when they are in the equilibrium configuration of the centrosymmetric molecule and as a result the operation $\hat{O}_i$ does not change the Euler angles. The vibronic coordinates are inverted and the nuclear spins of the pairs of nuclei AA', BB', CC', ..., NN' are interchanged by $\hat{O}_i$. Thus the molecular symmetry group operation $\hat{O}_i$ and the molecular point group operation i are related by [compare with Eq. (11-11c)]

$$\hat{O}_i = iR^0 p_{(AA')(BB')(CC') \cdots (NN')}, \qquad (11\text{-}16)$$

where $p_{(AA')(BB')(CC')\ldots(NN')}$ is the permutation of the nuclear spins. From this we see that the point group operation i is a true symmetry operation of the rovibronic Hamiltonian and not just of the vibronic Hamiltonian as are all other point group operations (see Problem 11-1, which follows). The operation i is not a true symmetry operation of the complete Hamiltonian because of the effect of nuclear spin coupling terms, but this is usually a very small effect. The point group operation i is not the same as the operation E^* and it does not give the parity label of the state; instead, it gives the g or u label. The g and u labels on molecular energy levels are near symmetry labels that are spoiled by interactions involving the nuclear spins.

Equation (11-10) can be reversed so that the molecular point group operation is the subject, i.e.,

$$O_a = OO_c^{-1}O_b^{-1}. \tag{11-17}$$

We can now generalize the concept of the molecular point group to nonrigid molecules that have no unique point group symmetry. We call the group that is a generalization of the molecular point group the *molecular vibronic group* and its elements are obtained as follows. Having obtained the molecular symmetry group (or, when necessary, the extended molecular symmetry group as we will discuss in Chapter 12), each element O is converted to an element of the molecular vibronic group by undoing (or just neglecting) the changes in Euler angles and the permutation of the nuclear spins that it causes. This is achieved in Eq. (11-17) where O_b^{-1} "removes" the Euler angle change and O_c^{-1} "removes" the nuclear spin permutation. For a rigid nonlinear molecule this would be the way to obtain the molecular point group, and Eq. (11-17) provides a good way of introducing the molecular point group after having defined the molecular symmetry group. In general the molecular vibronic group is used to classify vibrational and electronic states and to study vibronic interactions without having to bother about the Euler angles or the nuclear spins.

Before we leave the point group we must discuss the so-called "rotational subgroup of the point group" that is usually used in determining the nuclear spin statistical weights of the levels of rigid nonlinear molecules. The real rotational subgroup of the molecular point group consists only of the rotation operations from the point group, e.g., it is $\{E, C_{2x}\}$ for the C_{2v} group (see Table 11-3) for the water molecule. Such operations do not permute nuclei and so the statistical formulas cannot be applied to their result. However, what is called the "rotational subgroup of the point group" is really the permutation subgroup of the molecular symmetry group. The use of this group, and the use of the molecular symmetry group, in determining statistical weights is discussed in Chapter 10.

NEAR QUANTUM NUMBERS

To introduce the idea of near (and good) quantum numbers we return to the asymmetric top Hamiltonian as written in Eq. (8-114). In diagonalizing this Hamiltonian we use a $|J, k_a, m\rangle$ basis set where $\hat{J}_a^2 \; (= \hat{J}_z^2)$ has eigenvalues $K_a^2 \hbar^2$. If $(B - C)$ were zero then Eq. (8-114) would reduce to the symmetric top Hamiltonian and the functions $|J, k_a, m\rangle$ would be its eigenfunctions. As far as the symmetric top Hamiltonian is concerned the quantum number K_a is a *good quantum number* in that states having different values of K_a are not mixed by any terms in that Hamiltonian. The symmetric top Hamiltonian has the symmetry of the D_∞ molecular rotation group. For an asymmetric top if $(B - C)$ were small (compared to A, B, or C) then the D_∞ symmetry breaking effects of the term $[(B - C)/4](\hat{J}_m^{+2} + \hat{J}_m^{-2})$ would be slight and the asymmetric top states would largely consist of one K_a state with a small admixture of states having different K_a values. For example, for the ground state of the SO_2 molecule the rotational constants are approximately

$$A = 2.0\,\text{cm}^{-1}, \qquad B = 0.34\,\text{cm}^{-1}, \qquad \text{and} \qquad C = 0.29\,\text{cm}^{-1}.$$

Thus $(B - C) = 0.05\,\text{cm}^{-1}$ which is small compared to the rotational constants. From Eqs. (8-137)–(8-139) we deduce from these values of A, B, and C that the two $J = 2$ states of type E^+ can be written in terms of $|J, k_a, m\rangle$ functions as follows:

$$\Phi^-(J = 2, E^+) = 0.99992|2, 0, m\rangle - 0.00908(|2, 2, m\rangle + |2, -2, m\rangle), \quad (11\text{-}18)$$

$$\Phi^+(J = 2, E^+) = 0.01285|2, 0, m\rangle + 0.70705(|2, 2, m\rangle + |2, -2, m\rangle). \quad (11\text{-}19)$$

Clearly Φ^- is very closely a $K_a = 0$ state and Φ^+ is very closely a $K_a = 2$ state. The quantum number K_a is a useful *near quantum number* for determining which are the important perturbations of the states of an asymmetric top molecule, such as SO_2, that is a near prolate top. For a near oblate top $(A \cong B)$ K_c is a useful near quantum number. For an asymmetric top molecule having a large degree of asymmetry [i.e., $\kappa \cong 0$, see Eq. (8-143)] neither K_a nor K_c is a useful near quantum number in this regard. However, each does provide a convenient unambiguous label for the energy levels, and asymmetric top energy levels are labeled $J_{K_a K_c}$. The labels K_a and K_c also allow the symmetry in D_2 or in the MS group to be simply determined by consideration of their evenness or oddness.

The eigenfunctions of a one-dimensional harmonic oscillator Hamiltonian are labeled by the vibrational quantum number v. For a harmonic oscillator v is a good quantum number. For the low vibrational states of an anharmonic oscillator, v is a useful near quantum number in that each state consists mainly of one v state. For a two-dimensional harmonic oscillator l is an additional quantum number that is spoiled by anharmonicity and for a three-dimensional harmonic oscillator n is also used. The vibrational states of polyatomic molecules are labeled using the near quantum numbers v, l, and n as appropriate, e.g., in methane the vibrational states are labeled by the values of $v_1, v_2, v_3, v_4, l_2, l_3, l_4, n_3$, and n_4. Such labels are useful near quantum number labels as long as severe mixing between states having different labels does not take place. For example, in the CO_2 molecule the states ($v_1 = 0$, $v_2 = 2$, $v_3 = 0$) and $(1,0,0)$ for $l_2 = 0$ are heavily mixed so that for these states the quantum numbers v_1 and v_2 are not useful near quantum numbers. The relationship between the vibrational quantum numbers, the degeneracies of the levels, and an appropriate near symmetry group has been discussed many times [see, for example, Baker (1956) and Hwa and Nuyts (1966)].

For the rigid rotor rotational states of a symmetric top molecule the quantum number K is a true quantum number, but it is a near quantum number for the rotation–vibration or rovibronic states. This quantum number is spoiled by centrifugal distortion and Coriolis coupling effects. As a result of the fact that the Hamiltonian of a molecule commutes with the operation of time reversal (which sends every wavefunction into its complex conjugate; see Chapter 6) each eigenstate always involves the sum or difference of the $k = K$ and $k = -K$ wavefunctions. The *energy levels* should therefore be labeled using the unsigned quantum number K rather than the signed quantum number k. The quantum number J is a near quantum number label for the complete internal states E_{int} and it is spoiled by, for example, the nuclear spin coupling term $\hat{H}_{nsr}$. However, F is a true quantum number label for an isolated molecule in free space.

Usually near symmetry labels and near quantum numbers are just called symmetry labels and quantum numbers. In the rest of this chapter we will see that they provide very useful extra restrictions on allowed transitions and perturbations. While not rigorously true these restrictions do enable us to understand the main perturbations and transitions in a molecule.

NONVANISHING PERTURBATION TERMS

Symmetry can be used to show that certain perturbation terms in the Hamiltonian of a molecule vanish. Rotation–vibration perturbation terms (within a particular electronic state) are particularly amenable to such an analysis; these perturbations give rise to anharmonicity, centrifugal distortion, and Coriolis coupling effects and are written as [see Eqs. (8-28b)–(8-28d) and Eqs. (7-138) and (7-149)]

$$V_{\text{anh}} = \frac{1}{6} \sum_{r,s,t} \Phi_{rst} Q_r Q_s Q_t + \frac{1}{24} \sum_{r,s,t,u} \Phi_{rstu} Q_r Q_s Q_t Q_u + \cdots, \quad (11\text{-}20)$$

$$T_{\text{cent}} = -\frac{1}{2} \sum_{\alpha,\beta,r} \mu_{\alpha\alpha}^e a_r^{\alpha\beta} \mu_{\beta\beta}^e Q_r \hat{J}_\alpha \hat{J}_\beta$$

$$+ \frac{3}{8} \sum_{\alpha,\beta,\gamma,r,s} \mu_{\alpha\alpha}^e a_r^{\alpha\gamma} \mu_{\gamma\gamma}^e a_s^{\gamma\beta} \mu_{\beta\beta}^e Q_r Q_s \hat{J}_\alpha \hat{J}_\beta + \cdots, \quad (11\text{-}21)$$

and

$$T_{\text{Cor}} = -\sum_{\alpha,r,s} \mu_{\alpha\alpha}^e \zeta_{r,s}^\alpha Q_r \hat{P}_s \hat{J}_\alpha + \cdots. \quad (11\text{-}22)$$

Each of these terms is the product of a *coupling coefficient*, such as Φ_{rst} or $-\mu_{\alpha\alpha}^e \zeta_{r,s}^\alpha$ and a *coupling operator*, such as $Q_r Q_s Q_t$ or $Q_r \hat{P}_s \hat{J}_\alpha$. Since each of these terms must be totally symmetric in the molecular symmetry group of the molecule under consideration (each term is part of the complete Hamiltonian $\hat{H}$, and $\hat{H}$ is totally symmetric in the MS group), the coupling coefficient must vanish if the coupling operator is not totally symmetric. For example, in the water molecule the normal coordinates Q_1 and Q_3 are of species A_1 and B_2, respectively, in $C_{2v}(\text{M})$, so that $Q_1{}^2 Q_3$ is of species B_2. Thus Φ_{113} must be zero since the nontotally symmetric term $\Phi_{113} Q_1{}^2 Q_3$ cannot occur in the Hamiltonian. We can write the following rules, where we use the notation that $\Gamma(x)$ is the symmetry species of x in the MS group, and that $\Gamma^{(s)}$ is the totally symmetric representation of the MS group (called variously $A, A_1, A_g, A', A_{1g}, A_1', A_1{}^+, \Sigma^+$, or $\Sigma_g{}^+$ in the character tables in Appendix A): Φ_{rst} can be nonvanishing only if

$$\Gamma(Q_r) \otimes \Gamma(Q_s) \otimes \Gamma(Q_t) \supset \Gamma^{(s)}, \quad (11\text{-}23)$$

Φ_{rstu} can be nonvanishing only if

$$\Gamma(Q_r) \otimes \Gamma(Q_s) \otimes \Gamma(Q_t) \otimes \Gamma(Q_u) \supset \Gamma^{(s)}, \quad (11\text{-}24)$$

$a_r^{\alpha\beta}$ can be nonvanishing only if

$$\Gamma(Q_r) \otimes \Gamma(\hat{J}_\alpha) \otimes \Gamma(\hat{J}_\beta) \supset \Gamma^{(s)}, \quad (11\text{-}25)$$

or, equivalently, only if

$$\Gamma(\hat{J}_\alpha) \otimes \Gamma(\hat{J}_\beta) \supset \Gamma(Q_r), \quad (11\text{-}26)$$

and $\zeta_{r,s}^\alpha$ can be nonvanishing only if

$$\Gamma(Q_r) \otimes \Gamma(\hat{P}_s) \otimes \Gamma(\hat{J}_\alpha) \supset \Gamma^{(s)}, \quad (11\text{-}27)$$

or, equivalently, only if

$$\Gamma(Q_r) \otimes \Gamma(Q_s) \supset \Gamma(\hat{J}_\alpha), \quad (11\text{-}28)$$

since

$$\Gamma(\hat{P}_s) = \Gamma(Q_s). \qquad (11\text{-}29)$$

If the coupling operator is totally symmetric the coupling coefficient could still vanish, and this would be said to be an accident.

The Transformation Properties of the Rotational and Translational Coordinates

To apply the preceding rules for a particular molecule it is necessary to determine the representations of the MS group that are generated by the normal coordinates Q_r and by the rovibronic angular momentum components $\hat{J}_x, \hat{J}_y$, and $\hat{J}_z$. To determine the normal coordinate representation we determine the representation generated by the $3N$ Cartesian displacement coordinates and subtract the representation generated by the translational and rotational coordinates T_λ and R_λ, where $\lambda = x, y$, or z [see Eqs. (7-233)–(7-250)]. We therefore need to know the representation generated by the T_λ and the R_λ. In this subsection we show that for asymmetric and symmetric top molecules the transformation properties of R_λ, T_λ, and $\hat{J}_\lambda$ (as well as of the direction cosine matrix elements and vibrational and electronic angular momentum components, $\hat{p}_\lambda$ and $\hat{L}_\lambda$) under the effect of an element from the MS group follow once the equivalent rotation of the MS group element has been determined. For spherical top molecules the symmetry species of these factors is best determined by reduction of their species in the group $K(\text{mol})$.

From the results in Table 7-1 the transformation properties of $\hat{J}_x, \hat{J}_y$, and $\hat{J}_z$ under the effect of each element of an MS group follow once the equivalent rotations (R_z^β or R_α^π) of the elements have been determined. Similarly since the direction cosine matrix elements only involve the Euler angles [see Eq. (7-52)] their transformation properties in the MS group follow from the results in Table 7-1 once the equivalent rotations of the elements have been determined. It turns out that the transformation properties of $(\lambda_{x\tau}, \lambda_{y\tau}, \lambda_{z\tau})$ are the same as those of $(\hat{J}_x, \hat{J}_y, \hat{J}_z)$ regardless of whether $\tau = \xi, \eta$, or ζ. This result follows by application of the discussion that leads to Eq. (5-118), since we can write

$$\hat{J}_\tau = \lambda_{x\tau}\hat{J}_x + \lambda_{y\tau}\hat{J}_y + \lambda_{z\tau}\hat{J}_z, \qquad (11\text{-}30)$$

and $\hat{J}_\tau$ must be totally symmetric in the MS group, as we now show. $\hat{J}_\tau$ is given by [see, for example, Eq. (7-78)]

$$\hat{J}_\tau = \sum_{i,\mu,\nu} \varepsilon_{\tau\mu\nu}\mu_i\hat{P}_{\nu_i}, \qquad (11\text{-}31)$$

where $\tau, \mu, \nu = \xi, \eta$, or ζ, $\hat{P}_{\nu_i} = -i\hbar\,\partial/\partial\nu_i$, and $\varepsilon_{\tau\mu\nu}$ is defined in Eq. (7-90). A permutation operation simply interchanges terms ($\mu_i\hat{P}_{\nu_i}$) in the sum over i

and the sum, i.e., $\hat{J}_\tau$, is unaffected. The inversion E^* changes the sign of all μ_i and $\hat{P}_{v_i}$ so that $\hat{J}_\tau$ is invariant. As a result $\hat{J}_\tau$ is invariant to all elements of the MS group and is totally symmetric. Also following from Eq. (5-118) is the fact that the vibrational angular momenta $\hat{p}_\lambda$ and electronic angular momenta $\hat{L}_\lambda$ transform in the same way as $\hat{J}_\lambda$ (where $\lambda = x, y$, or z) since the terms $\mu^e_{\lambda\lambda}\hat{p}_\lambda\hat{J}_\lambda$ and $\mu^e_{\lambda\lambda}\hat{L}_\lambda\hat{J}_\lambda$ occur in the molecular Hamiltonian and must, therefore, be totally symmetric in the MS group. The representations generated by the three $\hat{J}_\lambda$ for all the MS groups have been determined from the results in Table 7-1 and are given in each of the character tables in Appendix A.

The transformation properties of the translational and rotational coordinates T_λ and R_λ can be determined from the equivalent rotations as follows. Suppose P is a permutation element of the MS group of an asymmetric top or symmetric top molecule. We write this element here as $P[R_a{}^b]$ where $R_a{}^b$ is the equivalent rotation of P (it is either $R_z{}^\beta$ or $R_\alpha{}^\pi$ in the notation of Table 7-1). Suppose that the operation $P[R_a{}^b]$ has the effect of replacing nucleus j by nucleus i [i.e., it contains the permutation $(\cdots ji \cdots)$], then we can write

$$P[R_a{}^b]\lambda_i = \lambda_i' = x_j f_{\lambda x}(R_a{}^b) + y_j f_{\lambda y}(R_a{}^b) + z_j f_{\lambda z}(R_a{}^b), \qquad (11\text{-}32)$$

where $\lambda = x, y$, or z, and $f_{\lambda\mu}(R_a{}^b)$ is a function of the angle α or β that defines $R_a{}^b$ (see Table 7-1). For the operation $R_z{}^\beta$ we have

$$\begin{aligned} P[R_z{}^\beta]x_i &= x_j \cos\beta + y_j \sin\beta, \\ P[R_z{}^\beta]y_i &= y_j \cos\beta - x_j \sin\beta, \\ P[R_z{}^\beta]z_i &= z_j, \end{aligned} \qquad (11\text{-}33)$$

so that

$$\begin{aligned} f_{xx}(R_z{}^\beta) &= f_{yy}(R_z{}^\beta) = \cos\beta, \\ f_{xy}(R_z{}^\beta) &= -f_{yx}(R_z{}^\beta) = \sin\beta, \\ f_{xz}(R_z{}^\beta) &= f_{yz}(R_z{}^\beta) = f_{zx}(R_z{}^\beta) = f_{zy}(R_z{}^\beta) = 0, \\ f_{zz}(R_z{}^\beta) &= 1. \end{aligned} \qquad (11\text{-}34)$$

For example, in the BF_3 molecule (see Fig. 10-1) the operation (123) has the equivalent rotation $R_z^{2\pi/3}$ and we have

$$(123)[R_z^{2\pi/3}]x_3 = x_3' = x_2 \cos(2\pi/3) + y_2 \sin(2\pi/3). \qquad (11\text{-}35)$$

For the operation $R_\alpha{}^\pi$ we determine that in general

$$\begin{aligned} f_{xx}(R_\alpha{}^\pi) &= -f_{yy}(R_\alpha{}^\pi) = \cos 2\alpha, \\ f_{xy}(R_\alpha{}^\pi) &= f_{yx}(R_\alpha{}^\pi) = \sin 2\alpha, \\ f_{xz}(R_\alpha{}^\pi) &= f_{yz}(R_\alpha{}^\pi) = f_{zx}(R_\alpha{}^\pi) = f_{zy}(R_\alpha{}^\pi) = 0, \\ f_{zz}(R_\alpha{}^\pi) &= -1. \end{aligned} \qquad (11\text{-}36)$$

Notice that the $f_{\lambda\mu}(R_a{}^b)$ are the $\lambda\mu$ elements of the three-dimensional representation matrix of the operation $R_a{}^b$ generated by $\hat{J}_x, \hat{J}_y,$ and $\hat{J}_z$. For a permutation–inversion element P^* (where P replaces j by i) we similarly write

$$P^*[R_a{}^b]\lambda_i = x_j f_{\lambda x}^*(R_a{}^b) + y_j f_{\lambda y}^*(R_a{}^b) + z_j f_{\lambda z}^*(R_a{}^b), \qquad (11\text{-}37)$$

and since a permutation–inversion operation changes the signs of all the coordinates we must have

$$f_{\lambda\mu}^*(R_a{}^b) = -f_{\lambda\mu}(R_a{}^b). \qquad (11\text{-}38)$$

We are now in a position to determine the general effect of the MS group elements $P[R_a{}^b]$ and $P^*[R_a{}^b]$ on the coordinates T_λ and R_λ.

First we look at the translational coordinates T_λ, and from Eq. (7-235) we have

$$T_\lambda = M_N^{-1/2} \sum_i m_i \Delta\lambda_i, \qquad (11\text{-}39)$$

and, therefore,

$$P[R_a{}^b]T_\lambda = M_N^{-1/2} \sum_i m_i \Delta\lambda_i' \qquad (11\text{-}40)$$

$$= M_N^{-1/2} \sum_j m_j \left(\sum_\mu f_{\lambda\mu}(R_a{}^b) \Delta\mu_j \right). \qquad (11\text{-}41)$$

Equation (11-41) follows from Eq. (11-40) since $m_i = m_j$ (P only permutes identical nuclei) and the sum ($\sum_i = \sum_j$) is over all nuclei in the molecule. We can write Eq. (11-41) as

$$P[R_a{}^b]T_\lambda = \sum_\mu f_{\lambda\mu}(R_a{}^b)T_\mu. \qquad (11\text{-}42)$$

Similarly

$$P^*[R_a{}^b]T_\lambda = \sum_\mu f_{\lambda\mu}^*(R_a{}^b)T_\mu, \qquad (11\text{-}43)$$

and from Eq. (11-38) we have

$$P^*[R_a{}^b]T_\lambda = -\sum_\mu f_{\lambda\mu}(R_a{}^b)T_\mu. \qquad (11\text{-}44)$$

From Eq. (11-42) we see that T_λ transforms in the same way as $\hat{J}_\lambda$ under a permutation operation of the MS group but, from Eq. (11-44), T_λ transforms with opposite sign to $\hat{J}_\lambda$ under a permutation–inversion operation. The transformation properties of $T_x, T_y,$ and T_z are given in Table 11-6 under the effect of MS group operations P and P^* as a function of the equivalent rotation of the operations. Using this table the representation of any MS group (of an asymmetric or symmetric top molecule) that is generated by the

Table 11-6

The transformation properties of the translational coordinates T_x, T_y, and T_z[a]
under the effect of permutations (P) and permutation–inversions (P^*) of an MS group[b]

$P[R_z{}^\beta]$	$P[R_\alpha{}^\pi]$	$P^*[R_z{}^\beta]$	$P^*[R_\alpha{}^\pi]$
T_x: $\quad T_x \cos\beta + T_y \sin\beta$	$T_x \cos 2\alpha + T_y \sin 2\alpha$	$-T_x \cos\beta - T_y \sin\beta$	$-T_x \cos 2\alpha - T_y \sin 2\alpha$
T_y: $\quad -T_x \sin\beta + T_y \cos\beta$	$T_x \sin 2\alpha - T_y \cos 2\alpha$	$T_x \sin\beta - T_y \cos\beta$	$-T_x \sin 2\alpha + T_y \cos 2\alpha$
T_z: $\quad T_z$	$-T_z$	$-T_z$	T_z

[a] See Eq. (7-235).

[b] The results are given as a function of the equivalent rotation of each element (see Table 7-1 for the definition of the equivalent rotations).

T_λ can be determined; these species are indicated in the character tables in Appendix A.

To appreciate how the rotational coordinates R_x, R_y, and R_z transform we look in detail at the transformation properties of R_x [see Eq. (7-236)]; R_x is given by

$$R_x = (\mu_{xx}^e)^{1/2} \sum_i m_i (y_i^e \Delta z_i - z_i^e \Delta y_i). \tag{11-45}$$

Using $m_i = m_j$, $\sum_i = \sum_j$, and the relationships between (y_i^e, z_i^e) and (y_j^e, z_j^e) we have

$$P[R_z{}^\beta]R_x = (\mu_{xx}^e)^{1/2} \sum_j m_j [(- \sin\beta x_j^e + \cos\beta y_j^e)\Delta z_j$$

$$- z_j^e(- \sin\beta \Delta x_j + \cos\beta \Delta y_j)]$$

$$= \cos\beta(\mu_{xx}^e)^{1/2} \sum_j m_j (y_j^e \Delta z_j - z_j^e \Delta y_j)$$

$$+ \sin\beta(\mu_{xx}^e)^{1/2} \sum_j m_j(z_j^e \Delta x_j - x_j^e \Delta z_j)$$

$$= R_x \cos\beta + (\mu_{xx}^e/\mu_{yy}^e)^{1/2} R_y \sin\beta. \tag{11-46}$$

For a symmetric top molecule $\mu_{xx}^e = \mu_{yy}^e$ and for an asymmetric top molecule we can only have $\beta = 0$ or π when $\sin\beta = 0$; thus we can effectively write

$$P[R_z{}^\beta]R_x = R_x \cos\beta + R_y \sin\beta \tag{11-47}$$

and R_x transforms in the same way as $\hat{J}_x$ under $P[R_z{}^\beta]$. Since R_x involves the products $y_i^e \Delta z_i$ and $z_i^e \Delta y_i$ the permutation–inversion $P^*[R_z{}^\beta]$ has the same effect as $P[R_z{}^\beta]$ since the sign change from Eq. (11-38) cancels out. Thus R_x transforms as $\hat{J}_x$ under $P[R_z{}^\beta]$ and $P^*[R_z{}^\beta]$. In a similar manner it is possible to show that R_x, R_y, and R_z transform in the same way as $\hat{J}_x, \hat{J}_y,$ and $\hat{J}_z$, respectively, under all the elements of the MS group. Hence the species of the rotational coordinates in an MS group are the same as those of $\hat{J}_x, \hat{J}_y$, and $\hat{J}_z$, and the species of the latter are indicated in the character tables in Appendix A.

For spherical top molecules the operators $(\hat{J}_x, \hat{J}_y, \hat{J}_z)$ transform as the representation $D^{(1)}$ of the group K(mol). From the correlation of the species of K(mol) with the species of the MS group the species of $(\hat{J}_x, \hat{J}_y, \hat{J}_z)$ in the MS group can be determined. The species of (R_x, R_y, R_z), $(\hat{p}_x, \hat{p}_y, \hat{p}_z)$, $(\hat{L}_x, \hat{L}_y, \hat{L}_z)$, and $(\lambda_{x\tau}, \lambda_{y\tau}, \lambda_{z\tau})$ are the same as those of $(\hat{J}_x, \hat{J}_y, \hat{J}_z)$. The representation generated by (T_x, T_y, T_z) is obtained from that generated by $(\hat{J}_x, \hat{J}_y, \hat{J}_z)$ by changing the signs of the characters for all permutation–inversion operations.

The transformation properties of these various terms in the molecular point group can also be determined. For nonlinear rigid molecules (R_x, R_y, R_z), (T_x, T_y, T_z), $(\hat{L}_x, \hat{L}_y, \hat{L}_z)$ and $(\hat{p}_x, \hat{p}_y, \hat{p}_z)$ transform in the same way in the point group as they do in the isomorphic MS group. However, since the elements of the molecular point group do not change the Euler angles then $(\hat{J}_x, \hat{J}_y, \hat{J}_z)$ and the direction cosine matrix elements are not transformed by the elements of the point group.

The results of this subsection on transformation properties are needed for the application of Eqs. (11-23)–(11-28) to determine the nonvanishing Φ, $a^{\alpha\beta}$, and ζ^α coefficients for a molecule. For molecules containing degenerate representations and degenerate normal coordinates it is necessary to choose a convention for the transformation properties of the individual members of each set of degenerate normal coordinates before the equations can be applied. In Problem 11-2 below such a case is discussed.

Problem 11-1. Determine the nonvanishing Φ, $a^{\alpha\beta}$, and ζ^α coefficients for the ground electronic state of the trans diimide molecule (N_2H_2). This molecule is shown in its equilibrium configuration in Fig. 11-6 and we assume that there is no torsional tunneling. The MS group of this molecule is C_{2h}(M) [see Table A-5 where the permutation (56) should be omitted].

Fig. 11-6. A nuclear labeled diimide molecule in the equilibrium configuration of its ground electronic state. The inertial axes (a, b, c) are also marked $(I_{aa} < I_{bb} < I_{cc})$.

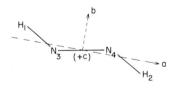

Answer. The six normal coordinates of this molecule have the symmetry species:

$$\Gamma(Q_1) = \Gamma(Q_2) = \Gamma(Q_3) = A_g,$$
$$\Gamma(Q_4) = A_u, \qquad (11\text{-}48)$$
$$\Gamma(Q_5) = \Gamma(Q_6) = B_u.$$

From these results and the rules of Eqs. (11-23) and (11-24) we deduce that the following Φ coefficients do not vanish by symmetry:

$$\Phi_{pqr}, \Phi_{p44}, \Phi_{ptu}, \Phi_{pqrs}, \Phi_{pq44}, \Phi_{pqtu}, \Phi_{44tu}, \Phi_{4444}, \text{ and } \Phi_{tuvw}, \quad (11\text{-}49)$$

where p, q, r, and $s = 1$, 2, or 3, and t, u, v, and $w = 5$ or 6. From Eq. (11-26) and the symmetries of $\hat{J}_a, \hat{J}_b, \hat{J}_c$ (given in Table A-5) and of the Q_r, we deduce that the following $a^{\alpha\beta}$ coefficients do not vanish by symmetry

$$a_r^{cc}, a_r^{ab} = a_r^{ba}, a_r^{aa}, \text{ and } a_r^{bb}, \quad (11\text{-}50)$$

where $r = 1$, 2, or 3. Note that for the u vibrations of a centrosymmetric molecule all $a_r^{\alpha\beta}$ vanish since the species of the components of $\hat{J}$ are always g (the operation $\hat{O}_i$ that defines the g or u character does not affect the Euler angles). As a result the rotational levels of g and u vibrational states cannot be mixed by the centrifugal distortion perturbation [see Eq. (11-16) and the discussion after it]. From Eq. (11-28) we see that the nonvanishing Coriolis coupling coefficients are

$$\zeta_{p,q}^c, \zeta_{5,6}^c, \zeta_{4,t}^a, \text{ and } \zeta_{4,t}^b, \quad (11\text{-}51)$$

where p, $q = 1$, 2, or 3 and $t = 5$ or 6. The rotational levels of g and u vibrational states cannot be mixed by the Coriolis coupling perturbation since the $\hat{J}_\alpha$ all have g species, and all $\zeta_{r,s}^\alpha$ involving one g and one u vibration must vanish [see Eq. (11-16) and the discussion after it]. In N_2H_2 the vibrational states $v_4 = 1$ and $v_6 = 1$ are close in energy and the perturbations from $\zeta_{4,6}^a$ and $\zeta_{4,6}^b$ coupling will be very important (see Problem 11-3).

Problem 11-2. Determine the nonvanishing Φ, $a^{\alpha\beta}$, and ζ^α coefficients for the CH_3F molecule in its ground electronic state. The CH_3F molecule is labeled in Fig. 11-7 and its MS group is the $C_{3v}(M)$ group given in Table A-8.

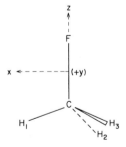

Fig. 11-7. A nuclear labeled methyl fluoride molecule in the equilibrium configuration of its ground electronic state. The (x, y, z) axes are the inertial axes $(I_{zz} < I_{xx} = I_{yy})$.

Answer. The symmetry species of the normal coordinates for CH_3F in $C_{3v}(M)$ are

$$\Gamma(Q_1) = \Gamma(Q_2) = \Gamma(Q_3) = A_1,$$
$$\Gamma(Q_{4a}, Q_{4b}) = \Gamma(Q_{5a}, Q_{5b}) = \Gamma(Q_{6a}, Q_{6b}) = E. \quad (11\text{-}52)$$

Table 11-7

The transformation properties of the normal coordinates and angular momentum components of CH_3F in the $C_{3v}(M)$ group[a]

	E	(123)	$(23)^*$	
Equiv. rot.	R^0	$R_z^{2\pi/3}$	$R_{\pi/2}^{\pi}$	
	Q_p	Q_p	$Q_p \quad :A_1$	
	Q_{ta}	$Q_{ta}c + Q_{tb}s$	$\left.\begin{array}{c} Q_{ta} \\ -Q_{tb} \end{array}\right\} :E$	
	Q_{tb}	$Q_{tb}c - Q_{ta}s$		
	Q_t	Q_t	Q_t	
	$\exp(i\alpha_t)$	$\omega^2 \exp(i\alpha_t)$	$\exp(-i\alpha_t)$	
	Q_t^+	$\omega^2 Q_t^+$	$\left.\begin{array}{c} Q_t^- \\ Q_t^+ \end{array}\right\} :E$	
	Q_t^-	ωQ_t^-		
	$\hat{J}_x$	$\hat{J}_x c + \hat{J}_y s$	$\left.\begin{array}{c} -\hat{J}_x \\ \hat{J}_y \end{array}\right\} :E$	
	$\hat{J}_y$	$\hat{J}_y c - \hat{J}_x s$		
	$\hat{J}_z$	$\hat{J}_z$	$-\hat{J}_z \quad :A_2$	
	$\hat{J}_m^+$	$\omega^2 \hat{J}_m^+$	$\left.\begin{array}{c} -\hat{J}_m^- \\ -\hat{J}_m^+ \end{array}\right\} :E$	
	$\hat{J}_m^-$	$\omega \hat{J}_m^-$		

[a] $p = 1$, 2, or 3; $t = 4$, 5, or 6; $c = \cos(2\pi/3)$; $s = \sin(2\pi/3)$; $\omega = \exp(2\pi i/3) = c + i\,s$; $\omega^2 = \exp(-2\pi i/3) = c - i\,s$. $Q_{ta} = Q_t \cos\alpha_t$ and $Q_{tb} = Q_t \sin\alpha_t$. $Q_t^\pm = Q_{ta} \pm iQ_{tb}$ and $\hat{J}_m^\pm = \hat{J}_x \pm i\hat{J}_y$.

In order to follow the convention of Di Lauro and Mills (1966) we choose the degenerate coordinates (Q_{ta}, Q_{tb}) to transform like (T_x, T_y); see Table 11-7. In the $C_{3v}(M)$ group we have

$$\begin{aligned} A_1 \otimes A_1 &= A_2 \otimes A_2 = A_1, \\ A_1 \otimes E &= A_2 \otimes E = E, \\ E \otimes E &= A_1 \oplus A_2 \oplus E, \\ [E]^2 &= A_1 \oplus E. \end{aligned} \tag{11-53}$$

From Eqs. (11-23) and (11-24), together with the species of the normal coordinates, given in Eq. (11-52), we deduce that the following anharmonic force constants can be nonvanishing:

$$\Phi_{pqr}, \Phi_{rst}, \Phi_{stu}, \Phi_{opqr}, \Phi_{qrst}, \Phi_{rstu}, \text{ and } \Phi_{stuv}, \tag{11-54}$$

where $o, p, q, r = 1$, 2, or 3, and $s, t, u, v = 4$, 5, or 6. For the degenerate vibrations we need to know how the a and b labels should be added, i.e., is Φ_{rst} to be taken as $\Phi_{rsata}, \Phi_{rsatb}, \Phi_{rsbta},$ or Φ_{rsbtb}? We determine this from the transformation properties of products such as $Q_r Q_{sa} Q_{ta}$ which can be obtained from the results in Table 11-7. We find that the following combination is totally symmetric:

$$(Q_r Q_{sa} Q_{ta} + Q_r Q_{sb} Q_{tb}) = Q_r (Q_s^+ Q_t^- + Q_s^- Q_t^+)/2, \qquad (11\text{-}55)$$

where $Q_s^{\pm}$ and $Q_t^{\pm}$ are related to (Q_{sa}, Q_{sb}) and (Q_{ta}, Q_{tb}) as discussed in Chapter 8 [see Eqs. (8-204)–(8-207) and (8-213)–(8-214)]. Thus

$$\Phi_{rsata} = \Phi_{rsbtb}(=\Phi_{rst}, \text{ say}) \qquad (11\text{-}56)$$

can be nonvanishing, but

$$\Phi_{rsatb} = \Phi_{rsbta} = 0. \qquad (11\text{-}57)$$

   The potential function, therefore, can contain the term

$$\Phi_{rst} Q_r (Q_{sa} Q_{ta} + Q_{sb} Q_{tb}) = \Phi_{rst} Q_r (Q_s^+ Q_t^- + Q_s^- Q_t^+)/2. \qquad (11\text{-}58)$$

The other nonvanishing cubic anharmonicity term that involves the degenerate coordinate is

$$\Phi_{stu}(Q_{sa} Q_{ta} Q_{ua} - Q_{sb} Q_{ta} Q_{ub} - Q_{sb} Q_{tb} Q_{ua} - Q_{sa} Q_{tb} Q_{ub})$$
$$= \Phi_{stu}(Q_s^+ Q_t^+ Q_u^+ + Q_s^- Q_t^- Q_u^-)/2. \qquad (11\text{-}59)$$

Quartic anharmonicity terms involving the degenerate normal coordinates that are totally symmetric in $C_{3v}(M)$ and which therefore can be nonvanishing are:

$$\begin{aligned}
&\Phi_{qrst} Q_q Q_r (Q_s^+ Q_t^- + Q_s^- Q_t^+)/2, \\
&\Phi_{rstu} Q_r (Q_s^+ Q_t^+ Q_u^+ + Q_s^- Q_t^- Q_u^-)/2, \\
&\Phi_{sstu} Q_s^+ Q_s^- (Q_t^+ Q_u^- + Q_t^- Q_u^+)/2, \\
&\Phi'_{sstu} (Q_s^+ Q_s^+ Q_t^- Q_u^- + Q_s^- Q_s^- Q_t^+ Q_u^+)/2.
\end{aligned} \qquad (11\text{-}60)$$

The nonvanishing $a^{\alpha\beta}$ coefficients are determined from Eq. (11-26) and Table 11-7. We see that $Q_r \hat{J}_z^2$ and $Q_r (\hat{J}_x^2 + \hat{J}_y^2)$ are totally symmetric, so

$$a_r^{zz} \neq 0 \qquad \text{and} \qquad a_r^{xx} = a_r^{yy} \neq 0, \qquad (11\text{-}61)$$

where $r = 1, 2,$ or 3. Further we have

$$\Gamma(\hat{J}_z) \otimes \Gamma(\hat{J}_x, \hat{J}_y) = \Gamma(Q_{ta}, Q_{tb}) \qquad (11\text{-}62)$$

and

$$\Gamma(\hat{J}_x, \hat{J}_y) \otimes \Gamma(\hat{J}_x, \hat{J}_y) \supset \Gamma(Q_{ta}, Q_{tb}), \qquad (11\text{-}63)$$

where $t = 4, 5,$ or 6. From a detailed consideration of the transformation properties of Q_{ta} and Q_{tb} (or of Q_t^+ and Q_t^-) in Table 11-7 we determine that the centrifugal coupling operator

$$[Q_{ta}(\hat{J}_x \hat{J}_z + \hat{J}_z \hat{J}_x) + Q_{tb}(\hat{J}_y \hat{J}_z + \hat{J}_z \hat{J}_y)] \qquad (11\text{-}64)$$

is totally symmetric. Thus

$$a_{ta}^{xz} = a_{ta}^{zx} = a_{tb}^{yz} = a_{tb}^{zy} (= a_t^{xz}, \text{say}) \neq 0 \tag{11-65}$$

in CH_3F where $t = 4, 5,$ or 6. We can write the appropriate centrifugal distortion term as

$$a_t^{xz}[Q_t^-(\hat{J}_m^+ \hat{J}_z + \hat{J}_z \hat{J}_m^+) + Q_t^+(\hat{J}_m^- \hat{J}_z + \hat{J}_z \hat{J}_m^-)]/2. \tag{11-66}$$

Similarly

$$a_{tb}^{xy} = a_{tb}^{yx} = a_{ta}^{yy} = -a_{ta}^{xx} (= -a_t^{xx}, \text{say}) \neq 0 \tag{11-67}$$

and the appropriate nonvanishing centrifugal distortion term can be written as

$$a_t^{xx}[Q_t^+(\hat{J}_m^+)^2 + Q_t^-(\hat{J}_m^-)^2]/2, \tag{11-68}$$

where $t = 4, 5,$ or 6. The symmetric Coriolis coupling operators are [where the terms arising from the antisymmetry of the ζ coefficients, see Eq. (7-139), are to be understood]

$$(Q_r \hat{P}_{sa} \hat{J}_y - Q_r \hat{P}_{sb} \hat{J}_x), \tag{11-69}$$

$$(Q_{sa} \hat{P}_{tb} - Q_{sb} \hat{P}_{ta}) \hat{J}_z, \tag{11-70}$$

and

$$(Q_{sa} \hat{P}_{ta} - Q_{sb} \hat{P}_{tb}) \hat{J}_y + (Q_{sa} \hat{P}_{tb} + Q_{sb} \hat{P}_{ta}) \hat{J}_x, \tag{11-71}$$

so that the nonvanishing Coriolis coupling coefficients are

$$\zeta_{r,sa}^y = -\zeta_{r,sb}^x (= \zeta_{r,s}^y, \text{say}), \tag{11-72}$$

$$\zeta_{sa,tb}^z = -\zeta_{sb,ta}^z (= \zeta_{s,t}^z, \text{say}), \tag{11-73}$$

and

$$\zeta_{sa,ta}^y = -\zeta_{sb,tb}^y = \zeta_{sa,tb}^x = \zeta_{sb,ta}^x (= \zeta_{s,t}^y, \text{say}), \tag{11-74}$$

where $r = 1, 2,$ or 3 and $s, t = 4, 5,$ or 6 [s can equal t in Eq. (11-73) but not in Eq. (11-74) because of the antisymmetry of the ζ coefficients]. The nonvanishing Coriolis coupling terms [note the minus sign in Eq. (11-22)] can be written

$$-(i/2)\mu_{yy}^e \zeta_{r,s}^y Q_r(\hat{P}_s^+ \hat{J}_m^- - \hat{P}_s^- \hat{J}_m^+), \tag{11-75}$$

$$-(i/2)\mu_{zz}^e \zeta_{s,t}^z (Q_s^+ \hat{P}_t^- - Q_s^- \hat{P}_t^+) \hat{J}_z, \tag{11-76}$$

and

$$-(i/2)\mu_{yy}^e \zeta_{s,t}^y (Q_s^- \hat{P}_t^- \hat{J}_m^- - Q_s^+ \hat{P}_t^+ \hat{J}_m^+), \tag{11-77}$$

where $r = 1, 2,$ or 3 and $s, t = 4, 5,$ or 6 [$s \neq t$ in Eq. (11-77)].

292   11. Near Symmetry and Perturbations

PERTURBATIONS BETWEEN STATES

The Hamiltonian of a molecule is totally symmetric in the MS group and in the group K(spatial). As a result the symmetry labels (Γ_{int} and F) that are obtained by classifying the approximate (zero order) wavefunctions $\Phi^0_{int} = \Phi_{nspin}\Phi_{rot}\Phi_{vib}\Phi_{elec}\Phi_{espin}$ in these groups are true symmetry labels; internal perturbations will only cause states of the same symmetry in these groups to interact, and each final (exact) wavefunction bears the same true symmetry labels as the approximate wavefunctions of which it is composed (see Chapter 6). As a result of this we can use the symmetry of the approximate wavefunctions to determine these final labels and to determine which zero order molecular energy levels can interact with each other.

The MS group label Γ_{int}, obtained by classifying the approximate wavefunctions Φ^0_{int}, is not the only MS group symmetry label that is useful when studying molecular interactions. The MS group labels on the basis set wavefunctions Φ_{nspin}, Φ_{rot}, Φ_{vib}, Φ_{elec}, and Φ_{espin} are Γ_{nspin}, Γ_{rot}, Γ_{vib}, Γ_{elec}, and Γ_{espin}, respectively. These labels, or combinations of them, are called basis symmetry labels and they are useful for studying various perturbations between Φ^0_{int} states: Configuration interaction perturbations can only occur between states having the same Γ_{elec} label; anharmonicity perturbations can only occur between states having the same Γ_{vib} label; rotation–vibration interactions can only occur between states having the same $\Gamma_{rv}(=\Gamma_{rot}\otimes\Gamma_{vib})$ label; vibronic interactions can only occur between states having the same $\Gamma_{ve}(=\Gamma_{vib}\otimes\Gamma_{elec})$ label; rovibronic interactions can only occur between states having the same $\Gamma_{rve}(=\Gamma_{rot}\otimes\Gamma_{vib}\otimes\Gamma_{elec})$ label; and so on. These rules result from the fact that each perturbation term involves only some of the coordinates in the molecule and can, therefore, only affect that part of the zero order wavefunction that involves those coordinates. Invariably only certain interactions are important in a particular case, and only certain basis symmetry labels are needed in the discussion of the interactions occurring. Usually Hund's case (b) is appropriate and we can neglect the interaction of the nuclear and electron spins with the other degrees of freedom; the basis symmetry labels we would use in these cases to determine which rovibronic states can interact would be Γ_{rve}, from the MS group, and J [or N for nonsinglet states in Hund's case (b)], from the group K(spatial). These are the symmetry labels to use for studying the most common interactions that occur between molecular energy levels.

Near symmetry labels and near quantum numbers are generally spoiled by perturbations, i.e., states having different near symmetry labels or near quantum number labels interact. However, a level labeled by certain near symmetry labels and near quantum number labels can only be mixed, by a particular perturbation, with levels having certain other near symmetry labels and near quantum number labels. As a result there are *selection rules* on the near symmetry labels and near quantum number labels for allowed interactions, and these are very useful.

For a symmetric top molecule in a singlet electronic state the near quantum numbers that are used to label the energy levels are I, J, K, $(\pm l)$, $(v_1, v_2, \ldots, v_s^{l_s}, v_t^{l_t}, \ldots)$ and the occupation numbers $(n_1, n_2, \ldots)$ of the molecular orbitals in the single configuration electronic wavefunction used to describe the state; for nonsinglet states in Hund's case (b) we use N (instead of J) and S. The use of the $(\pm l)$ quantum number label for the levels of a symmetric top molecular rotation group. The basis symmetry labels Γ_{nspin}, Γ_{rot}, Γ_{vib}, Γ_{elec}, For an asymmetric top molecule we use K_a and K_c instead of K, and there are no l_s quantum numbers. For a spherical top molecule we do not use K, and for each three-dimensional harmonic oscillator state we use the quantum numbers v, l, and n (see Chapter 8); there is also the quantum number R that will be discussed later. Each type of perturbation mixes states according to definite selection rules for these quantum numbers. Near symmetry labels on the energy levels are obtained by using the molecular point group and the molecular rotation group. The basis symmetry labels Γ_{nspin}, Γ_{rot}, Γ_{vib}, Γ_{elec}, and Γ_{espin}, discussed previously, are really the same as near symmetry labels, and states having different such labels can be mixed. For example, a rotation–vibration interaction (such as Coriolis interaction) mixes states having the same Γ_{rv} label but possibly different Γ_{rot} and Γ_{vib} labels; a level in H_2O having $\Gamma_{\text{rot}} = A_1$, and $\Gamma_{\text{vib}} = B_2$, can be mixed by a rotation–vibration interaction with a level having $\Gamma_{\text{rot}} = B_2$ and $\Gamma_{\text{vib}} = A_1$, since both have $\Gamma_{\text{rv}} = B_2$. The molecular symmetry group basis symmetry labels Γ_{vib}, Γ_{elec}, and Γ_{ve} are the same as the near symmetry labels from the molecular point group for a rigid nonlinear molecule.

We will now look in detail at perturbations arising from various terms in the molecular Hamiltonian. Each of these terms couples certain degrees of freedom that in zero order are separate. Terms that couple the electronic coordinates with the rotational and/or the vibrational coordinates give rise to the breakdown of the Born–Oppenheimer approximation, terms that couple the vibrational and rotational coordinates give rise to rotation–vibration interactions, and terms that couple the nuclear spins to the other degrees of freedom can give rise to so-called ortho–para mixing. Each of these perturbations will be discussed using the true symmetry labels as well as the basis symmetry labels and the near symmetry labels. On a first reading of this chapter it would probably be best to skip this discussion and go straight to the section on the optical selection rules.

The Breakdown of the Born–Oppenheimer Approximation

When there is a breakdown of the Born–Oppenheimer approximation the rovibronic levels of one electronic state are coupled to the rovibronic levels of another electronic state by one of the following kinetic energy terms in the molecular Hamiltonian[1]:

$$\hat{T}_{\text{vib}} = \frac{1}{2} \sum_r \hat{P}_r^{\ 2}, \tag{11-78}$$

$$\hat{T}_{\text{er}} = -\frac{1}{2} \sum_{\alpha,\beta} \mu_{\alpha\beta} (\hat{J}_\alpha \hat{L}_\beta + \hat{L}_\alpha \hat{J}_\beta), \tag{11-79}$$

and

$$\hat{T}_{\text{ev}} = \frac{1}{2} \sum_{\alpha,\beta} \mu_{\alpha\beta} (\hat{p}_\alpha \hat{L}_\beta + \hat{L}_\alpha \hat{p}_\beta). \tag{11-80}$$

See Eq. (7-150) for $\hat{T}_{\text{vib}}$ and Eq. (8-19) for $\hat{T}_{\text{er}} + \hat{T}_{\text{ev}}$. In these expressions the normal coordinates are taken to be those of one of the electronic states, Φ_{elec}, say, and the normal coordinates of the other electronic state, Φ'_{elec}, say, are expressed in terms of them. Similarly the elements of $\mu_{\alpha\beta}$ are expanded about the equilibrium configuration of the state Φ_{elec} in the normal coordinates of the state Φ_{elec}. Without making further approximations all we can say is that these terms couple states that have the same N ($= J$ for singlet states), I and S quantum numbers, and the same MS group species Γ_{rve}; the vibronic perturbation $\hat{T}_{\text{vib}}$ will mix states having the same Γ_{rot} and Γ_{ve}. The MS group is now that appropriate for the simultaneous treatment of the electronic states under consideration. The dominant interactions can be determined by making some appropriate approximations and by using near symmetry and near quantum numbers.

An approximation that is usually made in order to obtain an appreciation of the dominant interactions is to say that the two electronic states involved in the interaction have the same equilibrium nuclear geometry and harmonic force constants (i.e., identical normal coordinates). In these circumstances we can write the matrix element for the *vibronic* interaction caused by $\hat{T}_{\text{vib}}$ between the states

$$\Phi^0_{\text{int}} = \Phi_{\text{nspin}} \Phi_{\text{rot}} \Phi_{\text{vib}} \Phi_{\text{elec}} \Phi_{\text{espin}} \tag{11-81}$$

and

$$\Phi^{0'}_{\text{int}} = \Phi_{\text{nspin}} \Phi_{\text{rot}} \Phi'_{\text{vib}} \Phi'_{\text{elec}} \Phi_{\text{espin}} \tag{11-82}$$

(the nuclear spin, rotational, and electron spin states must be the same since $\hat{T}_{\text{vib}}$ does not involve these coordinates) as the sum of the following two

---

[1] In Hund's case (a) there will also be terms involving products such as $\hat{J}_\alpha \hat{S}_\beta$, $\hat{p}_\alpha \hat{S}_\beta$, and $\hat{L}_\alpha \hat{S}_\beta$; see footnote on page 249.

terms:

$$H_1 = \sum_r \langle \Phi_{\text{vib}} | (\langle \Phi_{\text{elec}} | \hat{P}_r | \Phi'_{\text{elec}} \rangle) \hat{P}_r | \Phi'_{\text{vib}} \rangle$$

$$= \sum_r \langle \Phi_{\text{vib}} | X_r^{\text{ee}'} \hat{P}_r | \Phi'_{\text{vib}} \rangle \qquad (11\text{-}83)$$

and

$$H_2 = \frac{1}{2} \sum_r \langle \Phi_{\text{vib}} | (\langle \Phi_{\text{elec}} | \hat{P}_r^{\,2} | \Phi'_{\text{elec}} \rangle) | \Phi'_{\text{vib}} \rangle$$

$$= \frac{1}{2} \sum_r \langle \Phi_{\text{vib}} | Y_r^{\text{ee}'} | \Phi'_{\text{vib}} \rangle. \qquad (11\text{-}84)$$

The term H_1 is much larger than H_2 since it is the matrix element of a function involving the vibrational momenta whereas H_2 is the matrix element of a function of vibrational coordinates; we neglect H_2.

If we neglect the dependence of $X_r^{\text{ee}'}$ on the normal coordinates (i.e., consider only the first term, $X_{r0}^{\text{ee}'}$, say, in the Taylor series expansion of $X_r^{\text{ee}'}$ about the equilibrium nuclear configuration) then it is totally symmetric in the MS group, and we must have

$$\Gamma_{\text{elec}} \otimes \Gamma(\hat{P}_r) \otimes \Gamma'_{\text{elec}} \supset \Gamma^{(s)},$$

i.e.,

$$\Gamma_{\text{elec}} \otimes \Gamma'_{\text{elec}} \supset \Gamma(Q_r), \qquad (11\text{-}85)$$

With $X_{r0}^{\text{ee}'}$ not depending on the normal coordinates we can write

$$H_1 = \sum_r X_{r0}^{\text{ee}'} \langle \Phi_{\text{vib}} | \hat{P}_r | \Phi'_{\text{vib}} \rangle \qquad (11\text{-}86)$$

and for the matrix element of $\hat{P}_r$ to be nonvanishing we must have

$$\Delta v_r = \pm 1 \qquad (11\text{-}87)$$

in the harmonic oscillator approximation (see Table 8-2).

The dominant vibronic interaction from $\hat{T}_{\text{vib}}$ will therefore be between vibronic states that are such that the electronic states have symmetry connected by that of a normal coordinate, and the vibrational states have vibrational quantum numbers that differ by one for a vibration whose normal coordinate has the species to connect the two electronic states. In the SCF approximation for the electronic states we have further restrictions on the changes in the MO occupation number n_i. The operator $\hat{P}_r$ is a one-electron operator [Brillouin (1934)] and its matrix elements will vanish if the electron configuration of the states differ by more than one orbital [see Sidis and Lefebvre-Brion (1971)]. For example, the configurations of the $\tilde{X}\,^2A_1$ and

$\tilde{B}\,^2B_2$ states of NO_2 are

$$\tilde{X}\,^2A_1: \quad \cdots (1a_2)^2(4b_2)^2(6a_1)^1 \tag{11-88}$$

and

$$\tilde{B}\,^2B_2: \quad \cdots (1a_2)^2(4b_2)^1(6a_1)^2, \tag{11-89}$$

which differ by one orbital and have species that are connected by that of a normal coordinate $[\Gamma(Q_3) = B_2]$. Hence there will be a large vibronic interaction matrix element satisfying $\Delta v_3 = \pm 1$ between vibrational levels of these states from the term H_1.

The dominant effect of the rotational kinetic energy coupling term $\hat{T}_{er}$ [see Eq. (11-79)] is understood if we neglect the dependence of $\mu_{\alpha\beta}$ and of the matrix elements of $\hat{L}_\lambda$ on the normal coordinates. In this circumstance $\hat{T}_{er}$ only involves the rotational and electronic coordinates so that Φ_{nspin}, Φ_{vib}, and Φ_{espin} (as well as the basis symmetry label Γ_{re}) must be the same for the states that are coupled. Thus the dominant interaction is a *rotation–electronic* interaction and the matrix element of $\hat{T}_{er}$ is taken as

$$-\sum_\alpha \mu_{\alpha\alpha}^e [\langle \Phi_{elec}|\hat{L}_\alpha|\Phi'_{elec}\rangle\langle \Phi_{rot}|\hat{J}_\alpha|\Phi'_{rot}\rangle], \tag{11-90}$$

where $\alpha = x$, y, or z. For the matrix element of $\hat{L}_\alpha$ between the electronic wavefunctions to be nonvanishing the electronic species must be connected by a rotation, (since $\hat{L}_\alpha$ transforms like $\hat{J}_\alpha$), i.e.,

$$\Gamma_{elec} \otimes \Gamma'_{elec} \supset \Gamma(\hat{L}_\alpha) = \Gamma(\hat{J}_\alpha), \tag{11-91}$$

where $\alpha = x$, y, or z. The rotational selection rules will depend on which of the three rotations is involved in Eq. (11-91). For a symmetric top molecule if the product of the electronic species is that of $\hat{J}_z$ then the electronic matrix element of $\hat{L}_z$ can be nonvanishing and the allowed interaction must have $\langle \Phi_{rot}|\hat{J}_z|\Phi'_{rot}\rangle \neq 0$. If Eq. (11-91) involves the x and y components for a symmetric top then the rotational states coupled must be such that $\langle \Phi_{rot}|\hat{J}_x$ or $\hat{J}_y|\Phi'_{rot}\rangle \neq 0$. In the molecular rotation group D_∞ the angular momenta $(\hat{J}_x, \hat{J}_y)$ transform as Π, and $\hat{J}_z$ as Σ^- (see Tables 7-1 and 11-1). As a result, by classifying the rigid rotor functions Φ_{rot} and Φ'_{rot} in D_∞ we see that, from the vanishing integral rule [Eq. (5-133)],

$$\langle \Phi_{rot}|\hat{J}_z|\Phi'_{rot}\rangle \neq 0 \tag{11-92}$$

implies that in D_∞

$$\Gamma(\Phi_{rot}) \otimes \Sigma^- \otimes \Gamma(\Phi'_{rot}) \supset \Sigma^+$$

i.e.,

$$\Gamma(\Phi_{rot}) \otimes \Gamma(\Phi'_{rot}) \supset \Sigma^-$$

or [see Herzberg (1966), p. 572]

$$K = K' \qquad (11\text{-}93)$$

(with $K > 0$ since $J = J'$). Similarly

$$\langle \Phi_{\text{rot}} | \hat{J}_x \text{ or } \hat{J}_y | \Phi'_{\text{rot}} \rangle \neq 0 \qquad (11\text{-}94)$$

implies [see Herzberg (1966), p. 572]

$$K = K' \pm 1. \qquad (11\text{-}95)$$

Thus for a symmetric top molecule the dominant interactions caused by $\hat{T}_{\text{er}}$ will be between electronic states having symmetry connected by a rotation, and rotational states coupled have the selection rule $\Delta K = 0$ or ± 1 depending on which rotational symmetry couples the electronic states. Centrifugal distortion or Coriolis coupling mixes states of different K within the same electronic state [see Eqs. (11-105) and (11-108)] and spoils the usefulness of the selection rules on K. For an asymmetric top molecule using the D_2 molecular rotation group we see that the rotational selection rules will be ($\Delta K_a = $ even, $\Delta K_c = $ odd), ($\Delta K_a = $ odd, $\Delta K_c = $ odd), and ($\Delta K_a = $ odd, $\Delta K_c = $ even) as the product of the species of the electronic states is that of $\hat{J}_a, \hat{J}_b$, and $\hat{J}_c$, respectively. If for the states concerned the molecule is a near prolate rotor (i.e., K_a is a useful near quantum number) then $\Delta K_a = $ even (or odd) can be replaced by $\Delta K_a = 0$ (or ± 1); for a near oblate rotor similar remarks apply for ΔK_c.

The term $\hat{T}_{\text{ev}}$, like $\hat{T}_{\text{vib}}$, causes vibronic perturbations and it will couple states, such as those represented in Eqs. (11-81) and (11-82), that differ in vibrational and electronic state only. Neglecting the dependence of $\mu_{\alpha\beta}$ and of the matrix elements of $\hat{L}_\lambda$ on the normal coordinates the dominant matrix element is

$$\sum_\alpha \mu^e_{\alpha\alpha} \langle \Phi_{\text{elec}} | \hat{L}_\alpha | \Phi'_{\text{elec}} \rangle \langle \Phi_{\text{vib}} | \hat{p}_\alpha | \Phi'_{\text{vib}} \rangle. \qquad (11\text{-}96)$$

The operators $\hat{p}_\alpha$ and $\hat{L}_\alpha$ transform like $\hat{J}_\alpha$ so that the symmetry conditions for this matrix element to be nonvanishing are

$$\Gamma_{\text{elec}} \otimes \Gamma'_{\text{elec}} \supset \Gamma(\hat{J}_\alpha) \qquad (11\text{-}97)$$

and

$$\Gamma_{\text{vib}} \otimes \Gamma'_{\text{vib}} \supset \Gamma(\hat{J}_\alpha), \qquad (11\text{-}98)$$

where $\alpha = x$, y, or z. Assuming the same harmonic force field for the two electronic states, and using Eq. (7-138) for $\hat{p}_\alpha$ we deduce the vibrational selection rules to be

$$\Delta v_r = \pm 1, \qquad \Delta v_s = \pm 1,$$

where

$$\Gamma(Q_r) \otimes \Gamma(Q_s) \supset \Gamma(\hat{J}_\alpha). \qquad (11\text{-}99)$$

Usually electronic matrix elements of $\hat{L}_\alpha$ are smaller than those of $\hat{P}_r$ so that $\hat{T}_{\text{vib}}$ is the dominant cause of the breakdown of the Born–Oppenheimer approximation. However, there is one case where perturbation by $\hat{T}_{\text{ev}}$ can be very important and that is for bent molecules, such as NH_2, that can vibrate through a linear configuration. The important interactions caused by $\hat{T}_{\text{ev}}$ in this case are between vibrational levels of two electronic states that become degenerate in the linear nuclear configuration. The reason for the importance of the interaction in this case is that such electronic states have nonnegligible electronic angular momentum about the linear (z) axis, and the vibrational levels so coupled can be very close in energy (because of the electronic degeneracy in the linear configuration). This effect is termed the Renner effect [Renner (1934), Jungen and Merer (1976)].

In the Renner effect the vibrational levels of two electronic states that become degenerate for the linear configuration are coupled. In polyatomic molecules which are not easily able to achieve a linear configuration a similar effect called the *Jahn–Teller effect* [Jahn and Teller (1937)] can be important. The dynamic Jahn–Teller effect [Herzberg (1966), page 54] is said to occur when there are interactions between the vibrational levels of two electronic states that are such that their nuclear potential energy surfaces intersect at a particular (symmetric) nuclear configuration. A pair of electronic states of a polyatomic molecule that are degenerate at a particular symmetric nuclear configuration (the degeneracy is a result of the symmetry of the electronic Hamiltonian at that nuclear configuration) will always have their energy split by some nuclear distortion [Jahn and Teller (1937)]. This is the static Jahn–Teller effect and the minima in the two potential surfaces so obtained will each be at an unsymmetrical nuclear configuration. The MS group to be used in studying interactions between levels of two such electronic states will be isomorphic to the point group of the symmetrical nuclear configuration at the electronic degeneracy.

Rotation–Vibration Perturbations

These perturbations within an electronic state are caused by the terms given in Eqs. (11-20)–(11-22). In a rigid rotor harmonic oscillator basis perturbation terms mix states according to definite selection rules on the vibrational quantum numbers v_i, l_i (for doubly degenerate vibrations) and n_i (for triply degenerate vibrations), and rotational quantum numbers K (for symmetric tops) or K_a and K_c (for asymmetric tops). The selection rules will now be discussed, and the way that the perturbations spoil these near quantum numbers should be appreciated. The rotation–vibration species Γ_{rv} is maintained in these perturbations.

Anharmonicity perturbations mix states of the same symmetry, Γ_{vib}, in the MS group, and with selection rules on the vibrational quantum numbers that can be easily deduced from the results in Tables 8-2 and 8-3. For example, for the CH_3F molecule (see Problem 11-2) the following terms are among the nonvanishing cubic and quartic anharmonicity terms

$$V_{\text{anh}}^{(1)} = \Phi_{1123}Q_1{}^2Q_2Q_3/2,$$
$$V_{\text{anh}}^{(2)} = \Phi_{145}Q_1(Q_4{}^+Q_5{}^- + Q_4{}^-Q_5{}^+)/2, \qquad (11\text{-}100)$$
$$V_{\text{anh}}^{(3)} = \Phi_{456}(Q_4{}^+Q_5{}^+Q_6{}^+ + Q_4{}^-Q_5{}^-Q_6{}^-)/2.$$

The nonvanishing matrix elements of these terms can be determined from the results in Eq. (8-183) and in Tables 8-2 and 8-3. We determine that $V_{\text{anh}}^{(1)}$ has nonvanishing matrix elements between vibrational states with the selection rules

$$\Delta v_1 = 0, \pm 2, \qquad \Delta v_2 = \pm 1, \qquad \Delta v_3 = \pm 1. \qquad (11\text{-}101)$$

$V_{\text{anh}}^{(2)}$ has nonvanishing matrix elements with selection rules

$$\Delta v_1 = \pm 1, \qquad \Delta v_4 = \pm 1, \qquad \Delta v_5 = \pm 1, \qquad \Delta l_4 = -\Delta l_5 = \pm 1, \quad (11\text{-}102)$$

and $V_{\text{anh}}^{(3)}$ has nonvanishing matrix elements with selection rules

$$\Delta v_4 = \pm 1, \qquad \Delta v_5 = \pm 1, \qquad \Delta v_6 = \pm 1, \qquad \Delta l_4 = \Delta l_5 = \Delta l_6 = \pm 1. \quad (11\text{-}103)$$

The centrifugal distortion perturbation terms [Eq. (11-21)] mix states in which both the vibrational and rotational quantum numbers change, although the vibration–rotation species Γ_{rv} must be maintained. The selection rules on the vibration and rotation quantum numbers can be derived from the results in Tables 8-1–8-3. For example, in the CH_3F molecule among the nonvanishing centrifugal distortion terms [see Eq. (11-68)] is

$$V_{\text{cent}}^{(1)} = a_4^{xx}[Q_4{}^+(\hat{J}_m{}^+)^2 + Q_4{}^-(\hat{J}_m{}^-)^2]/2. \qquad (11\text{-}104)$$

This term mixes rotation–vibration states according to the selection rules

$$\Delta v_4 = \pm 1$$

with

$$\Delta l_4 = +1 \quad \text{and} \quad \Delta k = -2 \qquad or \qquad \Delta l_4 = -1 \quad \text{and} \quad \Delta k = +2. \quad (11\text{-}105)$$

Clearly the first term in the centrifugal distortion term of Eq. (11-21) will mix vibration–rotation states with vibrational quantum numbers that differ by one because Q_r occurs in it. For a rigid molecule such as we are considering here none of the vibrational energy separations will be of the order of a rotational energy separation, and perturbation theory can reasonably be used to treat the effect of this term (unless $a_r^{\alpha\beta}$ is anomalously large). The second term in Eq. (11-21) has a coefficient of $Q_r Q_s \hat{J}_\alpha \hat{J}_\beta$ that is usually an order of magnitude smaller than the coefficient of $Q_r \hat{J}_\alpha \hat{J}_\beta$ in the first term. However the second term couples states that differ by one unit in two vibrational quantum numbers, i.e., the levels of the states $v_r = 1$ and $v_s = 1$ can be coupled by this term; it is possible for such coupled states to be close in energy (i.e., in *resonance*) and for the perturbation treatment of the effect of V_{cent} to fail. For example, in trans diimide the centrifugal distortion term

$$\tfrac{3}{4}\mu^e_{aa} a^{aa}_2 \mu^e_{aa} a^{aa}_3 \mu^e_{aa} Q_2 Q_3 \hat{J}_a^{\ 2} \tag{11-106}$$

is nonvanishing [see Eq. (11-50)] and the $v_2 = 1$ and $v_3 = 1$ states that are mixed by this term are close in energy. As a result of the resonance between levels that are coupled by this term, it can have a significant effect.

The selection rules on the vibrational and rotational quantum numbers for allowed Coriolis coupling interactions can likewise be obtained from the results in Tables 8-1–8-3. From Eq. (11-77) we see that a nonvanishing Coriolis coupling term in CH_3F is

$$-(i/2)\mu^e_{yy}\zeta^y_{4,5}[Q_4^{\ -}\hat{P}_5^{\ -}\hat{J}_m^{\ -} - Q_4^{\ +}\hat{P}_5^{\ +}\hat{J}_m^{\ +}] \tag{11-107}$$

and the selection rules for states to be coupled by this term are

$$\Delta v_4 = \pm 1 \qquad \text{and} \qquad \Delta v_5 = \pm 1$$

with

$$\Delta l_4 = \Delta l_5 = -\Delta k = \pm 1. \tag{11-108}$$

Another nonvanishing Coriolis coupling term for CH_3F is

$$-(i/2)\mu^e_{zz}\zeta^z_{s,s}(Q_s^{\ +}\hat{P}_s^{\ -} - Q_s^{\ -}\hat{P}_s^{\ +})\hat{J}_z, \tag{11-109}$$

where $s = 4$, 5, or 6, and this couples states with the selection rules

$$\Delta v_s = 0, \pm 2 \qquad \text{and} \qquad \Delta l_s = \Delta k = 0. \tag{11-110}$$

These terms have nonvanishing diagonal matrix elements and produce an important first order Coriolis splitting of levels having the same quantum number labels v_s and K.

In the absence of resonances the treatment of centrifugal distortion and Coriolis coupling by perturbation theory in the vibrational interaction leads to an *effective rotational Hamiltonian*, the *Watsonian* [Watson (1967, 1977)], in which the successive terms have quadratic, quartic, sextic, etc., powers of the angular momentum operators. The effective rotational Hamiltonian commutes with a molecular rotation group and, in the absence of resonances between states coupled by the centrifugal distortion or Coriolis coupling perturbations, K is a near quantum number for a symmetric top and the irreducible representations of the group D_2 are good labels for an asymmetric top. For spherical top molecules centrifugal distortion and Coriolis coupling have the important effect of causing a partial splitting of the $(2J + 1)$-fold k degeneracy of each level. The maximum splitting of this degeneracy that can occur is given by the total number of irreducible representations of the MS group that occur in Γ_{rv}. For example, the $J = 18$ level of the vibrational ground state of the methane molecule has MS group symmetry given by (see Table 10-14)

$$\Gamma_{rv} = 2A_1 \oplus 2A_2 \oplus 3E \oplus 4F_1 \oplus 5F_2 \qquad (11\text{-}111)$$

and this level can be split at most into 16 components (four nondegenerate, three doubly degenerate, and nine threefold degenerate) by centrifugal distortion and Coriolis coupling.

For spherical top molecules it has been found convenient to introduce the quantum number R when labeling the rotational levels of states in which one quantum of a triply degenerate vibration is excited [see, for example, Hougen (1976)]. The quantum number takes on the values $J + 1$, J, and $J - 1$ as the zero order rotation–vibration wavefunctions generate the representations $D^{(J+1)}$, $D^{(J)}$, and $D^{(J-1)}$. It is possible to define a first order vibration–rotation Hamiltonian which only connects states according to the selection rule $\Delta R = 0$. Even though the complete Hamiltonian is not diagonal in R this is often a useful near quantum number for a spherical top just as G is for a symmetric top.

The Quantum Numbers G and $(\pm l)$ Centrifugal distortion and Coriolis coupling interactions in a symmetric top can mix states of different K value [see, for example, Eqs. (11-105) and (11-108)]. If the interaction is strong it spoils the usefulness of K as a near quantum number. However, we can use symmetry to define the related quantum numbers G and G_v for the rotation–vibration states of a symmetric top molecule [Hougen (1962b)]. To introduce these quantum numbers we will use the CH_3F molecule as an example. A general zero order rotation–vibration wavefunction for CH_3F can be written

$$\psi_{vr} = \phi_{vr} \exp[i(l_4\alpha_4 + l_5\alpha_5 + l_6\alpha_6 + k\chi)], \qquad (11\text{-}112)$$

where ϕ_{vr} is a function of $Q_1, Q_2, \ldots, Q_6$, θ, and ϕ. From the transformation properties of Q_t and $\exp(i\alpha_t)$ under (123) given in Table (11-7), and knowing that $R_z^{2\pi/3}\chi = \chi + (2\pi/3)$, we deduce that

$$(123)\psi_{vr} = \exp[-2\pi i(l_4 + l_5 + l_6 - k)/3]\psi_{vr}. \qquad (11\text{-}113)$$

From Eq. (11-113) we see that the symmetry Γ_{rv} of the state will be E if $(l_4 + l_5 + l_6 - k)$ is ± 1, ± 2, ± 4, etc. (i.e., ± 1 mod 3), whereas it will be A_1 or A_2 if $(l_4 + l_5 + l_6 - k)$ is 0, ± 3, ± 6, etc. (i.e., 0 mod 3). The symmetry of the vibrational state is similarly related to $(l_4 + l_5 + l_6)$ in CH_3F. Therefore it proves useful to introduce the quantum numbers g and g_v which for CH_3F in its ground electronic state are given by

$$g_v = (l_4 + l_5 + l_6) \tag{11-114}$$

and

$$g = g_v - k. \tag{11-115}$$

The vibronic quantum number g_{ev} is defined for CH_3F by the equation

$$(123)\psi_{ev} = \exp(-2\pi i g_{ev}/3)\psi_{ev} \tag{11-116}$$

and the general definition of the quantum number g is that

$$(123)\psi_{evr} = \exp(-2\pi i g/3)\psi_{evr} \tag{11-117}$$

so that

$$g = g_{ev} - k. \tag{11-118}$$

We introduce the unsigned quantum numbers, for labeling energy levels,

$$G = |g|, \qquad G_{ev} = |g_{ev}|, \qquad \text{and} \qquad G_v = |g_v|.$$

By relating g_v to Γ_{vib} we see that anharmonicity perturbations in CH_3F can only mix vibrational states according to the selection rule

$$\Delta g_v = 0 \bmod 3,$$

i.e.,

$$\Delta(l_4 + l_5 + l_6) = 0, \pm 3, \pm 6, \text{ etc.} \tag{11-119}$$

[see Eqs. (11-102) and (11-103) for examples]. Similarly rotation–vibration perturbations in CH_3F mix rotation–vibration states with the selection rule

$$\Delta g = 0 \bmod 3,$$

i.e.,

$$\Delta(l_4 + l_5 + l_6 - k) = 0, \pm 3, \pm 6, \text{ etc.} \tag{11-120}$$

[see Eqs. (11-105) and (11-108) for examples]. The selection rules on g_{ev} and g for vibronic and rovibronic interactions follow in a similar manner.

For any electronic state of any symmetric top molecule the quantum numbers g, g_{ev}, and g_v (as well as G, G_{ev}, and G_v) are defined in a similar way to the above by considering the transformation properties of the rovibronic, vibronic, and vibrational wavefunctions under the effect of a single operation of the MS group [Hougen (1962b)]. This single operation is a permutation for which the equivalent rotation is of type R_z^β and it is that operation for which β ($= 2\pi/n$) has the smallest value (β_{min}) (or alternatively involves the largest value of n; this being the order of the primary axis of symmetry).

The groups D_{nd} where n is even and S_m where $m/2$ is even are exceptions in that the single operation used to define g, g_{ev}, and g_v is that permutation–inversion for which the equivalent rotation is R_z^{β} with the smallest positive β value. The selection rules for rovibronic, vibronic, and rotation–vibration perturbations can be formulated using these quantum numbers. The quantum numbers g, g_{ev}, and g_v contain no more information than the MS symmetry labels Γ_{rve}, Γ_{ve}, and Γ_{vib}, respectively, but they are useful, particularly for defining the $(\pm l)$ quantum number label. The $(\pm l)$ quantum number label is itself of particular use in discussing optical selection rules in symmetric tops when degenerate vibrational or vibronic states are involved.

To define the $(\pm l)$ quantum number label that is used for the levels of symmetric top molecules we will consider a vibrational state of the CH_3F molecule having $v_4 \neq 0$ and $v_5 = v_6 = 0$. The Coriolis coupling perturbation

$$-(i/2)\mu_{zz}^e \zeta_{4,4}^{yz}(Q_4^+ \hat{P}_4^- - Q_4^- \hat{P}_4^+)\hat{J}_z \qquad (11\text{-}121)$$

has nonvanishing first order matrix elements (i.e., $\Delta k = 0$ and $\Delta v_4 = \Delta l_4 = 0$) within any rotation–vibration state for which l_4 and k are nonzero. Using Eq. (11-121) together with the expressions for the matrix elements of the two-dimensional harmonic oscillator (see Table 8-3) we deduce that the first order energy of states for which $v_5 = v_6 = 0$ is given by (in cm^{-1})

$$E_{rv} = E_{rv}^0 - 2Ak\zeta_{4,4}^z l_4, \qquad (11\text{-}122)$$

where E_{rv}^0 is the rigid rotor harmonic oscillator energy which has a $2(v_4 + 1)$-fold degeneracy if $k \neq 0$. As a result the pair of levels with $k = +K$, $l_4 = +G_v$ and $k = -K$, $l_4 = -G_v$ have energy

$$E_{rv} = E_{rv}^0 - 2AK\zeta_{4,4}^z G_v \qquad (11\text{-}123)$$

and the pair of levels with $k = -K$, $l_4 = +G_v$ and $k = +K$, $l_4 = -G_v$ have energy

$$E_{rv} = E_{rv}^0 + 2AK\zeta_{4,4}^z G_v. \qquad (11\text{-}124)$$

Since the splitting is first order it is useful to have an extra label to distinguish the level with energy given by (11-123) from that with energy given by (11-124) for a given pair of (K, G_v) values. This label is the $(\pm l)$ label; one level is called the $(+l)$ level and the other is called the $(-l)$ level. The sign of the $(\pm l)$ label depends on the relative and absolute values of k and g_v and is given by [Hoy and Mills (1973)]

$$\mathrm{sgn}[k\sin(2\pi g_v/3)]. \qquad (11\text{-}125)$$

Levels for which Eq. (11-125) is positive are $(+l)$ levels and levels for which Eq. (11-125) is negative are $(-l)$ levels. In the $v_4 = 1$ state $g_v = l_4 = \pm 1$ and we see that levels with energy given by Eq. (11-123) are the $(+l)$ levels and those with energy given by Eq. (11-124) are the $(-l)$ levels. As another

Table 11-8
Symmetry of $(\pm l)$ levels of
CH_3F in an E vibronic state
with $K > 0^a$

	Γ_{rve}	
K:	$(+l)$	$(-l)$
$3n + 1$:	$A_1 \oplus A_2$	E
$3n + 2$:	E	$A_1 \oplus A_2$
$3n + 3$:	E	E

^{a} n is a nonnegative integer;
$K = |k|$.

example, in the $v_4 = 2$ state of CH_3F the pair of levels $k = +K$, $l_4 = +2$ and $k = -K$, $l_4 = -2$, with a first order energy shift of $-4AK\zeta_{4,4}^z$, are the $(-l)$ pair, and the pair $k = -K$, $l_4 = +2$ and $k = +K$, $l_4 = -2$, with a first order energy shift of $+4AK\zeta_{4,4}^z$, are the $(+l)$ pair. In the $v_4 = 2$ state the pair of $l_4 = 0$ levels do not suffer a first order Coriolis splitting and the $(\pm l)$ label is not used for them. In general for a vibrational state of CH_3F that has species E, the rotation–vibration species of the $(+l)$ and $(-l)$ levels depend on K as shown in Table 11-8. It is this symmetry that determines the $(\pm l)$ label and not the relative signs of k and l_4 (compare the $v_4 = 1$ and $v_4 = 2$ examples discussed above for CH_3F). To extend the $(\pm l)$ label to degenerate vibronic states we use g_{ev} instead of g_v in Eq. (11-125), and to extend it to other symmetric tops we rewrite the equation as [Hoy and Mills (1973)]

$$\text{sgn}[k \sin(2\pi g_{ev}/n)] \tag{11-126}$$

where n is the order of the primary axis of symmetry in the molecule.

The NO_2 Molecule

We will use the NO_2 molecule as an example and discuss the use of its MS group in determining which molecular energy levels can interact with each other. As well as considering the breakdown of the Born–Oppenheimer approximation, and rotation–vibration interactions, we will also discuss the effects of some perturbations that involve the coupling of the nuclear spins to the other coordinates in the molecule.

The MS group of NO_2 is the $C_{2v}(M)$ group given in Table A-4 and the oxygen nuclei are labeled 1 and 2. For $^{14}N^{16}O_2$ the complete internal wavefunction must be invariant to the permutation (12), since ^{16}O nuclei have spin zero, and therefore, it can only transform as A_1 or A_2 as its parity is $+$ or $-$, respectively. The ^{14}N nucleus has a spin of one, and the three nuclear spin states ($m_I = -1, 0,$ or $+1$) have symmetry

$$\Gamma^{\text{tot}}_{\text{nspin}} = 3A_1. \tag{11-127}$$

Thus rovibronic–electron spin states Φ_{rves} of symmetry B_1 or B_2 will be missing. The asymmetric top functions $|N_{K_aK_c}\rangle$ have the same symmetry species, as a function of the eveness and oddness of K_a and K_c, as the $|J_{K_aK_c}\rangle$ functions of H_2O given in Table 10-10. The three normal coordinates of NO_2 transform as those of H_2O so that

$$\Gamma(Q_1, Q_2, Q_3) = 2A_1 \oplus B_2. \tag{11-128}$$

The coordinates Q_1 and Q_2 of symmetry A_1 can be approximately described as the symmetric stretch and bend, and Q_3, of symmetry B_2, as the anti-symmetric stretch. A vibrational state Φ_{vib} with quantum numbers (v_1, v_2, v_3) has symmetry A_1 or B_2 as v_3 is even or odd and we can write this as

$$\Gamma_{\text{vib}} = [B_2]^{v_3}. \tag{11-129}$$

The species Γ_{elec} of the electronic orbital states are determined from the transformation properties of the atomic orbitals. We take a minimal basis set of 1s, 2s, and 2p orbitals on each nucleus; these 15 atomic orbitals generate the representation

$$7A_1 \oplus A_2 \oplus 2B_1 \oplus 5B_2, \tag{11-130}$$

where the (x, y, z) axes are located as in Fig. 11-1 (i.e., we use a I^r convention). The following 15 symmetry adapted orbitals can be constructed [given in approximate order of increasing orbital energy for the bent molecule; see, for example, Table II in Jackels and Davidson (1976)].

$(1a_1), (2a_1),$ and $(1b_2)$ from $1s(N), 1s(O_1),$ and $1s(O_2)$,

$(3a_1)$: $2s(N) + [2s(O_2) + 2s(O_1)]$,

$(2b_2)$: $2p_z(N) + [2s(O_2) - 2s(O_1)]$,

$(4a_1)$: $[2p_z(O_2) - 2p_z(O_1)]$,

$(5a_1)$: $2p_x(N) + [2p_x(O_2) + 2p_x(O_1)]$,

$(1b_1)$: $2p_y(N) + [2p_y(O_2) + 2p_y(O_1)]$,

$(3b_2)$: $[2p_z(O_2) + 2p_z(O_1)]$, $\qquad$ (11-131)

$(1a_2)$: $[2p_y(O_2) - 2p_y(O_1)]$,

$(4b_2)$: $[2p_x(O_2) - 2p_x(O_1)]$,

$(6a_1)$: $2p_x(N) - [2p_x(O_2) + 2p_x(O_1)]$,

$(2b_1)$: $2p_y(N) - [2p_y(O_2) + 2p_y(O_1)]$,

$(7a_1)$: $2s(N) - [2s(O_2) + 2s(O_1)]$,

$(5b_2)$: $2p_z(N) - [2s(O_2) - 2s(O_1)]$.

The ground electronic state configuration for the 23 electrons of the NO_2 molecule is

$$(1a_1)^2(1b_2)^2(2a_1)^2(3a_1)^2(2b_2)^2(4a_1)^2(5a_1)^2(1b_1)^2(3b_2)^2(1a_2)^2(4b_2)^2(6a_1)^1$$

$$(11\text{-}132)$$

with symmetry $\Gamma_{elec} = A_1$. Three excited electronic states have the configurations:

$$\cdots(1a_2)^2(4b_2)^2(2b_1)^1, \qquad \Gamma_{elec} = B_1,$$
$$\cdots(1a_2)^2(4b_2)^1(6a_1)^2, \qquad \Gamma_{elec} = B_2, \qquad (11\text{-}133)$$
$$\cdots(1a_2)^1(4b_2)^2(6a_1)^2, \qquad \Gamma_{elec} = A_2.$$

These four states are the $\tilde{X}\,^2A_1$, $\tilde{A}\,^2B_1$, $\tilde{B}\,^2B_2$ and $\tilde{C}\,^2A_2$ states of NO_2.

Before discussing the symmetries of the rovibronic states of NO_2 some discussion of the potential energy surfaces of the $\tilde{X}$, $\tilde{A}$, $\tilde{B}$, and $\tilde{C}$ states is necessary in order to show that $C_{2v}(M)$ used above is appropriate. The qualitative variations of the energies of the MO's of NO_2 with bond angle can be appreciated from rather simple bond strength arguments. We can draw "Walsh diagrams" [Walsh (1953)] showing the variation of orbital energies with bond angle and these are given for the $(1a_2)$, $(4b_2)$, $(6a_1)$, and $(2b_1)$ orbitals of NO_2 in Fig. 11-8. From the LCAO description of the MO's given in Eq. (11-131) the reader can appreciate why the $(6a_1)$ orbital is strongly bonding at the bent configuration but becomes weakly antibonding [and degenerate with the $(2b_1)$ orbital] at the linear configuration. From this diagram we see that the electronic states $\tilde{B}$ and $\tilde{C}$ [having two electrons in

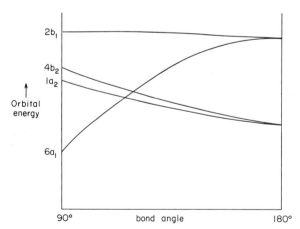

Fig. 11-8. The Walsh diagram for some MO's of NO_2. This diagram shows qualitatively how the orbital energies change with bond angle in NO_2.

the $(6a_1)$ orbital] should be strongly bent at equilibrium, the $\tilde{X}$ state [having one electron in the $(6a_1)$ orbital] should be bent (but not as much as in the $\tilde{B}$ and $\tilde{C}$ states) at equilibrium, and the $\tilde{A}$ state should be linear at equilibrium. Furthermore, at the linear configuration the $\tilde{A}$ and $\tilde{X}$ states become degenerate and the $\tilde{B}$ and $\tilde{C}$ states become degenerate. This is a result of the fact that the electronic Hamiltonian of the molecule has higher than C_{2v} symmetry when the nuclei are in the linear configuration (it has $D_{\infty h}$ symmetry; see Chapter 12) and degeneracies occur. However, the rovibronic Hamiltonian does not have symmetry that requires degeneracies and the simultaneous MS group of all four of these electronic states of NO_2 is the $C_{2v}(M)$ group; this happens to be the CNPI group of the molecule. It turns out, from *ab initio* calculations, that the $\tilde{B}$ state has its minimum at a bent configuration with *unsymmetrical bond lengths* [Jackels and Davidson (1976)]. However the small barrier between symmetrically equivalent forms permits tunneling, and the MS group of the $\tilde{B}$ electronic state is the $C_{2v}(M)$ group.

We now consider the symmetry classification of the rotational levels of these four electronic states and the interactions that are possible between them. We use a Hund's case (b) electron spin basis set so that (since $2S + 1 = 2$)

$$\Gamma_{espin} = 2A_1, \tag{11-134}$$

and we will consider an arbitrary rotational level as an example for the determination of the species of a rotational level. The level we consider is the $N_{K_aK_c} = 4_{21}$ level of the $(v_1,v_2,v_3) = (0,0,1)$ vibrational state of the $\tilde{X}^2A_1$ electronic state. Using a Hund's case (b) electron spin basis set the level has rovibronic–electron spin symmetry given by

$$\Gamma_{rves} = \Gamma_{rot} \otimes \Gamma_{vib} \otimes \Gamma_{elec} \otimes \Gamma_{espin}$$
$$= B_1 \otimes B_2 \otimes A_1 \otimes (2A_1) = 2A_2. \tag{11-135}$$

There are two A_2 irreducible representations because of the electron spin degeneracy of the level; the m_N degeneracy of 9 is not indicated. For $^{14}N^{16}O_2$ this state can be combined with the three totally symmetric nuclear spin states to give

$$\Gamma_{int} = 2A_2 \otimes 3A_1 = 6A_2. \tag{11-136}$$

From the group K(spatial) we have $I = 1$, $N = 4$, and $S = \frac{1}{2}$ so that $J = \frac{7}{2}$ or $\frac{9}{2}$, and $F = \frac{5}{2}, \frac{7}{2}, \frac{9}{2}, \frac{7}{2}, \frac{9}{2}$, or $\frac{11}{2}$ [the F value distinguishes the six states of Eq. (11-136)]. Thus the 4_{21} level of the $(0,0,1)\tilde{X}^2A_1$ state of $^{14}N^{16}O_2$ consists of six components with the F values just given, and each has MS group symmetry A_2 (i.e., $-$ parity). The electron spin interactions would split each rovibronic level into a doublet and these two levels would each be split into a triplet by nuclear hyperfine interactions.

Table 11-9

The rovibronic symmetries of the $\tilde{X}^2A_1$, $\tilde{A}^2B_1$, $\tilde{B}^2B_2$, and $\tilde{C}^2A_2$ states of NO_2[a]

	$\tilde{X}^2A_1$		$\tilde{A}^2B_1$		$\tilde{B}^2B_2$		$\tilde{C}^2A_2$	
	v_3		v_3		v_3		v_3	
K_aK_c:	Even	Odd	Even	Odd	Even	Odd	Even	Odd
ee:	A_1	(B_2)	(B_1)	A_2	(B_2)	A_1	A_2	(B_1)
oo:	A_2	(B_1)	(B_2)	A_1	(B_1)	A_2	A_1	(B_2)
eo:	(B_1)	A_2	A_1	(B_2)	A_2	(B_1)	(B_2)	A_1
oe:	(B_2)	A_1	A_2	(B_1)	A_1	(B_2)	(B_1)	A_2

[a] The states in parentheses are missing for $^{14}N^{16}O_2$.

The dominant interactions in the molecule will not involve the spins, and as a result we will label the states with Γ_{rve}, $N_{K_aK_c}$, v_1, v_2, v_3, Γ_{elec}, and Γ_{ve}. The symmetry label Γ_{rve} for a $N_{K_aK_c}$ level of any (v_1, v_2, v_3) state of each of the four electronic states considered here is given in Table 11-9; these symmetries depend on the evenness and oddness of v_3, as indicated, but not on the values of v_1 and v_2 [see Eq. (11-129)]. For $^{14}N^{16}O_2$ all rovibronic states of symmetry B_1 and B_2 will be missing as a result of the nuclear spin statistics and they are in parentheses in Table 11-9. States having the same labels Γ_{rve} and N can be mixed by terms in the spin free Hamiltonian. Thus any pair of rovibronic levels of this molecule that are close in energy and which have the same Γ_{rve} and N will interact as the result of being connected by *some* term in the spin free Hamiltonian. Certain interaction matrix elements will be larger than others and we now consider the interaction matrix elements that will be large. However, a small energy separation between two interacting states is as important as a large interaction matrix element, and states connected by a small matrix element, but having a small energy separation, can perturb each other significantly.

We will first consider possible vibronic interactions. The dominant vibronic interactions from H_1 [see Eqs. (11-83)–(11-87)] are between electronic states whose species are connected by a vibration, with a change of one in the quantum number of that vibration. The electronic states $\tilde{A}^2B_1$ and $\tilde{C}^2A_2$ are connected by the species of the antisymmetric stretching vibration but they have configurations that differ by two orbitals so that the vibronic coupling between them will be weak (and dependent on the extent of configuration interaction mixing); see remarks above Eq. (11-88). The electronic states $\tilde{X}^2A_1$ and $\tilde{B}^2B_2$ are also connected by the species of the antisymmetric stretching vibration and these states have configurations that only differ by one orbital. Thus there will be a large vibronic interaction matrix element between vibrational levels of the $\tilde{X}$ and $\tilde{B}$ states with the selection rule $\Delta v_3 = \pm 1$. Depending on the value of the energy separation between the vibronic states strong interactions can occur between close lying vibrational levels of the $\tilde{X}$ and $\tilde{B}$ states if they satisfy $\Delta v_3 = \pm 1$.

Vibronic interactions from $\hat{T}_{ev}$ [see Eq. (11-80)] can occur if the species of the electronic states are connected by the species of a rotation [see Eq. (11-97)] and will be important if the two electronic states become degenerate at the linear configuration. The pairs of electronic states $(\tilde{X}, \tilde{A})$ and $(\tilde{B}, \tilde{C})$ are each connected by $\Gamma(\hat{J}_a)$ and each pair becomes degenerate at the linear configuration.

The strongest rotation–electronic interactions from $\hat{T}_{er}$ occur between the levels of electronic states having electronic species connected by the species of a rotation [see Eq. (11-91)]. The pairs $(\tilde{X}, \tilde{A})$ and $(\tilde{B}, \tilde{C})$ each have species connected by that of $\hat{J}_a$, the pairs $(\tilde{X}, \tilde{C})$ and $(\tilde{A}, \tilde{B})$ each have species connected by that of $\hat{J}_b$, and the pairs $(\tilde{X}, \tilde{B})$ and $(\tilde{A}, \tilde{C})$ each have species connected by that of $\hat{J}_c$. The dominant coupling matrix elements will have rotational selection rules $(\Delta K_a = 0, \Delta K_c = \pm 1)$, $(\Delta K_a = \pm 1, \Delta K_c = \pm 1)$, or $(\Delta K_a = \pm 1, \Delta K_c = 0)$, since the product of the electronic species is B_1 $(= \Gamma(\hat{J}_a))$, $A_2 (= \Gamma(\hat{J}_b))$, or $B_2 (= \Gamma(\hat{J}_c))$, respectively. These levels (as they should) all have the same Γ_{rve} species. Rovibronic interactions can also occur between rotational levels of these states, and as an example, the levels of the $\tilde{X}$ and $\tilde{A}$ states can be coupled with selection rules $\Delta v_3 = \pm 1, \Delta K_a = \pm 1, \Delta K_c = \pm 1$.

Rotation–vibration interactions within each electronic state, caused by centrifugal distortion and Coriolis coupling, mix rotational levels that have the same Γ_{rv} species. The selection rules will be either $\Delta v_3 = $ even and $\Delta K_a = 0$, or $\Delta v_3 = $ odd and $\Delta K_a = \pm 1$ (both with $\Delta K_c = 0$ and with Δv_1 and Δv_2 not restricted by symmetry arguments). Purely vibrational perturbations caused by anharmonicity terms in the potential function of each electronic state will mix levels of the same Γ_{vib}. Thus this interaction can only occur with $\Delta v_3 = $ even. There will be no purely electronic interaction between these states since they all have different electronic symmetry. However, each electronic state can be mixed with other, more highly excited, electronic states by configuration interaction.

We have so far neglected some interesting coupling interactions caused by the terms in $\hat{H}_{ns}$ (see Table 6-1); these terms involve the nuclear spins and we will illustrate their effect by considering the $^{14}N^{17}O_2$ molecule. An ^{17}O nucleus has a spin of $\frac{5}{2}$ and the pair of oxygen nuclei (labeled 1 and 2) give rise to 36 nuclear spin states having symmetry in $C_{2v}(M)$ given by

$$21A_1 \oplus 15B_2. \tag{11-137}$$

The states with symmetry A_1 have I_{12} (where $I_{12} = I_1 + I_2$) values of 5, 3, and 1, and the states with symmetry B_2 have $I_{12} = 4, 2,$ and 0. Multiplying by the ^{14}N nuclear spin species $(3A_1)$ we obtain

$$\Gamma^{\text{tot}}_{\text{nspin}} = 63A_1 \oplus 45B_2. \tag{11-138}$$

Since ^{17}O nuclei are fermions the complete internal wavefunction Φ_{int} can only be of species $\Gamma_{\text{int}} = B_1$ or B_2 in C_{2v}(M) (as the parity of Φ_{int} is $-$ or $+$, respectively). Thus rovibronic states of symmetry A_1 or A_2 (for which $K_a K_c = $ ee or oo in a totally symmetric vibronic state) can be combined with any of the 45 nuclear spin states of symmetry B_2, and rovibronic states of symmetry B_1 or B_2 (for which $K_a K_c = $ eo or oe in a totally symmetric vibronic state) can be combined with any of the 63 nuclear spin states of symmetry A_1. The levels with the higher statistical weight (i.e., I_{12} odd) are called ortho levels and the other levels (I_{12} even) are called para levels. What is of interest here is the possible perturbation interactions between the ortho and para levels, i.e., interactions that spoil the near quantum number I_{12}. We call such an interaction a $\Delta I_{12} \neq 0$ interaction or an ortho–para interaction. For an ortho and a para level to interact they must have the same true symmetry labels (i.e., F value and Γ_{int} symmetry in the MS group). Within an electronic state the term $\hat{H}_{\text{nsr}}$ (see Table 6-1) can cause ortho–para interactions. From the symmetries in Table 10-10 we see that the $K_a K_c$ pairs (ee with oe) and (eo with oo) can interact with each other when they have the same F value. The pair of levels N_{0N} and N_{1N} within an electronic state of ^{14}N^{17}O$_2$ (such as the pair 1_{01} and 1_{11}) would generally be the closest pairs of ortho–para levels that could perturb each other in this way. Such interactions have not been observed [but see Curl, Kasper, and Pitzer (1967)]. Ortho–para interactions between levels of different electronic states can be caused by the terms $\hat{H}_{\text{nso}}$ and $\hat{H}_{\text{nses}}$ in Table 6-1. This interaction matrix element can be large and if such a pair of coupled levels were in close resonance the effect of an ortho–para interaction would be observed. Thus, for example, in ^{14}N^{17}O$_2$ the interaction between an ee level in an A_1 vibronic state (a para level for which $\Gamma_{\text{rve}} = A_1$) and an ee level in a B_2 vibronic state (an ortho level for which $\Gamma_{\text{rve}} = B_2$) could be observable if they have the same F value and lie close in energy. For any molecule having two or more identical nuclei of nonzero spin there is a possibility of observing such ortho–para interactions [see Raich and Good (1964) and Smirnov (1974)].

An esoteric point involving the nuclear spins and missing levels might as well be mentioned at this point. In ^{14}N^{16}O$_2$ rovibronic states of species B_1 or B_2 are missing since oxygen nuclei have spin $I = 0$ and are bosons. However, since each oxygen nucleus in the molecule is not in a perfectly spherical environment the quantum number I is not perfectly good; excited nuclear states having I nonzero could be very slightly mixed into the $I = 0$ state by the nonspherical charge distribution around each nucleus. As a result the missing levels would not be completely missing. Since excited nuclear states are typically at energies of about 1 MeV this effect will be exceedingly small.[2]

Problem 11-3. In the trans diimide molecule N_2H_2 the states $v_4 = 1$ and $v_6 = 1$ are close together in energy [see Carlotti, Johns, and Trombetti (1974)]. Using the results obtained in the solution to Problem 11-1 determine the direct interactions that can occur between the rotational levels of these states.

Answer. The states $v_4 = 1$ and $v_6 = 1$ have vibrational symmetry $\Gamma_{vib} = A_u$ and B_u, respectively. Since they have different vibrational symmetries they cannot be mixed by anharmonicity perturbations. The first term in the centrifugal distortion perturbation [see Eq. (11-21)] contains the coupling operator $Q_r \hat{J}_\alpha \hat{J}_\beta$, and hence couples rotation–vibration states according to the vibrational selection rule $\Delta v_r = \pm 1$. This term cannot give rise to a perturbation between states, such as the $v_4 = 1$ and $v_6 = 1$ states, that differ in two vibrational quantum numbers. The second term in T_{cent} [see Eq. (11-21)] contains the coupling operator $Q_r Q_s \hat{J}_\alpha \hat{J}_\beta$ and this term couples states that differ in two vibrational quantum numbers. In $C_{2h}(M)$ the combinations $Q_4 Q_6 \hat{J}_a \hat{J}_c$ and $Q_4 Q_6 \hat{J}_b \hat{J}_c$ are totally symmetric and we might expect levels of the $v_4 = 1$ and $v_6 = 1$ states to be coupled by such terms. However, the coefficients of $Q_4 Q_6 \hat{J}_a \hat{J}_c$ and $Q_4 Q_6 \hat{J}_b \hat{J}_c$ vanish for trans diimide, since they involve the parameters $a_4^{\alpha\beta}$ and $a_6^{\alpha\beta}$ which were shown to be zero in the solution of Problem 11-1 [see Eq. (11-50)]. In the solution to Problem 11-1 we saw that for any centrosymmetric molecule $a_r^{\alpha\beta}$ vanishes if Q_r is of u species, and hence the rotational levels of two vibrational states cannot be connected by T_{cent} if the levels have different levels of excitation of any u vibrations. This contrasts with the fact that the rotational levels for N_2H_2 of the $v_2 = 1$ and $v_3 = 1$ vibrational states (both of A_g symmetry) can be mixed by terms such as $Q_1 Q_3 J_a^2$ in V_{cent} since $a_r^{\alpha\alpha}$ and a_r^{ab} do not vanish for $r = 1$ or 3.

The rotational levels of the $v_4 = 1$ and $v_6 = 1$ states can be mixed by the Coriolis coupling perturbation

$$-\sum_\alpha \mu_{\alpha\alpha}^e \zeta_{4,6}^\alpha (Q_4 \hat{P}_6 - Q_6 \hat{P}_4) \hat{J}_\alpha. \qquad (11\text{-}139)$$

In particular $\zeta_{4,6}^a$ and $\zeta_{4,6}^b$ are nonvanishing [see Eq. (11-51)] so that the two nonvanishing coupling terms are (in cm^{-1})

$$V_{Cor}^{(a)} = -A\hbar^{-2}\zeta_{4,6}^a (Q_4 \hat{P}_6 - Q_6 \hat{P}_4) \hat{J}_a \qquad (11\text{-}140)$$

and

$$V_{Cor}^{(b)} = -B\hbar^{-2}\zeta_{4,6}^b (Q_4 \hat{P}_6 - Q_6 \hat{P}_4) \hat{J}_b. \qquad (11\text{-}141)$$

---

[2] Missing levels involving "nonspherical" nuclear spin states (i.e., $I \neq 0$), such as the A_1' and A_1'' states of NH_3, may be more likely to occur.

These perturbations can only connect levels of the $v_4 = 1$ and $v_6 = 1$ states that have the same J value and Γ_{rv} symmetry. There are two possible Γ_{rv} symmetries for these levels (see Table 10-13): A_u and B_u. The A_u levels are the $J_{K_a K_c}$ levels of the $v_4 = 1$ state having K_c even and the levels of the $v_6 = 1$ state having K_c odd; these levels can be mixed by $V_{Cor}^{(a)} + V_{Cor}^{(b)}$. The K_c odd levels in the $v_4 = 1$ state and the K_c even levels of the $v_6 = 1$ state are of B_u symmetry, and they can also be mixed together. Using a I^r prolate rotor basis set (trans diimide is a near prolate rotor) the nonvanishing matrix elements are (from Tables 8-1 and 8-2)

$$\langle v_4 = 1, J, k_a | V_{Cor}^{(a)} | v_6 = 1, J, k_a \rangle = iA\zeta_{4,6}^{ra}[\sqrt{\gamma_6/\gamma_4} + \sqrt{\gamma_4/\gamma_6}]k_a/2 \quad (11\text{-}142)$$

and

$$\langle v_4 = 1, J, k_a \pm 1 | V_{Cor}^{(b)} | v_6 = 1, J, k_a \rangle$$
$$= iB\zeta_{4,6}^{rb}[\sqrt{\gamma_6/\gamma_4} + \sqrt{\gamma_4/\gamma_6}]\tfrac{1}{4}\sqrt{J(J+1) - k_a(k_a \pm 1)}. \quad (11\text{-}143)$$

We see that states having high K_a and $K_a \cong J$ will be strongly coupled by $V_{Cor}^{(a)}$ and weakly coupled by $V_{Cor}^{(b)}$, whereas those with high J and $K_a \cong 0$ will be strongly coupled by $V_{Cor}^{(b)}$ and weakly coupled by $V_{Cor}^{(a)}$. However, the effect of $V_{Cor}^{(a)}$ will generally be the more important since A ($\cong 10$ cm^{-1}) is much larger than B ($\cong 1.3$ cm^{-1}) in N_2H_2 [see Carlotti, Johns, and Trombetti (1974)].

OPTICAL SELECTION RULES AND FORBIDDEN TRANSITIONS

For an isolated molecule a transition with the absorption or emission of electric dipole radiation can only occur between certain pairs of energy levels. The restrictions defining the pairs of energy levels between which such transitions can occur are called selection rules. We can determine rigorous selection rules on the true symmetry labels Γ_{int} and F by using the MS group and the K(spatial) group. By making appropriate approximations we can determine much more restrictive selection rules on near quantum numbers and near symmetry labels. Transitions that are not allowed by these restrictive selection rules, but that are allowed if no approximations are made (or when magnetic dipole and electric quadrupole radiation are considered), are called *forbidden transitions*. Forbidden transitions are usually weak and the main features of the electromagnetic spectrum of a molecule can usually be understood without considering them. We will first derive the rigorous selection rules and then discuss the approximations that are made to obtain the approximate selection rules. The effect of the breakdown of the approximations in allowing forbidden transitions to occur will be discussed.

A transition between two states Φ''_{int} and Φ'_{int} will occur with the absorption or emission of electromagnetic radiation if the matrix element between the states of the electric dipole moment operator along a space fixed direction does not vanish [see, for example, Section 8c in Eyring, Walter, and Kimball (1944)]. Choosing this direction as the ζ direction we have the condition

$$\langle \Phi'_{int} | M_\zeta | \Phi''_{int} \rangle \neq 0, \tag{11-144}$$

if $\Phi''_{int} \rightarrow \Phi'_{int}$ is to be an allowed electric dipole transition where the component of the electric dipole moment is given by

$$M_\zeta = \sum_j e_j \zeta_j \tag{11-145}$$

and e_j and ζ_j are the charge and ζ coordinate of the jth particle in the molecule, where j runs over all nuclei and electrons in the molecule. For the integral in Eq. (11-144) not to vanish by symmetry the product of the symmetries of Φ'^*_{int} and Φ''_{int} must contain the symmetry of M_ζ from the vanishing integral rule [Eq. (5-133)]. We will use the MS group and the K(spatial) group to determine symmetry selection rules. From the definition of M_ζ in Eq. (11-145) we see that M_ζ is invariant to the permutation of identical nuclei (since identical nuclei have the same charge) and that it is changed in sign by the inversion operation E^* (since $E^* \zeta_j = -\zeta_j$). Thus M_ζ (like the translation T_ζ) transforms as that one-dimensional representation of the MS group that has character $+1$ under all permutations and character -1 under all permutation inversions; we call this the *antisymmetric* representation Γ^* of the MS group. Clearly M_ξ and M_η also each transform as Γ^* in the MS group. We can classify M_ξ, M_η, and M_ζ in the K(spatial) group and they transform as $D^{(1)}$. Thus, for the transition to be an allowed electric dipole transition we must have

$$\Gamma''_{int} \otimes \Gamma'_{int} \supset \Gamma^* \quad \text{in the MS group}, \tag{11-146}$$

where Γ''_{int} and Γ'_{int} are the symmetries of Φ''_{int} and Φ'^*_{int}, and

$$D''_{int} \otimes D'_{int} \supset D^{(1)} \quad \text{in the } K\text{(spatial) group}, \tag{11-147}$$

where D''_{int} and D'_{int} are the symmetries of Φ''_{int} and Φ'^*_{int}. The selection rules can be written in terms of the parity of Φ_{int} and the F quantum number labels as

$$+ \leftrightarrow -, \qquad + \nleftrightarrow +, \qquad - \nleftrightarrow - \tag{11-148}$$

(which means that a transition between a $+$ and a $-$ level can occur but a transition between two $+$ levels or between two $-$ levels cannot occur) and from Eq. (11-147) [see Eq. (6-43)]

$$\Delta F = 0, \pm 1 \quad (F' = 0 \nleftrightarrow F'' = 0). \tag{11-149}$$

These constitute rigorous selection rules for electric dipole transitions of an isolated molecule in field free space.

We now discuss the further restrictions obtained by making approximations. The approximations we make are those required to obtain separable wavefunctions which we write as $\Phi_{\text{nspin}}\Phi_{\text{rot}}\Phi_{\text{vib}}\Phi_{\text{elec}}\Phi_{\text{espin}}$, and we also make the approximations of neglecting electrical anharmonicity and axis switching; these two approximations will be explained when they arise.

Using separable wavefunctions Eq. (11-144) can be written as

$$\langle \Phi'_{\text{nspin}}\Phi'_{\text{rot}}\Phi'_{\text{vib}}\Phi'_{\text{elec}}\Phi'_{\text{espin}}|M_\zeta|\Phi''_{\text{nspin}}\Phi''_{\text{rot}}\Phi''_{\text{vib}}\Phi''_{\text{elec}}\Phi''_{\text{espin}}\rangle \neq 0. \quad (11\text{-}150)$$

The dipole moment operator is independent of the spin coordinates so we can separate the spin variables in the integral and write it as

$$\langle \Phi'_{\text{nspin}}|\Phi''_{\text{nspin}}\rangle\langle \Phi'_{\text{espin}}|\Phi''_{\text{espin}}\rangle\langle \Phi'_{\text{rot}}\Phi'_{\text{vib}}\Phi'_{\text{elec}}|M_\zeta|\Phi''_{\text{rot}}\Phi''_{\text{vib}}\Phi''_{\text{elec}}\rangle \neq 0. \quad (11\text{-}151)$$

We write M_ζ in terms of components along the molecule fixed (x, y, z) axes by using direction cosine elements $\lambda_{\alpha\zeta}$ [see Eq. (7-52)]:

$$M_\zeta = \lambda_{x\zeta}M_x + \lambda_{y\zeta}M_y + \lambda_{z\zeta}M_z. \quad (11\text{-}152)$$

The direction cosine elements involve the Euler angles and the M_α involve the vibronic coordinates so that we can separate the rotational and vibronic terms in Eq. (11-151) to obtain

$$\langle \Phi'_{\text{nspin}}|\Phi''_{\text{nspin}}\rangle\langle \Phi'_{\text{espin}}|\Phi''_{\text{espin}}\rangle \sum_{\alpha=xyz} \langle \Phi'_{\text{rot}}|\lambda_{\alpha\zeta}|\Phi''_{\text{rot}}\rangle\langle \Phi'_{\text{vib}}\Phi'_{\text{elec}}|M_\alpha|\Phi''_{\text{vib}}\Phi''_{\text{elec}}\rangle \neq 0.$$

$$(11\text{-}153)$$

We can integrate over the electronic coordinates in the vibronic matrix element by writing

$$\langle \Phi'_{\text{vib}}\Phi'_{\text{elec}}|M_\alpha|\Phi''_{\text{vib}}\Phi''_{\text{elec}}\rangle = \langle \Phi'_{\text{vib}}|[\langle \Phi'_{\text{elec}}|M_\alpha|\Phi''_{\text{elec}}\rangle]|\Phi''_{\text{vib}}\rangle$$
$$= \langle \Phi'_{\text{vib}}|M_\alpha(e', e'')|\Phi''_{\text{vib}}\rangle, \quad (11\text{-}154)$$

where the factor $M_\alpha(e', e'')$ will depend on the nuclear coordinates. We expand this factor in terms of the normal coordinates of one of the electronic states, about its value at the nuclear equilibrium configuration of that state, by writing

$$M_\alpha(e', e'') = M_\alpha^{(o)}(e', e'') + \sum_r M_\alpha^{(r)}(e', e'')Q_r + \frac{1}{2}\sum_{r,s} M_\alpha^{(r,s)}(e', e'')Q_r Q_s + \cdots.$$

$$(11\text{-}155)$$

If $\Phi'_{\text{elec}} \neq \Phi''_{\text{elec}}$ we neglect all but the first term in Eq. (11-155) and if $\Phi'_{\text{elec}} = \Phi''_{\text{elec}}$ we neglect all but the first two terms; the latter is termed the neglect of *electrical anharmonicity*, by analogy to the neglect of mechanical anharmonicity which is the neglect of V_{anh} [see Eq. (11-20)]. Substituting Eq. (11-155) into Eq. (11-154) and Eq. (11-154) into Eq. (11-153) we obtain the final equation defining an *allowed transition* as

$$\langle \Phi'_{nspin}|\Phi''_{nspin}\rangle \langle \Phi'_{espin}|\Phi''_{espin}\rangle \left[\langle \Phi'_{vib}|\Phi''_{vib}\rangle \sum_\alpha \langle \Phi'_{rot}|\lambda_{\alpha\zeta}|\Phi''_{rot}\rangle M_\alpha^{(o)}(e',e'')\right.$$

$$\left. + \sum_\alpha \langle \Phi'_{rot}|\lambda_{\alpha\zeta}|\Phi''_{rot}\rangle \sum_r \langle \Phi'_{vib}|Q_r|\Phi''_{vib}\rangle M_\alpha^{(r)}(e',e'')\right] \neq 0, \qquad (11\text{-}156)$$

where the second term in the square brackets is neglected if $\Phi'_{elec} \neq \Phi''_{elec}$. We will use this equation to determine the selection rules for allowed transitions, but first it is necessary to discuss the effect of axis switching since it has been neglected in the above derivation.

If the equilibrium geometries of the two electronic states Φ'_{elec} and Φ''_{elec} are different then the molecule fixed (x, y, z) axes would be differently oriented in the two states for a given instantaneous nuclear arrangement. This is because the orientation of the axes is determined by using the Eckart conditions and the equilibrium geometry enters into these [see Eqs. (7-127)–(7-135)]. This effect is called *axis switching* [Hougen and Watson (1965)]. As a result we must choose the equilibrium configuration of one of the electronic states as *the* equilibrium geometry to use in the Eckart equations in order to define uniquely a set of (x, y, z) axes so that the $\lambda_{\alpha\zeta}$ and M_α in Eq. (11-152) can be unambiguously determined. The rotational wavefunctions of the other electronic state would then have to be written in terms of rotational wavefunctions that are functions of Euler angles defined by these new axes in order that the rotational matrix elements of $\lambda_{\alpha\zeta}$ involve only one set of angles. The result of this would be that certain "extra" rotational transitions called axis switching transitions would become allowed, and they do not obey the selection rules on K (or K_a and K_c) that we derive below. This effect would also need to be considered in the comparison of *ab initio* and experimental vibronic matrix elements of M_α. Axis switching transitions are not of common occurrence and are generally weak.

We now return to a determination of the selection rules required by Eq. (11-156). From the orthogonality of the spin wavefunctions it is clear that we must have

$$\Phi'_{nspin} = \Phi''_{nspin} \qquad (11\text{-}157)$$

and

$$\Phi'_{espin} = \Phi''_{espin}, \qquad (11\text{-}158)$$

so that we obtain the selection rules on the spin quantum numbers

$$\Delta I = 0 \qquad (11\text{-}159)$$

and

$$\Delta S = 0. \tag{11-160}$$

Two results that are useful and that will be used below are that in the MS group

$$\Gamma(M_\alpha) = \Gamma(T_\alpha) \tag{11-161}$$

and

$$\Gamma(\lambda_{\alpha\zeta}) = \Gamma(\hat{J}_\alpha), \tag{11-162}$$

where $\alpha = x, y$, or z, T_α is the translational coordinate [see Eq. (11-39)] and $\hat{J}_\alpha$ is a component of the total angular momentum along a molecule fixed direction.

Electronic Transitions

When $\Phi'_{elec} \neq \Phi''_{elec}$ we only consider the first term in the square brackets of Eq. (11-156), and this is nonvanishing [from Eqs. (11-154), (11-155), and (5-133)] if

$$\Gamma'_{elec} \otimes \Gamma''_{elec} \supset \Gamma(M_\alpha) = \Gamma(T_\alpha) \tag{11-163}$$

and

$$\Gamma'_{vib} = \Gamma''_{vib}. \tag{11-164}$$

Thus allowed electronic transitions are such that the species of the electronic wavefunctions in the MS group[3] must be connected by a translation. Further the vibrational states involved in the transition must have the same symmetry species in the MS group. For an absorption transition from the vibronic ground state of a molecule, the vibrational ground state wavefunction is totally symmetric so that only totally symmetric vibrations of the excited state can be excited. If there are vibronic interactions mixing $\Phi'_{elec}\Phi'_{vib}$ and/or $\Phi''_{elec}\Phi''_{vib}$ with other vibronic levels of other electronic states [Herzberg and Teller (1933)], or a strong dependence of the electronic transition moment $M_\alpha(e', e'')$ on the nuclear coordinates, we are left with the following symmetry selection rule for vibronically allowed (but electronically forbidden) transitions [from Eqs. (11-154) and (5-133)]:

$$\Gamma'_{ve} \otimes \Gamma''_{ve} \supset \Gamma(M_\alpha) = \Gamma(T_\alpha), \tag{11-165}$$

where $\alpha = x, y$, or z. Rovibronic interactions can make additional electronically forbidden transitions allowed [see Eq. (11-177)].

---

[3] The MS group is that appropriate for the simultaneous treatment of the electronic states under consideration.

Vibrational Transitions

When $\Phi'_{elec} = \Phi''_{elec}$ and $\Phi'_{vib} \neq \Phi''_{vib}$ the first term in the square brackets of Eq. (11-156) will vanish (since the vibrational wavefunctions of a single electronic state are orthogonal to each other). For the second term to be nonvanishing we must have

$$\langle \Phi'_{vib} | Q_r | \Phi''_{vib} \rangle \neq 0 \qquad (11\text{-}166)$$

and [from Eqs. (11-154) and (11-155) with $\Phi'_{elec} = \Phi''_{elec}$]

$$\Gamma(Q_r) = \Gamma(M_\alpha) = \Gamma(T_\alpha). \qquad (11\text{-}167)$$

For Eq. (11-166) to be true in the harmonic oscillator approximation we have the selection rule

$$\Delta v_r = \pm 1, \qquad (11\text{-}168)$$

where, from Eq. (11-167), the only allowed vibrational transitions are those for which the normal coordinate transforms like a translation. For vibrational transitions from the vibrational ground state only those $v_r = 1$ states are accessible (giving rise to *fundamental bands* in the spectrum) whose normal coordinates transform like a translation. Since the number of normal coordinates that transform like a translation differs if we make different assumptions about the equilibrium symmetry of the molecule, we can use the observed number of allowed fundamental bands to help us determine the equilibrium symmetry. Electrical and mechanical anharmonicity will give rise to forbidden transitions that do not satisfy Eq. (11-168) but

$$\Gamma'_{vib} \otimes \Gamma''_{vib} \supset \Gamma(T_\alpha) \qquad (11\text{-}169)$$

is still required. Rotation–vibration and rovibronic interactions will make additional forbidden transitions allowed [see Eq. (11-177)].

Rotational Transitions

Rotational transitions will accompany the vibrational and vibronic transitions just discussed and they must satisfy [see Eq. (11-156)]:

$$\langle \Phi'_{rot} | \lambda_{\alpha\zeta} | \Phi''_{rot} \rangle \neq 0. \qquad (11\text{-}170)$$

For a symmetric top molecule we deduce from the form of the symmetric top wavefunctions [see Eqs. (8-64) and (8-67)] and from the explicit expressions for the $\lambda_{\alpha\zeta}$ [see Eq. (7-52)] that Eq. (11-170) implies [see also Eqs. (11-92)–(11-95) and Eq. (11-162)]

$$\Delta K = 0 \qquad \text{if} \quad \alpha = z \qquad (11\text{-}171)$$

and

$$\Delta K = \pm 1 \qquad \text{if} \quad \alpha = x \text{ or } y \qquad (11\text{-}172)$$

both with $\Delta J = 0,^4 \pm 1$ [$\Delta N = 0, \pm 1$ for nonsinglet states in Hund's case (b)]. If the rotational transitions within a vibrational or vibronic band satisfy Eq. (11-171) the band has a characteristic appearance and is called a *parallel band*; Eq. (11-172) leads to a *perpendicular band*. This provides a useful diagnostic for the vibrational or vibronic states involved in the transition, since it tells us the symmetry relationship between them [from Eqs. (11-165)–(11-169)]. If the transition involves a degenerate vibronic state then selection rules on the $(\pm l)$ label can be derived and are very useful [Herzberg (1945)]. In particular, transitions from an A vibronic state to an E or E_1 vibronic state (in which the vibronic symmetry species are connected by the species of T_x and T_y) are such that the $\Delta K = +1$ transitions are to the $(+l)$ levels and $\Delta K = -1$ transitions are to the $(-l)$ levels. For an asymmetric top molecule the rotational selection rules are derived from Eq. (11-170) and Table 11-2 to be

$$\Delta K_a = \text{even} \quad \Delta K_c = \text{odd} \quad \text{if} \quad \alpha = a,$$
$$\Delta K_a = \text{odd} \quad \Delta K_c = \text{odd} \quad \text{if} \quad \alpha = b, \qquad (11\text{-}173)$$
$$\Delta K_a = \text{odd} \quad \Delta K_c = \text{even} \quad \text{if} \quad \alpha = c,$$

all with $\Delta J = 0, \pm 1$ [or $\Delta N = 0, \pm 1$ for nonsinglet states in Hund's case (b)] giving rise to type A, B, or C bands. If, for the states concerned, the molecule is a near prolate rotor (i.e., K_a is a useful near quantum number) then $\Delta K_a =$ even (or odd) can be replaced by $\Delta K_a = 0$(or ± 1); for a near oblate rotor similar remarks apply to ΔK_c. Transitions obeying Eq. (11–173) but not obeying these more restrictive selection rules on ΔK_a or ΔK_c, as appropriate, can also be called forbidden. For spherical top molecules we just have ΔJ[or ΔN in Hund's case (b)] $= 0, \pm 1$ as the rotational selection rule.

Pure rotational transitions (in which the vibronic state does not change, i.e., $\Phi_{\text{vib}}' = \Phi_{\text{vib}}''$, and $\Phi_{\text{elec}}' = \Phi_{\text{elec}}''$) can occur with the rotational selection rules given above if the vibronic state is such that [for Eq. (11-154) to be nonvanishing]

$$\Gamma_{\text{ve}} \otimes \Gamma_{\text{ve}} \supset \Gamma(M_\alpha) = \Gamma(T_\alpha). \qquad (11\text{-}174)$$

If Eq. (11-174) is satisfied the molecule can have a nonvanishing *permanent dipole moment* given by

$$\langle \Phi_{\text{vib}}\Phi_{\text{elec}}|M_\alpha|\Phi_{\text{vib}}\Phi_{\text{elec}}\rangle = \mu_\alpha, \qquad (11\text{-}175)$$

where $\alpha = x, y,$ or z. (Note that the diagonal matrix element of M_ζ for a state for which Φ_{int} is nondegenerate must vanish since M_ζ only has nonvanishing matrix elements between states of opposite parity.) In a symmetric

---

[4] $\Delta J = 0$ is forbidden for $K = 0 \rightarrow 0$ [see Eq. (11-93) and Table 11-1].

top molecule of C_{nv} symmetry, for example, the translation T_z is totally symmetric and the vibronic ground state is almost always totally symmetric, so that such molecules can have a nonvanishing permanent dipole moment in the z direction. In these circumstances pure rotational transitions are allowed with selection rules $\Delta K = 0, \Delta J = 0, \pm 1$. For an asymmetric top the selection rules on K_a and K_c depend on the direction of the permanent dipole moment [i.e., on the values of α in Eq. (11-174)]. If the direction of the permanent dipole moment is along an axis that is not a principal inertial axis then allowed transitions occur with selection rules given by two, or all three, of Eq. (11-173) depending on the projection of that axis direction on the a, b, and c axes.

Rotation–vibration interactions caused by centrifugal distortion and Coriolis coupling spoil K as a quantum number for a symmetric top and hence give rise to forbidden transitions that do not satisfy Eqs. (11-171) and (11-172). Also Γ_{vib} and Γ_{ve} are spoiled as symmetry labels so that Eqs. (11-169) and (11-174) are spoiled. In these circumstances the selection rule for allowed rotation–vibration transitions within an electronic state is relaxed to

$$\Gamma'_{rv} \otimes \Gamma''_{rv} \supset \Gamma^*, \tag{11-176}$$

where Γ^* is the antisymmetric representation, but still with $\Delta J = 0, \pm 1$ [or $\Delta N = 0, \pm 1$ for nonsinglet states in Hund's case (b)]. Rovibronic interactions relax the rovibronic selection rule to

$$\Gamma'_{rve} \otimes \Gamma''_{rve} \supset \Gamma^* \tag{11-177}$$

but still with ΔJ [or ΔN for nonsinglet states in Hund's case (b)] $= 0, \pm 1$. Electron spin–orbit coupling spoils a Hund's case (b) basis and spoils S and N as quantum numbers; the selection rules become further relaxed to

$$\Gamma'_{rves} \otimes \Gamma''_{rves} \supset \Gamma^* \tag{11-178}$$

with $\Delta J = 0, \pm 1$ (but now $\Delta S \neq 0$ and $\Delta N \neq 0, \pm 1$ transitions become allowed). Finally nuclear spin interactions relax the selection rules completely to those of Eqs. (11-146)–(11-149) when ortho–para type transitions become allowed. Forbidden transitions are particularly important in circumstances in which their occurrence allows otherwise inaccessible energy level separations in a molecule to be determined [as, for example, in the determination of A_0 from the infrared spectrum of a symmetric top molecule; Olson (1972), Sarka (1976)].

In summary, the strongest vibronic transitions will be those that satisfy Eqs. (11-163) and (11-164) and the strongest vibrational transitions will be those that satisfy Eqs. (11-167) and (11-168). The strongest rotational lines accompanying these transitions will satisfy Eqs. (11-171)–(11-173). The strongest pure rotational transitions will occur within vibronic states that satisfy Eq. (11-174) with rotational selection rules as given by Eqs. (11-171)–(11-173). All transitions that do not satisfy these selection rules are called forbidden transitions. Vibronic interactions and the dependence of the electronic transition moment on the nuclear coordinates give rise to forbidden electronic transitions that do not satisfy Eq. (11-163) but which still satisfy Eq. (11-165). Mechanical and electrical anharmonicity give rise to forbidden vibrational transitions that do not satisfy Eqs. (11-166)–(11-168) but which satisfy Eq. (11-169), and rotation–vibration interactions and axis-switching spoil the selection rules of Eqs. (11-171)–(11-173) although Eq. (11-176) must be satisfied [rotation–vibration interaction can also give rise to forbidden rotational transitions within a vibronic state for which Eq. (11-174) is not satisfied]. Electron spin–orbit coupling spoils the selection rules on S and N, and the coupling of the nuclear spins with the other degrees of freedom in the molecule reduces the selection rules to those of Eqs. (11-146)–(11-149). Except in the case of resonances (when the interaction spoiling the selection rules is between levels that are close lying in energy) the forbidden transitions will be much weaker than the allowed transitions.

Centrosymmetric molecules again merit special attention. For a centrosymmetric molecule the species of the translations T_α are always u since the operation[5] $\hat{O}_i$ inverts the vibronic coordinates in the molecule fixed axis system. Thus allowed vibrational or vibronic transitions must obey the selection rule u $\leftrightarrow$ g. The rotational levels within a g (u) vibronic state are all of g (u) symmetry since the rotational wavefunctions must all be of g symmetry. Since the symmetry of M_ζ (i.e., the irreducible representation Γ^*) must be u ($\hat{O}_i$ is a permutation–inversion operation and must reverse M_ζ), no rotational transitions within any vibronic state of a centrosymmetric molecule can occur as a result of rotation–vibration or rovibronic interactions. Electron spin functions in either a case (a) or case (b) basis must be of g species so that spin–orbit coupling cannot induce rotational transitions. Only interactions involving the nuclear spins can permit pure rotational transitions to occur in a centrosymmetric molecule. Such transitions will be ortho–para type transitions, and the presence of identical nuclei with nonzero spin is required. These forbidden transitions are most likely to be observed in one of a pair of close lying electronic states if one is of g symmetry and the other of u symmetry, when a magnetic interaction between the nuclear spins and the electrons (from $\hat{H}_{nso} + \hat{H}_{nses}$ in Table 6-1) mixes levels and spoils the g/u classification. In molecules that are not centrosymmetric, but which do not have a permanent dipole moment, rotation–vibration, or rovibronic, interactions can give rise to an allowed rotational spectrum as long as Eq. (11-176) or (11-177) is obeyed. The forbidden rotational spectrum of

---

[5] See Eq. (11-12) for the definition of $\hat{O}_i$.

methane is an example of this and is caused by rotation–vibration inter-actions; the observed transitions satisfy Eq. (11-176) [see, for example, Hougen (1976)]. The forbidden rotational spectrum of HD is caused by rovibronic interactions and Eq. (11-177) is satisfied by the observed transitions [see, for example, Bunker (1973a)].

MAGNETIC DIPOLE AND ELECTRIC QUADRUPOLE TRANSITIONS

A transition between the states Φ'_{int} and Φ''_{int} can occur with the (weak) absorption or emission of electromagnetic radiation even if the electric dipole matrix element [Eq. (11-144)] is zero as long as the matrix element of the magnetic dipole moment operator or electric quadrupole moment operator is nonvanishing (higher multipole transitions are possible but have not been observed). In general magnetic dipole and electric quadrupole transition probabilites are about 10^{-5} and 10^{-8}, respectively, of electric dipole transition probabilities. Such transitions are called forbidden since they are forbidden as electric dipole transitions.

For the states Φ'_{int} and Φ''_{int} to be connected by magnetic dipole radiation we must have [see, for example, Y.-N. Chiu (1965)]

$$\langle \Phi'_{int} | D_\zeta | \Phi''_{int} \rangle \neq 0, \tag{11-179}$$

where the ζ component of the magnetic dipole moment operator is given by

$$D_\zeta = \sum_j \frac{e_j}{2m_j c} (\hat{l}_{\zeta j} + g_j s_{\zeta j}), \tag{11-180}$$

the ζ component of the orbital angular momentum of particle j is given by

$$\hat{l}_{\zeta j} = -i\hbar \left(\xi_j \frac{\partial}{\partial \eta_j} - \eta_j \frac{\partial}{\partial \xi_j} \right), \tag{11-181}$$

and the ζ component of the spin angular momentum of particle j is $s_{\zeta j}$ ($= I_{\zeta j}$ for a nucleus). For each electron g_j is nearly 2 [see, for example, Moss (1973)]. D_ζ is invariant to any operation of the MS group (it transforms like $\hat{J}_\zeta$) and so the rigorous magnetic dipole transition selection rules are

$$\Gamma'_{int} \otimes \Gamma''_{int} \supset \Gamma^{(s)}, \quad \text{i.e.,} \quad + \leftrightarrow +, \quad - \leftrightarrow -, \quad \text{and} \quad + \nleftrightarrow -. \tag{11-182}$$

The operator D_ζ transforms as $D^{(1)}$ in K(spatial) so the selection rule on F is

$$\Delta F = 0, \pm 1 \qquad (F' = 0 \nleftrightarrow F'' = 0). \tag{11-183}$$

As a result of the fact that the masses of the particles occur in Eq. (11-180) we see that the terms involving the orbital and spin magnetic moments of the electrons will be about 10^3 times larger than the terms involving the orbital and spin magnetic moments of the nuclei. Magnetic dipole transitions involving a reorientation of the electron orbital or spin magnetic moment are the only transitions (excepting NMR) so far observed [see, for example, Herzberg (1946), Herzberg and Herzberg (1947), Babcock and Herzberg (1948), Arpigny (1966), and Brown, Cole, and Honey (1972)]. Magnetic dipole rotation–vibration transitions could well provide useful information that would complement the information obtained from the electric dipole rotation–vibration spectrum of a molecule but such transitions have not yet been observed. We can resolve D_ζ along the molecule fixed axes [as done for M_ζ in Eq. (11-152)] and since the D_α transform like R_α (or equivalently like $\hat{J}_\alpha$) the vibronic symmetry selection rules [Eqs. (11-163), (11-165), (11-167), (11-169), and (11-174)] would all involve the species of R_α rather than the species of T_α. The rotational selection rules involve the matrix elements of the direction cosines and are the same as given in Eqs. (11-171)–(11-173).

Allowed electric quadrupole transitions involve the matrix element of the electric quadrupole moment of the molecule $Q_{\zeta\eta}$, say [see, for example, Y.-N. Chiu (1965)]. This transforms as $T_\zeta \times T_\eta$, i.e., as $\Gamma^* \otimes \Gamma^* = \Gamma^{(s)}$, in the MS group, and as $D^{(2)}$ in K(spatial). Thus the rigorous selection rules for allowed electric quadrupole transitions are

$$\Gamma'_{\text{int}} \otimes \Gamma''_{\text{int}} \supset \Gamma^{(s)}, \qquad \text{i.e.,} \qquad + \leftrightarrow +, \quad - \leftrightarrow -, \quad \text{and} \quad + \not\leftrightarrow -, \qquad (11\text{-}184)$$

and from Eq. (6-41)

$$\Delta F = 0, \pm 1, \pm 2 \qquad (0 \not\leftrightarrow 0), \quad (\tfrac{1}{2} \not\leftrightarrow \tfrac{1}{2}), \quad (0 \not\leftrightarrow 1). \qquad (11\text{-}185)$$

We can express the elements of $Q_{\zeta\eta}$ in terms of molecule fixed components Q_{xy}, etc., which transform like $T_x T_y$, etc. (the species of the product of two translations). The species of $T_x T_y$ is that of the electric polarizability component α_{xy} [see Eq. (11-190) that follows] and these species are given in the character tables in Appendix A. Thus electric quadrupole transitions occur between vibronic states whose species are connected by the species of a component of the electric polarizability tensor. The rotational transitions accompanying an α_{xy} vibronic transition, say, must be such that the rotational matrix element of, for example, $\lambda_{x\zeta}\lambda_{y\eta}$ is nonvanishing. The most famous example of an electric quadrupole rotation–vibration spectrum is that of the hydrogen molecule [Herzberg (1949, 1950b)].

MULTIPHOTON PROCESSES AND THE RAMAN EFFECT

The selection rules for multiphoton processes, such as occur in Raman scattering, can be obtained by expressing the intensity of the multiphoton process as the sum of products of one-photon electric dipole matrix elements. For example, a two-photon transition from a state i to a state k can take place only if there exists some third state j such that both transitions $i \to j$ and $j \to k$ are electric dipole allowed (and each, therefore, obeys the electric dipole selection rules just derived). If states i and k are such that there are no states j for which both $i \to j$ and $j \to k$ are electric dipole allowed then the transition $i \to k$ is forbidden as a two-photon process. The strict selection rules [obtained from the strict one-photon selection rules of Eqs. (11-146)–(11-149)] are that the transition between the states Φ''_{int} and Φ'_{int} is forbidden as an n-photon electric dipole transition unless

$$\Gamma'_{\text{int}} \otimes \Gamma''_{\text{int}} \supset [\Gamma^*]^n \qquad (11\text{-}186)$$

and[6]

$$\Delta F = 0, \pm 1, \pm 2, \ldots, \pm n. \qquad (11\text{-}187)$$

Since

$$[\Gamma^*]^n = \Gamma^{(s)} \qquad \text{if} \quad n \text{ is even}$$

and   (11-188)

$$[\Gamma^*]^n = \Gamma^* \qquad \text{if} \quad n \text{ is odd},$$

then for n even we have $+ \leftrightarrow +$, $- \leftrightarrow -$, and $+ \not\leftrightarrow -$ whereas for n odd we have $+ \leftrightarrow -$, $+ \not\leftrightarrow +$, and $- \not\leftrightarrow -$. Further selection rules on the electronic and vibrational symmetries, and on the vibrational and rotational quantum numbers, can be obtained from the one-photon selection rules previously discussed.

In Raman scattering radiation of frequency v (usually visible light) is incident on a molecule that is in a state i with energy E_i. Scattered radiation of frequency $v - v_{ki}$ is detected, where $h v_{ki} = E_k - E_i$, and the molecule undergoes a two-photon transition to the state k (which can be of greater or lesser energy than the state i). Usually the states i and k are different rotation–vibration levels of the ground electronic state of the molecule and $v \gg |v_{ki}|$. More restrictive selection rules for $i \to k$ to be allowed in the Raman effect, over and above those derived by considering the selection rules for the successive electric dipole transitions $i \to j$ and $j \to k$, are derived by making the *polarizability approximation*. This approximation is only applicable if i and k are ground electronic state rotation–vibration energy levels, if $h v \gg |h v_{ki}|$, and if the excited electronic states j are at energies much higher than $h v$. The polarizability approximation arises as follows. The intensity of the scattered radiation in the Raman effect [Placzek (1934)] depends on the intensity of the incident radiation, on the frequency $(v - v_{ki})$, and on the square of the two-photon transition moment C^{β}_{ik}, where C^{β}_{ik} has components

---

[6] Further restrictions on ΔF for low values of F will occur [see, for example, Eq. (11-185)].

$C_{ik}^{\alpha\beta}$ ($\alpha = \xi, \eta$, or ζ), given by [see Mills (1964), Eq. (2)]

$$C_{ik}^{\alpha\beta} = \sum_j \left[\frac{M_{ij}^{\beta} M_{jk}^{\alpha}}{h\nu_{ji} - h\nu} + \frac{M_{ij}^{\alpha} M_{jk}^{\beta}}{h\nu_{jk} + h\nu} \right], \tag{11-189}$$

where β is the direction of polarization of the electric vector of the incident radiation (i.e., $\beta = \xi, \eta$, or ζ) and M_{ij}^{γ} and M_{jk}^{γ} are electric dipole matrix elements of M_γ ($\gamma = \alpha$ or β) such as given in Eq. (11-144) for $\gamma = \zeta$. If we make the approximation of a separable wavefunction [see Eq. (11-150)], assume that $\nu \ll \nu_{ji}$, and that $\nu \gg |\nu_{ki}|$, then the intensity of a rotation–vibration transition, within a nondegenerate ground electronic state Φ_{elec}, depends on the square of

$$\sum_{\gamma,\delta} \langle \Phi_{\text{rot}}^i | \lambda_{\gamma\alpha} \lambda_{\delta\beta} | \Phi_{\text{rot}}^k \rangle \langle \Phi_{\text{vib}}^i | \alpha_{\gamma\delta} | \Phi_{\text{vib}}^k \rangle, \tag{11-190}$$

where $\alpha_{\gamma\delta}$ is a component of the static electric polarizability and it has the same symmetry as

$$\left\langle \Phi_{\text{elec}} \left| \left(\sum_m e^2 \gamma_m \, \delta_m \right) \right| \Phi_{\text{elec}} \right\rangle, \tag{11-191}$$

where γ and $\delta = x, y$, or z, and m runs over all the electrons in the molecule. As a result of the approximations we can omit the explicit introduction of the intermediate electronic states j.

The polarizability tensor occurring in Eq. (11-190) is symmetric and the six independent components transform like the symmetrized square of the representation of the MS group generated by the electric dipole moment operator components M_x, M_y, and M_z. We see now that the allowed transitions are more restricted by demanding Eq. (11-190) to be nonvanishing than by demanding Eq. (11-189) to be nonvanishing [see, for example, Mills (1964)]. Equation (11-190) is nonvanishing if $\langle \Phi_{\text{rot}}^i | \lambda_{\gamma\alpha} \lambda_{\delta\beta} | \Phi_{\text{rot}}^k \rangle \neq 0$ (giving the selection rules on the rotational quantum numbers) and if the product of the vibrational species contains the symmetrized square of the species of the dipole moment operator components (M_x, M_y, and M_z). The vibrational part of Eq. (11-189) is nonvanishing if the product of the vibrational species contains the complete square of the species of (M_x, M_y, and M_z). For example, in a C_{3v} molecule M_x, M_y, and M_z transform as $A_1 \oplus E$ the square of which is $2A_1 \oplus A_2 \oplus 3E$, and the symmetrized square of which is $2A_1 \oplus 2E$. Using polarizability theory a vibrational transition of the type $A_1 \to A_2$ would be forbidden in the Raman effect whereas using the less approximate theory demanding that Eq. (11-189) be nonvanishing this transition would be allowed in the Raman effect ($A_1 \to E \to A_2$ are both dipole allowed). The polarizability approximation is a very good one in practice when the criteria for its applicability ($v \gg |v_{ik}|$, and hv much less than the electronic excitation energies) are satisfied.

The species of the polarizability tensor components $\alpha_{\gamma\delta}$ are the same as the species of the products $(T_{\gamma}T_{\delta})$ and these species are indicated in the MS tables in Appendix A. A vibrational transition will be Raman active if the vibrational species are connected by the species of a polarizability tensor component. We can expand the electronic polarizability tensor as a Taylor series in the normal coordinates in the same way that $M_{\alpha}(e', e'')$ is expanded in Eq. (11-155). As a result the most intense vibrational Raman bands that can be excited from the ground vibrational state will be the fundamentals (for which $\Delta v_r = 1$) and each Raman active state must have a normal coordinate that transforms like a component of the polarizability tensor. In centrosymmetric molecules the polarizability tensor is of g symmetry (it is the product of two T_{α} which are u) and only fundamental bands involving g vibrations can be active. This is in contrast to the infrared active fundamentals which must be u. Thus in centrosymmetric molecules there is an exclusion principle that no fundamental band can be simultaneously infrared and Raman active. The determination of the number of infrared and Raman active fundamentals in a molecule is a great help in determining the symmetry of the equilibrium geometry of a rigid molecule.

THE ZEEMAN EFFECT

In the Zeeman effect a magnetic field $\boldsymbol{B}$ is applied to the molecule under study. Choosing $\boldsymbol{B}$ to be along the ζ axis, the field free molecular Hamiltonian is augmented by the term [see, for example, Chapter 11 in Townes and Schawlow (1955)]

$$\hat{H}_{\text{Zeeman}} = -D_{\zeta}B_{\zeta} \tag{11-192}$$

where D_{ζ} is the ζ component of the magnetic dipole moment operator for the molecule; see Eq. (11-180). From Eq. (11-180) we see that D_{ζ} involves both the orbital and spin angular momenta of the particles in the molecule.

We are interested in knowing how this perturbation changes the energy levels and wavefunctions of the molecule from those of the molecule in free space. We can appreciate this qualitatively from the symmetry of $\hat{H}_{\text{Zeeman}}$ in the MS group and in the K(spatial) group. Since identical particles have the same charge, mass, and spin, $\hat{H}_{\text{Zeeman}}$ is invariant to any permutation of identical particles. Also the inversion, which simultaneously changes the sign of all $\xi_i, \eta_i,$ and ζ_i (but which does not affect the spins), will leave $\hat{H}_{\text{Zeeman}}$ invariant. Thus the term $\hat{H}_{\text{Zeeman}}$ is totally symmetric in the MS

Table 11-10
The character table of
the C_∞ group

	E	$2R_\zeta^\varepsilon$	$\cdots$
Σ:	1	1	$\cdots$
Π $\begin{cases} \\ \end{cases}$	1 1	$e^{i\varepsilon}$ $e^{-i\varepsilon}$	$\cdots$ $\cdots$
Δ $\begin{cases} \\ \end{cases}$	1 1	$e^{2i\varepsilon}$ $e^{-2i\varepsilon}$	$\cdots$ $\cdots$
Φ $\begin{cases} \\ \end{cases}$	1 1	$e^{3i\varepsilon}$ $e^{-3i\varepsilon}$	$\cdots$ $\cdots$
$\vdots$	$\vdots$	$\vdots$	$\cdots$

group. This means that $\hat{H}_{\text{Zeeman}}$ can only cause states of the same MS group symmetry to interact and that the Hamiltonian matrix (*including* Zeeman terms) can be factored into blocks by using the MS group symmetry species. The term $\hat{H}_{\text{Zeeman}}$ is not, however, totally symmetric in the group K(spatial) but it is invariant to any rotation about the direction of B (the ζ direction in our example). The reader can see that twofold rotations about the ξ or η directions will change the sign of the orbital part of $\hat{H}_{\text{Zeeman}}$ but that any rotation about the ζ axis will leave it invariant. Thus the rotational symmetry is reduced to C_∞(spatial) and the C_∞ character table is given in Table 11-10. In the group C_∞(spatial) the molecular states span the representations Σ, Π, Δ, etc., as $m_F = 0$, ± 1, ± 2, etc. (note that the Π, Δ, etc., representations are separably degenerate), and F does not remain a good quantum number. In the presence of a magnetic field time reversal symmetry is lost, since $\hat{H}_{\text{Zeeman}}$ is not invariant to the reversal of all momenta and spins, and separably degenerate states are not required by symmetry to be degenerate. As a result each level that in field free space is characterized by the quantum number F, with a $(2F + 1)$-fold m_F degeneracy, is split into $(2F + 1)$ levels each of which has a definite m_F value and the same MS group symmetry label. In a weak field F would be a useful near quantum number and the most intense electric dipole transitions for the molecule would satisfy $\Delta F = 0$, ± 1. The strictly allowed electric dipole transitions would be all those that satisfy $\Delta m_F = 0$, ± 1 ($m_F'' = 0 \nleftrightarrow m_F' = 0$) with MS group species connected by Γ^*.

THE STARK EFFECT

In the Stark effect an electric field E is applied to the molecules under study. Choosing E to be along the ζ axis, the field free molecular Hamiltonian is augmented by the term [see, for example, Chapter 10 in Townes and

Schawlow (1955)]

$$\hat{H}_{\text{Stark}} = -\left(\sum_j e_j \zeta_j\right) E = -M_\zeta E. \tag{11-193}$$

This has the symmetry of the electric dipole moment operator M_ζ and hence has symmetry species Γ^* in the MS group and species $D^{(1)}$ in the group K(spatial). Thus the Stark effect mixes states connected by the symmetry species Γ^* and $D^{(1)}$; the selection rules governing which states can be mixed by the application of an electric field are the same as the selection rules for electric dipole transitions since the matrix element involved is that of M_ζ in both cases. The Stark effect mixes those states between which electric dipole transitions are allowed. Note that $\hat{H}_{\text{Stark}}$ is invariant to time reversal since it is unaffected by the reversal of momenta and spins.

The Hamiltonian of a molecule in an electric field is invariant to any permutation of identical nuclei but not to the inversion E^* (since $\hat{H}_{\text{Stark}}$ is changed in sign by E^*). As a result the permutation subgroup of the MS group (the PSMS group) can be used to label the states. The correlation table of the MS group and its PSMS group can be used to determine the PSMS group labels from the MS group labels of the levels of the molecule in free space. States having different PSMS group labels cannot be mixed by $\hat{H}_{\text{Stark}}$ and the Hamiltonian matrix can be factorized by using this symmetry.

The K(spatial) symmetry is spoiled by $\hat{H}_{\text{Stark}}$, and F is no longer a good quantum number. However, the Stark Hamiltonian is invariant to the operations of the C_∞(spatial) group, where the rotations are about the field direction, and time reversal symmetry is not spoiled by an electric field. As a result $M_F (= |m_F|)$ is a good quantum number and the $(2F + 1)$ degeneracy of each F state is resolved into $(F + 1)$ different M_F components in an electric field. Each level can be labeled by M_F and the PSMS group label.

The electric dipole spectrum of a molecule in an electric field contains the normal electric dipole transitions (possibly shifted and split by the field) as well as additional transitions made allowed by interactions caused by $\hat{H}_{\text{Stark}}$. Since these interactions satisfy electric dipole transition selection rules the additional transitions will be those that constitute allowed two-photon electric dipole transitions of the isolated molecule.

SUMMARY

As a result of using true symmetry, basis symmetry, and near symmetry, as well as true and near quantum numbers, we label molecular energy levels. The most useful labels are Γ_{int} (or parity), F, Γ_{rve}, J, I, S, N, the vibrational quantum numbers v_r, and the rotational quantum numbers $(K,(\pm l))$ for a symmetric top, (K_a, K_c) for an asymmetric top, and R for a spherical top. For certain purposes we may also wish to use the basis symmetry labels Γ_{rot}, Γ_{vib}, Γ_{elec}, Γ_{rv}, and Γ_{ve} from the MS group. We can use these labels to specify which levels can be mixed by perturbation terms and to specify which levels are connected by electric dipole transitions. Among the more important selection rules for perturbations are that anharmonicity perturbations connect levels having the same Γ_{vib} labels, centrifugal distortion and Coriolis coupling connect levels having the same Γ_{rv} labels, and vibronic interactions connect states having the same Γ_{ve} labels. Selection rules on the vibrational and rotational quantum numbers can also be derived. The selection rules for electric dipole transitions can be formulated in terms of the vibrational, rotational, and electronic quantum numbers and symmetries; transitions that occur which do not satisfy these restrictions are called forbidden transitions. The appearance of forbidden transitions result from the effects of anharmonicity, centrifugal distortion, Coriolis coupling, vibronic, rotation–electronic, and rovibronic interactions, as well as from electrical anharmonicity, axis switching, and changes in the magnetic dipole and electric quadrupole moments. The energy level labels are also useful in specifying the Zeeman and Stark effects as well as for determining the selection rules for multiphoton processes.

BIBLIOGRAPHICAL NOTES

Near Symmetry

McIntosh (1971).
Wulfman (1971).

Optical Selection Rules and Forbidden Transitions

Herzberg (1945).   Chapter III.
Herzberg (1966).   Chapter II.
Herzberg (1969).
Bunker (1974).   This paper discusses the determination of the optical selection rules for some ionic species.
Oka (1976).

Zeeman and Stark Effects

Bunker (1973b).
Watson (1974).
Watson (1975).

12

Linear Molecules and Nonrigid Molecules

This chapter deals with the application of the MS group to linear molecules and to nonrigid molecules. For linear molecules we first introduce the isomorphic Hamiltonian which is much more convenient to work with than the true Hamiltonian. The classification of the eigenfunctions of the isomorphic Hamiltonian in the MS group is discussed, and it is shown that the molecular point group, but not the MS group, can be used to classify the vibronic eigenfunctions. As usual the MS group, but not the molecular point group, is used to classify the rovibronic and complete internal eigenfunctions. The extended MS group of a linear molecule is introduced, and this group, like the MS group of a nonlinear rigid molecule, is isomorphic to the molecular point group and can be used to classify both vibronic and rovibronic eigenfunctions. Nonrigid molecules have one or more large amplitude internal coordinates and the derivation of the rotation–vibration Hamiltonian discussed in Chapters 7 and 8 must be generalized. The rotation–vibration coordinates and eigenfunctions of nonrigid molecules are introduced, and the technique used in determining their transformation properties in the MS group is discussed. For nonrigid molecules having identical coaxial internal rotors (such as hydrogen peroxide and dimethylacetylene) it is not possible to classify the vibronic eigenfunctions in the MS group, just as for linear molecules, and it is necessary to introduce an extended MS group for these molecules also. The extended MS group can be used for classifying rovibronic and vibronic states. The electron spin double groups for linear molecules and nonrigid molecules are also introduced.

LINEAR MOLECULES

The Isomorphic Hamiltonian

For a linear molecule there are only two rotational degrees of freedom corresponding to the two Euler angles θ and ϕ required to specify the direction of the molecular axis (the z axis) in space, and this lack of a third Euler angle causes some complications. [At this point the reader may find it useful to read again the section on the diatomic molecule in Chapter 7, Eqs. (7-65)–(7-116).] The x and y coordinates of the nuclei vanish when a linear molecule is in its equilibrium configuration, and from Eqs. (7-127)–(7-129) we see that there are only two Eckart equations for a linear molecule:

$$\sum_{i=1}^{N} m_i z_i^e y_i = 0 \tag{12-1}$$

and

$$\sum_{i=1}^{N} m_i z_i^e x_i = 0. \tag{12-2}$$

These equations enable us to determine θ and ϕ, and thus the spatial orientation of the z axis, from the instantaneous coordinates of the nuclei in space. There is no Eckart condition that specifies the Euler angle χ since the orientation of the x and y axes is immaterial from the point of view of minimizing the vibrational angular momentum given in Eq. (7-122). The Euler angle χ is usually chosen as a constant and in the derivation of the diatomic molecule Hamiltonian in Chapter 7 the choice $\chi = 0°$ was made. The most general choice would be to choose χ as a function of θ and ϕ. Because of this the direction cosine matrix elements [see Eq. (7-52)] will only depend on the two independent variables θ and ϕ. As a result of the absence of χ as a rotational variable the components of the angular momentum in the molecule fixed axis system for a linear molecule do not obey the commutation relations given in Eq. (7-147). Instead the commutation relations are more complicated [see, for example, Eqs. (7-84) and (7-85)], and the matrix elements involving these operators and the rotational eigenfunctions are not the same as for a nonlinear molecule (given in Table 8-1). The rotational kinetic energy operator for a linear molecule is also more complicated than that of a nonlinear molecule [see Eq. (7-137)] owing to the presence of extra angular factors [see Eq. (7-94) and note the presence of the $\sin\theta$ factors occurring in the second term].

A method has been devised for introducing the Euler angle χ as an independent variable for a linear molecule, and this is by the use of Hougen's isomorphic Hamiltonian [Hougen (1962a), Bunker and Papousek (1969), Watson (1970), Howard and Moss (1971)]. The isomorphic Hamiltonian has one more degree of freedom than the true Hamiltonian and it has many eigenvalues that are not eigenvalues of the true Hamiltonian. However, it is easy to use a particular set of basis functions to diagonalize the isomorphic Hamiltonian so that these extra eigenvalues are not obtained. With the isomorphic Hamiltonian the components of the angular momentum along the molecule fixed axes have the commutation relations given in Eq. (7-147), and the rotational kinetic energy operator is simplified. Therefore, as a result of using the isomorphic Hamiltonian the linear molecule is treated in much the same way as the nonlinear molecule.

To obtain the isomorphic Hamiltonian for a linear molecule, χ is introduced as an independent variable, and the coordinates of the particles in the molecule are referred to an (x', y', z') axis system having Euler angles (θ, ϕ, χ) in the (ξ, η, ζ) axis system. As a result we introduce components of the rovibronic, electronic, and vibrational angular momentum, $\hat{\boldsymbol{J}}$, $\hat{\boldsymbol{L}}$, and $\hat{\boldsymbol{p}}$, respectively, along these axes (these components will be called $\hat{J}_{x}{}'$, $\hat{J}_{y}{}'$, etc.). If we initially choose $\chi = 0°$ in the true Hamiltonian (as for the diatomic molecule Hamiltonian discussed in Chapter 7) the (x', y', z') axes are rotated from the (x, y, z) axes about the $z\,(=z')$ axis through the angle χ (see Fig. 12-1). We have

$$(\hat{J}_x - \hat{p}_x - \hat{L}_x) = \cos\chi(\hat{J}_x{}' - \hat{p}_x{}' - \hat{L}_x{}') - \sin\chi(\hat{J}_y{}' - \hat{p}_y{}' - \hat{L}_y{}'), \quad (12\text{-}3)$$

$$(\hat{J}_y - \hat{p}_y - \hat{L}_y) = \cos\chi(\hat{J}_y{}' - \hat{p}_y{}' - \hat{L}_y{}') + \sin\chi(\hat{J}_x{}' - \hat{p}_x{}' - \hat{L}_x{}'), \quad (12\text{-}4)$$

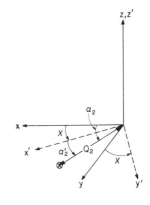

Fig. 12-1. Molecule fixed axes for a linear molecule. The (x, y, z) axes have Euler angles $(\theta, \phi, 0)$ in the (ξ, η, ζ) axis system, and the (x', y', z') axes used with the isomorphic Hamiltonian have Euler angles (θ, ϕ, χ) where χ is arbitrary. The bending vibration is described by an amplitude Q_2 and an angle $\alpha_2 = (\chi + \alpha_2{}')$ in the normal Hamiltonian, and by the coordinates Q_2 and $\alpha_2{}'$ in the isomorphic Hamiltonian.

and

$$(\hat{J}_z - \hat{p}_z - \hat{L}_z) = (\hat{J}_z' - \hat{p}_z' - \hat{L}_z') = 0. \qquad (12\text{-}5)$$

The $\hat{J}_x', \hat{J}_y',$ and $\hat{J}_z'$ operators are the same as $\hat{J}_x, \hat{J}_y,$ and $\hat{J}_z$ given in Eqs. (7-144)–(7-146). The rotational kinetic energy operator (in cm^{-1}) of a linear molecule is transformed by this coordinate change to [see Eq. (83) of Watson (1970)]

$$\hat{T}_{\text{rot}}^{\text{iso}} = Bh^{-2}[(\hat{J}_x' - \hat{p}_x' - \hat{L}_x')^2 + (\hat{J}_y' - \hat{p}_y' - \hat{L}_y')^2], \qquad (12\text{-}6)$$

where the awkward angular factors have disappeared. The term U that is present for the nonlinear molecule [see Eq. (7-137)] is not present for the linear molecule. The coordinate change, from (x, y, z) to (x', y', z'), will not affect the form of the electronic kinetic energy operator $\hat{T}_e$, as we showed generally in the work leading to Eqs. (7-58) and (7-60), since the definition of χ is independent of the electronic coordinates. Also, since the vibrational coordinates are unaffected by the transformation, the vibrational kinetic energy operator $(\frac{1}{2}\sum_r \hat{P}_r^2)$ is also unchanged. The angular momenta $\hat{J}_x', \hat{J}_y',$ and $\hat{J}_z'$ in the isomorphic Hamiltonian satisfy the commutation relations given in Eq. (7-147) and, as for nonlinear molecules, they all commute with the components $\hat{p}_\alpha'$ and $\hat{L}_\alpha'$ of the vibrational and electronic angular momenta because they involve different sets of variables.

The isomorphic Hamiltonian $\hat{H}^{\text{iso}}$ commutes with $(\hat{J}_z' - \hat{p}_z' - \hat{L}_z')$ so that there exist simultaneous eigenfunctions of $\hat{H}^{\text{iso}}$ and $(\hat{J}_z' - \hat{p}_z' - \hat{L}_z')$. From Eq. (12-5) we see that the only eigenvalues of $\hat{H}^{\text{iso}}$ that are eigenvalues of the true Hamiltonian are those for which the corresponding eigenfunctions have eigenvalue zero for $(\hat{J}_z' - \hat{p}_z' - \hat{L}_z')$. Thus in setting up a zero order set of basis functions as products of rigid rotor (Φ_{rot}), harmonic oscillator (Φ_{vib}), and electron orbital (Φ_{elec}) functions we only use those basis functions for which $k = \Lambda + l$ where

$$\hat{J}_z'\Phi_{\text{rot}} = kh\Phi_{\text{rot}}, \qquad (12\text{-}7)$$

$$\hat{p}_z'\Phi_{\text{vib}} = lh\Phi_{\text{vib}}, \qquad (12\text{-}8)$$

and

$$\hat{L}_z'\Phi_{\text{elec}} = \Lambda h\Phi_{\text{elec}}. \qquad (12\text{-}9)$$

We now look at the form of these basis functions so that we can investigate their symmetry transformation properties.

The Rovibronic Wavefunctions

The (isomorphic) rigid rotor rotational Hamiltonian of a linear molecule is (in cm^{-1})

$$\hat{H}_{\text{rot}}^{\text{iso}} = Bh^{-2}(\hat{J}_x'^2 + \hat{J}_y'^2) = Bh^{-2}(\hat{J}^2 - \hat{J}_z'^2), \qquad (12\text{-}10)$$

so that the eigenvalues are

$$E_{\text{rot}}^{\text{iso}} = B[J(J + 1) - k^2],$$ (12-11)

and the eigenfunctions are symmetric top functions [see Eq. (8-64)] which we write as [see Eq. (8-73b)]

$$\Phi_{\text{rot}}^{\text{iso}} = [1/(2\pi)]^{1/2} S_{Jkm}(\theta, \phi) \exp(ik\chi).$$ (12-12)

The vibrational Hamiltonian of a linear molecule involves $(3N - 5)$ normal coordinates (there being three translational and two rotational degrees of freedom) and there are $(N - 1)$ nondegenerate normal coordinates and $(N - 2)$ doubly degenerate bending normal coordinates. As a result the zero order harmonic oscillator wavefunctions are the products of $(N - 1)$ one-dimensional and $(N - 2)$ two-dimensional harmonic oscillator functions which we write as

$$\Phi_{\text{vib}}^{\text{iso}} = \Phi_{v_1, \ldots, v_{2N-3}}(Q_1, \ldots, Q_{2N-3}) \exp(il_1\alpha_1') \exp(il_2\alpha_2') \cdots \exp(il_{N-2}\alpha_{N-2}')$$
(12-13)

where Q_1 to Q_{N-2} are two-dimensional and Q_{N-1} to Q_{2N-3} are one-dimensional normal coordinates. The angles α_r' are as the α introduced in Eqs. (8-204)–(8-207) for a two-dimensional harmonic oscillator and here they are measured from the x' axis in the $x'y'$ plane (see Figs. 12-1 and 8-2). The total vibrational angular momentum about the linear axis (i.e., the eigenvalue of $\hat{p}_z'$) is given by

$$l\hbar = \sum_{r=1}^{N-2} l_r \hbar.$$ (12-14)

The electronic Hamiltonian of a linear molecule is solved in the same way as for a nonlinear molecule (see Chapter 8) and the zero order electronic wavefunction is a product of MO's each of which depends parametrically on the vibrational coordinates. When the nuclei are in the linear configuration the electronic Hamiltonian will commute with $\hat{L}_z'$, and Λ will be a good quantum number for the eigenstates. In this circumstance we can write

$$\Phi_{\text{elec}}^{\text{iso}} = \psi_{\text{elec}} \exp(i\Lambda\chi_e')$$ (12-15)

where ψ_{elec} is a function of interelectron distances, and χ_e' (a cylindrical angle about the z' axis) is the coordinate conjugate to $\hat{L}_z'$.

The rovibronic basis functions that we use to diagonalize $\hat{H}^{\text{iso}}$ are the products of the functions in Eqs. (12-12), (12-13), and (12-15) with the restriction that

$$k = \Lambda + l.$$ (12-16)

This ensures that the eigenvalues obtained will only be those of the true Hamiltonian.

The Symmetry Classification of the Basis Set Wavefunctions

In order to simplify the equations we will not look at the symmetry classification of the general basis set function just discussed but rather at the rotation–vibration wavefunction of HCN in its ground electronic state (this is a $\Lambda = 0$ state). This will enable us to see the relationship of the MS group to the molecular point group for a linear molecule fairly simply.

The rotation–vibration basis set functions for the isomorphic Hamiltonian of the HCN molecule in its ground electronic state are

$$\Phi_{rv}^{iso} = [1/(2\pi)]^{1/2} S_{Jkm}(\theta, \phi) \exp(ik\chi) \Phi_{v_1 v_2 v_3}(Q_1, Q_2, Q_3) \exp(il\alpha_2'), \quad (12\text{-}17)$$

where the two-dimensional bending vibration is numbered 2 in keeping with the established convention. We only use basis functions having $k = l$ [see Eq. (12-16)] which we can write as

$$\Phi_{rv}^{iso} = [1/(2\pi)]^{1/2} S_{JlM}(\theta, \phi) \exp(il\chi) \Phi_{v_1 v_2 v_3}(Q_1, Q_2, Q_3) \exp(il\alpha_2'), \quad (12\text{-}18)$$

and this is identical to

$$\Phi_{rv}^{iso} = [1/(2\pi)]^{1/2} S_{JlM}(\theta, \phi) \exp(il\alpha_2) \Phi_{v_1 v_2 v_3}(Q_1, Q_2, Q_3), \quad (12\text{-}19)$$

where we have introduced the angle α_2 given by

$$\alpha_2 = \alpha_2' + \chi \quad (12\text{-}20)$$

and α_2 is measured in the xy plane from the x axis [using the $\chi = 0°$ convention to define the (x, y, z) axis system] as shown in Fig. (12-1). We can also introduce the angle χ_e by writing

$$\chi_e = \chi_e' + \chi, \quad (12\text{-}21)$$

and the rovibronic wavefunction of a $\Lambda \neq 0$ state would involve the product $\exp(il\alpha_2)\exp(i\Lambda\chi_e)$.

Before discussing the symmetry classification of this rotation–vibration function we must introduce the symmetry groups of linear molecules. The point group of a linear molecule is $D_{\infty h}$ if it is centrosymmetric, otherwise it is $C_{\infty v}$. Both of these groups have an infinite number of elements. The MS group of a molecule such as HCN that has $C_{\infty v}$ point group symmetry is called $C_{\infty v}(M)$ and it consists of the two elements E and E^*. The MS group of a molecule such as CO_2 that has $D_{\infty h}$ point group symmetry is called $D_{\infty h}(M)$ and it consists of the four elements E, (p), E^* and $(p)^*$, where (p) is the permutation of all pairs of identical nuclei that, in the equilibrium configuration, are located symmetrically about the nuclear center of mass. The operation $(p)^*$ is therefore identical to the operation $\hat{O}_i$ introduced in Eq. (11-12). The character table of the group $C_{\infty v}(M)$ is given in Table A-12 and its two irreducible representations are labeled $+$ (or Σ^+) and $-$ (or Σ^-) as the character under E^* (i.e., the parity) is $+1$ or -1. The character table of the group $\boldsymbol{D}_{\infty h}(M)$ is given in Table A-13. The irreducible representations are labeled $+s$ (or Σ_g^+), $+a$ (or Σ_u^+), $-s$ (or Σ_u^-) and $-a$ (or Σ_g^-), where the $+$ or $-$ depends on the character under E^*, the s or a on the character under (p), and the g or u on the character under $(p)^*$. We see that for a linear rigid molecule the MS group and the molecular point group are not isomorphic, unlike the situation for a nonlinear rigid molecule. We will now discuss the use of the MS group and molecular point group to classify the rotation–vibration wavefunction of HCN given in Eq. (12-19).

The rotation–vibration wavefunction of HCN given in Eq. (12-19) can be classified in the MS group $\boldsymbol{C}_{\infty v}(M)$. We can do this because the coordinates occurring in the wavefunction, θ, ϕ, α_2, Q_1, Q_2, and Q_3, all have well-defined transformation properties under the elements of the group $\boldsymbol{C}_{\infty v}(M)$. The transformation properties of the coordinates under the effect of E^* are [see Fig. 12-2 and Eq. (7-115)]

$$E^*(\theta, \phi) = (\theta_{\text{new}}, \phi_{\text{new}}) = (\pi - \theta, \phi + \pi), \qquad (12\text{-}22)$$

$$E^*\alpha_2 = (\alpha_2)_{\text{new}} = (\pi - \alpha_2), \qquad (12\text{-}23)$$

and

$$E^*(Q_1, Q_2, Q_3) = (Q_1, Q_2, Q_3). \qquad (12\text{-}24)$$

The z axis, the y axis, and the direction of the bending displacement in space (but not the x axis from which α_2 is measured) are all reversed by E^*. The coordinates Q_1 and Q_3 transform like $(\Delta z_H - \Delta z_C)$ and $(\Delta z_C - \Delta z_N)$, and are

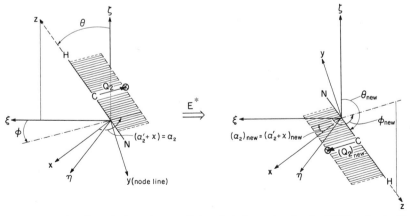

Fig. 12-2. The effect of E^* on the coordinates θ, ϕ, Q_2, and α_2 for the HCN molecule. The directions of the z and y axes in space are reversed by E^*.

unaffected by E^*; Q_2 is the amplitude of the degenerate vibration and is unaffected by a symmetry operation. For a $D_{\infty h}$ linear molecule such as CO_2 it can be shown that [see Eqs. (7-112) and (7-116)]

$$(p)(\theta, \phi, \alpha_2) = (\pi - \theta, \phi + \pi, -\alpha_2) \tag{12-25}$$

and

$$(p)^*(\theta, \phi, \alpha_2) = (\theta, \phi, \alpha_2 + \pi). \tag{12-26}$$

From Eqs. (12-22)–(12-24) we deduce that the effect of E^* on the rotation–vibration wavefunction of HCN is

$$E^*\Phi_{rv}^{iso} = [1/(2\pi)]^{1/2} S_{Jlm}(\pi - \theta, \phi + \pi) \exp[il(\pi - \alpha_2)]\Phi_{v_1 v_2 v_3}(Q_1, Q_2, Q_3). \tag{12-27}$$

By analogy with the effect of $R_{\pi/2}^\pi$ on a symmetric top function [see Table 7-1 and Eq. (10-22)] we deduce that

$$E^*\Phi_{rv}^{iso} = (-1)^{J+l}[1/(2\pi)]^{1/2} S_{J-lm}(\theta, \phi) \exp(-il\alpha_2)\Phi_{v_1 v_2 v_3}(Q_1, Q_2, Q_3). \tag{12-28}$$

Thus for $l = 0$ states the rovibronic states with J even are $+$ and those with J odd are $-$. For $l \neq 0$ states we must form sum and difference functions in order that the functions transform irreducibly, i.e., we form (using an obvious notation)

$$\Phi_{rv}^\pm = (|J, |l|, m, v_1, v_2, v_3\rangle \pm |J, -|l|, m, v_1, v_2, v_3\rangle)/\sqrt{2}. \tag{12-29}$$

The Φ_{rv}^+ (or Φ_{rv}^-) function is $+$ (or $-$) for $J + l$ even and $-$ (or $+$) for $J + l$ odd. In general the rovibronic states of a $C_{\infty v}$ molecule can be classified as being $+$ or $-$ in its MS group, and for a $D_{\infty h}$ molecule as $+s$, $-s$, $+a$, or $-a$ in its MS group.

As well as classifying the rotation–vibration wavefunctions we wish to classify the rotation and vibration wavefunctions separately. The rotation and vibration wavefunctions of HCN in the ground electronic state are [from Eqs. (12-12), (12-13), and (12-18)]

$$\Phi_{rot}^{iso} = [1/(2\pi)]^{1/2} S_{Jlm}(\theta, \phi) \exp(il\chi) \tag{12-30}$$

and

$$\Phi_{vib}^{iso} = \Phi_{v_1 v_2 v_3}(Q_1, Q_2, Q_3) \exp(il\alpha_2'). \tag{12-31}$$

To determine the transformation properties of these functions in the MS group we must determine the transformation properties of the rotational and vibrational coordinates θ, ϕ, χ, Q_1, Q_2, Q_3, and α_2' under the effect of the elements E and E^* of the MS group. The transformation properties of θ, ϕ, Q_1, Q_2, and Q_3 present no problem but χ and α_2' do present a problem. This is because there is no Eckart condition that defines χ. Hence, after performing E or E^* we do not know where to locate the $x'y'$ axes and we do not know what value χ and α_2' have. The transformation properties of the angle $\alpha_2 = (\chi + \alpha_2')$ are well defined, see Eq. (12-23), since this does not depend on the orientation of the $x'y'$ axes. The rotation and vibration functions of HCN do not, therefore, have separately determinable parities. This problem occurs for all linear molecules since the rotational angle χ and the vibronic variables α_r' and χ_e' do not have determinable transformation properties under the effect of a nuclear permutation or permutation–inversion. On the other hand the Euler angle χ has well-defined transformation properties under the elements of the molecular point group. This is because the elements of the molecular point group are defined as doing nothing at all to the Euler angles. Since χ is unchanged after a point group operation the location of the $x'y'$ axes after a point group operation is defined and the transformation properties of the vibronic variables α_2' and χ_e' are defined. Thus the vibronic wavefunctions of a linear molecule can be classified in the appropriate molecular point group. For a $\boldsymbol{C}_{\infty v}$ molecule the vibronic states can be of species Σ^+, Σ^-, Π, Δ, etc., and for a $\boldsymbol{D}_{\infty h}$ molecule there is an additional g or u label. Just as for nonlinear molecules the molecular point group is not a symmetry group of the rovibronic Hamiltonian (since the components $\hat{J}_\alpha'$ are unaffected by its elements whereas $\hat{p}_\alpha'$ and $\hat{L}_\alpha'$ are transformed), and it is not the correct group to use for classifying rovibronic states.

We now discuss the classification of vibrational and electronic wavefunctions in the molecular point group for a linear molecule. The elements of the $\boldsymbol{D}_{\infty h}$ point group are

$$E\,(= C_\infty{}^0),\ C_\infty{}^\varepsilon,\ C_2^{(\varepsilon/2)},\ \sigma_v^{(\varepsilon/2)},\ \text{and}\ S_\infty{}^\varepsilon, \tag{12-32}$$

where

$C_\infty{}^\varepsilon$ is a rotation (in a right handed sense) of the vibronic variables about the z' axis through the angle ε,

$C_2^{(\varepsilon/2)}$ is a twofold rotation of the vibronic variables about an axis that makes an angle $(\varepsilon/2)$ (measured in a right handed sense about the z' axis) with the x' axis,

$\sigma_v^{(\varepsilon/2)}$ is a reflection of the vibronic variables in a plane containing the z' axis and the axis just described that makes an angle of $(\varepsilon/2)$ with the x' axis, and

$S_\infty{}^\varepsilon$ is the product of $C_\infty{}^\varepsilon$ and the reflection (σ_h) of the vibronic variables in the $x'y'$ plane, i.e.,

$$S_\infty{}^\varepsilon = C_\infty{}^\varepsilon \sigma_h. \tag{12-33}$$

Note that in the $D_{\infty h}$ point group

$$S_\infty{}^0 = \sigma_h \qquad (12\text{-}34)$$

and

$$S_\infty{}^\pi = i. \qquad (12\text{-}35)$$

In the preceding definitions the angles ε can assume any value (mod 2π) in the range $0 \le \varepsilon \le 2\pi$ independently of each other, and for the $C_{\infty v}$ point group the elements $C_2^{(\varepsilon/2)}$ and $S_\infty{}^\varepsilon$ do not occur.

The classification of the vibrational wavefunctions of a linear molecule in the appropriate point group is straightforward. The effects of the operations of Eq. (12-32) on the vibrational angles α_r' for the degenerate vibrations are

$$C_\infty{}^\varepsilon \alpha_r' = S_\infty{}^\varepsilon \alpha_r' = \alpha_r' + \varepsilon \qquad (12\text{-}36)$$

and

$$C_2^{(\varepsilon/2)} \alpha_r' = \sigma_v^{(\varepsilon/2)} \alpha_r' = -\alpha_r' + \varepsilon. \qquad (12\text{-}37)$$

Electronic cylindrical angles (measured from the x' axis) will transform in the same way. Using the point group we can classify the vibrational and electronic wavefunctions of a linear molecule. The two-dimensional vibrational wavefunctions of the HCN molecule [see Eq. (12-31)] are transformed as

$$C_\infty{}^\varepsilon \Phi_{v_2}(Q_2) \exp(il\alpha_2') = \exp(il\varepsilon) \Phi_{v_2}(Q_2) \exp(il\alpha_2') \qquad (12\text{-}38)$$

and

$$\sigma_v^{(\varepsilon/2)} \Phi_{v_2}(Q_2) \exp(il\alpha_2') = \exp(il\varepsilon) \Phi_{v_2}(Q_2) \exp(-il\alpha_2'). \qquad (12\text{-}39)$$

Thus the pair of functions $(|v_2, l\rangle, |v_2, -l\rangle)$ have character $2\cos l\varepsilon$ (i.e., $e^{il\varepsilon} + e^{-il\varepsilon}$) under $C_\infty{}^\varepsilon$, and they have a character of 0 if $l \ne 0$ under $\sigma_v^{(\varepsilon/2)}$. For a $D_{\infty h}$ molecule such as CO_2 we have

$$S_\infty{}^\pi |v_2, l\rangle = (-1)^l |v_2, l\rangle, \qquad (12\text{-}40)$$

so that functions with l even are g and those with l odd are u, and the species of the functions depends on l as given in Table 12-1 (see Table A-15 for the $D_{\infty h}$ character table). The symmetry classification of the vibrational states associated with the nondegenerate vibrations presents no special problems.

The point group classification of the electronic wave functions of a linear molecule has an interesting feature which we will now discuss. The electronic wavefunctions are obtained by solving the electronic wave equation for a particular nuclear configuration [see Eq. (8-2)], and we also obtain the electronic eigenvalues V_{elec}. By solving this equation at many different nuclear configurations we can determine the variation of V_{elec} with the nuclear coordinates and, after adding the nuclear–nuclear repulsion term

Table 12-1

Species of bending
wavefunctions of CO_2
in $D_{\infty h}$ point group[a]

l	$\Gamma(|v_2, l\rangle)$
0	Σ_g^+
± 1	Π_u
± 2	Δ_g
± 3	Φ_u
$\vdots$	$\vdots$

[a] For a $C_{\infty v}$ molecule
such as HCN the g and
u subscripts should be
omitted.

V_{nn}, we can obtain the nuclear potential energy function V_N, for each electronic state, as a function of the nuclear coordinates [see Eq. (8-5)]. For HCN in the linear configuration the ground electronic state is a Σ state and the first excited electronic state is a Π state. However, when the molecule is bent it has C_s point group symmetry and no electronic states are required by symmetry to be degenerate. The electronic Π state of linear HCN splits into an A' and an A'' electronic state of the bent molecule and the potential energy curves are shown schematically in Fig. 12-3. For any linear molecule, degenerate electronic states of the linear configuration are not required by symmetry to remain degenerate when the molecule is bent and the potential energy curves that are degenerate in the linear configuration will split apart. This splitting of a degenerate electronic state, and the further interaction between the vibrational levels of the states caused by the term $\hat{T}_{ev}$ [see Eq. (11-80) and Fig. 12-3], is called the Renner effect. Such a splitting can occur in one of the three ways indicated in Fig. 12-4. To determine the species of the electronic states of the bent molecule from those of the linear molecule the appropriate correlation table of $D_{\infty h} \to C_{2v}$ and $C_{\infty v} \to C_s$ (given in Table B-3 of Appendix B) can be used.

For linear molecules we classify the rovibronic states and vibronic states in different symmetry groups, using the MS group and molecular point group, respectively. However, we can introduce an extended molecular symmetry (EMS) group [Bunker and Papousek (1969)] which can be used to classify both functions. The classification obtained reproduces the point group classification of the vibronic states, (i.e., Σ, Π, Δ, etc., with g or u in addition for $D_{\infty h}$ molecules) and the MS group classification of the rovibronic states (i.e., $+$ or $-$, with a or s in addition for $D_{\infty h}$ molecules). No new

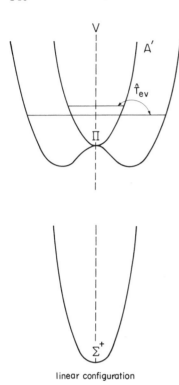

Fig. 12-3.   The lowest potential energy curves of the singlet states of HCN. The abscissa is the bending coordinate. A pair of bending energy levels coupled by the vibronic kinetic energy $\hat{T}_{ev}$ are marked to indicate schematically a Renner effect interaction.

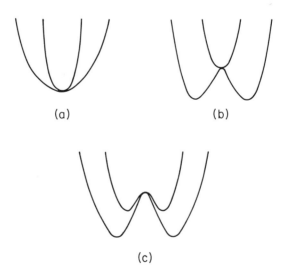

Fig. 12-4.   The potential curves of triatomic molecules. The three ways that an electronic degeneracy at the linear nuclear configuration can be resolved when the molecule is bent.

classifications emerge but this does enable the symmetry classification of linear molecule wavefunctions and the vibronic and rovibronic selection rules to be treated using a single group, the EMS group, in the same way that nonlinear molecule wavefunctions are treated using the MS group.

The elements of the EMS group for a molecule of $D_{\infty h}$ point group symmetry are

$$E\,(= E_0),\, E_\varepsilon,\, (p)_\varepsilon,\, E_\varepsilon{}^*,\, \text{and}\, (p)_\varepsilon{}^*, \tag{12-41}$$

and those for a molecule of $C_{\infty v}$ point group symmetry are

$$E\,(= E_0),\, E_\varepsilon,\, \text{and}\, E_\varepsilon{}^*, \tag{12-42}$$

where the subscripts ε can assume any value (mod 2π) in the range $0 \leq \varepsilon \leq 2\pi$ independently of each other. A general element O_ε of the EMS group is defined as follows: (i) The effect of O_ε on the spatial coordinates of the nuclei and electrons in the molecule [i.e., on the (ξ_i, η_i, ζ_i) coordinates] is the same as that of the element O of the MS group. (ii) The effect of O_ε on the Euler angles θ and ϕ is the same as the effect of O of the MS group. (iii) The effect of O_ε on the Euler angle χ is defined by the value of ε according to the following rules:

$$E_\varepsilon \chi = \chi - \varepsilon, \tag{12-43}$$

$$(p)_\varepsilon \chi = 2\pi - \chi - \varepsilon, \tag{12-44}$$

$$E_\varepsilon{}^* \chi = \pi - \chi - \varepsilon, \tag{12-45}$$

$$(p)_\varepsilon{}^* \chi = \pi + \chi - \varepsilon. \tag{12-46}$$

This completely defines the elements of the EMS group.

The transformation properties of χ have been defined in such a manner that each element of the EMS group has the same effect on the vibronic variables $\alpha_r{}'$ and $\chi_e{}'$ as an element in the molecular point group. From parts (i) and (ii) of the definition of O_ε we see that the effect of O_ε on the rovibronic angles α_r and χ_e must be the same as the effect[1] of O. The effect of O_ε on χ is defined by part (iii) of the definition of O_ε, and thus we can determine the transformation properties of the vibronic coordinates $\alpha_r{}'\,(= \alpha_r - \chi)$ and $\chi_e{}'\,(= \chi_e - \chi)$. The results for $\alpha_r{}'$ and $\chi_e{}'$ are the same, and for $\alpha_r{}'$ we have

$$E_\varepsilon \alpha_r{}' = \alpha_r{}' + \varepsilon, \tag{12-47}$$

$$(p)_\varepsilon \alpha_r{}' = -\alpha_r{}' + \varepsilon, \tag{12-48}$$

$$E_\varepsilon{}^* \alpha_r{}' = -\alpha_r{}' + \varepsilon, \tag{12-49}$$

$$(p)_\varepsilon{}^* \alpha_r{}' = \alpha_r{}' + \varepsilon. \tag{12-50}$$

---

[1] Given in Eqs. (12-23), (12-25), and (12-26).

As can be seen from the definition of the effect of O_ε, together with the results of Eqs. (12-47)–(12-50), each element of the EMS group can be written as the product of a molecular point group operation that affects only the vibronic variables $(Q_r, \alpha_r', \text{and } \chi_e')$, a rotation operation that affects only the Euler angles (see Table 7-1), and a nuclear spin permutation. We can, therefore, write each element of the EMS group as follows:

$$E_\varepsilon = C_\infty{}^\varepsilon R_z^{-\varepsilon} p_0, \tag{12-51}$$

$$(p)_\varepsilon = C_2^{(\varepsilon/2)} R_{\varepsilon/2}^\pi p_{ns}, \tag{12-52}$$

$$E_\varepsilon{}^* = \sigma_v^{(\varepsilon/2)} R_{(\pi+\varepsilon)/2}^\pi p_0, \tag{12-53}$$

$$(p)_\varepsilon{}^* = S_\infty{}^\varepsilon R_z^{\pi-\varepsilon} p_{ns} \quad [(p)_\pi{}^* = S_\infty{}^\pi R_z{}^0 p_{ns} = i p_{ns}], \tag{12-54}$$

where p_{ns} is the permutation of the nuclear spin coordinates of the nuclei that are permuted by (p). Each of these equations is of the form of Eq. (11-10) which was developed for nonlinear molecules. Since the EMS group and the point group of a linear molecule are isomorphic [the 1:1 relation between the elements follows from Eqs. (12-51)–(12-54)] we use the same character table and irreducible representation labels for each (see Tables A-14 and A-15). Further we can develop the electron spin double groups of the EMS groups of linear molecules as shown in Tables A-14 and A-15. Since each element of the EMS group transforms the vibronic variables in the same way as its partner in the molecular point group the vibronic wavefunctions can equally well be classified in the EMS group as in the molecular point group and the same results (such as in Table 12-1) will be obtained. The rotational wavefunctions of a linear molecule can also be classified in the EMS group and the results for a $D_{\infty h}$ molecule are given in Table 12-2. For a $C_{\infty v}$ molecule the g subscripts should be omitted. The species of the rovibronic functions are obtained by multiplying the rotational and vibronic species together bearing in mind the restriction imposed by Eq. (12-16).

The effect of a general element O_ε of the EMS group of a linear molecule on a rovibronic wavefunction will be exactly the same as the effect of the element O of the MS group. This follows from part (i) of the definition of O_ε. This means that the elements of the EMS group leave the rovibronic Hamiltonian invariant and that we can classify the rovibronic states in the EMS group. We deduce that the rovibronic wavefunctions of a $D_{\infty h}$ molecule can only transform as one of the four one-dimensional irreducible representations $\Sigma_g{}^+$, $\Sigma_u{}^+$, $\Sigma_g{}^-$, and $\Sigma_u{}^-$ of the EMS group $D_{\infty h}$(EM). Similarly for a $C_{\infty v}$ molecule the rovibronic wavefunctions can only transform as Σ^+ or Σ^- of $C_{\infty v}$(EM). The classification of a rovibronic wavefunction of a $D_{\infty h}$ molecule as $\Sigma_g{}^+$, $\Sigma_u{}^+$, $\Sigma_g{}^-$, and $\Sigma_u{}^-$ is equivalent to the standard (MS group) notation of $+s$, $+a$, $-a$, and $-s$, respectively; for a $C_{\infty v}$ molecule Σ^+ and

Table 12-2
Species of rotational
wavefunctions of
$D_{\infty h}$ molecule in
the EMS group $D_{\infty h}(EM)^a$

k		$\Gamma(|Jkm\rangle)$
0	(J even)	Σ_g^+
	(J odd)	Σ_g^-
± 1		Π_g
± 2		Δ_g
± 3		Φ_g
$\vdots$		$\vdots$

^{a} For a molecule of $C_{\infty v}$ symmetry the classification in $C_{\infty v}(EM)$ gives the same results with the g subscripts omitted.

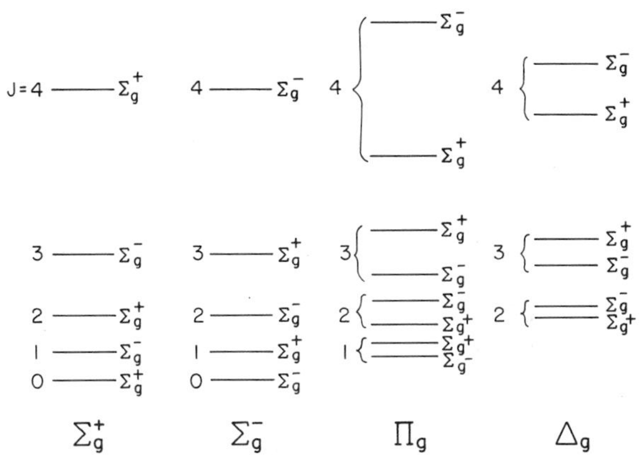

Fig. 12-5. The symmetries of the rovibronic states (rotational levels) of Σ_g^+, Σ_g^-, Π_g, and Δ_g vibronic states of a $D_{\infty h}$ molecule. For u vibronic states the g subscripts should be replaced by u subscripts; for a $C_{\infty v}$ molecule the subscripts should be omitted. [See Herzberg (1966), p. 572.]

Σ^- are equivalent to the standard notations + and −. In Fig. 12-5 examples of the application of this classification scheme to rovibronic states are given.

A component of the dipole moment operator along a space fixed (ξ, η, ζ) direction transforms according to the representation $\Gamma^* = \Sigma_u^-$ of $D_{\infty h}(EM)$ and the representation Σ^- of $C_{\infty v}(EM)$. Thus we have the following selection rules for allowed electric dipole transitions between rovibronic states

$$\Sigma_g^+ \leftrightarrow \Sigma_u^- \qquad \text{and} \qquad \Sigma_u^+ \leftrightarrow \Sigma_g^-, \tag{12-55}$$

where the g and u labels are omitted for a $C_{\infty v}$ molecule. These selection rules are equivalent to the standard selection rules $+ \leftrightarrow -$, $s \leftrightarrow s$ and $a \leftrightarrow a$ [see Herzberg (1966), Eqs. (II, 46) and (II, 47)].

NONRIGID MOLECULES

A nonrigid molecule is a molecule that is in an electronic state that has more than one accessible minimum in its potential energy surface V_N. As a result the molecule can contort (i.e., vibrate with large amplitude) from one conformation to another in a time scale short enough to be experimentally detectable, and observable energy level splittings or shifts are produced. As we saw in Chapter 9 the order of the MS group of a nonrigid molecule is usually higher than that of any one conformer. To use this group to symmetry label the molecular energy levels we must first determine an appropriate set of molecular coordinates that lead to a useful zero order Hamiltonian, and the classification of the eigenfunctions of the zero order Hamiltonian in the MS group then gives the desired symmetry labels.

To determine the zero order wavefunctions of a nonrigid molecule we begin, just as for a rigid molecule, by neglecting spin coupling terms and by making the Born–Oppenheimer approximation. As a result the zero order nuclear and electron spin functions and the zero order electron orbital functions are all as for a rigid molecule and require no further discussion. The classification of these wavefunctions in the MS group of the nonrigid molecule proceeds as for a rigid molecule, and to classify the electron orbital functions, and case (a) electron spin functions, it is necessary to know how the molecule fixed (x, y, z) axes (i.e., Euler angles) are transformed by the elements of the MS group. For a few nonrigid molecules (those having identical coaxial internal rotors on a linear frame) the transformation properties of the Euler angle χ in the MS group are ambiguous, and, just as for linear molecules, an extended MS group is required. This special problem will be discussed when we consider the dimethylacetylene molecule at the end of this chapter. The zero order rotation–vibration Hamiltonian of a nonrigid molecule is not the same as for a rigid molecule because the Taylor series expansions of both $\mu_{\alpha\beta}$ [see Eq. (7-149)] and V_N [see Eq. (8-26)] cannot, in general, be satisfactorily approximated by their leading terms when one or more of the vibrational coordinates is of large amplitude. A special discussion is therefore required of the rotation–vibration Hamiltonian, the rotation–vibration coordinates, and the rotation–vibration wavefunctions of a nonrigid molecule. We will call any large amplitude vibrational coordinate of a nonrigid molecule a *contortional coordinate*, and to emphasize its presence the zero order rotation–vibration wavefunctions will be called rotation–contortion–vibration wavefunctions. In all that follows in this chapter we will only consider the situation in which there is one contortional coordinate but the generalization to situations in which there is more than one is in principle straightforward.

The Rotation–Contortion–Vibration Problem

To appreciate the approach used in determining a useful zero order basis set of rotation–contortion–vibration wavefunctions for a nonrigid molecule we first summarize the method used for rigid molecules. As discussed in Chapters 7 and 8 the zero order rotation–vibration Hamiltonian of a nonlinear rigid molecule is obtained as follows:

(i) We obtain the rotation–vibration Hamiltonian, $\hat{H}_{rv} = (\hat{T}_N + V_N)$, in the coordinates $(\xi_2, \eta_2, \zeta_2, \ldots, \xi_N, \eta_N, \zeta_N)$ for the electronic state of interest after having made the Born–Oppenheimer approximation.

(ii) We change the coordinates in $\hat{H}_{rv}$ to $(\theta, \phi, \chi, Q_1, \ldots, Q_{3N-6})$, where the expressions for the Euler angles (θ, ϕ, χ) in terms of the (ξ_i, η_i, ζ_i) coordinates are obtained from the Eckart equations, and the expressions for the normal coordinates Q_r in terms of the (ξ_i, η_i, ζ_i) coordinates are obtained by using the l matrix; the kinetic energy expression obtained is given in Eq. (7-137).

(iii) We expand $\mu_{\alpha\beta}$ and V_N as Taylor series, about their values at equilibrium, in the normal coordinates Q_r [see Eqs. (7-149) and (8-26)].

(iv) The zero order rotation–vibration Hamiltonian is then obtained by neglecting all but the leading terms ($\mu_{\alpha\beta}^e$ and $\frac{1}{2}\sum \lambda_r Q_r^2$) in the Taylor series expansions of $\mu_{\alpha\beta}$ and V_N and by neglecting the vibronic angular momenta $\hat{p}_\alpha$ and $\hat{L}_\alpha$.

The zero order Hamiltonian obtained is the sum of a three-dimensional rigid rotor Hamiltonian and of $3N - 6$ one-dimensional harmonic oscillator Hamiltonians. Using the Eckart conditions the $\hat{p}_\alpha$ are minimized so that their neglect is not usually a bad approximation, and the l matrix is such that the leading term in V_N contains no cross terms $\Phi_{rs}Q_rQ_s$ to prevent the zero order separation of the $3N - 6$ harmonic oscillators. The eigenfunctions of this Hamiltonian are easily determined as functions of the Euler angles and normal coordinates, and they can be classified in the MS group once the transformation properties of the coordinates have been determined.

The approach used to obtain the zero order rotation–contortion–vibration Hamiltonian of a nonrigid molecule is slightly different. This is because in this case we cannot usually approximate both $\mu_{\alpha\beta}$ and V_N by their leading terms in a Taylor series expansion about equilibrium; the contortional coordinate is of large amplitude and can prevent the convergence of either or both of the series (using rigid molecule formalism the lack of convergence of $\mu_{\alpha\beta}$ and

V_{N} would give rise to a large centrifugal rotation–contortion interaction and a large anharmonic contortion–vibration interaction, respectively). An outline of the approach used is as follows:

(i)   We obtain the rotation–contortion–vibration Hamiltonian $\hat{H}_{\mathrm{rcv}}$ $(= \hat{T}_{\mathrm{N}} + V_{\mathrm{N}})$ in the coordinates $(\xi_2, \eta_2, \zeta_2, \ldots, \xi_N, \eta_N, \zeta_N)$ after having made the Born–Oppenheimer approximation; this Hamiltonian will be basically the same whether or not the molecule in the electronic state under consideration is rigid or nonrigid.

(ii)   We change coordinates in $\hat{H}_{\mathrm{rcv}}$ to $(\theta, \phi, \chi, \rho, Q_1, \ldots, Q_{3N-7})$, where ρ is the contortional coordinate. The expressions for the Euler angles (θ, ϕ, χ) in terms of the (ξ_i, η_i, ζ_i) coordinates are obtained from the Eckart equations, the expression for ρ in terms of the (ξ_i, η_i, ζ_i) coordinates is obtained from an equation called a *Sayvetz* equation (discussed later) [Sayvetz (1939)], and the expressions for the $3N - 7$ normal coordinates Q_r are obtained from the l matrix. The elements of the l matrix will now depend on ρ.

(iii)   We expand $\mu_{\alpha\beta}$ and V_{N} about their values when all $(3N - 7)$ Q_r are zero as a Taylor series in the $(3N - 7)$ Q_r with coefficients that depend on ρ. These expansions can be written as

$$\mu_{\alpha\beta} = \mu_{\alpha\beta}^{\mathrm{ref}} - \mu_{\alpha\gamma}^{\mathrm{ref}} a_r^{\gamma\delta} \mu_{\delta\beta}^{\mathrm{ref}} Q_r + \cdots \tag{12-56}$$

and

$$V_{\mathrm{N}} = V_0(\rho) + \Phi_r Q_r + \tfrac{1}{2}\lambda_r Q_r^2 + \tfrac{1}{6}\Phi_{rst} Q_r Q_s Q_t + \cdots, \tag{12-57}$$

where $\alpha, \beta, \gamma, \delta = x, y,$ or z, and $r, s, t = 1, 2, \ldots, 3N - 7$ and summation over repeated indices is assumed; $\mu_{\alpha\beta}^{\mathrm{ref}}$ and the $a_r^{\gamma\delta}$ are functions of ρ, $V_0(\rho)$ is the pure contortional potential function, and λ_r and the Φ are also functions of ρ. The term linear in Q_r occurs in V_{N} because Φ_r is a function of ρ, and in the standard (rigid molecule) expansion this term would become (expanding only to quartic terms)

$$[\Phi_{r\rho}\rho + \tfrac{1}{2}\Phi_{r\rho\rho}\rho^2 + \tfrac{1}{6}\Phi_{r\rho\rho\rho}\rho^3]Q_r. \tag{12-58}$$

(iv)   The zero order rotation–contortion–vibration Hamiltonian is then obtained by neglecting all but the leading terms ($\mu_{\alpha\beta}^{\mathrm{ref}}$ and $V_0(\rho) + \tfrac{1}{2}\sum_r \lambda_r Q_r^2$) in the Taylor series expansions, and by neglecting the vibronic angular momenta that occur.

The zero order rotation–contortion–vibration Hamiltonian obtained is the sum of a four-dimensional rotation–contortion Hamiltonian $\hat{H}_{\mathrm{rc}}$ and of $3N - 7$ harmonic oscillator Hamiltonians. Neglecting some terms that are usually small, the zero order rotation–contortion Hamiltonian is given by [see, for example, Eq. (7) in Bunker and Stone (1972)]

$$\hat{H}_{\rm rc} = \frac{1}{2} \sum_{\alpha,\beta} \mu_{\alpha\beta}^{\rm ref} \hat{J}_{\alpha} \hat{J}_{\beta} + \frac{1}{2} \mu_{\rho\rho} \hat{J}_{\rho}{}^2 + V_0(\rho), \qquad (12\text{-}59)$$

where $\alpha, \beta = x, y$, or z, $\hat{J}_{\rho} = -i\hbar\, \partial/\partial\rho$, and $\mu_{\rho\rho}$ is the inverse of the contortional reduced mass. The quantities $\mu_{\alpha\beta}^{\rm ref}, \mu_{\rho\rho}$, and $V_0(\rho)$ all depend on ρ. The $3N - 7$ harmonic oscillator Hamiltonians are each a parametric function of ρ (just as the electronic Hamiltonian of a molecule is a parametric function of the nuclear coordinates in the Born–Oppenheimer approximation) because of the dependence of the λ_r on ρ. The normal coordinates will therefore vary with the ρ value and this ρ dependence can be determined explicitly if the ρ dependence of the molecular force field is known.

The preceding outline of the derivation of the zero order rotation–contortion–vibration Hamiltonian is presented in order to bring out the similarities and differences to the rigid molecule Hamiltonian. The details of the derivation will differ from molecule to molecule. However, the idea of a *reference configuration* is common to all nonrigid molecules and is worth explaining in some detail.

The reference configuration of a nonrigid molecule is the analog of the equilibrium configuration of a rigid molecule. A nonrigid molecule is in its reference configuration when all the small amplitude vibrational displacements are set to zero but when the Euler angles and ρ coordinate are arbitrary. The inverse moment of inertia tensor of the reference configuration has the elements $\mu_{\alpha\beta}^{\rm ref}$ just discussed [see Eq. (12-56)]. Thus, for example, in the hydrogen peroxide molecule the reference configuration is that molecular configuration in which the O–H and O–O bond lengths and H–O–O angles are fixed at their equilibrium values but in which the torsional angle and rotational orientation in space are arbitrary. The energy levels of a molecule in its reference configuration as it rotates and contorts are obtained from its rotation–contortion Hamiltonian [see Eq. (12-59)]. The convention used to locate the molecule fixed (x, y, z) axes in the reference configuration is a matter of convenience (as it is for their location in the equilibrium configuration of a rigid molecule), and their location as the principal inertial axes is one possibility. Having defined the reference configuration, the location of the (x, y, z) axes in the reference configuration, and the contortional coordinate ρ of the reference configuration, it is easy to use the reference configuration to determine the effect of the elements of the MS group on the coordinates θ, ϕ, χ, and ρ.

In the reference configuration just discussed the molecule can be said to *rigidly contort*, i.e., the ordinary bond lengths and angles remain fixed as the molecule contorts. For a triatomic molecule this approach leads to the *rigid bender Hamiltonian* [Bunker and Stone (1972)]. An improvement is obtained by allowing the ordinary bond lengths and angles to change as the molecule contorts; in this new reference configuration the molecule *semirigidly contorts*. For triatomic molecules and four atomic molecules such as HCNO, this leads to the *semirigid bender Hamiltonian* [Bunker and Landsberg (1977) and Bunker, Landsberg, and Winnewisser (1978)].

The rotation–contortion Hamiltonian is a four-dimensional Hamiltonian but it is greatly simplified if the ρ dependence of the $\mu_{\alpha\beta}^{\mathrm{ref}}$ elements can be neglected. In this circumstance the rotation–contortion Hamiltonian can be separated into an ordinary rigid rotor Hamiltonian and a one-dimensional contortional Hamiltonian. The contortional Hamiltonian can be diagonalized analytically for certain forms of $V_0(\rho)$, and in the general case it can be diagonalized numerically. For the dimethylacetylene and methanol molecules the moments of inertia are independent of the torsional angle and there is complete separation of rotation and torsion to zero order. For the ammonia and hydrogen peroxide molecules the moments of inertia are only weakly dependent on the contortional coordinate, and for the purpose of appreciating the general energy level pattern and symmetry labels we can ignore this dependence as a useful first approximation.

We now discuss the Eckart and Sayvetz conditions for a nonrigid molecule. Having chosen a convention for locating the (x, y, z) axes in the reference configuration, a nucleus, i, say, will have coordinates x_i^{r}, y_i^{r}, and z_i^{r} in that axis system.[2] The reference configuration coordinates of all the nuclei in a molecule are known, once the reference configuration is defined, in the same way that the equilibrium nuclear coordinates x_i^{e}, y_i^{e}, and z_i^{e} of a rigid molecule are known. The reference configuration coordinates of each nucleus will not be constants (as are the equilibrium configuration coordinates of a rigid molecule) but they will depend on the contortional angle ρ. The Eckart equations now become

$$\sum_{i=1}^{N} m_i(x_i^{r}y_i - y_i^{r}x_i) = 0, \tag{12-60}$$

$$\sum_{i=1}^{N} m_i(y_i^{r}z_i - z_i^{r}y_i) = 0, \tag{12-61}$$

and

$$\sum_{i=1}^{N} m_i(z_i^{r}x_i - x_i^{r}z_i) = 0, \tag{12-62}$$

----

[2] The superscript r stands for reference configuration.

and the Sayvetz condition is

$$\sum_{\alpha,i} m_i (\alpha_i - \alpha_i^{\,r}) \frac{\partial \alpha_i^{\,r}}{\partial \rho} = 0, \tag{12-63}$$

where $\alpha = x, y,$ or z. The $\alpha_i^{\,r}$ and $\partial \alpha_i^{\,r}/\partial \rho$ are known functions of ρ, and the α_i can be expressed in terms of the direction cosine elements (i.e., the Euler angles) and the known (ξ_i, η_i, ζ_i) nuclear coordinates as done in Eqs. (7-132)– (7-134). We will then have four equations in the four unknowns $(\theta, \phi, \chi,$ and ρ) which can be solved simultaneously for any given set of (ξ_i, η_i, ζ_i) coordinates, i.e., for any given nuclear configuration in space.[3] We also obtain the (x_i, y_i, z_i) coordinates of the nuclei once we know the Euler angles. These equations could be used to determine how $\theta, \phi, \chi,$ and ρ are transformed by elements of the MS group but it is easier to study the transformation properties of the reference configuration for this purpose [i.e., we determine the transformation properties of the coordinates $\theta, \phi, \chi,$ and ρ when all $(3N - 7)$ Q_r's are zero]. In the same way that the Eckart conditions define "rotation" so as to minimize the terms that couple rotation and vibration in the kinetic energy, the Sayvetz condition defines "contortion" so as to minimize the term that couples contortion and vibration in the kinetic energy. There is one Sayvetz condition for each contortional coordinate. Loosely speaking Eq. (12-63) defines ρ so that the change in nuclear coordinates when ρ is changed, $\partial \alpha_i/\partial \rho$, is orthogonal to the change in nuclear coordinates, $\alpha_i - \alpha_i^{\,r}$, that results from the vibrations.

The $3N - 7$ normal coordinates and the vibrational wavefunctions of a nonrigid molecule involve the contortional angle as a parameter. This makes the determination of the symmetry species of the normal coordinates, and hence of the vibrational wavefunctions, a little more complicated than for a rigid molecule. As we allow the contortional coordinate to vary over its allowed range the point group symmetry of the instantaneous nuclear configuration may change, e.g., as the dimethylacetylene molecule twists the instantaneous point group symmetry changes from D_{3h} through D_3 to D_{3d}. For each different point group geometry we can determine the symmetry species, in that point group, of the normal coordinates using the technique discussed in Chapter 7 [see Eq. (7-238)]. Excluding the species of the contortional coordinate from these we obtain the species of the $3N - 7$ normal coordinates in each possible instantaneous point group. We now use the correlation tables of these point groups (or really of the rigid molecule MS groups with which they are isomorphic) to the nonrigid molecule MS group (or, where necessary, the EMS group) to determine the species of the normal coordinates in the nonrigid molecule MS (or EMS) group. The

---

[3] See Fig. 2 and Eq. 8 of Hougen, Bunker, and Johns (1970).

species of the normal coordinates obtained is not always unique and the actual species will depend on the ρ dependence of the force field. It is often a useful approximation to completely neglect the ρ dependence of the force field or to include only a specific part of it. This will be discussed in detail for the nitromethane and dimethylacetylene molecules in the following.

Optical Selection Rules

The rigorous selection rules obtained in Eqs. (11-146)–(11-149) apply to all molecules, rigid, nonrigid, and linear, as do the spin quantum number selection rules of Eqs. (11-159) and (11-160) in the absence of strong spin interactions. The rotation, vibration, and electronic selection rules obtained in Chapter 11 must be reexamined for nonrigid molecules since the zero order wavefunction is separated differently. Assuming the separation of rotation and contortion, the zero order *roconvibronic* wavefunctions are of the form

$$\Phi_{\text{rcve}} = \Phi_{\text{rot}}(\theta\phi\chi)\Phi_{\text{con}}(\rho)\Phi_{\text{vib}}\Phi_{\text{elec}}. \tag{12-64}$$

The rotational quantum numbers involved will be J, k, and m for symmetric or spherical tops and J, K_a, K_c, and m for asymmetric tops; we will use k_i for the contortional quantum number. The selection rules are obtained by considering the conditions under which the following is true [from Eq. (11-151)]

$$\langle \Phi'_{\text{rot}}\Phi'_{\text{con}}\Phi'_{\text{vib}}\Phi'_{\text{elec}}|M_\zeta|\Phi''_{\text{rot}}\Phi''_{\text{con}}\Phi''_{\text{vib}}\Phi''_{\text{elec}}\rangle \neq 0. \tag{12-65}$$

We follow the arguments of Eqs. (11-152)–(11-155), but now $M_\alpha^{(o)}(e', e'')$, $M_\alpha^{(r)}(e', e'')$, etc. are functions of the contortional coordinate ρ, and we obtain the equation that defines an allowed transition as [compare with Eq. (11-156)]

$$\sum_\alpha \langle \Phi'_{\text{rot}}|\lambda_{\alpha\zeta}|\Phi''_{\text{rot}}\rangle \langle \Phi'_{\text{con}}|M_\alpha^{(o)}(e', e'')\langle \Phi'_{\text{vib}}|\Phi''_{\text{vib}}\rangle|\Phi''_{\text{con}}\rangle$$

$$+ \sum_\alpha \langle \Phi'_{\text{rot}}|\lambda_{\alpha\zeta}|\Phi''_{\text{rot}}\rangle \sum_r \langle \Phi'_{\text{con}}|M_\alpha^{(r)}(e', e'')\langle \Phi'_{\text{vib}}|Q_r|\Phi''_{\text{vib}}\rangle|\Phi''_{\text{con}}\rangle \neq 0, \tag{12-66}$$

where we neglect the second term if $\Phi'_{\text{elec}} \neq \Phi''_{\text{elec}}$. The rotational selection rules will depend on α and are the same as discussed in Chapter 11 [see Eqs. (11-170)–(11-173)]. The general contortion–vibronic symmetry selection rule is

$$\Gamma'_{\text{cve}} \otimes \Gamma''_{\text{cve}} \supset \Gamma(T_\alpha), \tag{12-67}$$

which is similar to Eq. (11-165).

For electronic transitions we deduce from the first term in Eq. (12-66) that we must have

$$\Gamma'_{\text{elec}} \otimes \Gamma''_{\text{elec}} \otimes \Gamma'_{\text{con}} \otimes \Gamma''_{\text{con}} \supset \Gamma(T_\alpha) \tag{12-68}$$

and

$$\Gamma'_{vib} = \Gamma''_{vib}. \tag{12-69}$$

The vibrational symmetry rule is unchanged from that for a rigid molecule but now more electronic transitions are allowed since we do not neglect the ρ dependence of $M_\alpha^{(o)}(e', e'')$. For a given electronic transition only certain contortional transitions can occur [satisfying Eq. (12-68)] from which we obtain selection rules on k_i. The contortional transitions give rise to contortional fine structure (like rotational fine structure) in the observed bands if the contortional energy level separations are small.

Similarly for a vibrational transition within an electronic state we deduce the general symmetry rule for an allowed transition as

$$\Gamma'_{vib} \otimes \Gamma''_{vib} \otimes \Gamma'_{con} \otimes \Gamma''_{con} \supset \Gamma(T_\alpha), \tag{12-70}$$

which is similar to Eq. (11-169). Using the second term in Eq. (12-66) we obtain the conditions for an allowed contortion–vibration transition as

$$\langle \Phi'_{vib} | Q_r | \Phi''_{vib} \rangle \neq 0 \tag{12-71}$$

together with Eq. (12-70). In the harmonic oscillator approximation Eq. (12-71) leads to

$$\Delta v_r = \pm 1. \tag{12-72}$$

We see that for a nonrigid molecule, when we allow for contortional fine structure, vibrational fundamentals are allowed for normal modes that do not have the species of a translation. This arises from the fact that we are not neglecting the ρ dependence of $M_\alpha^{(r)}(e'', e'')$. This point is perhaps made clearer by reference to Fig. 12-6 which depicts the allowed contortional lines in a fundamental band for which the species of the normal coordinate is not that of a translation. Thus the $k_i = 0 \leftarrow 0$ line, in the $v_r = 1 \leftarrow 0$ band shown, is forbidden. It is supposed that the contortional symmetries are such that $\Delta k_i = \pm 1$ transitions accompanying the $v_r = 1 \leftarrow 0$ band are allowed; these would be sum and difference bands if the molecule were rigid and the contortional energy separation large. Thus for the nonrigid molecule with small contortional energy level separations the $v_r = 1 \leftarrow 0$ band is allowed; it consists of contortional fine structure that satisfies $\Delta k_i = \pm 1$. The additional fine structure from the rotational transitions is ignored in Fig. 12-6.

We will now briefly discuss the application of these ideas to the nonrigid molecules ammonia, nitromethane, and dimethylacetylene. The ammonia and nitromethane molecules illustrate the effects of inversion tunneling and torsional tunneling, respectively, and the dimethylacetylene molecule illustrates how an EMS group arises for a nonrigid molecule.

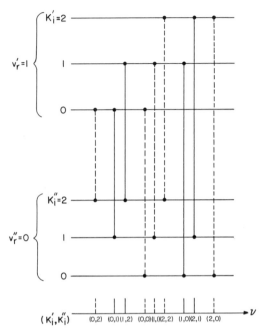

Fig. 12-6. The contortional fine structure ($\Delta K_i = \pm 1$) on a $v_r = 1 \leftarrow 0$ fundamental band for which $\Gamma(Q_r) \neq \Gamma(T_z)$ but for which $\Gamma(Q_r) \otimes \Gamma[\Phi_{con}(k_i)] \otimes \Gamma[\Phi_{con}(k_i \pm 1)] \supset \Gamma(T_z)$. The full lines are allowed transitions and the dotted lines are forbidden transitions. $K_i = |k_i|$.

The Ammonia Molecule

An ammonia molecule in its equilibrium configuration is shown in Fig. 12-7, and inversion between the two configurations shown in the figure gives rise to observable splittings in the spectrum. The inversion potential and inversion energy levels are shown in Fig. 12-8 [see Papousek, Stone, and Spirko (1973) and Bunker (1975) and references therein]. If the effects of inversion tunneling were not observable the energy level situation would be

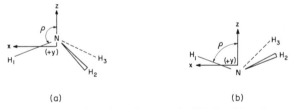

(a)   (b)

Fig. 12-7. The reference configuration of NH_3. The NH bond lengths are equal to their equilibrium values and the three HNH angles are equal to each other. The molecule fixed (x, y, z) axes are the principal inertial axes.

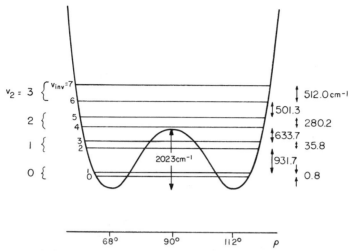

Fig. 12-8. The inversion energy levels of NH_3. The energy level splittings are in cm^{-1}.

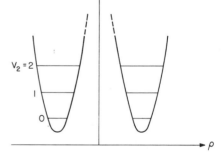

Fig. 12-9. The potential function and umbrella vibration energy levels of a rigidly C_{3v} molecule such as NF_3. The abscissa is the inversion coordinate.

as shown in Fig. 12-9. NF_3 is an example of such a case and the states of the umbrella vibration are labeled $v_2 = 0, 1, 2, \ldots$. The MS group of NF_3 is $C_{3v}(M)$ and that of inverting NH_3 is $D_{3h}(M)$; the character table of the $D_{3h}(M)$ group is given in Table A-9. In Fig. 12-8 we number the inversion states by the v_2 value for the rigid molecule level with which they correlate and also with the quantum number v_{inv}. The use of the quantum number v_{inv} (which gives the total number of nodes in the inversion wavefunction) is preferable to v_2 for NH_3, particularly for the higher vibrational states, and it allows us to treat NH_3 as a planar molecule with a highly anharmonic out-of-plane vibration. Using this quantum number the vibrational and rotational selection rules for allowed optical transitions and perturbations are easy to formulate.

To classify the rotation–inversion wavefunctions it is first necessary to define the reference configuration, the orientation of the molecule fixed axes in the reference configuration, and the inversion angle ρ of the reference configuration. The reference configuration of the ammonia molecule is defined as having all three NH bond lengths equal to their equilibrium values and all three HNH angles equal to each other. The molecule fixed (x, y, z) axes are chosen to be right handed and to be the principal axes of the reference configuration, as indicated in Fig. 12-7. The positive z axis direction is defined as that direction a right handed screw would travel if twisted in the direction $H_1 \rightarrow H_2 \rightarrow H_3$, and the x axis is in the z–N–H_1 plane. The inversion angle in the reference configuration is the angle between the positive z axis direction and an NH bond. This completes the definition of the reference configuration. For an arbitrarily oriented ammonia molecule in its reference configuration we can use this definition to determine θ, ϕ, χ, and ρ, and hence we can determine the transformation properties of these angles in the $D_{3h}(M)$ group. When the molecule is not in its reference configuration the Eckart and Sayvetz conditions would be used to obtain θ, ϕ, χ, and ρ.

Neglecting the dependence of the moments of inertia on the inversion angle the zero order rotation–inversion Hamiltonian is

$$\hat{H}_{ri} = \hat{H}_{rot} + \hat{H}_{inv}, \tag{12-73}$$

where

$$\hat{H}_{rot} = (\mu^e_{xx}/2)(\hat{J}_x{}^2 + \hat{J}_y{}^2) + (\mu^e_{zz}/2)\hat{J}_z{}^2 \tag{12-74}$$

and

$$\hat{H}_{inv} = (\mu_{\rho\rho}/2)\hat{J}_\rho{}^2 + V_0(\rho). \tag{12-75}$$

The rotational Hamiltonian is that of a symmetric top, and the transformation properties of its eigenfunctions in $D_{3h}(M)$ can be determined using the equivalent rotations indicated in Table A-9. The inversion Hamiltonian can be diagonalized for any shape of $V_0(\rho)$ by numerical methods. For determining the symmetry of the eigenfunctions Φ_{inv} we can take $V_0(\rho)$ as being harmonic and centered on the planar configuration. Doing this we see [from Eq. (10-30) and the results after it] that the states with v_{inv} even are totally symmetric in $D_{3h}(M)$ and that the states with v_{inv} odd have the same symmetry as ρ; ρ has the symmetry A_2''. We can therefore write the species of the inversion states as

$$\Gamma(\Phi_{inv}) = [A_2'']^{v_{inv}} \tag{12-76}$$

The species of the rotational states are given in Table 12-3. By determining the normal coordinate representation of the rigid D_{3h} molecule and correlating with the $D_{3h}(M)$ group we see that for the small amplitude vibrations the normal coordinates generate the representation:

$$\Gamma(Q) = A_1' \oplus 2E' \tag{12-77}$$

Table 12-3
Symmetry species of
rotational states of
NH_3 in $D_{3h}(M)^a$

see p. 236 & 237

K		Γ_{rot}
0	(J even)	A_1'
	(J odd)	A_2'
$6n \pm 1$		E''
$6n \pm 2$		E'
$6n \pm 3$		$A_1'' \oplus A_2''$
$6n \pm 6$		$A_1' \oplus A_2'$

^{a} n is a nonnegative integer;
$K \geq 0$.

in $D_{3h}(M)$. No other species is possible for the normal coordinates in this case since the MS group and the group of a rigid conformer (D_{3h}) are isomorphic and the species correlation is 1:1. The electron spin–orbit wavefunction of the ground state of NH_3 is totally symmetric in $D_{3h}(M)$ and the rotation–inversion–vibronic species of the ground electronic state levels can be obtained from the results in Tables 12-3 and Eqs. (12-76) and (12-77).

The statistical weights of NH_3 exhibit an interesting feature. The nuclear spin states span the representation (see Problem 6-2)

$$\Gamma_{nspin}^{tot} = 12A_1' + 6E', \qquad (12\text{-}78)$$

and from the Pauli exclusion principle the overall wavefunction can have either A_2' or A_2'' species. As a result the rotation–inversion–vibration levels of the ground electronic state have statistical weights as given in Table 12-4. We see that states of species A_1' and A_1'' have a statistical weight of zero. As a result half the $K = 0$ energy levels in any nondegenerate vibrational state will be missing (see Fig. 12-10).

Having determined the symmetry species of the rotational, inversion, and vibrational states of the molecule the selection rules for allowed electric

Table 12-4
Statistical weights of rovibronic states^a of NH_3

Γ_{rve}	Statistical weight	Γ_{rve}	Statistical weight
A_1'	0	A_1''	0
A_2'	12	A_2''	12
E'	6	E''	6

^{a} Inversion states are included.

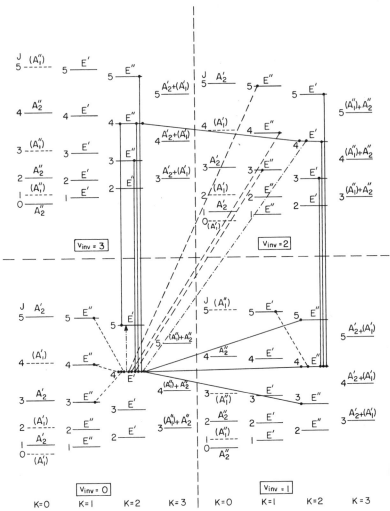

Fig. 12-10.   The lower rotational energy levels of the $v_{inv} = 0, 1, 2,$ and 3 states of NH_3. The $D_{3h}(M)$ symmetry labels Γ_{rve} have been added and rovibronic states forbidden by nuclear spin statistics are in parentheses. Some of the allowed transitions are marked and a full line is used; these transitions satisfy $\Delta v_{inv} =$ odd and $\Delta K = 0$. Using a dashed line some of the forbidden transitions made allowed by vibration–rotation interactions are marked and these have $\Delta v_{inv} =$ even, $\Delta K \neq 0$. Both of these types of transition occur with $\Delta J = 0$ or ± 1. Forbidden transitions made allowed by applying an external electric field (these transitions are allowed two-photon transitions) have selection rules $\Delta v_{inv} =$ even, $\Delta K = 0$, and $\Delta J = 0, \pm 1,$ and ± 2; two of these transitions are marked by a dot–dash line.

dipole transitions can be determined. The selection rules for transitions between the rotation–inversion states are particularly important. From Table A-9 we see that the species of M_z is A_2'', that of (M_x, M_y) is E', and that of Γ^* is A_1''. Thus the E' fundamentals will be infrared active with selection rules $\Delta K = \pm 1$ and $\Delta J = 0, \pm 1$, and the rotation–inversion spectrum will satisfy the selection rules $\Delta K = 0$, $\Delta v_{inv} =$ odd with $\Delta J = 0, \pm 1$. Since the $v_{inv} = 1$ inversion state is very close to the $v_{inv} = 0$ state, hot transitions from it will be as important as those from the ground ($v_{inv} = 0$) state. In Fig. 12-10 the lowest rotational levels of the $v_{inv} = 0, 1, 2$, and 3 states are drawn, and some of the rotation–inversion transitions allowed in the electric dipole absorption spectrum are marked by solid lines. The $v_{inv} = 3 \leftarrow 0$ and $v_{inv} = 2 \leftarrow 1$ bands are completely on top of each other in the infrared spectrum, and these bands correlate with the $v_2 = 1 \leftarrow 0$ band of the rigid nonplanar molecule. In the microwave spectrum $v_{inv} = 1 \leftarrow 0$ and $v_{inv} = 0 \leftarrow 1$ transitions occur in absorption, and three of these types of transition are also marked in Fig. 12-10; these correlate with pure rotation transitions of the rigidly nonplanar molecule. Rotational transitions within the $v_{inv} = 0$ or 1 states are forbidden, but vibration–rotation interactions (spoiling the goodness of K and of the vibrational quantum numbers as labels) can make such forbidden transitions weakly allowed; four of these transitions are indicated by a dashed line in Fig. 12-10 and the selection rules are $K =$ odd $\leftrightarrow$ even with $K = 3n \pm 1$ in both states *or* $K = 3n$ in both states [see, for example, Oka (1976)]. Obviously the $v_{inv} = 2 \leftarrow 0$ and $3 \leftarrow 1$ forbidden bands also gain intensity (with the same selection rules on K) by such interactions and three of these transitions are also marked by a dashed line in the figure. In an electric field (i.e., in the Stark effect) rovibrational states connected by the species A_1'' with $\Delta J = 0$ or ± 1 are mixed. In particular states with the same K value and with v_{inv} values differing by one will be mixed by an electric field, and as a result forbidden transitions with Δv_{inv} even will steal intensity from allowed transitions having Δv_{inv} odd; such transitions will have $\Delta K = 0$. Two examples of such a forbidden transition made allowed by applying an electric field are indicated by a dot–dash line in Fig. 12-10.

The Nitromethane Molecule

The nitromethane molecule in its reference configuration is shown in Fig. 12-11 [see Papousek, Sarka, Spirko, and Jordanov (1971) and Bunker (1975)]. All bond lengths and angles have their equilibrium values (except the torsional angle ρ) and the molecule fixed axes are located as principal axes in the reference configuration; the z axis is in the $C \rightarrow N$ direction and the x axis is in the CNO_4 plane. The torsional angle ρ is the angle between the CH_1 and NO_4 bonds projected onto the xy plane and measured from CH_1 in the right handed sense about the z axis. This molecule has practically free internal

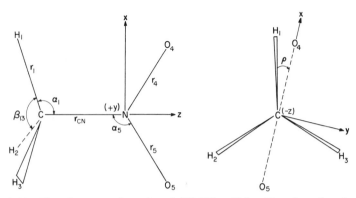

Fig. 12-11. The reference configuration of CH_3NO_2. All bond lengths and angles, except the torsional angle ρ, have their equilibrium values, and the molecule fixed (x, y, z) axes are the principal inertial axes.

rotation, and its MS group is the group G_{12}; the character table of G_{12} and of the electron spin double group G_{12}^2 is shown in Table A-17. The group G_{12} is isomorphic to the D_{3h} group.

The zero order rotational Hamiltonian is that of an asymmetric top ($xyz \equiv bca$) and from the equivalent rotations given in Table A-17 we deduce the asymmetric rotor species to be those in Table 12-5. This molecule is a near oblate top ($I_{aa} \cong I_{bb} < I_{cc}$), and the asymmetric top rotational energy levels can be determined in the usual way using a $|J, k_c\rangle$ basis. The zero order torsional Hamiltonian is obtained by setting the barrier hindering internal rotation to zero and its eigenfunctions and eigenvalues (in cm^{-1}) are given by:

$$\Phi_{\text{tor}} = \exp(ik_i\rho) \qquad (12\text{-}79)$$

and

$$E_{\text{tor}} = Dk_i^2, \qquad (12\text{-}80)$$

where $k_i = 0, \pm 1, \pm 2, \ldots$, and D is the rotational constant of the CH_3 group about the CN axis. The transformation properties of ρ under the effect of the elements of G_{12} are determined, from the transformations of the reference

Table 12-5
Species of asymmetric top
rotational wavefunctions of
CH_3NO_2 in the group G_{12}

K_aK_c	Γ_{rot}	K_aK_c	Γ_{rot}
ee	A_1'	oe	A_1''
eo	A_2'	oo	A_2''

configuration, to be

$$(123)\rho = \rho - (2\pi/3),$$
$$(23)^*\rho = 2\pi - \rho,$$
$$(45)\rho = \rho + \pi, \tag{12-81}$$
$$(123)(45)\rho = \rho + (\pi/3),$$
$$(23)(45)^*\rho = \pi - \rho.$$

The effect of $(23)^*$ is depicted in Fig. 12-12. The species of the torsional functions depend on K_i as shown in Table 12-6. The introduction of a small torsional barrier does not affect these symmetry considerations but we see

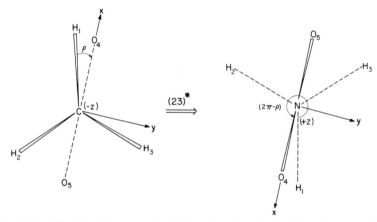

Fig. 12-12. The effect of the operation $(23)^*$ on the torsional angle of the reference configuration of CH_3NO_2.

Table 12-6
Species of torsional
wavefunctions, $\exp(ik_i\rho)$,
of CH_3NO_2 in the
MS group $\mathbf{G}_{12}{}^a$

K_i	Γ
0	A_1'
$6n \pm 1$	E''
$6n \pm 2$	E'
$6n \pm 3$	$A_1'' \oplus A_2''$
$6n \pm 6$	$A_1' \oplus A_2'$

^{a} n is a nonnegative integer and $K_i = |k_i| > 0$.

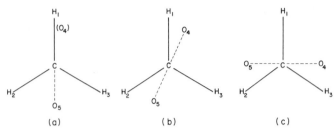

Fig. 12-13. The three rigid conformers of CH_3NO_2 that have different MS group symmetries.

that the degeneracy of the states with $K_i = 6n \pm 3$ and $6n \pm 6$ can be lifted by the effect of the barrier since they are not necessarily degenerate in G_{12}.

To determine the species of the normal coordinates in G_{12} we consider the species of the normal coordinates of the molecule when it is rigidly in each of the three conformations shown in Fig. 12-13. Conformer (a) in Fig. 12-13 has MS group $G_a = \{E, (23)^*\}$, conformer (b) has MS group $G_b = \{E\}$, and conformer (c) has MS group $G_c = \{E, (23)(45)^*\}$. The character tables of G_a and G_c are the same as that of $C_s(M)$ given in Table A-1. Classifying the normal modes of each of these rigid conformers in its MS group (using the technique discussed in Chapter 7) and omitting the torsional species we obtain

$$G_a: \quad \Gamma(Q) = 10A' \oplus 4A'',$$
$$G_b: \quad \Gamma(Q) = 14A, \tag{12-82}$$
$$G_c: \quad \Gamma(Q) = 9A' \oplus 5A''.$$

The correlation table of G_{12} with G_a, G_b and G_c is given in Table 12-7, and by using this table with Eq. (12-82) we determine that the species of the normal

Table 12-7

Correlation table of
MS group G_{12} of
torsionally tunneling $CH_3NO_2{}^a$

G_{12}:	G_a	G_b	G_c
A_1':	A'	A	A'
A_2':	A''	A	A''
E':	$A' \oplus A''$	$2A$	$A' \oplus A''$
A_1'':	A'	A	A''
A_2'':	A''	A	A'
E'':	$A' \oplus A''$	$2A$	$A' \oplus A''$

^{a} The groups G_a, G_b, and G_c are for the rigid conformers of Fig. 12-13.

coordinates of free rotor CH_3NO_2 in the group G_{12} are given by

$$\Gamma(Q) = (10 - l - m - n)A_1' \oplus (5 - l - m - n)A_2' \oplus lA_1'' \oplus (l - 1)A_2''$$
$$\oplus mE' \oplus nE'', \tag{12-83}$$

where l, m, and n are integers satisfying

$$l \geq 1, \qquad m \geq 0, \qquad n \geq 0, \qquad (l + m + n) \leq 5. \tag{12-84}$$

The restrictions of Eq. (12-84) result from the fact that none of the coefficients in Eq. (12-83) can be negative. As a result of this there are 35 different possible normal coordinate species in this case. This is a rather extreme case of the species of the normal coordinates being variable, but the general result is not atypical if the rigid conformer groups are all of much smaller order than the MS group of the nonrigid molecule. We proceed by neglecting the ρ dependence of the normal coordinates or by assuming a particular form of it. Perturbation theory can then be used to allow for the neglected ρ dependencies. If we completely neglect the ρ dependence of the normal coordinates then the normal coordinate representation of CH_3NO_2 is obtained as the species of the ρ independent symmetry coordinates given in Table 12-8. This species is

$$\Gamma^0(Q) = 5A_1' \oplus 2A_1'' \oplus A_2'' \oplus 3E' \tag{12-85}$$

[i.e., this has $l = 2$, $m = 3$, and $n = 0$ in Eq. (12-83)]. The double degeneracies of the methyl group vibrations S_9, S_{10}, and S_{11} will be split by interactions with the NO_2 frame into in-plane $S_r^{(i)}$ and out-of-plane $S_r^{(o)}$ modes, and if we wished to allow for this in zero order we would use symmetry coordinates

$$S_r^{(i)} = \cos \rho\, S_{ra} - \sin \rho\, S_{rb}, \qquad S_r^{(o)} = \sin \rho\, S_{ra} + \cos \rho\, S_{rb} \tag{12-86}$$

Table 12-8

A set of symmetry coordinates of CH_3NO_2 not involving the torsional angle[a]

A_1'	$S_1 = (\Delta r_1 + \Delta r_2 + \Delta r_3)/\sqrt{3}$
	$S_2 = \Delta r_{CN}$
	$S_3 = (\Delta r_4 + \Delta r_5)/\sqrt{2}$
	$S_4 = (\Delta \alpha_4 + \Delta \alpha_5)/\sqrt{2}$
	$S_5 = (\Delta \alpha_1 + \Delta \alpha_2 + \Delta \alpha_3)/\sqrt{3}$
A_1''	$S_6 = (\Delta r_4 - \Delta r_5)/\sqrt{2}$
	$S_7 = (\Delta \alpha_4 - \Delta \alpha_5)/\sqrt{2}$
A_2''	$S_8 = \Delta \lambda$ (out-of-plane NO_2 wag)
E'	$S_{9a} = (2\Delta r_1 - \Delta r_2 - \Delta r_3)/\sqrt{6}, S_{9b} = (\Delta r_2 - \Delta r_3)/\sqrt{2}$
	$S_{10a} = (2\Delta \beta_{23} - \Delta \beta_{13} - \Delta \beta_{12})/\sqrt{6}, S_{10b} = (\Delta \beta_{13} - \Delta \beta_{12})/\sqrt{2}$
	$S_{11a} = (2\Delta \alpha_1 - \Delta \alpha_2 - \Delta \alpha_3)/\sqrt{6}, S_{11b} = (\Delta \alpha_2 - \Delta \alpha_3)/\sqrt{2}$

[a] The bond and angle coordinates are defined in Fig. 12-11.

for $r = 9$, 10, and 11. $S_r^{(i)}$ is of species A_1'' and $S_r^{(o)}$ is of species A_2'' so that the symmetry coordinate representation is now

$$5A_1' \oplus 5A_1'' \oplus 4A_2'' \qquad (12\text{-}87)$$

[i.e., $l = 5$ and $m = n = 0$ in Eq. (12-83)]. Symmetry coordinates with the species given by Eq. (12-87) are the best to use for interpreting the rotation–vibration spectrum quantitatively since with them the G matrix is independent of ρ and the ρ dependence of the F matrix can be added by perturbation theory just as the torsional barrier can.[4] The effects of these perturbations will be to alter the rotational and torsional energy level pattern accompanying each vibrational state.

In this low barrier molecule the torsional energy levels will accompany each vibrational state just like extra rotational structure, and we will need selection rules on K_a, K_c, and K_i to interpret the rotational–torsional fine structure that will accompany each (ordinary) vibrational transition. In the infrared spectrum a transition can occur if the states satisfy the symmetry requirement

$$\Gamma_{\text{rtv}}' \otimes \Gamma_{\text{rtv}}'' \supset \Gamma^* = A_2'. \qquad (12\text{-}88)$$

We will use the normal coordinate representation of Eq. (12-87), together with the rotational and torsional species of Tables 12-5 and 12-6, to elaborate the selection rules in detail. The five A_1' fundamentals will be infrared active with accompanying rotational–torsional transitions satisfying

$$\Gamma_{\text{rt}}' \otimes \Gamma_{\text{rt}}'' \supset A_2' \qquad (12\text{-}89)$$

and, from Tables 12-5 and 12-6, the most intense of the rotational–torsional transitions will satisfy the selection rules

$$\Delta K_i = 0, \qquad \Delta K_a = 0, \qquad \Delta K_c = \pm 1. \qquad (12\text{-}90)$$

The five A_1'' fundamentals will be infrared active with accompanying rotational–torsional transitions satisfying

$$\Gamma_{\text{rt}}' \otimes \Gamma_{\text{rt}}'' \supset A_2'' \qquad (12\text{-}91)$$

for which the most intense transitions will satisfy

$$\Delta K_i = 0, \qquad \Delta K_a = \pm 1, \qquad \Delta K_c = \pm 1. \qquad (12\text{-}92)$$

For the four A_2'' fundamentals the infrared selection rules from

$$\Gamma_{\text{rt}}' \otimes \Gamma_{\text{rt}}'' \supset A_1'' \qquad (12\text{-}93)$$

---

[4] See Hougen (1965), Bunker (1967), and Bunker and Hougen (1967). The F and G matrices are discussed in Section 4-3 of Wilson, Decius, and Cross (1955).

are, for the most intense transitions,

$$\Delta K_i = 0, \qquad \Delta K_a = \pm 1, \qquad \Delta K_c = 0. \tag{12-94}$$

The CH_3NO_2 and CH_3BF_2 molecules to which all these results apply are near oblate tops so that K_c, but not K_a, is a useful near quantum number. As a result of asymmetric top effects forbidden transitions with $\Delta K_a = \pm 2$ will occur in the A_1' fundamentals, and with $\Delta K_a = \pm 3$ will occur in the A_1'' and A_2'' fundamentals. Torsional transitions with $\Delta K_i = 6n \pm 3$ and $\Delta K_i = 6n \pm 6$ are allowed by symmetry in these bands, but these will be very weak since they depend on rather high terms in the expansion of $M_\alpha^{(r)}(e'', e'')$ in ρ.

Dimethylacetylene

The dimethylacetylene molecule in its reference configuration is shown in Fig. 12-14 [see Bunker (1975) and references therein]. This molecule has a very low barrier to internal rotation, and its MS group is the group G_{36}. The MS group of acetone is isomorphic to that of dimethylacetylene and its character table is given in Table A-19. The transformation properties of the $\hat{J}_a, T_a, \alpha_{ab}$, etc., given in Table A-19 apply to acetone but not to dimethylacetylene. The species of Γ^* is A_3.

The molecule fixed z axis of the reference configuration is defined as pointing from C_7 to C_8, and its orientation in space is given by the Euler

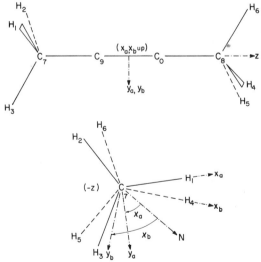

Fig. 12-14. The reference configuration of dimethylacetylene and the definition of the rotor fixed axes (x_a, y_a, z) and (x_b, y_b, z). These axes have origin at the nuclear center of mass O and the x_a and x_b axes are in the $H_1C_7C_8$ and $H_4C_8C_7$ planes, respectively. ON is the node line from which the Euler angles χ_a and χ_b are measured (see Fig. 7-1).

angles θ and ϕ. We introduce $x_a y_a$ and $x_b y_b$ axes that rotate with the $C_7(H_1 H_2 H_3)$ and $C_8(H_4 H_5 H_6)$ rotors, respectively; x_a being in the $C_8 C_7 H_1$ plane and x_b being in the $C_7 C_8 H_4$ plane as indicated in Fig. 12-14. The orientation of these axes is defined by the two Euler angles χ_a and χ_b. Assuming no torsional barrier the zero order rotational–torsional Hamiltonian is (in cm^{-1})

$$[B(\hat{J}_x{}^2 + \hat{J}_y{}^2) + A(\hat{J}_{za}^2 + \hat{J}_{zb}^2)]\hbar^{-2}, \qquad (12\text{-}95)$$

where A is the rotational constant for each CH_3 group and $\hat{J}_{za}$ and $\hat{J}_{zb}$ are the z components of the angular momenta of each CH_3 group. The zero order rotational–torsional wavefunctions are

$$\Phi_{rt} = [(1/(2\pi)]S_{Jkm}(\theta, \phi) \exp(ik_a\chi_a) \exp(ik_b\chi_b), \qquad (12\text{-}96)$$

where k, k_a, and k_b are positive or negative integers, and $k = k_a + k_b$. The transformation properties of θ, ϕ, χ_a, and χ_b under the effect of some elements of G_{36} are given in Table 12-9; from these results the effect of any operation of G_{36} on the angles can be deduced. The symmetry classification of the rotational–torsional wavefunctions can be determined from the results in Table 12-9, and the results are given in Table 12-10.

Table 12-9

The transformation properties of the Euler angles of dimethylacetylene under the effect of elements of G_{36}

E	(123)	(456)	(23)(56)*	(14)(26)(35)(78)(90)*
θ	θ	θ	$\pi - \theta$	θ
ϕ	ϕ	ϕ	$\phi + \pi$	ϕ
χ_a	$\chi_a + 2\pi/3$	χ_a	$\pi - \chi_a$	$\chi_b + \pi$
χ_b	χ_b	$\chi_b - 2\pi/3$	$\pi - \chi_b$	$\chi_a + \pi$

Table 12-10

The symmetry classification of the rotational–torsional functions of dimethylacetylene in G_{36}[a]

$|k_a - k_b|$	$|k_a + k_b|$		
	0	$3n \pm 1$	$3n \pm 3$
0:	A_1 (J even) A_2 (J odd)	E_1	$A_1 \oplus A_2$
$3m \pm 1$:	E_3	G	$E_3 \oplus E_4$
$3m \pm 3$:	$A_1 \oplus A_3$ (J even) $A_2 \oplus A_4$ (J odd)	$E_1 + E_2$	$A_1 \oplus A_2 \oplus A_3 \oplus A_4$

[a] n, m are nonnegative integers.

In the approximation of ignoring the dependence of the vibrational kinetic and potential energies on the torsional angle (i.e., the dependence of the elements of the 23×23 F and G matrices on the torsional angle) we deduce that the normal coordinate species is [Hougen (1965)]

$$\Gamma^0(Q) = 4A_1 \oplus 3A_4 \oplus 4G. \tag{12-97}$$

We deduce that the three A_4 fundamental bands are infrared active with selection rules

$$\Delta|k_a + k_b| = \Delta|k_a - k_b| = 0, \qquad \text{i.e.,} \qquad \Delta K_a = \Delta K_b = 0, \tag{12-98}$$

and the four G fundamentals are infrared active with selection rules

$$\Delta|k_a + k_b| \text{ and } \Delta|k_a - k_b| = \pm 1, \quad \text{i.e.,} \quad (\Delta K_a, \Delta K_b) = (0, \pm 1) \text{ or } (\pm 1, 0). \tag{12-99}$$

In both cases the weak symmetry allowed transitions having ΔK_a and/or ΔK_b greater than one are ignored.

A detailed understanding of the spectrum and energy levels of this molecule demands that we allow in zero order for the end-to-end coupling of the perpendicular vibrations, into cis and trans vibrations (the G matrix for such symmetry coordinates is independent of the torsional angle), and that the zero order wavefunction be separable into rotational, torsional, and vibrational parts. The rotational wavefunctions will involve the three Euler angles θ, ϕ, and χ upon which the intramolecular potential function does not depend. The zero order wavefunctions are

$$\Phi_{\text{rtv}} = [1/(2\pi)]S_{Jkm}(\theta, \phi)\exp(ik\chi)\exp(ik_i\gamma)\Phi_{\text{vib}}[Q_s(\gamma)], \tag{12-100}$$

where

$$\chi = (\chi_a + \chi_b)/2, \qquad \gamma = (\chi_a - \chi_b)/2,$$

and the 23 normal coordinates $Q_s(\gamma)$ are allowed to depend parametrically on the torsional angle γ.

We now wish to classify the rotational, torsional, and vibrational wavefunctions separately where these functions are

$$\Phi_{\text{rot}} = [1/(2\pi)]^{1/2}S_{Jkm}(\theta, \phi)\exp(ik\chi),$$
$$\Phi_{\text{tor}} = [1/(2\pi)]^{1/2}\exp(ik_i\gamma), \tag{12-101}$$
$$\Phi_{\text{vib}} = \Phi_{\text{vib}}[Q_s(\gamma)].$$

Once we have achieved this we can determine the selection rules on K and K_i for the vibrational transitions. However, to make the separate classification of Φ_{rot}, Φ_{tor}, and Φ_{vib} in G_{36} we run into the same difficulty of double valuedness that we encountered in Chapter 10 [see Eq. (10-59)] when we tried to classify angular momentum wavefunctions with half-integral angular momentum quantum numbers. For example,

$$(123)\Phi_{\text{rot}} = (123)[1/(2\pi)]^{1/2}S_{Jkm}(\theta, \phi)\exp[ik(\chi_a + \chi_b)/2] \quad (12\text{-}102)$$

and if we use the transformation properties of θ, ϕ, χ_a, and χ_b given in Table 12-9 we obtain

$$(123)\Phi_{\text{rot}} = \exp(i\pi k/3)\Phi_{\text{rot}}. \quad (12\text{-}103)$$

However, if we rotate χ_a or χ_b through 2π (which does nothing to the spatial orientation of the molecule) we obtain

$$(123)\Phi_{\text{rot}} = \exp(4i\pi k/3)\Phi_{\text{rot}} = (-1)^k \exp(i\pi k/3)\Phi_{\text{rot}}, \quad (12\text{-}104)$$

and the character generated by Φ_{rot} under (123) is double valued if k is odd. The same kind of thing happens for Φ_{tor} or Φ_{vib} and for all elements of G_{36} (including E). We overcome this difficulty, in a way similar to that used to introduce the electron spin double groups, by bringing in the operation E' which is the operation of increasing χ_a (or, equivalently, χ_b) by 2π radians and by fixing the effects of the operations (123), (456), etc., on χ and γ to be as obtained from Table 12-9. By introducing E' we obtain a group of twice the order of the group G_{36}, and we call it $G_{36}(\text{EM})$ by analogy with the similar group required for the separate classification of the rotational and vibronic wavefunctions of a linear molecule. The character table of $G_{36}(\text{EM})$ is given in Table A-25, and the extra representations are labeled with a d subscript (for double valued) [Hougen (1964)]. The overall wavefunction cannot belong to a d representation.

In the EMS group $G_{36}(\text{EM})$ the transformation properties of θ, ϕ, χ, and γ are definite and are given in Table 12-11. Notice that $(123)^3 \equiv (456)^3 \equiv E'$. We can classify the rotational, torsional, and vibrational wavefunctions in the group. The rotational and torsional species are given in Table 12-12. The normal coordinates that involve γ, with cis and trans coupled degenerate vibrations, give a G matrix that is independent of γ and the coordinates have

Table 12-11

The transformation properties of θ, ϕ, χ, and γ for dimethylacetylene under the effect of operations of $G_{36}(\text{EM})^a$

E	(123)	(456)	(23)(56)*	(14)(26)(35)(78)(90)*	E'
θ	θ	θ	$\pi - \theta$	θ	θ
ϕ	ϕ	ϕ	$\phi + \pi$	ϕ	ϕ
χ	$\chi + 2\pi/6$	$\chi - 2\pi/6$	$\pi - \chi$	$\chi + \pi$	$\chi + \pi$
γ	$\gamma + 2\pi/6$	$\gamma + 2\pi/6$	$-\gamma$	$-\gamma$	$\gamma + \pi$

^{a} ϕ, χ, and γ are mod 2π and $0 \leq \theta \leq \pi$.

Table 12-12

The separate classification of the rotational[a] and torsional[b] wavefunctions of dimethylacetylene in $G_{36}(EM)$

K	Γ_{rot}	K_i	Γ_{tor}
0 (J even)	A_1	0	A_1
0 (J odd)	A_2		
$6n \pm 1$	E_{2d}	$6m \pm 1$	E_{3d}
$6n \pm 2$	E_1	$6m \pm 2$	E_3
$6n \pm 3$	$A_{3d} \oplus A_{4d}$	$6m \pm 3$	$A_{1d} \oplus A_{3d}$
$6n \pm 6$	$A_1 \oplus A_2$	$6m \pm 6$	$A_1 \oplus A_3$

[a] n is a nonnegative integer; $K \geq 0$.
[b] m is a nonnegative integer; $K_i \geq 0$.

Table 12-13

The selection rules[a] on the fundamentals of dimethylacetylene using the $G_{36}(EM)$ group

No.	Approximate description	Species	Infrared ($\Delta J = 0, \pm 1$)	Raman ($\Delta J = 0, \pm 1, \pm 2$)
1	C—H stretch			
2	C≡C stretch	A_1	Inactive[a]	$\Delta K = \Delta K_i = 0$
3	CH$_3$ deformation			
4	C—C stretch			
6	C—H stretch			
7	CH$_3$ deformation	A_4	$\Delta K = \Delta K_i = 0$	Inactive[a]
8	C—C stretch			
9	C—H stretch			
10	CH$_3$ deformation	E_{1d}	$\Delta K = \pm 1, \Delta K_i = 0$	Inactive[a]
11	CH$_3$ rock			
12	Skeletal bend			
13	C—H stretch			
14	CH$_3$ deformation	E_{2d}	Inactive[a]	$\Delta K = \pm 1, \Delta K_i = 0$
15	CH$_3$ rock			
16	Skeletal bend			

[a] Neglecting the very weak transitions with ΔK or ΔK_i larger than one. ν_5 is the torsion.

symmetry [Bunker (1967)]

$$4A_1 \oplus 3A_4 \oplus 4E_{1d} \oplus 4E_{2d} \tag{12-105}$$

in $G_{36}(EM)$. The selection rules on the fundamentals of these vibrations are as given in Table 12-13.

Other Nonrigid Molecules

Nonrigid molecules that require an EMS group for the separate classification of the rotational and vibronic wavefunctions are those that have identical coaxial internal rotors on a linear framework. Hydrogen peroxide and internally rotating ethylene would be examples of such molecules. The EMS group of hydrogen peroxide is G_4(EM) and its character table is given in Table A-21 [see, also, Dellepiane, Gussoni, and Hougen (1973) and Yamada, Nakagawa, and Kuchitsu (1974)]. The EMS group for a torsionally tunneling ethylene molecule is the group G_{16}(EM), and its character table is given in Table A-24 [see Papousek, Sarka, Spirko, and Jordanov (1971) and Merer and Watson (1973)]. For the nonrigid molecules acetone, methyl silane, and methanol no EMS group is required and the MS groups are G_{36}, G_{18}, and G_6, the character tables of which are given in Tables A-19, A-18, and A-16, respectively [see Bunker (1975) and references therein].

Problem 12-1.   Determine the character table of the electron spin double group $G_6{}^2$ of a molecule with an odd number of electrons that has the geometrical structure of the CH_3OH molecule and for which the odd electron spin is coupled to the COH part of the molecule (e.g., perhaps CH_3SeH^+). Also determine the character table if the odd electron spin is coupled to the CH_3 part of the molecule (e.g., perhaps, SiH_3OH^+); see Eqs. (10-68)–(10-84) for the technique used in determining the character table of a spin double group.

Answer.   The CH_3OH molecule and its principal axes of inertia are indicated in Table A-16. The COH plane is the ab plane. The orientation of the (a, b, c) axes in space is given by the Euler angles (θ, ϕ, χ) where we use a I^r convention $(abc \equiv zxy)$. The effect of each of the elements of the group $G_6{}^2$ on the Euler angles [see Eqs. (10-64) and (10-65)] can be determined and is given in Table 12-14. Using the results in Table 12-14 we can determine the group multiplication table, and this is given in Table 12-15. Using the

Table 12-14

Transformation properties of Euler angles for abc ($\equiv zxy$) principal axes of CH_3OH under the effect of elements of the $G_6{}^2$ group

E	$(12)^*$	R	$R(12)^*$
(123)	$(23)^*$	$R(123)$	$R(23)^*$
(132)	$(13)^*$	$R(132)$	$R(13)^*$
θ	$\pi - \theta$	θ	$\pi - \theta$
ϕ	$\phi + \pi$	ϕ	$\phi + \pi$
χ	$\pi - \chi$	$\chi + 2\pi$	$3\pi - \chi$

Table 12-15
Part[a] of the multiplication table of the elements of
the group $G_6{}^2$

	E	(123)	(132)	(12)*	(23)*	(13)*
E:	E	(123)	(132)	(12)*	(23)*	(13)*
(123):	(123)	(132)	E	(13)*	(12)*	(23)*
(132):	(132)	E	(123)	(23)*	(13)*	(12)*
(12)*:	(12)*	(23)*	(13)*	R	R(123)	R(132)
(23)*:	(23)*	(13)*	(12)*	R(132)	R	R(123)
(13)*:	(13)*	(12)*	(23)*	R(123)	R(132)	R

[a] Since $R^2 = E$ and since R commutes with all the elements of the group G_6 the rest of the multiplication table can be easily constructed.

Table 12-16
The class structure of the $G_6{}^2$ group

C_1	C_2	C_3	C_4	C_5	C_6
E	(123)	(12)*	R	R(123)	R(12)*
	(132)	(23)*		R(132)	R(23)*
		(13)*			R(13)*

multiplication table [see Eq. (4-47)] we can determine the class structure of $G_6{}^2$ and this is given in Table 12-16. The class products can be determined [see Eq. (4-61)] and those that we use in constructing the character table are

$$C_2{}^2 = 2C_1 + C_2,$$
$$C_4{}^2 = C_1,$$
$$C_2 C_4 = C_5,$$
$$C_6{}^2 = 3C_4 + 3C_5,$$
$$C_3 C_4 = C_6.$$
(12-106)

Using Eqs. (4-62) and (12-106) we see that the characters of the two-dimensional representations of $G_6{}^2$ satisfy

$$\chi_1 = 2,$$
$$\chi_2{}^2 = 2 + \chi_2,$$
$$\chi_4{}^2 = 4,$$
$$\chi_2 \chi_4 = 2\chi_5,$$
$$3\chi_6{}^2 = 2(\chi_4 + 2\chi_5),$$
$$\chi_3 \chi_4 = 2\chi_6.$$
(12-107)

Table 12-17

Characters of two-dimensional
representations of group $G_6{}^2$ as
determined from Eq. (12-107)

C_1	C_2	C_3	C_4	C_5	C_6
2	2	2	2	2	2
2	2	-2	2	2	-2
2	2	$2i$	-2	-2	$-2i$
2	2	$-2i$	-2	-2	$2i$
2	-1	0	2	-1	0
2	-1	0	-2	1	0

The characters for the two-dimensional representations of $G_6{}^2$ that we obtain from these equations are given in Table 12-17. The characters of the irreducible representations follow directly from these results, and the character table is given in Table A-16.

If the electron spin is coupled to the CH_3 group then we must introduce an axis system fixed to the CH_3 rotor. Doing this we obtain Euler angle transformations as given in Table 10-21, and thus we deduce that the character table of the spin double group is that of the $C_{3v}(M)^2$ group given in Table A-8. The characters in the half-integral representations depend on which rotor fixed axis system the odd electron is coupled to by the spin–orbit coupling operator. For nonrigid molecules the character tables of the spin

Table 12-18

The character table of the spin double group for CH_3NO_2 obtained when the
axes are fixed to the CH_3 rotor as in CH_3F

	E	(123)	(12)*	(45)	(123)(45)	(12)(45)*	R	R(123)	R(45)	R(123)(45)	
	1	2	3[6]	1	2	3[6]	1	2	1	2	
Equiv. rot.:	R^0	$R_z^{2\pi/3}$	$R_{\pi/6}^\pi$	R^0	$R_z^{2\pi/3}$	$R_{\pi/6}^\pi$	$R^{2\pi}$	$R_z^{8\pi/3}$	$R^{2\pi}$	$R_z^{8\pi/3}$	
A_1':	1	1	1	1	1	1	1	1	1	1	
A_2':	1	1	-1	1	1	-1	1	1	1	1	
E':	2	-1	0	2	-1	0	2	-1	2	-1	
A_1'':	1	1	1	-1	-1	-1	1	1	-1	-1	
A_2'':	1	1	-1	-1	-1	1	1	1	-1	-1	
E'':	2	-1	0	-2	1	0	2	-1	-2	1	
$E_{1/2}'$:	2	1	0	2	1	0	-2	-1	-2	-1	
$E_{1/2}''$:	2	-1	0	-2	1	0	-2	1	2	-1	
$E_{3/2}'$:	2	-2	0	2	-2	0	-2	2	-2	2	: sep
$E_{3/2}''$:	2	-2	0	-2	2	0	-2	2	2	-2	: sep

double groups, given in Tables A-16–A-18, A-21, and A-24, are all constructed under the assumption that the odd electron is coupled to the axis system that is shown in the figure at the head of each table. The character table for the spin double group of the MS group of CH_3NO_2 obtained using an axis system fixed to the CH_3 rotor can also be derived. This is given in Table 12-18.

The Molecular Rotation–Contortion Group

As well as introducing the vibronic group as a generalization of the molecular point group for the classification of the vibronic states of nonrigid molecules we can also introduce the molecular rotation–contortion group as a generalization of the molecular rotation group. The molecular rotation–contortion group is a group of elements each of which transforms the rotational and contortional variables and which commutes with the rotation–contortion Hamiltonian $\hat{H}_{rc}$ [see Eq. (12-59)]. This group can be obtained from the MS (or EMS) group by neglecting the effect of each element on the vibronic coordinates and nuclear spin coordinates. Such groups have been used by microwave spectroscopists for many years [see, for example, Myers and Wilson (1960)] although no general discussion was given of their determination. Recently such groups have been discussed in a general fashion [Bauder, Meyer, and Günthard (1974)] and these authors have called them *isometric* groups.

DISCUSSION AND SUMMARY

In this chapter the MS group has been applied to linear molecules and to nonrigid molecules. For linear molecules, and some nonrigid molecules, we have seen that an extended MS group is required for the separate classification of rotational and vibronic states. In Chapter 11 the vibronic symmetry group was introduced [see Eq. (11-17)], and for rigid molecules this is the molecular point group. For a rigid nonlinear molecule the elements of the vibronic group (molecular point group) are obtained from the elements of the MS group by neglecting the effect of each of the MS group elements on the Euler angles and nuclear spin coordinates. For a linear molecule the vibronic group ($D_{\infty h}$ or $C_{\infty v}$) is obtained from the EMS group in a similar manner; the EMS group is required since it is this group, and not the MS group, that can be used for classifying vibronic states. For a nonrigid molecule the vibronic group can be obtained similarly from the EMS or MS group, where appropriate; the vibronic group of a nonrigid molecule is not isomorphic, in general, to any point group. The rotation–contortion group is obtained in a similar manner from the MS or EMS group by neglecting the effects of the elements on the vibronic and nuclear spin coordinates.

The spin double groups of linear molecules and of nonrigid molecules have been obtained, and these are required for molecules having an odd number of electrons in Hund's case (a). For a $C_{\infty v}$ molecule in a case (a) $^2\Pi$ electronic state the electron spin symmetry is obtained from Table B-2(ii) as $E_{1/2}$; since $\Pi \otimes E_{1/2} = E_{1/2} \oplus E_{3/2}$ the spin–orbit states obtained have symmetry $E_{1/2}$ and $E_{3/2}$. The traditional notation for these states is $^2\Pi_{1/2}$ and $^2\Pi_{3/2}$. Similarly, for example, a $^4\Delta$ state gives rise, in case (a), to spin–orbit states of species $E_{1/2}$, $E_{3/2}$, $E_{5/2}$, and $E_{7/2}$. The correlation of the case (a) and case (b) limiting energy level patterns for CH_3F^+ and $CH_3BF_2^+$, and of the symmetry species labels obtained using axes attached to each of the internal rotors in $CH_3BF_2^+$, are discussed by Bunker (1979).

A collection of colliding or reacting molecules can, in principle, be described by a Hamiltonian, and this Hamiltonian will be symmetric under the effect of the permutation of identical nuclei and the inversion E^*. Thus we can set up the CNPI group for such a system of nuclei. By generalizing the concept of feasibility the MS groups for these systems can be set up and they enable useful symmetry limitations for interactions and transitions to be determined which are unaffected by the details of the Hamiltonian (these details frequently being unknown); a recent discussion of this has been given by Quack (1977). Matrix isolated molecules in static crystal fields have also been treated using the idea of the MS group [Miller and Decius (1973)].

APPENDIX **A**

The Character Tables

This Appendix gives the character tables of the spin double groups of the most common molecular symmetry (MS) groups and extended molecular symmetry (EMS) groups. The MS group is defined in Chapter 9 and the EMS group is defined in Chapter 12. The EMS group is necessary for the classification of the vibronic states of linear molecules and of nonrigid molecules that have identical coaxial internal rotors; in the latter case the special operation E', which is the rotation of one internal rotor through 2π radians, is introduced. The spin double group is necessary for classifying states that involve half-integral angular momenta, and this group is defined in Chapter 10; the special operation R, which is the rotation of the whole molecule through 2π radians, is used in this group. Above and to the left of the dashed lines the standard MS or EMS group (in which R does not occur) is given. Separably degenerate representations are treated as being degenerate in giving the class structure of the groups and in giving the characters (see Tables 5-4 and 6-2); such representations are denoted sep.

For a rigid nonlinear molecule the MS group is isomorphic to the molecular point group, and in such a case the name of the MS group is taken to be that of the point group followed by (M), e.g., the MS group of CH_3F is called $C_{3v}(M)$. For a linear rigid molecule the EMS group is isomorphic to the molecular point group and is called $C_{\infty v}$(EM) or $D_{\infty h}$(EM) as appropriate; the MS group of a linear molecule is called $C_{\infty v}$(M) or $D_{\infty h}$(M) but these are not isomorphic to the $C_{\infty v}$ or $D_{\infty h}$ point groups. For a nonrigid molecule the MS group is called G_n, where n is the order of the group, and the EMS group for a molecule whose MS group is G_n is called G_n(EM). The spin double groups are indicated by a 2 superscript on the MS or EMS group name.

In each character table one element from each class is given and the number of elements in the class is indicated underneath the element (the number within the square brackets applies to the spin double group if there is a change). If the MS or EMS group is isomorphic to the molecular point group (this only happens for rigid molecules), the appropriate elements for each class of the molecular point group are given (this shows the effects of the MS group element on the vibronic variables), and the names of the irreducible representations are taken from the molecular point group. The equivalent rotation (equiv. rot.) of the MS or EMS group element given in each class is also given (R_a^{π}, R_b^{π}, and R_c^{π} are rotations about the a, b, and c axes through π radians, and the other notation is defined in Table 7-1; for convenience $R^{2\pi}R_z^{\beta}$ is written $R_z^{2\pi+\beta}$ and $R^{2\pi}R_{\alpha}^{\pi}$ is written $R_{\alpha}^{3\pi}$) so that the effect of the element on the Euler angles and angular momentum components $\hat{J}_x, \hat{J}_y$, and $\hat{J}_z$ can be determined (from the results in Table 7-1). The species of the $\hat{J}_{\alpha}$ so obtained are indicated to the right in each table. The translational coordinate T_{α} [see Eq. (7-235)] transforms in the same way as $\hat{J}_{\alpha}$ under a nuclear permutation but with opposite sign under a permutation–inversion (see Table 11-6). The species of the T_{α} and of the components $\alpha_{\gamma\delta}$ of the electronic polarizability, given by the species of $T_{\gamma}T_{\delta}$ [see Eq. (11-191)], are also given to the right of each table. Finally the representation Γ^* that has character $+1$ for the permutations and -1 for the permutation–inversions is indicated; the use of this representation is discussed in Chapter 11 [see Eq. (11-146)]. The rotational coordinate R_{α} [see Eq. (7-236)] transforms in the same way as $\hat{J}_{\alpha}$ under permutations and permutation–inversions.

The definition adopted here for the molecular point group is such that its elements transform the vibronic variables [including case (a) electron spins] but do not transform the Euler angles or nuclear spins. The definition accords with that given in Section 5-5 of Wilson, Decius, and Cross (1955); see also Bunker (1975), p. 21.

It should be apparent that no uniform system has been devised here for naming the MS or EMS groups and their irreducible representations. In particular there are often many nonisomorphic groups having the same number of elements and the G_n notation for the MS groups of nonrigid molecules would not distinguish between such groups. Fortunately no ambiguities have arisen for the nonrigid molecules studied so far using these groups.

The ground electronic states of the molecules depicted in each table are singlet states and the spin double groups would not be required for them. The spin double groups would be required for molecules having an odd number of electrons and strong spin–orbit coupling [i.e., Hund's case (a)]. The spin double groups of use for nonrigid molecules having the geometrical structure of CH_3OH, CH_3NO_2, or SiH_3CH_3, with an odd number of electrons and strong spin–orbit coupling, are given in Tables A-16–A-18 under the assumption that the electron spin is coupled to the COH, NO_2, and SiH_3 parts of the molecules, respectively. The half-integral representations of these spin double groups would be different if the electron spin were coupled to the CH_3 part of each of these molecules (see Problem 12-1).

Table A-1

The group $C_s(M)^2$

Example: HN_3

$C_s(M)^2$:	E	E^*	R		
	1	1[2]	1		
C_s:	E	σ_{ab}	–		
Equiv. rot.:	R^0	R_c^π	$R^{2\pi}$		
A':	1	1	1	:	$T_a, T_b, \hat{J}_c, \alpha_{aa}, \alpha_{bb}, \alpha_{cc}, \alpha_{ab}$
A'':	1	−1	1	:	$T_c, \hat{J}_a, \hat{J}_b, \alpha_{ac}, \alpha_{bc}, \Gamma^*$
$E_{1/2}$:	2	0	−2	:	sep

Table A-2
The group $C_i(M)^2$
Example: Trans C(HIF)CHIF (without torsional tunneling)

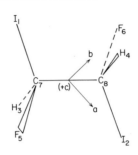

$C_i(M)^2$:	E	$(12)(34)(56)(78)^*$	R	$R(12)(34)(56)(78)^*$	
	1	1	1	1	
C_i:	E	i	—	—	
Equiv. rot.:	R^0	R^0	$R^{2\pi}$	$R^{2\pi}$	
A_g:	1	1	1	1	$:\ \hat{J}_a, \hat{J}_b, \hat{J}_c, \boldsymbol{\alpha}$
A_u:	1	-1	1	-1	$:\ T_a, T_b, T_c, \Gamma^*$
$B_{g/2}$:	1	1	-1	-1	
$B_{u/2}$:	1	-1	-1	1	

Table A-3
The group $C_2(M)^2$
Example: Hydrogen persulfide (without torsional tunneling)

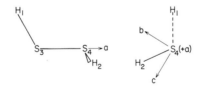

$C_2(M)^2$:	E	$(12)(34)$	R	
	1	$1[2]$	1	
C_2:	E	C_{2b}	—	
Equiv. rot.:	R^0	R_b^π	$R^{2\pi}$	
A:	1	1	1	$:\ T_b, \hat{J}_b, \alpha_{aa}, \alpha_{bb}, \alpha_{cc}, \alpha_{ac}, \Gamma^*$
B:	1	-1	1	$:\ T_a, T_c, \hat{J}_a, \hat{J}_c, \alpha_{ab}, \alpha_{bc}$
$E_{1/2}$:	2	0	-2	$:\ $ sep

Table A-4
The group $C_{2v}(M)^2$
Example: Water

$C_{2v}(M)^2$:	E 1	(12) 1[2]	E^* 1[2]	(12)* 1[2]	R 1	
C_{2v}:	E	C_{2b}	σ_{ab}	σ_{bc}	—	
Equiv. rot.:	R^0	$R_b{}^\pi$	$R_c{}^\pi$	$R_a{}^\pi$	$R^{2\pi}$	
A_1:	1	1	1	1	1	: $T_b, \alpha_{aa}, \alpha_{bb}, \alpha_{cc}$
A_2:	1	1	−1	−1	1	: $\hat{J}_b, \alpha_{ac}, \Gamma^*$
B_1:	1	−1	−1	1	1	: $T_c, \hat{J}_a, \alpha_{bc}$
B_2:	1	−1	1	−1	1	: $T_a, \hat{J}_c, \alpha_{ab}$
$E_{1/2}$:	2	0	0	0	−2	

Table A-5
The group $C_{2h}(M)^2$
Example: Trans-difluoroethylene (without torsional tunneling)

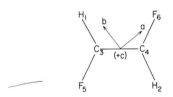

$C_{2h}(M)^2$:	E 1	(12)(34)(56) 1[2]	E^* 1[2]	(12)(34)(56)* 1	R 1	$R(12)(34)(56)^*$ 1	
C_{2h}:	E	C_{2c}	σ_{ab}	i	—	—	
Equiv. rot.:	R^0	$R_c{}^\pi$	$R_c{}^\pi$	R^0	$R^{2\pi}$	$R^{2\pi}$	
A_g:	1	1	1	1	1	1	: $\hat{J}_c, \alpha_{aa}, \alpha_{bb}, \alpha_{cc}, \alpha_{ab}$
A_u:	1	1	−1	−1	1	−1	: T_c, Γ^*
B_g:	1	−1	−1	1	1	1	: $\hat{J}_a, \hat{J}_b, \alpha_{ac}, \alpha_{bc}$
B_u:	1	−1	1	−1	1	−1	: T_a, T_b
$E_{g/2}$:	2	0	0	2	−2	−2	: sep
$E_{u/2}$:	2	0	0	−2	−2	2	: sep

Table A-6

The group $D_{2h}(M)^2$

Example: Ethylene (without torsional tunneling)

$D_{2h}(M)^2$:	E	$(12)(34)$ $1[2]$	$(13)(24)(56)$ $1[2]$	$(14)(23)(56)$ $1[2]$	E^* $1[2]$	$(12)(34)^*$ $1[2]$	$(13)(24)(56)^*$ $1[2]$	$(14)(23)(56)^*$ 1	R 1	$R(14)(23)(56)^*$ 1	
D_{2h}:	E	C_{2a}	C_{2b}	C_{2c}	σ_{ab}	σ_{ac}	σ_{bc}	i	–	–	
Equiv. rot.:	R^0	R_a^{π}	R_b^{π}	R_c^{π}	R_c^{π}	R_b^{π}	R_a^{π}	R^0	$R^{2\pi}$	$R^{2\pi}$	
A_g:	1	1	1	1	1	1	1	1	1	1	: $\alpha_{aa},\alpha_{bb},\alpha_{cc}$
A_u:	1	1	1	1	-1	-1	-1	-1	1	-1	: Γ^*
B_{1g}:	1	1	-1	-1	1	1	-1	-1	1	1	: $\hat{J}_a,\alpha_{bc}$
B_{1u}:	1	1	-1	-1	-1	-1	1	1	1	-1	: T_a
B_{2g}:	1	-1	1	-1	1	-1	1	-1	1	1	: $\hat{J}_b,\alpha_{ac}$
B_{2u}:	1	-1	1	-1	-1	1	-1	1	1	-1	: T_b
B_{3g}:	1	-1	-1	1	1	-1	-1	1	1	1	: $\hat{J}_c,\alpha_{ab}$
B_{3u}:	1	-1	-1	1	-1	1	1	-1	1	-1	: T_c
$E_{g/2}$:	2	0	0	0	0	0	0	2	-2	-2	
$E_{u/2}$:	2	0	0	0	0	0	0	-2	-2	2	

Table A-7

The group $D_{2d}(M)^2$

Example: Allene (without torsional tunneling)

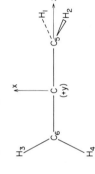

$D_{2d}(M)^2$:	E	(1324)(56)* 2	(12)(34) 1[2]	(14)(23)(56) 2[4]	(12)* 2[4]	R 1	R(1324)(56)* 2		
D_{2d}:	E	$2S_4$	C_2	$2C_2'$	$2\sigma_d$	—	—		
Equiv. rot.:	R^0	$R_z^{\pi/2}$	R_z^{π}	$R_{\pi/4}^{\pi}$	$R_{\pi/2}^{\pi}$	$R^{2\pi}$	$R_z^{5\pi/2}$		
A_1:	1	1	1	1	1	1	1	:	$\alpha_{xx} + \alpha_{yy},\ \alpha_{zz}$
A_2:	1	1	1	-1	-1	1	1	:	$\hat{J}_z$
B_1:	1	-1	1	1	-1	1	-1	:	$\alpha_{xx} - \alpha_{yy},\ \Gamma^*$
B_2:	1	-1	1	-1	1	1	-1	:	$T_z,\ \alpha_{xy}$
E:	2	0	-2	0	0	2	0	:	$(T_x, T_y),(\hat{J}_x, \hat{J}_y),(\alpha_{xz}, \alpha_{yz})$
$E_{1/2}$:	2	$\sqrt{2}$	0	0	0	-2	$-\sqrt{2}$		
$E_{3/2}$:	2	$-\sqrt{2}$	0	0	0	-2	$\sqrt{2}$		

Table A-8
The group $C_{3v}(M)^2$
Example: Methyl fluoride

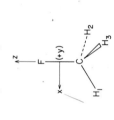

$C_{3v}(M)^2$:	E	(123)	$(23)^*$	R	$R(123)$	
	1	2	3[6]	1	2	
C_{3v}:	E	$2C_3$	$3\sigma_v$	—	—	
Equiv. rot.:	R^0	$R_z^{2\pi/3}$	$R_{\pi/2}^{\pi}$	$R^{2\pi}$	$R_z^{8\pi/3}$	
A_1:	1	1	1	1	1	: $T_z, \alpha_{zz}, \alpha_{xx}+\alpha_{yy}$
A_2:	1	1	-1	1	1	: $\hat{J}_z, \Gamma^*$
E:	2	-1	0	2	-1	: $(T_x, T_y), (\hat{J}_x, \hat{J}_y), (\alpha_{xz}, \alpha_{yz}), (\alpha_{xx}-\alpha_{yy}, \alpha_{xy})$
$E_{1/2}$:	2	1	0	-2	-2	: sep
$E_{3/2}$:	2	-2	0	-2	2	:

Table A-9

The group $\boldsymbol{D}_{3h}(M)^2$

Example: Boron trifluoride

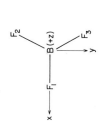

$D_{3h}(M)^2$:	E	(123)	(23)	E*	(123)*	(23)*	R	R(123)	R(123)*
	1	2	3[6]	1[2]	2	3[6]	1	2	2
D_{3h}:	E	$2C_3$	$3C_2$	σ_h	$2S_3$	$3\sigma_v$	–	–	–
Equiv. rot.:	R^0	$R_z^{2\pi/3}$	R_0^π	R_z^π	$R_z^{-\pi/3}$	$R_{\pi/2}^\pi$	$R^{2\pi}$	$R_z^{8\pi/3}$	$R_z^{5\pi/3}$
A_1'	1	1	1	1	1	1	1	1	1	: $\alpha_{zz}, \alpha_{xx}+\alpha_{yy}$
A_2'	1	1	−1	1	1	−1	1	1	1	: $\hat{J}_z$
E'	2	−1	0	2	−1	0	2	−1	−1	: $(T_x, T_y), (\alpha_{xx}-\alpha_{yy}, \alpha_{xy})$
A_1''	1	1	1	−1	−1	−1	1	1	−1	: Γ^*
A_2''	1	1	−1	−1	−1	1	1	1	−1	: T_z
E''	2	−1	0	−2	1	0	2	−1	1	: $(\hat{J}_x, \hat{J}_y), (\alpha_{xz}, \alpha_{yz})$
$E_{1/2}$	2	1	0	0	$\sqrt{3}$	0	−2	−1	−$\sqrt{3}$	
$E_{3/2}$	2	−2	0	0	0	0	−2	2	0	
$E_{5/2}$	2	1	0	0	−$\sqrt{3}$	0	−2	−1	$\sqrt{3}$	

Table A-10
The group $D_{3d}(M)^2$
Example: Ethane (without torsional tunneling in the staggered conformation)

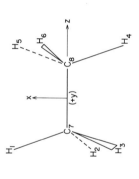

$D_{3d}(M)^2$:	E	(123)(465)	(16)(24)(35)(78)	(14)(26)(35)(78)*	(163425)(78)*	(12)(46)*	R	R(123)(465)	R(14)(26)(35)(78)*	R(163425)(78)*	
	1	2	3[6]	1	2	3[6]	1	2	1	2	
D_{3d}:	E	$2C_3$	$3C_2$	i	$2S_6$	$3\sigma_d$					
Equiv. rot.:	R^0	$R_z^{2\pi/3}$	$R_{\pi/6}^{\pi}$	R^0	$R_z^{2\pi/3}$	$R_{\pi/6}^{\pi}$	$R^{2\pi}$	$R_z^{8\pi/3}$	$R^{2\pi}$	$R_z^{8\pi/3}$	
A_{1g}:	1	1	1	1	1	1	1	1	1	1	$\alpha_{zz},\ \alpha_{xx}+\alpha_{yy}$
A_{2g}:	1	1	−1	1	1	−1	1	1	1	1	$\hat{J}_z$
E_g:	2	−1	0	2	−1	0	2	−1	2	2	$(\hat{J}_x,\hat{J}_y),\ (\alpha_{xz},\alpha_{yz})$ / $(\alpha_{xx}-\alpha_{yy},\ \alpha_{yy},\ \alpha_{xy})$
A_{1u}:	1	1	1	−1	−1	−1	1	1	−1	−1	Γ^*
A_{2u}:	1	1	−1	−1	−1	1	1	1	−1	−1	T_z
E_u:	2	−1	0	−2	1	0	2	−1	−2	−2	(T_x,T_y)
$E_{g/2}$:	2	0	0	2	0	0	−2	−1	−2	−1	
$E_{u/2}$:	2	0	0	−2	0	0	−2	−1	2	1	
$E_{3g/2}$:	2	−2	0	2	−2	0	−2	2	−2	2	: sep
$E_{3u/2}$:	2	−2	0	−2	2	0	−2	2	2	−2	: sep

382

Table A-11
The group $T_d(M)^2$
Example: Methane

$T_d(M)^2$:	E	(123)	(14)(23)	(1423)*	(23)*	R	R(123)	R(1423)*		
	1	8	3[6]	6	6[12]	1	8	6		
T_d:	E	$8C_3$	$3C_2$	$6S_4$	$6\sigma_d$					
A_1:	1	1	1	1	1	1	1	1	:	$\alpha_{xx} + \alpha_{yy} + \alpha_{zz}$
A_2:	1	1	1	-1	-1	1	1	-1	:	Γ^*
E:	2	-1	2	0	0	2	-1	0	:	$(\alpha_{xx} + \alpha_{yy} - 2\alpha_{zz}, \alpha_{xx} - \alpha_{yy})$
F_1:	3	0	-1	1	-1	3	0	1	:	$(\hat{J}_x, \hat{J}_y, \hat{J}_z)$
F_2:	3	0	-1	-1	1	3	0	-1	:	$(T_x, T_y, T_z), (\alpha_{xy}, \alpha_{yz}, \alpha_{xz})$
$E_{1/2}$:	2	1	0	$\sqrt{2}$	0	-2	-1	$-\sqrt{2}$		
$E_{5/2}$:	2	1	0	$-\sqrt{2}$	0	-2	-1	$\sqrt{2}$		
$G_{3/2}$:	4	-1	0	0	0	-4	1	0		

Table A-12
The group $C_{\infty v}(M)$
Example: Hydrogen cyanide

H ——— C ——— N

$C_{\infty v}(M)$:	E	E^*	
$\Sigma^+, +$:	1	1	
$\Sigma^-, -$:	1	-1 : Γ^*	

Table A-13
The group $D_{\infty h}(M)$
Example: Carbon dioxide

O_1 ——— C ——— O_2

$D_{\infty h}(M)$:	E	(12)	E^*	(12)*	
$\Sigma_g^+, +s$:	1	1	1	1	
$\Sigma_u^-, -s$:	1	1	-1	-1 : Γ^*	
$\Sigma_g^-, -a$:	1	-1	-1	1	
$\Sigma_u^+, +a$:	1	-1	1	-1	

Table A-14

The group^a $C_{\infty v}(EM)^2$

Example: Hydrogen cyanide

$$
\begin{array}{c}
x \\
\uparrow \\
\text{H} \longrightarrow \text{C} \longrightarrow \text{N} \longrightarrow z \\
(+y)
\end{array}
$$

$C_{\infty v}(EM)^2$:	E_0	⋯	E_ε	⋯	∞E_ε^*	⋯	R	RE_ε	⋯	
	1		2		$\infty[\times 2]$		1	2		
$C_{\infty v}$:	E	⋯	$2C_\infty^\varepsilon$	⋯	$\infty\sigma_v^{(\varepsilon/2)}$	⋯		—		
Equiv. rot.:	R^0		$R_z^{-\varepsilon}$		$R_{(\pi+\varepsilon)/2}^{\pi}$		$R^{2\pi}$	$R_z^{2\pi-\varepsilon}$		
$(+)\Sigma^+$	1	⋯	1	⋯	1	⋯	1	1	⋯	$T_z,\ \alpha_{xx}+\alpha_{yy},\ \alpha_{zz}$
$(-)\Sigma^-$	1	⋯	1	⋯	-1	⋯	1	1	⋯	$\hat{J}_z,\ \Gamma^*$
Π	2	⋯	$2\cos\varepsilon$	⋯	0	⋯	2	$2\cos\varepsilon$	⋯	$(T_x, T_y),\ (\hat{J}_x, \hat{J}_y),\ (\alpha_{xz}, \alpha_{yz})$
Δ	2	⋯	$2\cos 2\varepsilon$	⋯	0	⋯	2	$2\cos 2\varepsilon$	⋯	$(\alpha_{xx}-\alpha_{yy},\ \alpha_{xy})$
⋯	⋯		⋯		⋯		⋯	⋯		
$E_{1/2}$	2	⋯	$2\cos(\varepsilon/2)$	⋯	0	⋯	-2	$-2\cos(\varepsilon/2)$	⋯	
$E_{3/2}$	2	⋯	$2\cos(3\varepsilon/2)$	⋯	0	⋯	-2	$-2\cos(3\varepsilon/2)$	⋯	
⋯	⋯		⋯		⋯		⋯	⋯		

^{a} The xyz axes are the $x'y'z'$ axes introduced in Chapter 12 for the isomorphic Hamiltonian.

Table A-15
The group[a] $D_{\infty h}(EM)^2$
Example: Carbon dioxide

$O_1 \;\text{——}\; \underset{(+y)}{C} \;\text{——}\; O_2 \longrightarrow z$, with x axis upward.

$D_{\infty h}(EM)^2$:	E_0	E_ε	...	∞E_ε^*	$(12)_\pi^*$	$(12)_{\pi+\varepsilon}^*$	...	$\infty(12)_\varepsilon$	R	RE_ε	...	$R(12)_\pi^*$	$R(12)_{\pi+\varepsilon}^*$	...	
	1	2	...	$\infty[\times2]$	1	2	...	$\infty[\times2]$	1	2	...	1	2	...	
$D_{\infty h}$:	E	$2C_\infty^\varepsilon$	...	$\infty\sigma_v^{(\varepsilon/2)}$	i	$2S_\infty^{\pi+\varepsilon}$	...	$\infty C_2^{(\varepsilon/2)}$	–	–	...	–	–	...	
Equiv. rot.:	R^0	$R_z^{-\varepsilon}$	...	$R_{(\pi+\varepsilon)/2}^{\pi}$	R^0	$R_z^{-\varepsilon}$	...	$R_{\varepsilon/2}^{\pi}$	$R^{2\pi}$	$R_z^{2\pi-\varepsilon}$	...	$R_z^{2\pi}$	$R_z^{2\pi-\varepsilon}$	...	
$(+s)\Sigma_g^+$	1	1	...	1	1	1	...	1	1	1	...	1	1	...	$\alpha_{xx}+\alpha_{yy},\ \alpha_{zz}$
$(-a)\Sigma_g^-$	1	1	...	–1	1	1	...	–1	1	1	...	1	1	...	$\hat{J}_z$
$(+a)\Sigma_u^+$	1	1	...	1	–1	–1	...	–1	1	1	...	–1	–1	...	T_z
$(-s)\Sigma_u^-$	1	1	...	–1	–1	–1	...	1	1	1	...	–1	–1	...	Γ^*
Π_g	2	$2\cos\varepsilon$	...	0	2	$2\cos\varepsilon$	...	0	2	$2\cos\varepsilon$	...	2	$2\cos\varepsilon$	...	$(\hat{J}_x,\hat{J}_y),\ (\alpha_{xz},\alpha_{yz})$
Π_u	2	$2\cos\varepsilon$	...	0	–2	$-2\cos\varepsilon$	...	0	2	$2\cos\varepsilon$	...	–2	$-2\cos\varepsilon$	...	(T_x,T_y)
Δ_g	2	$2\cos2\varepsilon$	...	0	2	$2\cos2\varepsilon$	...	0	2	$2\cos2\varepsilon$	...	2	$2\cos2\varepsilon$	...	$(\alpha_{xx}-\alpha_{yy},\ \alpha_{xy})$
Δ_u	2	$2\cos2\varepsilon$	...	0	–2	$-2\cos2\varepsilon$	...	0	2	$2\cos2\varepsilon$	...	–2	$-2\cos2\varepsilon$	...	
...	...	...	...	...	...	...	...	...	...	...	...	...	...	...	
$E_{g/2}$	2	$2\cos(\varepsilon/2)$	...	0	2	$2\cos(\varepsilon/2)$	...	0	–2	$-2\cos(\varepsilon/2)$	...	–2	$-2\cos(\varepsilon/2)$	...	
$E_{u/2}$	2	$2\cos(\varepsilon/2)$	...	0	–2	$-2\cos(\varepsilon/2)$	...	0	–2	$-2\cos(\varepsilon/2)$	...	2	$2\cos(\varepsilon/2)$	...	
$E_{3g/2}$	2	$2\cos(3\varepsilon/2)$	...	0	2	$2\cos(3\varepsilon/2)$	...	0	–2	$-2\cos(3\varepsilon/2)$	...	–2	$-2\cos(3\varepsilon/2)$	...	
$E_{3u/2}$	2	$2\cos(3\varepsilon/2)$	...	0	–2	$-2\cos(3\varepsilon/2)$	...	0	–2	$-2\cos(3\varepsilon/2)$	...	2	$2\cos(3\varepsilon/2)$	...	
...	...	...	...	...	...	...	...	...	...	...	...	...	...	...	

[a] The xyz axes are the $x'y'z'$ axes introduced in Chapter 12 for the isomorphic Hamiltonian.

Table A-16

The group $G_6{}^2$

Example: Methanol (with torsional tunneling) (see Problem 12-1)

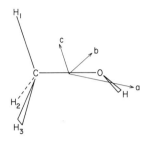

$G_6{}^2$:	E	(123)	(23)*	R	R(123)	
	1	2	3[6]	1	2	
Equiv. rot.:	R^0	R^0	$R_c{}^\pi$	$R^{2\pi}$	$R^{2\pi}$	
A_1:	1	1	1	1	1	: $T_a, T_b, \hat{J}_c, \alpha_{aa}, \alpha_{bb}, \alpha_{cc}, \alpha_{ab}$
A_2:	1	1	-1	1	1	: $T_c, \hat{J}_a, \hat{J}_b, \alpha_{ac}, \alpha_{bc}, \Gamma^*$
E:	2	-1	0	2	-1	
$E_{a/2}$:	2	-1	0	-2	1	
$E_{1/2}$:	2	2	0	-2	-2	: sep

Table A-17
The group G_{12}^2
Example: Nitromethane (with torsional tunneling) (see also Table 12-18)

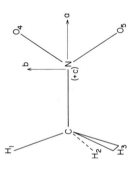

G_{12}^2:	E	(123)	$(23)^*$	(45)	$(123)(45)$	$(23)(45)^*$	R	$R(123)$	$R(123)(45)$	
	1	2	3[6]	1[2]	2	3[6]	1	2	2	
Equiv. rot.:	R^0	R^0	R_c^π	R_a^π	R_a^π	R_b^π	$R^{2\pi}$	$R^{2\pi}$	$R_a^{3\pi}$	
A_1':	1	1	1	1	1	1	1	1	1	$T_a, \alpha_{aa}, \alpha_{bb}, \alpha_{cc}$
A_2':	1	1	-1	1	1	-1	1	1	1	$J_a, \alpha_{bc}, \Gamma^*$
E':	2	-1	0	2	-1	0	2	-1	-1	
A_1'':	1	1	1	-1	-1	-1	1	1	-1	T_b, J_c, α_{ab}
A_2'':	1	1	-1	-1	-1	1	1	1	-1	T_c, J_b, α_{ac}
E'':	2	-1	0	-2	1	0	2	-1	1	
$E_{a/2}$:	2	-1	0	0	$\sqrt{3}$	0	-2	$+1$	$-\sqrt{3}$	
$E_{1/2}$:	2	2	0	0	0	0	-2	-2	0	
$E_{b/2}$:	2	-1	0	0	$-\sqrt{3}$	0	-2	$+1$	$\sqrt{3}$	

Table A-18
The group G_{18}^2
Example: Methylsilane (with torsional tunneling)

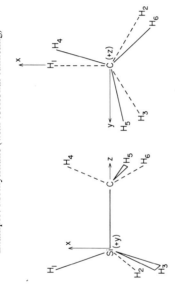

G_{18}^2	E	(456)	(123)	(123)(456)	(123)(465)	(23)(56)*	R	R(456)	R(123)	R(123)(456)	R(123)(465)	
	1	2	2	2	2	9[18]	1	2	2	2	2	
Equiv. rot.:	R^0	R^0	$R_z^{2\pi/3}$	$R_z^{2\pi/3}$	$R_z^{2\pi/3}$	$R_{\pi/2}^{\pi}$	$R^{2\pi}$	$R^{2\pi}$	$R_z^{8\pi/3}$	$R_z^{8\pi/3}$	$R_z^{8\pi/3}$	
A_1	1	1	1	1	1	1	1	1	1	1	1	$T_z, \alpha_{zz}, \alpha_{xx}+\alpha_{yy}$
A_2	1	1	1	1	1	-1	1	1	1	1	1	$\hat{J}_z, \Gamma^*$
E_1	2	2	-1	-1	-1	0	2	2	-1	-1	-1	$(T_x, T_y), (\hat{J}_x, \hat{J}_y), (\alpha_{xz}, \alpha_{yz}), (\alpha_{xx}-\alpha_{yy}, \alpha_{xy})$
E_2	2	-1	2	-1	-1	0	2	-1	2	-1	-1	
E_3	2	-1	-1	2	-1	0	2	-1	-1	2	-1	
E_4	2	-1	-1	-1	2	0	2	-1	-1	-1	2	
$E_{1/2}$	2	2	1	1	1	0	-2	-2	-1	-1	-1	
$E_{a/2}$	2	-1	-2	1	1	0	-2	1	2	-1	-1	
$E_{b/2}$	2	-1	1	-2	1	0	-2	1	-1	2	-1	
$E_{c/2}$	2	-1	1	1	-2	0	-2	1	-1	-1	2	
$E_{3/2}$	2	2	-2	-2	-2	0	-2	-2	2	2	2	sep

Table A-19
The group G_{36}
Example: Acetone (with torsional tunneling)

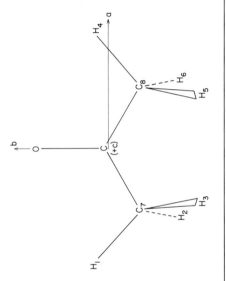

G_{36}:	E	(123)(456)	(14)(26)(35)|(78)*	(123)(465)	(132)	(142635)(78)*	(14)(25)(36)|(78)	(142536)|(78)	(23)|(56)*	
	1	2	3	2	4	6	3	6	9	
Equiv. rot.:	R^0	R^0	R_a^π	R^0	R^0	R_a^π	R_b^π	R_b^π	R_c^π	
A_1	1	1	1	1	1	1	1	1	1	: $T_b, \alpha_{aa}, \alpha_{bb}, \alpha_{cc}$
A_2	1	1	1	1	1	1	-1	-1	-1	: $T_c, \hat{J}_a, \alpha_{bc}$
A_3	1	1	-1	1	1	-1	1	1	-1	: $\hat{J}_b, \alpha_{ac}, \Gamma^*$
A_4	1	1	-1	1	1	-1	-1	-1	1	: $T_a, \hat{J}_c, \alpha_{ab}$
E_1	2	2	2	-1	-1	1	0	0	0	
E_2	2	2	-2	-1	-1	-1	0	0	0	
E_3	2	-1	0	2	-1	0	2	-1	0	
E_4	2	-1	0	2	-1	0	-2	1	0	
G	4	-2	0	-2	1	0	0	0	0	

Table A-20
The group G_4
Example: Hydrogen peroxide
(with torsional tunneling)

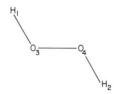

G_4:	E	$(12)(34)$	E^*	$(12)(34)^*$	
	1	1	1	1	
A_1:	1	1	1	1	
A_2:	1	1	-1	-1	$:\ \Gamma^*$
B_1:	1	-1	-1	1	
B_2:	1	-1	1	-1	

Table A-21
The group $G_4(EM)^2$
Example: Hydrogen peroxide (with torsional tunneling)

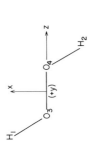

$G_4(EM)^2$:	E	$(12)(34)$	E^*	$(12)(34)^*$	E'	$(12)(34)E'$	E^*E'	$(12)(34)^*E'$	R	$R(12)(34)^*E'$	
	1	1[2]	1[2]	1[2]	1[2]	1[2]	1[2]	1	1	1	
Equiv. rot.:	R^0	$R^\pi_{\pi/2}$	R^π_π	R^π_z	R^π_z	R^π_π	$R^\pi_{3\pi/2}$	$R^{2\pi}$	$R^{2\pi}$	R^0	
A_1:	1	1	1	1	1	1	1	1	1	1	: $\alpha_{xx},\alpha_{yy},\alpha_{zz}$
A_2:	1	1	-1	-1	1	1	-1	-1	1	-1	: Γ^*
B_1:	1	-1	-1	1	1	-1	-1	1	1	1	: $\hat{J}_z,\alpha_{xy}$
B_2:	1	-1	1	-1	1	-1	1	-1	1	-1	: T_z
A_{1d}:	1	1	1	1	-1	-1	-1	-1	1	-1	: T_y
A_{2d}:	1	1	-1	-1	-1	-1	1	1	1	1	: $\hat{J}_y,\alpha_{xz}$
B_{1d}:	1	-1	-1	1	-1	1	1	-1	1	-1	: T_x
B_{2d}:	1	-1	1	-1	-1	1	-1	1	1	1	: $\hat{J}_x,\alpha_{yz}$
$E_{u/2}$:	2	0	0	0	0	0	0	2	-2	-2	
$E_{g/2}$:	2	0	0	0	0	0	0	-2	-2	2	

Table A-22
The group G_8
Example: CNPI group of H_2O_2

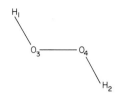

G_8:	E	$(12)(34)$	E^*	$(12)(34)^*$	(12)	(34)	$(12)^*$	$(34)^*$	
A_1':	1	1	1	1	1	1	1	1	
A_2':	1	1	-1	-1	1	1	-1	-1	: Γ^*
B_1':	1	-1	-1	1	1	-1	-1	1	
B_2':	1	-1	1	-1	1	-1	1	-1	
A_1'':	1	1	1	1	-1	-1	-1	-1	
A_2'':	1	1	-1	-1	-1	-1	1	1	
B_1'':	1	-1	-1	1	-1	1	1	-1	
B_2'':	1	-1	1	-1	-1	1	-1	1	

Table A-23
The group G_{16}
Example: Ethylene (with torsional tunneling)[a]

H$_1$ — C$_5$ — H$_2$
C$_5$ = C$_6$
H$_3$ — C$_6$ — H$_4$

G_{16}:	E	$(1423)(56)^*$	$(12)(34)$	$(13)(24)(56)$	$(34)^*$	E^*	$(1423)(56)$	$(12)(34)^*$	$(14)(23)(56)^*$	(34)
	1	2	1	2	2	1	2	1	2	2
$A_1^{\,+}$:	1	1	1	1	1	1	1	1	1	1
$A_2^{\,+}$:	1	1	1	-1	-1	1	1	1	-1	-1
$B_1^{\,+}$:	1	-1	1	1	-1	1	-1	1	1	-1
$B_2^{\,+}$:	1	-1	1	-1	1	1	-1	1	-1	1
$E^{\,+}$:	2	0	-2	0	0	2	0	-2	0	0
$A_1^{\,-}$:	1	-1	1	1	-1	-1	1	-1	-1	1
$A_2^{\,-}$:	1	-1	1	-1	1	-1	1	-1	1	-1
$B_1^{\,-}$:	1	1	1	1	1	-1	-1	-1	-1	-1
$B_2^{\,-}$:	1	1	1	-1	-1	-1	-1	-1	1	1
$E^{\,-}$:	2	0	-2	0	0	-2	0	2	0	0

1 : Γ^*

Table A-24
The group $G_{16}(EM)^2$
Example: Ethylene (with torsional tunneling)[a]

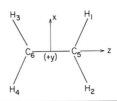

$G_{16}(EM)^2$:	E	a	a^2	b	ab	c	ac	a^2c	bc	abc
	1	4	1[2]	2[4]	4[8]	2[4]	4[8]	2[4]	2	4
Equiv. rot.:	R^0	$R_z^{\pi/2}$	R_z^{π}	R_{π}^{π}	$R_{3\pi/4}^{\pi}$	R_{π}^{π}	$R_{3\pi/4}^{\pi}$	$R_{\pi/2}^{\pi}$	$R^{2\pi}$	$R_z^{5\pi/2}$
A_1^+:	1	1	1	1	1	1	1	1	1	1
A_2^+:	1	1	1	-1	-1	1	1	1	-1	-1
B_1^+:	1	-1	1	1	-1	1	-1	1	1	-1
B_2^+:	1	-1	1	-1	1	1	-1	1	-1	1
E^+:	2	0	-2	0	0	2	0	-2	0	0
A_1^-:	1	1	1	1	1	-1	-1	-1	-1	-1
A_2^-:	1	1	1	-1	-1	-1	-1	-1	1	1
B_1^-:	1	-1	1	1	-1	-1	1	-1	-1	1
B_2^-:	1	-1	1	-1	1	-1	1	-1	1	-1
E^-:	2	0	-2	0	0	-2	0	2	0	0
E_{1d}:	2	0	2	2	0	0	0	0	0	0
E_{2d}:	2	0	2	-2	0	0	0	0	0	0
E_{3d}:	2	0	-2	0	0	0	0	0	2	0
E_{4d}:	2	0	-2	0	0	0	0	0	-2	0
$E_{u/2}$:	2	$\sqrt{2}$	0	0	0	0	0	0	2	$\sqrt{2}$
$E_{3u/2}$:	2	$-\sqrt{2}$	0	0	0	0	0	0	2	$-\sqrt{2}$
$E_{g/2}$:	2	$\sqrt{2}$	0	0	0	0	0	0	-2	$-\sqrt{2}$
$E_{3g/2}$:	2	$-\sqrt{2}$	0	0	0	0	0	0	-2	$\sqrt{2}$
$G_{a/2}$:	4	0	0	0	0	0	0	0	0	0

E'	a^2E'	bE'	a^2bc	R	Ra	Rbc	$Rabc$	Ra^2E'	
1[2]	1	2[4]	2[4]	1	4	2	4	1	
R_z^{π}	$R^{2\pi}$	$R_{3\pi/2}^{\pi}$	$R_z^{3\pi}$	$R^{2\pi}$	$R_z^{5\pi/2}$	R^0	$R_z^{\pi/2}$	R^0	
1	1	1	1	1	1	1	1	1	: $\alpha_{zz},(\alpha_{xx}+\alpha_{yy})$
1	1	-1	-1	1	1	-1	-1	1	
1	1	1	1	1	-1	1	-1	1	: $(\alpha_{xx}-\alpha_{yy})$
1	1	-1	-1	1	-1	-1	1	1	: T_z
2	-2	0	0	2	0	0	0	-2	
1	1	1	-1	1	1	-1	-1	1	
1	1	-1	1	1	1	1	1	1	: $\hat{J}_z$
1	1	1	-1	1	-1	-1	1	1	: Γ^*
1	1	-1	1	1	-1	1	-1	1	: α_{xy}
2	-2	0	0	2	0	0	0	-2	
-2	-2	-2	0	2	0	0	0	-2	
-2	-2	2	0	2	0	0	0	-2	
-2	2	0	-2	2	0	2	0	2	: $(\hat{J}_x,\hat{J}_y),(\alpha_{xz},\alpha_{yz})$
-2	2	0	2	2	0	-2	0	2	: (T_x,T_y)
0	-2	0	0	-2	$-\sqrt{2}$	-2	$-\sqrt{2}$	2	
0	-2	0	0	-2	$\sqrt{2}$	-2	$\sqrt{2}$	2	
0	-2	0	0	-2	$-\sqrt{2}$	2	$\sqrt{2}$	2	
0	-2	0	0	-2	$\sqrt{2}$	2	$-\sqrt{2}$	2	
0	4	0	0	-4	0	0	0	-4	

[a] $a = (1423)(56)^*$, $b = (13)(24)(56)$, $c = E^*$.

Table A-25

The group $G_{36}(EM)$
Example: Dimethylacetylene (with torsional tunneling)[a,b]

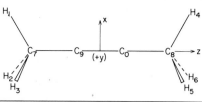

$G_{36}(EM)$:	E	ab	c	ab^5	a^4	b^2c	cd	b^4cd	d
	1	2	3	2	4	6	3	6	9
Equiv. rot.:	R^0	R^0	R_z^π	$R_z^{2\pi/3}$	$R_z^{4\pi/3}$	$R_z^{\pi/3}$	R_0^π	$R_{2\pi/3}^\pi$	$R_{\pi/2}^\pi$
A_1:	1	1	1	1	1	1	1	1	1
A_2:	1	1	1	1	1	1	-1	-1	-1
A_3:	1	1	-1	1	1	-1	1	1	-1
A_4:	1	1	-1	1	1	-1	-1	-1	1
E_1:	2	2	2	-1	-1	-1	0	0	0
E_2:	2	2	-2	-1	-1	1	0	0	0
E_3:	2	-1	0	2	-1	0	2	-1	0
E_4:	2	-1	0	2	-1	0	-2	1	0
G:	4	-2	0	-2	1	0	0	0	0
A_{1d}:	1	1	1	1	1	1	1	1	1
A_{2d}:	1	1	1	1	1	1	-1	-1	-1
A_{3d}:	1	1	-1	1	1	-1	1	1	-1
A_{4d}:	1	1	-1	1	1	-1	-1	-1	1
E_{1d}:	2	2	2	-1	-1	-1	0	0	0
E_{2d}:	2	2	-2	-1	-1	1	0	0	0
E_{3d}:	2	-1	0	2	-1	0	2	-1	0
E_{4d}:	2	-1	0	2	-1	0	-2	1	0
G_d:	4	-2	0	-2	1	0	0	0	0

E'	$E'ab$	$E'c$	$E'ab^5$	$E'a^4$	$E'b^2c$	$E'cd$	$E'b^4cd$	$E'd$		
1	2	3	2	4	6	3	6	9		
R_z^π	R_z^π	R^0	$R_z^{5\pi/3}$	$R_z^{\pi/3}$	$R_z^{4\pi/3}$	$R_{\pi/2}^\pi$	$R_{\pi/6}^\pi$	R_0^π		
1	1	1	1	1	1	1	1	1	:	$(\alpha_{xx}+\alpha_{yy}), \alpha_{zz}$
1	1	1	1	1	1	-1	-1	-1	:	$\hat{J}_z$
1	1	-1	1	1	-1	1	1	-1	:	Γ^*
1	1	-1	1	1	-1	-1	-1	1	:	T_z
2	2	2	-1	-1	-1	0	0	0	:	$(\alpha_{xx}-\alpha_{yy}, \alpha_{xy})$
2	2	-2	-1	-1	1	0	0	0		
2	-1	0	2	-1	0	2	-1	0		
2	-1	0	2	-1	0	-2	1	0		
4	-2	0	-2	1	0	0	0	0		
-1	-1	-1	-1	-1	-1	-1	-1	-1		
-1	-1	-1	-1	-1	-1	1	1	1		
-1	-1	1	-1	-1	1	-1	-1	1		
-1	-1	1	-1	-1	1	1	1	-1		
-2	-2	-2	1	1	1	0	0	0	:	(T_x, T_y)
-2	-2	2	1	1	-1	0	0	0	:	$(\hat{J}_x, \hat{J}_y), (\alpha_{xz}, \alpha_{yz})$
-2	1	0	-2	1	0	-2	1	0		
-2	1	0	-2	1	0	2	-1	0		
-4	2	0	2	-1	0	0	0	0		

[a] $a = (123)$, $b = (456)$, $c = (14)(26)(35)(78)(90)^*$, $d = (23)(56)^*$.
[b] For definitions of Euler angle transformations see Table 12-11.

Table A-26
The group G_{48}
Example: CNPI group of CH_4

(Structure of CH_4: central C bonded to H_4, H_1, H_2, H_3.)

G_{48}:	E	(123)	(14)(23)	(1423)*	(23)*	E*	(123)*	(14)(23)*	(1423)	(23)
	1	8	3	6	6	1	8	3	6	6
$A_1^{\,+}$	1	1	1	1	1	1	1	1	1	1
$A_2^{\,+}$	1	1	1	−1	−1	1	1	1	−1	−1
E^{+}	2	−1	2	0	0	2	−1	2	0	0
$F_1^{\,+}$	3	0	−1	1	−1	3	0	−1	1	−1
$F_2^{\,+}$	3	0	−1	−1	1	3	0	−1	−1	1
$A_1^{\,-}$	1	1	1	1	1	−1	−1	−1	−1	−1
$A_2^{\,-}$	1	1	1	−1	−1	−1	−1	−1	1	1
E^{-}	2	−1	2	0	0	−2	1	−2	0	0
$F_1^{\,-}$	3	0	−1	1	−1	−3	0	1	−1	1
$F_2^{\,-}$	3	0	−1	−1	1	−3	0	1	1	−1

(Margin annotation: $1 : \Gamma^{*}$)

The Correlation Tables

In this appendix there are four types of correlation tables.

In Table B-1 the representations $D^{(0)}$ through $D^{(3)}$ of the spin double group of the three-dimensional molecular rotation group $K(\text{mol})^2$ are reduced onto the molecular rotation groups D_∞^2 and D_2^2. The rotational states of a spherical top molecule can be labeled $D^{(J)}$ in $K(\text{mol})^2$ according to the value of J. The rotational states of a symmetric top molecule can be labeled Σ^+ (or Σ^-), Π, Δ, ..., in D_∞^2 as $K = 0$ with J even (or $K = 0$ with J odd), $K = 1$, $K = 2$, ..., respectively, and the rotational states of an asymmetric top molecule can be labeled A, B_a, B_b, and B_c in D_2^2 as $K_a K_c$ is ee, eo, oo, and oe (o = odd and e = even). Tables 11-1 and 11-2 were used in these reductions.

In Table B-2 the representations $D^{(0)}$ through $D^{(3)}$ of $K(\text{mol})^2$ are reduced onto the representations of the groups given in Appendix A by using the equivalent rotations of the elements of the group. We first reduce $K(\text{mol})$ onto the equivalent rotation group of the MS (or EMS) group, and then make use of the isomorphism or homomorphism of the MS (or EMS) group onto the equivalent rotation group in order to obtain the final result. The equivalent rotation group is the group of all distinct equivalent rotations of the MS (or EMS) group concerned; e.g., for the $C_{3v}(\text{M})$ group it is $D_3 =$

$\{E, R_z^{2\pi/3}, R_z^{4\pi/3}, R_{\pi/6}^{\pi}, R_{\pi/2}^{\pi}, R_{5\pi/6}^{\pi}\}$. From this table the symmetry Γ_{rot} of the rotational states with $J \leq 3$ of each molecule in its MS (or EMS) group can be obtained. The species of Hund's case (a) electron spin functions in these groups can be obtained from this table by identifying the representations of $K(mol)^2$ with $D^{(S)}$.

In Table B-3 the correlations of the representations of the molecular point groups of bent and linear molecules are given for use in determining the correlation of the symmetry of the electronic states of a linear triatomic molecule with those obtained by bending the molecule.

In Table B-4 reverse correlation tables, with statistical weights, are given for some common nonrigid molecules. For a rigid methanol molecule the $C_s(M)$ group is $\{E,(23)^*\}$ if we take the rigid conformer as having $O-C-H_1$ coplanar with the $C-O-H$ group. For a rigid eclipsed or staggered methyl silane molecule the $C_{3v}(M)$ group of a conformer is $\{E, (123)(465), (132)(456), (12)(46)^*, (13)(45)^*, (23)(56)^*\}$. For a rigid acetone molecule in the conformation indicated in Table A-19 the $C_{2v}(M)$ group is $\{E, (14)(25)(36)(78), (23)(56)^*, (14)(26)(35)(78)^*\}$. The results in Table B-4 are of use when the tunneling splittings are small, and the statistical weights should help in the assignments of the states.

Table B-1
The correlation table of $K(mol)^2$ with the molecular rotation groups D_∞^2 and D_2^2

$K(mol)^2$	D_∞^2	D_2^2
$D^{(0)}$	Σ^+	A
$D^{(1/2)}$	$E_{1/2}$	$E_{1/2}$
$D^{(1)}$	$\Sigma^- \oplus \Pi$	$B_a \oplus B_b \oplus B_c$
$D^{(3/2)}$	$E_{1/2} \oplus E_{3/2}$	$2E_{1/2}$
$D^{(2)}$	$\Sigma^+ \oplus \Pi \oplus \Delta$	$2A \oplus B_a \oplus B_b \oplus B_c$
$D^{(5/2)}$	$E_{1/2} \oplus E_{3/2} \oplus E_{5/2}$	$3E_{1/2}$
$D^{(3)}$	$\Sigma^- \oplus \Pi \oplus \Delta \oplus \Phi$	$A \oplus 2B_a \oplus 2B_b \oplus 2B_c$

Table B-2
The correlation of $K(\text{mol})^2$ with the spin double groups of MS and EMS groups

(i) Asymmetric top molecules

$K(\text{mol})^2$	$C_s(M)^2$	$C_i(M)^2$	$C_2(M)^2$	$C_{2v}(M)^2$	$C_{2h}(M)^2$	$D_{2h}(M)^2$
$D^{(0)}$	A'	A_g	A	A_1	A_g	A_g
$D^{(1/2)}$	$E_{1/2}$	$2B_{g/2}$	$E_{1/2}$	$E_{1/2}$	$E_{g/2}$	$E_{g/2}$
$D^{(1)}$	$A' \oplus 2A''$	$3A_g$	$A \oplus 2B$	$A_2 \oplus B_1 \oplus B_2$	$A_g \oplus 2B_g$	$B_{1g} \oplus B_{2g} \oplus B_{3g}$
$D^{(3/2)}$	$2E_{1/2}$	$4B_{g/2}$	$2E_{1/2}$	$2E_{1/2}$	$2E_{g/2}$	$2E_{g/2}$
$D^{(2)}$	$3A' \oplus 2A''$	$5A_g$	$3A \oplus 2B$	$2A_1 \oplus A_2 \oplus B_1 \oplus B_2$	$3A_g \oplus 2B_g$	$2A_g \oplus B_{1g} \oplus B_{2g} \oplus B_{3g}$
$D^{(5/2)}$	$3E_{1/2}$	$6B_{g/2}$	$3E_{1/2}$	$3E_{1/2}$	$3E_{g/2}$	$3E_{g/2}$
$D^{(3)}$	$3A' \oplus 4A''$	$7A_g$	$3A \oplus 4B$	$A_1 \oplus 2A_2 \oplus 2B_1 \oplus 2B_2$	$3A_g \oplus 4B_g$	$A_g \oplus 2B_{1g} \oplus 2B_{2g} \oplus 2B_{3g}$

(ii) Symmetric top, spherical top, and linear molecules

$K(mol)^2$	$D_{2d}(M)^2$	$C_{3v}(M)^2$	$D_{3h}(M)^2$	$D_{3d}(M)^2$	$T_d(M)^2$	$C_{\infty v}(EM)^2$	$D_{\infty h}(EM)^2$
$D^{(0)}$	A_1	A_1	A_1'	A_{1g}	A_1	Σ^+	Σ_g^+
$D^{(1/2)}$	$E_{1/2}$	$E_{1/2}$	$E_{1/2}$	$E_{g/2}$	$E_{1/2}$	$E_{1/2}$	$E_{g/2}$
$D^{(1)}$	$A_2 \oplus E$	$A_2 \oplus E$	$A_2' \oplus E''$	$A_{2g} \oplus E_g$	F_1	$\Sigma^- \oplus \Pi$	$\Sigma_g^- \oplus \Pi_g$
$D^{(3/2)}$	$E_{1/2} \oplus E_{3/2}$	$E_{1/2} \oplus E_{3/2}$	$E_{1/2} \oplus E_{3/2}$	$E_{g/2} \oplus E_{3g/2}$	$G_{3/2}$	$E_{1/2} \oplus E_{3/2}$	$E_{g/2} \oplus E_{3g/2}$
$D^{(2)}$	$A_1 \oplus B_1 \oplus B_2 \oplus E$	$A_1 \oplus 2E$	$A_1' \oplus E' \oplus E''$	$A_{1g} \oplus 2E_g$	$E \oplus F_2$	$\Sigma^+ \oplus \Pi \oplus \Delta$	$\Sigma_g^+ \oplus \Pi_g \oplus \Delta_g$
$D^{(5/2)}$	$E_{1/2} \oplus 2E_{3/2}$	$2E_{1/2} \oplus E_{3/2}$	$E_{1/2} \oplus E_{3/2} \oplus E_{5/2}$	$2E_{g/2} \oplus E_{3g/2}$	$G_{3/2} \oplus E_{5/2}$	$E_{1/2} \oplus E_{3/2} \oplus E_{5/2}$	$E_{g/2} \oplus E_{3g/2} \oplus E_{5g/2}$
$D^{(3)}$	$A_2 \oplus B_1 \oplus B_2 \oplus 2E$	$A_1 \oplus 2A_2 \oplus 2E$	$A_1' \oplus A_2' \oplus A_2'' \oplus E' \oplus E''$	$A_{1g} \oplus 2A_{2g} \oplus 2E_g$	$A_2 \oplus F_1 \oplus F_2$	$\Sigma^- \oplus \Pi \oplus \Delta \oplus \Phi$	$\Sigma_g^- \oplus \Pi_g \oplus \Delta_g \oplus \Phi_g$

(iii) Nonrigid molecules

$K(mol)^2$	G_6^2	G_{12}^2	G_{18}^2	G_{36}^2	$G_4(EM)^2$	$G_{16}(EM)^2$	$G_{36}(EM)^2$
$D^{(0)}$	A_1	A_1'	A_1	A_1	A_1	A_1^+	A_1
$D^{(1/2)}$	$E_{1/2}$	$E_{1/2}$	$E_{1/2}$	$E_{1/2}$	$E_{g/2}$	$E_{g/2}$	$E_{1/2}$
$D^{(1)}$	$A_1 \oplus 2A_2$	$A_2' \oplus A_1'' \oplus A_2''$	$A_2 \oplus E_1$	$A_2 \oplus A_3 \oplus A_4$	$B_1 \oplus A_{2d} \oplus B_{2d}$	$A_2^- \oplus E_{3d}$	$A_2 \oplus E_{2d}$
$D^{(3/2)}$	$2E_{1/2}$	$2E_{1/2}$	$E_{1/2} \oplus E_{3/2}$	$2E_{1/2}$	$2E_{g/2}$	$E_{g/2} \oplus E_{3g/2}$	$E_{1/2} \oplus E_{3/2}$
$D^{(2)}$	$3A_1 \oplus 2A_2$	$2A_1' \oplus A_2' \oplus A_1'' \oplus A_2''$	$A_1 \oplus 2E_1$	$2A_1 \oplus A_2 \oplus A_3 \oplus A_4$	$2A_1 \oplus B_1 \oplus A_{2d} \oplus B_{2d}$	$A_1^+ \oplus B_1^+ \oplus B_2^- \oplus E_{3d}$	$A_1 \oplus E_1 \oplus E_{2d}$
$D^{(5/2)}$	$3E_{1/2}$	$3E_{1/2}$	$2E_{1/2} \oplus E_{3/2}$	$3E_{1/2}$	$3E_{g/2}$	$E_{g/2} \oplus 2E_{3g/2}$	$E_{1/2} \oplus E_{3/2} \oplus E_{5/2}$
$D^{(3)}$	$3A_1 \oplus 4A_2$	$A_1' \oplus 2A_2' \oplus 2A_1'' \oplus 2A_2''$	$A_1 \oplus 2A_2 \oplus 2E_1$	$A_1 \oplus 2A_2 \oplus 2A_3 \oplus 2A_4$	$A_1 \oplus 2B_1 \oplus 2A_{2d} \oplus 2B_{2d}$	$A_2^- \oplus B_1^+ \oplus B_2^- \oplus 2E_{3d}$	$A_2 \oplus A_{3d} \oplus A_{4d} \oplus E_1 \oplus E_{2d}$

401

Table B-3

Correlation of species of electronic states of
a linear triatomic molecule[a] with species of
electronic states obtained when
molecule is bent[b]

$D_{\infty h}$	C_{2v}	$C_{\infty v}$	C_s
Σ_g^+	A_1	Σ^+	A'
Σ_g^-	B_1	Σ^-	A''
Σ_u^+	B_2	Π	$A' \oplus A''$
Σ_u^-	A_2	Δ	$A' \oplus A''$
Π_g	$A_2 \oplus B_2$	$\vdots$	$\vdots$
Π_u	$A_1 \oplus B_1$		
Δ_g	$A_1 \oplus B_1$		
Δ_u	$A_2 \oplus B_2$		
$\vdots$	$\vdots$		

[a] In $D_{\infty h}$ or $C_{\infty v}$.
[b] In C_{2v} or C_s.

Table B-4

Reverse correlation tables for nonrigid molecules
with statistical weights (C = ^{12}C, N = ^{14}N, and O = ^{16}O)

(i) NH$_3$[a]

C_{3v}(M)	D_{3h}(M)
A_1(12)	A_1'(0) $\oplus$ A_2''(12)
A_2(12)	A_2'(12) $\oplus$ A_1''(0)
E(12)	E'(6) $\oplus$ E''(6)

(ii) CH$_3$OH[b]

C_s(M)	G_6
A'(8)	A_1(4) $\oplus$ E(4)
A''(8)	A_2(4) $\oplus$ E(4)

(iii) SiH$_3$CH$_3$[c]

C_{3v}(M)	G_{18}
A_1(24)	A_1(16) $\oplus$ E_4(8)
A_2(24)	A_2(16) $\oplus$ E_4(8)
E(40)	E_1(16) $\oplus$ E_2(16) $\oplus$ E_3(8)

(iv) CH$_3$COCH$_3$[d]

C_{2v}(M)	G_{36}
A_1(28)	A_1(6) $\oplus$ E_1(4) $\oplus$ E_3(2) $\oplus$ G(16)
A_2(28)	A_3(6) $\oplus$ E_2(4) $\oplus$ E_3(2) $\oplus$ G(16)
B_1(36)	A_2(10) $\oplus$ E_1(4) $\oplus$ E_4(6) $\oplus$ G(16)
B_2(36)	A_4(10) $\oplus$ E_2(4) $\oplus$ E_4(6) $\oplus$ G(16)

(v) H$_2$O$_2$[e]

C_2(M)	G_4	G_8
A(2)	A_1(1) $\oplus$ A_2(1)	A_1'(1) $\oplus$ A_1''(0) $\oplus$ A_2'(1) $\oplus$ A_2''(0)
B(6)	B_1(3) $\oplus$ B_2(3)	B_1'(0) $\oplus$ B_1''(3) $\oplus$ B_2'(0) $\oplus$ B_2''(3)

(vi) C$_2$H$_4$[f]

D_{2h}(M)	G_{16}
A_g(7)	$A_1{}^+$(1) $\oplus$ $B_1{}^+$(6)
A_u(7)	$A_1{}^-$(6) $\oplus$ $B_1{}^-$(1)
B_{1g}(3)	$A_2{}^-$(0) $\oplus$ $B_2{}^-$(3)
B_{1u}(3)	$A_2{}^+$(3) $\oplus$ $B_2{}^+$(0)
B_{2g}(3)	E^-(3)
B_{2u}(3)	E^+(3)
B_{3g}(3)	E^+(3)
B_{3u}(3)	E^-(3)

(vii) C$_2$H$_6$[g]

D_{3d}(M)	G_{36}
A_{1g}(8)	A_1(6) $\oplus$ E_3(2)
A_{1u}(8)	A_3(6) $\oplus$ E_3(2)
A_{2g}(16)	A_2(10) $\oplus$ E_4(6)
A_{2u}(16)	A_4(10) $\oplus$ E_4(6)
E_g(20)	E_1(4) $\oplus$ G(16)
E_u(20)	E_2(4) $\oplus$ G(16)

(viii) CH$_4$[h]

T_d(M)	G_{48}
A_1(5)	$A_1{}^+$(0) $\oplus$ $A_1{}^-$(5)
A_2(5)	$A_2{}^+$(5) $\oplus$ $A_2{}^-$(0)
E(2)	E^+(1) $\oplus$ E^-(1)
F_1(3)	$F_1{}^+$(3) $\oplus$ $F_1{}^-$(0)
F_2(3)	$F_2{}^+$(0) $\oplus$ $F_2{}^-$(3)

[a] See Tables A-8 and A-9.
[b] See Tables A-1 and A-16.
[c] See Tables A-8 and A-18.
[d] See Tables A-4 and A-19.

[e] See Tables A-3, A-20, and A-22.
[f] See Tables A-6, A-23, and 10-2.
[g] See Tables A-10 and A-25.
[h] See Tables A-11 and A-26.

References

C. Arpigny (1966). *Astrophys. J.* **144**, 424.

H. D. Babcock and L. Herzberg (1948). *Astrophys. J.* **108**, 167.

P. E. G. Baird, M. W. S. M. Brimicombe, G. J. Roberts, P. G. H. Sandars, D. C. Soreide, E. N. Fortson, L. L. Lewis, and E. G. Lindahl (1976). *Nature* **264**, 528.

G. A. Baker (1956). *Phys. Rev.* **103**, 1119.

A. Bauder, R. Meyer, and H. H. Günthard (1974). *Molecular Phys.* **28**, 1305.

H. Bethe (1929). *Ann. Physik.* (5) **3**, 133.

M. Born (1951). *Nachr. Akad. Wiss. Göttingen* **1**.

M. Born and K. Huang (1956). "Dynamical Theory of Crystal Lattices," Oxford Univ. Press, New York.

M. Born and R. Oppenheimer (1927). *Ann. Physik.* **84**, 457.

M. A. Bouchiat and L. Pottier (1977). *In* "Laser Spectroscopy III" (J. L. Hall and J. L. Carlsten, eds.), page 9, Springer-Verlag, New York.

L. Brillouin (1934). "Les Champs 'Self-consistents' de Hartree–Fock," Hermann, Paris.

J. M. Brown, A. R. H. Cole, and F. R. Honey (1972). *Molecular Phys.* **23**, 287.

P. R. Bunker (1964). *Molecular Phys.* **8**, 81.

P. R. Bunker (1965). *Molecular Phys.* **9**, 257.

P. R. Bunker (1967). *J. Chem. Phys.*, **47**, 718; **48**, 2832.

P. R. Bunker (1973a). *J. Mol. Spectrosc.* **46**, 119.

P. R. Bunker (1973b). *J. Mol. Spectrosc.* **48**, 181.

P. R. Bunker (1974). *Chem. Phys. Lett.* **27**, 322.

P. R. Bunker (1975). "Vibrational Spectra and Structure," Vol. 3 (J. R. Durig, ed.), Chapter 1, Dekker, New York.

P. R. Bunker (1979). The Permutation Group in Physics and Chemistry, *in* "Lecture Notes in Chemistry" (H. Hartmann, ed.), Springer-Verlag, Berlin.

P. R. Bunker and J. T. Hougen (1967). *Can. J. Phys.* **45**, 3867.

P. R. Bunker and B. M. Landsberg (1977). *J. Mol. Spectrosc.* **67**, 374.

P. R. Bunker and D. Papousek (1969). *J. Mol. Spectrosc.* **32**, 419.

P. R. Bunker and J. M. R. Stone (1972). *J. Mol. Spectrosc.* **41**, 310.

P. R. Bunker, B. M. Landsberg, and B. P. Winnewisser (1978). *J. Mol. Spectrosc.* (in press).

M. Carlotti, J. W. C. Johns, and A. Trombetti (1974), *Can. J. Phys.* **52**, 340.

Y.-N. Chiu (1965). *J. Chem. Phys.* **42**, 2671.

F. E. Close (1976). *Nature* **264**, 505.

E. U. Condon and G. H. Shortley (1959). "The Theory of Atomic Spectra," Cambridge Univ. Press, London.

R. F. Curl, J. V. V. Kasper, and K. S. Pitzer (1967). *J. Chem. Phys.* **46**, 3220.

B. T. Darling and D. M. Dennison (1940). *Phys. Rev.* **57**, 128.

G. Dellepiane, M. Gussoni, and J. T. Hougen (1973). *J. Mol. Spectrosc.* **47**, 515.

C. DiLauro and I. M. Mills (1966). *J. Mol. Spectrosc.* **21**, 386.

T. R. Dyke (1977). *J. Chem. Phys.* **66**, 492.

C. Eckart (1935). *Phys. Rev.* **47**, 552.

A. R. Edmonds (1960). "Angular Momentum in Quantum Mechanics," Princeton Univ. Press, Princeton, New Jersey.

H. Eyring, J. Walter, and G. E. Kimball (1944). "Quantum Chemistry," Wiley, New York.

G. R. Gunther-Mohr, C. H. Townes, and J. H. Van Vleck (1954). *Phys. Rev.* **94**, 1191.

M. Hamermesh (1962). "Group Theory and Its Application to Physical Problems," Addison-Wesley, Reading, Massachusetts.

G. Herzberg (1945). "Molecular Spectra and Molecular Structure, II. Infrared and Raman Spectra of Polyatomic Molecules," Van Nostrand Rheinhold, Princeton, New Jersey.

G. Herzberg (1946). *Phys. Rev.* **69**, 362.

G. Herzberg (1949). *Nature* **163**, 170.

G. Herzberg (1950a). "Molecular Spectra and Molecular Structure, I. Spectra of Diatomic Molecules," Van Nostrand Rheinhold, Princeton, New Jersey.

G. Herzberg (1950b). *Can. J. Res., A* **28**, 144.

G. Herzberg (1966). "Molecular Spectra and Molecular Structure, III. Electronic Spectra and Electronic Structure of Polyatomic Molecules," Van Nostrand Rheinhold, Princeton, New Jersey.

G. Herzberg (1969). *Mém. Soc. Roy. Sci. Liège Coll., ser. V.* **XVI**, 121.

G. Herzberg and E. Teller (1933). *Z. Phys. Chem. Abt. B* **21**, 410.

L. Herzberg and G. Herzberg (1947). *Astrophys. J.* **105**, 353.

J. T. Hougen (1962a). *J. Chem. Phys.* **36**, 519.

J. T. Hougen (1962b). *J. Chem. Phys.* **37**, 1433.

J. T. Hougen (1963). *J. Chem. Phys.* **39**, 358.

J. T. Hougen (1964). *Can. J. Phys.* **42**, 1920.

J. T. Hougen (1965). *Can. J. Phys.* **43**, 935.

J. T. Hougen (1976). Methane symmetry operations, *in* "MTP International Reviews of Science, Physical Chemistry," Ser. 2, Vol. 3 (D. A. Ramsay, ed.), Butterworth, London.

J. T. Hougen, P. R. Bunker, and J. W. C. Johns (1970). *J. Mol. Spectrosc.* **34**, 136.

J. T. Hougen and J. K. G. Watson (1965). *Can J. Phys.* **43**, 298.

B. J. Howard and R. E. Moss (1971). *Molecular Phys.* **20**, 147.

A. R. Hoy and I. M. Mills (1973). *J. Mol. Spectrosc.* **46**, 333.

A. R. Hoy, I. M. Mills, and G. Strey (1972). *Molecular Phys.* **24**, 1265.

R. C. Hwa and J. Nuyts (1966). *Phys. Rev.* **145**, 1188.

C. F. Jackels and E. R. Davidson (1976). *J. Chem. Phys.* **65**, 2941.

H. A. Jahn and E. Teller (1937). *Proc. Roy. Soc. Ser. A* **161**, 220

C. Jungen and A. J. Merer (1976). *In* "Molecular Spectrocopy: Modern Research," Vol. II (K. N. Rao, ed.), Chapter 3, Academic Press, New York.

G. W. King, R. M. Hainer, and P. C. Cross (1943). *J. Chem. Phys.* **11**, 27.

L. D. Landau and E. M. Lifschitz (1958). "Quantum Mechanics," Pergamon, Oxford.

H. C. Longuet-Higgins (1963). *Molecular Phys.* **6**, 445.

J. D. Louck (1976). *J. Mol. Spectrosc.* **61**, 107.

H. Margenau and G. M. Murphy (1956). "The Mathematics of Physics and Chemistry," 2nd ed., Van Nostrand Rheinhold, Princeton, New Jersey.

F. A. Matsen (1964). *Advan. Quantum Chem.* **1**, 59.

H. V. McIntosh (1971). Symmetry and degeneracy *in* "Group Theory and Its Applications," Vol. II (E. M. Loebl, ed.), Academic Press, New York.

A. J. Merer and L. Schoonveld (1969). *Can. J. Phys.* **47**, 1731.

A. J. Merer and J. K. G. Watson (1973). *J. Mol. Spectrosc.* **47**, 499.

R. E. Miller and J. C. Decius (1973). *J. Chem. Phys.* **59**, 4871.

I. M. Mills (1964). *Molecular Phys.* **8**, 363.

W. Moffitt and A. D. Liehr (1957). *Phys. Rev.* **106**, 1195.

R. E. Moss (1973). "Advanced Molecular Quantum Mechanics," Chapman & Hall, London.

R. S. Mulliken (1975). *In* "Selected Papers of Robert S. Mulliken," Part VII (D. A. Ramsay and J. Hinze, eds.), Univ. of Chicago Press, Chicago.

F. D. Murnaghan (1938). "The Theory of Group Representations," Johns Hopkins Press, Baltimore, Maryland.

R. J. Myers and E. B. Wilson (1960). *J. Chem. Phys.* **33**, 186.

H. H. Nielsen (1959). "Encyclopedia of Physics," Vol. XXXVII/1 (S. Flugge, ed.), p. 173, Springer-Verlag, Berlin.

T. Oka (1973). *J. Mol. Spectrosc.* **48**, 503.

T. Oka (1976). *In* "Molecular Spectroscopy: Modern Research," Vol. II (K. N. Rao, ed.), Chapter 5.1, Academic Press, New York.

T. Oka and Y. Morino (1961). *J. Mol. Spectrosc.* **6**, 472.

W. B. Olson (1972). *J. Mol. Spectrosc.* **43**, 190.

W. Opechowski (1940). *Physica* **VII**, 522.

D. Papousek, K. Sarka, V. Spirko, and B. Jordanov (1971). *Collect. Czech. Chem. Commun.* **36**, 890.

D. Papousek, J. M. R. Stone, and V. Spirko (1973). *J. Mol. Spectrosc.* **48**, 17.

W. Pauli (1940). *Phys. Rev.* **58**, 716.

L. Pauling and E. B. Wilson (1935). "Introduction to Quantum Mechanics," McGraw-Hill, New York.

G. Placzek (1934). "Handbuch der Radiologie" Vol. 6, p. 205, Akademische Verlagsgesellschaft, Leipzig.

B. Podolsky (1928). *Phys. Rev.* **32**, 812.

M. Quack (1977). *Molecular Phys.* **34**, 477.

J. C. Raich and R. H. Good (1964). *Astrophys. J.* **139**, 1004.

R. W. Redding (1971). *J. Mol. Spectrosc.* **38**, 396.

R. Renner (1934). *Z. Physik.* **92**, 172.

K. Ruedenberg and R. D. Poshusta (1972). *Advan. Quantum Chem.* **6**, 267.

P. G. H. Sandars (1977). *In* "Laser Spectroscopy III" (J. L. Hall and J. L. Carlsten, eds.), page 21, Springer-Verlag, New York.

K. Sarka (1976). *Collect. Czech. Chem. Commun.* **41**, 2817.

A. Sayvetz (1939). *J. Chem. Phys.* **7**, 383.

V. Sidis and H. Lefebvre-Brion (1971). *J. Phys. B.* **4**, 1040.

V. A. Smirnov (1974). *Opt. Spectrosc.* **37**, 498.

T. Smith (1934). *Proc. Phys. Soc.* **46**, 344.

M. Tinkham (1964). "Group Theory and Quantum Mechanics," McGraw-Hill, New York.

C. H. Townes and A. L. Schawlow (1955). "Microwave Spectroscopy," McGraw-Hill, New York.

J. H. Van Vleck (1951). *Rev. Modern Phys.* **23**, 213.

T. E. H. Walker and J. I. Musher (1974). *Molecular Phys.* **27**, 1651.

A. D. Walsh (1953). *J. Chem. Soc.*, p. 2266.

J. K. G. Watson (1965). *Can. J. Phys.* **43**, 1996.

J. K. G. Watson (1967). *J. Chem. Phys.* **46**, 1935.

J. K. G. Watson (1968). *Molecular Phys.* **15**, 479.

J. K. G. Watson (1970). *Molecular Phys.* **19**, 465.

J. K. G. Watson (1974). *J. Mol. Spectrosc.* **50**, 281.

J. K. G. Watson (1975). *Can J. Phys.* **53**, 2210.

J. K. G. Watson (1977). "Vibrational Spectra and Structure," Vol. 6 (J. R. Durig, ed.), Chapter 1, Dekker, New York.

E. P. Wigner (1959). "Group Theory," Academic Press, New York.

E. B. Wilson, J. C. Decius, and P. C. Cross (1955). "Molecular Vibrations," McGraw-Hill, New York.

E. B. Wilson and J. B. Howard (1936). *J. Chem. Phys.* **4**, 260.

R. S. Winton and G. Winnewisser (1970). Columbus, Ohio, Symposium on Molecular Structure and Spectroscopy, Paper Q3.

C. M. Woodman (1970). *Molecular Phys.* **19**, 753.

C. E. Wulfman (1971). Dynamical groups in atomic and molecular physics *in* "Group Theory and Its Applications," Vol. II (E. M. Loebl, ed.), Academic Press, New York.

K. Yamada, T. Nakagawa, and K. Kuchitsu (1974). *J. Mol. Spectrosc.* **51**, 399.

Index

This index includes, in addition to the usual material of an index, all *symbols* and *molecules* discussed or mentioned. The symbols appear at the beginning of the appropriate letter section of the index in the following order: capitals before lower case before Greek letters, with Roman type before italic type before boldface type. Each molecule is listed, as in the volumes by Herzberg (1945, 1950a, 1966), under its formula considered as a word; for example, CH_3F as Chf and H_2O as Ho.

A

A-type band, 318

a, s label, diatomic molecule, 137

(a, b, c) axis system, 144

$a_r^{z\beta}$, moment of inertia factor, 143, 282, 287–291, 299–300

(α, β), spin functions, 101–102, 105, 109

(α, β, γ), Euler angles, 94

$\boldsymbol{\alpha}$, electric polarizability tensor, 322, 324, 374

Abelian group, 15, 53

Accidental degeneracy, 65, 112, 197, 203, 206, 221

Accidentally vanishing interaction, 283

Active picture, 90

Allowed transition, 315, 318, 320

Angular momentum

commutation relations, 132, 142, 182, 332

electronic, 130–133, 142–143

electron spin, 100–101, 103, 107–108, 250, 253

ladder operators, 180–184, 235, 263

matrix elements, 184–185

nuclear spin, 104, 107–108, 110

quantum number, 98, 107–108, 176–177, 180–181, 184–186, 226, 236–240, 250, 313, 315–316, 321, 323, 326–327

rovibronic, 107–108, 131, 137, 141–142, 250

total, 95–96, 107, 226